QUATERNARY

ENVIRONMENTS

QUATERNARY ◆ ENVIRONMENTS

Second Edition

MARTIN WILLIAMS

MAWSON GRADUATE CENTRE FOR ENVIRONMENTAL STUDIES
UNIVERSITY OF ADELAIDE, AUSTRALIA

DAVID DUNKERLEY

DEPARTMENT OF GEOGRAPHY, MONASH UNIVERSITY, AUSTRALIA

PATRICK DE DECKKER

DEPARTMENT OF GEOLOGY, AUSTRALIAN NATIONAL UNIVERSITY

PETER KERSHAW

DEPARTMENT OF GEOGRAPHY, MONASH UNIVERSITY, AUSTRALIA

AND

JOHN CHAPPELL

RESEARCH SCHOOL OF EARTH SCIENCES, AUSTRALIAN NATIONAL UNIVERSITY

ARNOLD

A member of the Hodder Headline Group
LONDON

First published in Great Britain in 1993
Second edition published in 1998, reprinted 2003 by
Arnold, a member of the Hodder Headline Group,
338 Euston Road, London NW1 3BH

http://www.arnoldpublishers.com

Distributed in the United States of America by
Oxford University Press Inc.,
198 Madison Avenue, New York, NY10016

British Library Cataloguing in Publication Data
A catalogue entry for this book is available from the British Library

Library of Congress Gataloging in Publication Data
A catalog entry for this book is available from the Library of Congress

ISBN 0 340 69151 4 (pb)

6 7 8 9 10

Composition by J&L Composition Ltd, Filey, North Yorkshire
Printed and bound in India by Replika Press Pvt. Ltd.

CONTENTS

THE AUTHORS

Martin Williams is a Foundation Professor of Environmental Studies in the Mawson Graduate Centre for Environmental Studies, University of Adelaide, Australia. He has a PhD degree from the Australian National University and an ScD degree from the University of Cambridge. He has carried out extensive fieldwork in Australia, Africa, India and China, and is the author of over 120 research papers on landscape evolution, climatic change and prehistoric environments in Australia, Africa and India. He co-edited *Evolution of Australasian Landforms* (ANU Press, Canberra, 1978); *The Sahara and the Nile* (Balkema, Rotterdam, 1980); *A Land Between Two Niles: Quaternary Geology and Biology of the Central Sudan* (Balkema, Rotterdam, 1982); *Monsoonal Australia* (Balkema, Rotterdam, 1991); *The Cainozoic in Australia* (Geological Society of Australia, Sydney, 1991); co-authored the first edition of *Quaternary Environments* (Edward Arnold, London, 1993); and is co-author of *Interactions of Desertification and Climate* (Arnold, with UNEP and WMO, London, 1996).

David Dunkerley is a Senior Lecturer in Geography and Environmental Science at Monash University, Australia. He has monitored chemical and physical denudational processes throughout Australia as well as in the equatorial lowlands of Papua New Guinea and Borneo; and has particular interests in global tectonic processes, karst geochemistry and modern fluvial and hillslope erosional and depositional processes in semi-arid areas. He was a co-author of the first edition of *Quaternary Environments* (Edward Arnold, London, 1993).

Patrick De Deckker is a Reader in Geology at the Australian National University, Canberra. He has studied lakes in Australia, Antarctica, South America, Europe, Africa and China, and has helped to pioneer the quantitative evaluation of water temperature and salinity using ostracods in marine and non-marine sediments. His current research is on late Quaternary marine environments in Australasia. He co-edited *Limnology in Australia* (CSIRO, Melbourne and Junk, Dordercht, 1986), *Ostracoda in the Earth Sciences* (Elsevier, Amsterdam, 1988) and *The Cainozoic in Australia* (Geological Society of Australia, Sydney, 1991). He is author of over 100 scientific papers covering numerous aspects of Quaternary and modern terrestrial and marine environments and biota. He edited a Special Issue of *Palaeogeography, Palaeoclimatology, Palaeoecology* entitled 'The Late Quaternary Palaeoceanography of the Australasian Region' (Elsevier, Amsterdam, 1997), and was a co-author of the first edition of *Quaternary Environments* (Edward Arnold, London, 1993).

Peter Kershaw is Director of the Centre for Palynology and Palaeoecology and Professor of Geography and Environmental Science at Monash University, Australia. His research on the Quaternary and Tertiary vegetation history of Australia has resulted in three books and over 90 papers. He is President of the Palynological and Palaeobotanical Association of Australia, and has also been involved in joint palynological research in China and the former USSR.

John Chappell is Professor of Quaternary Geology and Geomorphology in the Research School of Earth Sciences at the Institute of Advanced Studies, Australian National University, and a Fellow of the Australian Academy of Science. He was previously Professor in the Department of Biogeography and Geomorphology (later renamed the Division of Archaeology and Natural History) at the ANU. He has published over 200 research papers on all aspects of Quaternary geology, and is perhaps best known for his contributions to our knowledge of Quaternary sea-level history and Australian coastal studies. He has worked extensively throughout Australia as well as in China and Papua New Guinea.

PREFACE TO THE 1ST EDITION

As University teachers and active researchers we have long been aware of the need for an up-to-date text dealing with the global and regional environmental changes associated with the Quaternary glaciations. This book is primarily aimed at second and third year undergraduate classes, but it will also be useful for graduate students seeking to enlarge their understanding of global change. We have tried out on our own students, many of the ideas scattered through this book. Our bibliography does not purport to be exhaustive, and is mainly confined to sources in English, but does point the reader to some of the more comprehensive recent studies. Ultimately, the best way to learn is to go out and discover for yourself through fieldwork. It is also a lot more fun. As always, an ounce of practice is worth more than a pound of theory . . . but a little theory can and does help!

To attempt a global overview of Quaternary environments is a daunting task. Twenty years have elapsed since Richard Foster Flint wrote his unsurpassed *Glacial and Quaternary Geology* (Wiley 1972) and Karl Butzer his magisterial *Environment and Archaeology* (Methuen 1971). We understand all too well why so few have followed the difficult trail they blazed, but difficulties seldom disappear unless confronted.

Between us, we have carried out field research on every continent, including Antarctica. In this book we have tried to adopt a world view. There remain some inevitable gaps in our coverage but our preference has been for a selective rather than an encyclopaedic approach.

Many people have helped to enhance our appreciation of Quaternary environmental fluctuations, on land and sea, around the globe. Before and during the conception of this volume, we have enjoyed the benefit of stimulating discussions with friends and colleagues around the world. They include Don Adamson, Stan Ambrose, Mike Barbetti, Raymond Bonnefille, Jim Bowler, Karl Butzer, John Chappell, Desmond Clark, Jack Davies, Tom Dunne, Gerry Eck, Hugues Faure, Leon Follmer, Jean-Charles Fontes, Bob Galloway, Françoise Gasse, Alan Gillespie, John Gowlett, Dick Grove, Bernard Hallet, Jack Harris, Don Johnson, Hilt Johnson, Pete Lamb, Estella Leopold, Liu Tungsheng, Dan Livingstone, Virendra Misra, Nicole Petit-Maire, Steve Porter, S.N. Rajaguru, Pierre Rognon, Roman Schild, Asher Schick, Geoff Spaulding, Alayne Street-Perrott, Minze Stuiver, Marice Taïeb, Mike Talbot, Claudio Vita-Finzi, Donald Walker, Andrew Warren, Link Washburn, Bob Wasson, Fred Wendorf, Tim White, Herb Wright, Karl-Heinz Wyrwoll, and Aaron Yaïr. We thank them all.

Special thanks go to Gary Swinton, who drafted every figure; to Jan Liddicut, who handled numerous drafts with aplomb; and to Tim Barta, Sharon Davis, Alan Fried, Kate Harle, Kim Newbury and Helen Quilligan for their cheerful help with figure compilation and reference checking.

Martin Williams, David Dunkerley, Patrick De Deckker,
Peter Kershaw, Tonia Stokes
Melbourne, March 1992.

PREFACE TO THE 2ND EDITION

Given the rapid changes in our understanding of the interactions between land, sea and ice brought about by tenacious researchers in all parts of the globe, it is gratifying to be able to revise and expand upon many sections in this book. We have added two new chapters, a new appendix on dating methods, and have updated each of the remaining chapters.

We are as ever indebted to our friends and colleagues from around the world who have kept us informed of recent advances and have so generously proffered advice and reprints. Kris James and Alex Blood were towers of strength with word-processing help and bibliographies. Any errors of omission or commission remain ours. If this work proves useful, then the effort will have been worthwhile.

M.A.J. Williams
Adelaide, December 1997

ACKNOWLEDGEMENTS

The publishers would like to thank the following for permission to use copyright material:

A. A. Balkema for figures 1, 2, 6, and 7 reprinted from Zinderen Bakker, E. M. van (ed.), Antarctic glacial history and word palaeonenvironments – Proceedings of a symposium held on 17th August 1977 during the Xth INQUA Congress at Birmingham, UK, 1978. 180 pp. Hfl. 230 – A. A. Balkema, PO Box 1675, Rotterdam, Netherlands; Academic Press Ltd (Sydney) for Adamson, D. A. and Pickford, J. 'Cainozoic History of the Vestfold Hills' in Pickard, J. (ed.) *Antarctic Oasis Terrestrial Environments and the History of the Vestfold Hills* fig 3.20; Academic Press Ltd for Ruddiman, W. F. (1984) 'The Last Interglacial Ocean' *Quaternary Research* 21, fig 56 and Walcott, R. I. (1972) 'Past Sea Levels, Eustacy and Deformation of the Earth' *Quaternary Research* 2, fig 1; Annual Reviews Inc. for fig 1, reproduced with permission from the *Annual Review of Earth and Planetary Sciences* Vol 4 © 1976 by Annual Reviews Inc; Belhaven Press for fig 11.1, reproduced from Thomas, D. S. G. (ed.), *Arid Zone Geomorphology*, Belhaven Press, 1989. All rights reserved; Blackwell Scientific Publications Ltd for Fleet, A. J., Kelts, K. R. and Talbot, M. R. (eds), *Lacustrine Petroleum Source Rocks*, Special Publication of the Geological Society of London No 41 figs 2 and 8; Cambridge University Press for figs 6.2 and 6.3 Funnell, B. M. and Riedel, W. R. (eds), *Micropalaeontology of the Oceans* and Van Andel, T. H. (1985) *New Views on an Old Planet*, pages 171, 175 and 235; Croom Helm for Keya, M. I. (1986), 'Electron Spin Resonance' in Zimmerman, M. R. and Angel, J. L. (eds), *Dating and Age Determination of Biological Materials*; Elsevier Science Publishers for fig 5 from De Deckker, P., Collin, J. P. and Peypouquet, J. P., *Ostracada in the Earth Sciences* (1988) and for fig 2, reprinted from *Geochimica Cosmochimica Acta* 58, McCulloch, M. T. *et al.*, A high-resolution Sr/Ca and delta 180 coral record from the Great Barrier Reef, Australia, and the 1982–3 El Niño, 2749 © 1994; Pergamon Press Plc for fig 14 reprinted from *Physics and Chemistry of the Earth* 4, Fairbridge, R. W. 'Eustatic Changes in Sea level' pages 99–185, copyright (1961) with permission from Pergamon Press Ltd, Headington Hill Hall, Oxford, OX3 0BW, UK; Gebrüder Borntraeger Verslagsbuchhandlung for fig 6 from Hillaire-Marcel, C. and Occhietti, S. (1980) 'Chronology, palaeogeography and palaeoclimatic significance of the late and post glacial events in eastern Canada' in *Zeitschrift für Geomorphologie* 24(4); John Wiley & Sons, Inc. for Strahler, A. N. and Strahler A. H., *Modern Physical Geography*, fig 9.32, Plate D and pages 226–227, Copyright © 1987 by Arthur N. Strahler, for Flint, R. F., *Glacial and Quaternary Geology* figs 3.7 and 18.8, Copyright © 1971 for Mörner, N. A. (1980) 'The Fennoscandian Uplift: Geological Data and their Geodynamical Implication' fig 6, in Mörner, N. A. (ed.) *Earth Rheology, Isostasy and Eustacy* and for fig 8.2 from Denton, G. H. and Hughes, T. J., *Last Great Ice Sheets* Copyright © 1981, reprinted by permission of John Wiley & Sons, Inc., all rights reserved; Kluwer Academic Publishers for Berger, A. L., Imbrie, J., Hays, J., Kukla, G. and Saltzman, B. (eds), *Milankovitch and Climate: Understanding the Response to Astronomical Forcing* (1984) figs 2 and 29, Bach, W., *Our Threatened Climate* (1984), fig 11.2 and Loffler, H. (ed.), *Paleolimnology IV*, fig 2. Reprinted by permission by Kluwer Academic Publishers; Longman Group UK for Tricart, J. and Cailleux, A., *Introduction to Climatic Geomorphology* (1972), fig 3 and for West, R. G. *Pleistocene Geology and Biology* (1977), figs 2.7, 3.6a and 8.7b; National Academy Press for fig 3.1 and table A. 1, reprinted with permission from *Understanding Climatic Change: A Program for Action*, 1975, National Academy Press, Washington, D.C.; Oxford University Press for Ivanovitch, M. and Harmon, R. S. (eds) *Uranium-Series Disequilibrium: Applications to Environmental Problems* (1980), table 3.2, for Stratham, I., *Earth Surface Sediment Transport* (1977), fig 2.13 and for Goudie, A., *Environmental Change: Second Edition* (1983), fig 3.1 and fig 7.2 by permission of Oxford University Press; Pergamon Press for fig 3.4 reprinted with permission from *Ocean Chemistry and Deep Sea Sediments* (1989) Pergamon Press plc; Scientific American, Inc. for illustrations by Sally Black on pages 4–5 from Siever, R., *The Dynamic Earth* (1983); Society for Sedimentary Geology for fig 1, from *Journal of Sedimentary Petrology* 55(2) Thom, B. G. and Roy, P. S., 'Relative Sea Levels and Coastal Sedimentation in South-East Australia in the Holocene' page 258; Springer, Verlag for Spencer, R. J. *et al.* 1984, *Contributions to Mineralogy and Petrology*; The American Association for the Advancement of Science for fig 2, Thompson *et al.*, *Science* Vol. 234, Copyright 1986 by the AAAS and fig 3 from COHMAP Members 'Climatic Changes of the last 18000 years: Observations and model Simulations' *Science* 241, Copyright 1988 by the AAAS; The Macquarie Library Pty Ltd for *The Macquarie Illustrated World Atlas* (1984) pages 46 and 47; Editions Odile Jacob © 1996 for Quand *L'Ocean Se Face*, Duplessy, J.-C., p. 219.

Every effort has been made to trace copyright holders of material reproduced in this book. Any rights not acknowledged here will be acknowledged in subsequent printings if notice is given to the publisher.

CHAPTER 1

QUATERNARY ENVIRONMENTS: AN INTRODUCTION

Il n'y a pas de fait pur; mais toute expérience, si objective semble-t-elle, s'enveloppe inévitablement d'un système d'hypothèse dès que le savant cherche à la formuler.

Teilhard de Chardin (1881–1955)
Preface to *Le Phénomène Humain* (1947)

At a time of widespread apprehension over the impact of human activities upon our biosphere and atmosphere, it is helpful to consider the history of past environmental fluctuations and to draw some relevant lessons from careful study of former human interactions with their natural surroundings. At the present time, there is growing concern that the burning of coal and oil over the last 200 years and the more recent but rapidly accelerating clearance of tropical forests may alter the balance between incoming solar radiation and outgoing terrestrial radiation in ways as yet hard to predict in detail, but which are more likely to lead to global warming than otherwise. The desire to predict possible future climatic changes likely to be triggered by the anthropogenic increase in the atmospheric concentration of carbon dioxide, methane, nitrous oxide and other greenhouse-enhancing gases has spawned an enormous literature (see Graedel and Crutzen 1995, and Houghton *et al.* 1995, 1996, for useful summaries). These attempts to gaze into the future have also re-invigorated studies of the geologically recent past, particularly the Quaternary period, which is the topic of this book.

The Quaternary period spans roughly the last 2 million years of geological time (Fig. 1.1) and is of critical importance in Earth history. It was during this period of remarkably frequent and rapid changes in world climate (Fig. 1.2) (Flint 1971; Butzer 1974; West and Sparks 1977; Bowen 1978; Goudie 1983; Anderson and Borns 1994; Ehlers 1996) that bipedal, toolmaking, fire-using hominids emerged from Africa and gradually moved out to occupy Eurasia, Australia and the Americas, as well as distant oceanic islands throughout the globe. The Quaternary is thus not simply the coda to 4.5 billion years

of Earth history, but is also the time during which we became fully human.

One lesson we are slowly learning after the long saga of continuous human interaction with our environment is that we ourselves are an integral part of that same environment, and that we are the custodians rather than the owners of the lands we now inhabit (Suzuki 1990; Ponting 1991; Tolba 1992). We return to this theme in the final chapter of this book when we consider the past, present and possible future impact of our species upon the air we breathe, the water we drink, and the land and the sea which sustain the plant and animal life upon which we depend for our survival (Mungall and McLaren 1991; McMichael 1993; Holland and Petersen 1995; Simmons 1996).

The aims of this book are to examine some of the global environmental fluctuations of the last 2 million years, to analyse some of the more important evidence

Cainozoic time-scale (Ma BP)

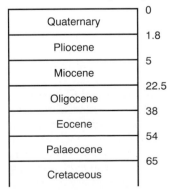

	Ma BP
Quaternary	0
	1.8
Pliocene	
	5
Miocene	
	22.5
Oligocene	
	38
Eocene	
	54
Palaeocene	
	65
Cretaceous	

Fig. 1.1 Cainozoic time-scale (Ma). (Modified from Cowie and Bassett 1989)

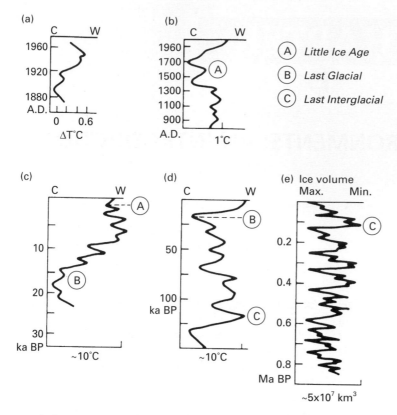

Fig. 1.2 Climatic variability at different time-scales during the last 0.9 Ma of the Quaternary. (Adapted from Australian Academy of Science 1976)

used in reconstructing Quaternary environments (Bowen 1978; Lowe and Walker 1984; Bradley 1985), and to consider some of the ways in which living organisms (including humans) have responded to past environmental changes. We also believe that a knowledge of the past, besides being intrinsically interesting, is also our only real guide to what may befall us in the future.

PRELUDE TO THE QUATERNARY

An accurate long-term perspective on global climatic change has now become possible owing to recent advances in our understanding of world tectonic history. The combined evidence from deep-sea drilling, seismic surveys and palaeomagnetic studies has allowed reconstruction of sea-floor spreading history, and of continental apparent polar-wandering curves. The data from land and sea are impressive and persuasive. The timing of late Cainozoic ice build-up in the two hemispheres is now reasonably well known, as are some of the associated changes in oceanic and atmospheric circulation, which are in turn related to the origin and expansion of the deserts and the contraction of the tropical rainforests. A proper understanding of Quaternary climatic changes therefore requires some appreciation of the legacy of the Tertiary.

The Tertiary and Quaternary periods together comprise the Cainozoic era and embrace the past 65 million years (Fig. 1.1). The present geographical distribution of land, sea and ice (Fig. 1.3) and of the corresponding morphoclimatic regions shown in Fig. 1.4 are the end-product of Mesozoic and Cainozoic lithospheric plate movements and sea-floor spreading. A number of major regional episodes, including Himalayan uplift, Antarctic ice accumulation, closure of the Panama isthmus, build-up of the North American ice sheets, intertropical cooling and desiccation, and expansion of savanna at the expense of tropical rainforest, were all closely linked with the global tectonic events of the Tertiary (Ruddiman and Kutzbach 1991; Quade *et al.* 1995; Derbyshire 1996; Liu *et al.* 1996; Ramstein *et al.* 1997) and are the subject of Chapter 2.

QUATERNARY GLACIATIONS

Ice began to accumulate on Antarctica well over 20 million years ago. Ice build-up came much later in the northern hemisphere, and it was not until 2.4 million years ago that major ice sheets began to grow rapidly in North America. For reasons which remain obscure, but which appear to be closely related to cyclical changes in the Earth's orbital path around the sun and in the tilt of the Earth's axis (Jiang and

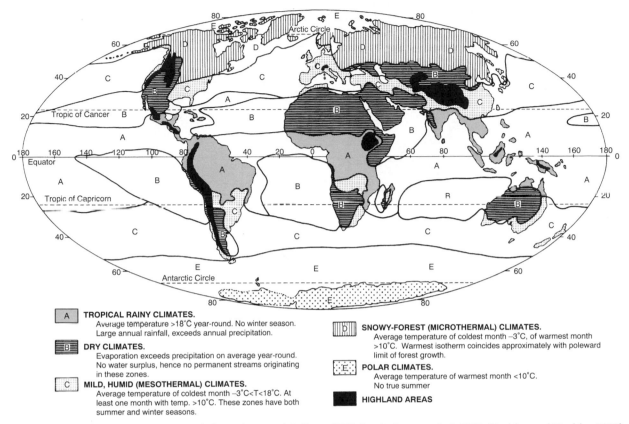

Fig. 1.3 Present-day global climates. (After Tricart and Cailleux 1972; Bartholomew *et al.* 1980; Strahler and Strahler 1987)

A	TROPICAL RAINY CLIMATES.

Average temperature >18°C year-round. No winter season. Large annual rainfall, exceeds annual precipitation.

B	DRY CLIMATES.

Evaporation exceeds precipitation on average year-round. No water surplus, hence no permanent streams originating in these zones.

C	MILD, HUMID (MESOTHERMAL) CLIMATES.

Average temperature of coldest month –3°C<T<18°C. At least one month with temp. >10°C. These zones have both summer and winter seasons.

D	SNOWY-FOREST (MICROTHERMAL) CLIMATES.

Average temperature of coldest month –3°C, of warmest month >10°C. Warmest isotherm coincides approximately with poleward limit of forest growth.

E	POLAR CLIMATES.

Average temperature of warmest month <10°C. No true summer

HIGHLAND AREAS

Peltier 1996), the great ice sheets of the northern hemisphere in particular developed a characteristic cycle of slow build-up to full glacial conditions, followed by rapid ice melting and deglaciation. These topics are the focus of Chapters 3 to 5.

QUATERNARY SEA-LEVEL CHANGES

The larger of the Quaternary ice caps were up to 4 km thick. As the ice caps slowly built up to attain their maximum thickness, the underlying bedrock was progressively depressed beneath the weight of accumulating ice. When the ice melted, the crust slowly rose again to its preglacial level. These *isostatic readjustments* to the waxing and waning of the great Quaternary ice sheets caused changes in the relative levels of land and sea. During glacial maxima, roughly 5.5% of the world's water was locked up in the form of ice (the corresponding value today is 1.7%). As the ice sheets grew, so the level of the world's oceans fell by up to 150 m, depending upon total ice volume. With deglaciation and rapid melting of the ice caps, sea-level rose once more to about

present levels. These *glacio-eustatic* sea-level fluctuations are analysed in Chapter 6, together with the influence of *isostasy* and other tectonic movements upon global and local sea-levels (Warrick *et al.* 1993).

EVIDENCE FROM THE OCEANS

Reconstruction of past sea-level fluctuations can throw useful light on the rate of accumulation and the rate of melting of global ice, but well-dated Quaternary sea-level histories only extend back some 250 000 years, so that the first 90% of the record must be sought elsewhere, most notably from deep-sea cores.

Inferences about changes or fluctuations in ocean circulation patterns used to depend very largely upon sedimentological and microfossil studies. Analysis of the oxygen isotopic composition of the calcareous tests of suitable benthonic and planktonic foraminifera now provides an additional and powerful means of assessing changes in ocean water temperature and salinity at depth and near the surface (see Chapter 7). After allowing for local effects, it is also possible to use this technique to estimate changes in

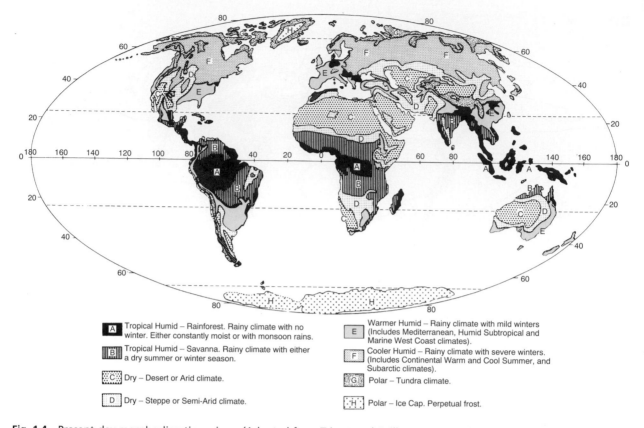

Fig. 1.4 Present-day morphoclimatic regions. (Adapted from Tricart and Cailleux 1972; Strahler and Strahler 1987)

global ice volume. Deduced changes in regional surface salinity can also indicate changes in runoff from major rivers, changes in evaporation, and changes in the amount of seasonal rainfall (De Deckker 1997).

The record from deep-sea cores has the double advantage of good global coverage and of spanning much of the Cainozoic. There are comparatively few such long, continuous terrestrial records, and those that do exist are usually confined to particular types of lake basin. It is still too soon to say whether or not the long ice cores collected from Antarctica and Greenland represent a continuous sequence of ice accumulation, although present evidence seems to indicate an unbroken record spanning the last interglacial in Greenland and over 50 000 years beyond that in Antarctica. The Chinese loess record also appears to provide an unbroken record matching that of the Greenland ice cores in its remarkable detail (Porter and An 1995; An and Porter 1997).

RIVERS, LAKES AND GROUNDWATER

Although the oceanic record can provide unrivalled information about the pattern and tempo of global climatic fluctuations in the Quaternary, it is often more useful to know about the direct changes to the landscape caused by local and regional hydrological fluctuations. Our increased understanding of the global linkages or teleconnections between historic floods, droughts and sea surface temperature anomalies, epitomised by the climatic variations associated with El Niño–Southern Oscillation events (Allan *et al.* 1996), demonstrates the very practical relevance of such studies. A further and still unresolved issue is the nature of the interactions between climatic variability (including short-term droughts and longer-term climatic desiccation) and desertification processes (Williams and Balling 1996). Such former hydrological changes are evident in the Quaternary depositional legacy of rivers large and small, as well as in the ever-changing response of lakes to local fluctuations in evaporation, precipitation and groundwater levels. Unfortunately for our purposes, the alluvial history of most rivers can only be pieced together from fragmentary and often poorly dated suites of sediments. However, as Chapter 8 points out, rivers and lakes together can yield highly informative accounts of how certain regions responded to the environmental vicissitudes of the Quaternary, and of how they may well respond in the future (Costa *et al.* 1995; Gregory *et al.* 1995).

EVIDENCE FROM THE DESERTS

A growing body of evidence from deep sea cores, lake deposits and ice cores shows that times of lowest world temperature during the Quaternary (*glacial maxima*) were times of greatest aridity on land, with massive export of desert dust offshore, and even to central Antarctica (Yung *et al.* 1996). Deserts are excellent geological and geomorphological museums, for the very aridity to which they owe their existence has minimised the destructive impact of fluvial erosion and has helped to conserve an array of river, lake and wind-blown deposits. These deposits sometimes contain remarkably well preserved and occasionally, as in certain semi-arid rift valleys in Africa, or the loess plateau in China, a nearly-continuous fossil vertebrate and invertebrate record spanning most of the late Pliocene and Quaternary. Chapter 9 enlarges on these topics.

EVIDENCE FROM NON-MARINE FLORA AND FAUNA

The emergence of the plants and animals upon which humans have long depended for food and shelter took place against the environmental changes of the late Tertiary and was finally accomplished during the Quaternary. Changes in the non-marine plant and animal record provide an invaluable adjunct to the purely physical evidence furnished by landforms and sediments, and can be used to reconstruct former temperature and rainfall fluctuations with great precision and accuracy. Some organisms are inherently sensitive to local changes in habitat, and may respond rapidly to external disturbance. Perhaps the most versatile and certainly one of the best tested methods used in Quaternary environmental reconstruction is the technique of *pollen analysis*, which is considered in some detail in Chapter 10, along with other more circumscribed techniques.

HUMAN ORIGINS, INNOVATIONS AND MIGRATIONS

As the great Quaternary ice caps waxed and waned, and deserts expanded and contracted, a small-brained vegetarian hominid left its footprints clearly visible in a carbonatite ash which was laid down during a volcanic eruption near Laetoli in Tanzania nearly 4 million years ago. This creature, *Australopithecus afarensis,* was fully bipedal, and may well be the ancestor from which later hominids, including the genus *Homo*, were to derive. Chapter 11 describes

the slow progression from user of tools to toolmaker, discusses the development and refinement of stone knapping techniques, and concludes with a short analysis of the origins of plant and animal domestication. The food-producing economy of the Neolithic saw the virtual demise of most hunter-gatherer societies around the world, and the inception of modern urban civilisation.

ATMOSPHERIC CIRCULATION DURING THE QUATERNARY

The cultural development of our human forebears took place against a background of ever-changing global climate. In the intertropical zone, for instance, cold, dry and windy glacial maxima alternated with warm, wet interglacials. Regions delineated as arid in Figs 1.3 and 1.4 were sometimes studded with deep freshwater lakes; areas now under rainforest were sometimes covered in savanna, or partly mantled with wind-blown sand.

Chapter 12 is an attempt to explore some of the changes in global atmospheric circulation patterns during the Quaternary, particularly the terminal Pleistocene towards 18 000 years ago, and the early Holocene around 9000 years ago. We do this for two very good reasons. First, the last 20 000 years contain the best-dated, best-preserved and most abundant palaeoclimatic evidence with which to test global atmospheric circulation models. Second, the two time-spikes considered coincide, respectively, with the last glacial maximum (18 000 years ago) and the so-called early Holocene 'climatic optimum' of 9000 years ago, which we prefer to regard as simply the postglacial antithesis of the full glacial climate. Between them, they encompass a substantial component of climatic range of the most recent glacial–postglacial cycle (Bell and Walker 1992; Wright *et al.* 1993).

ENVIRONMENTAL CHANGES: PAST, PRESENT, FUTURE

Throughout the Quaternary there has been a prolonged series of interactions between hominids (ancestral humans) and their environment. Stone toolmaking dates back to about 2.5 million years ago, and fire was being used in Africa about a million years later. The question of how far prehistoric hunters contributed to the demise of certain species of animals is a vexed one, as is the related question of the role of burning in bringing about plant extinctions. With the advent of Neolithic food production, and accelerated clearing of the natural vegetation, the degree of human impact upon the biosphere and hydrosphere began to increase dramatically (Turner *et al.* 1990; Mannion and Bowlby 1992; Roberts 1994; Middleton 1995; Brown 1997). By altering plant cover, we may increase runoff, and thereby accelerate soil erosion. There is a delicate balance between the different components of the hydrosphere (Fig. 1.5) and the atmosphere (Fig. 1.6). Since the Industrial Revolution, in particular, we have begun to interfere with that balance by unwittingly altering some of the feedback loops which are an integral part of the global climate system (Fig. 1.5). Chapter 13 discusses these issues in greater detail.

QUATERNARY CHRONOLOGY

There has long been controversy over the exact duration of the Quaternary. Some workers espouse a long chronology starting as early as 3.5 Ma. Others prefer a shorter chronology, beginning at 2, 1.8 or 1.6 Ma. We tentatively opt for 1.8 Ma (Fig. 1.1), which also coincides reasonably well with the Olduvai palaeomagnetic event, an interval with normal magnetic polarity bracketed by K–Ar dates of 1.87 and 1.67 Ma (see Appendix). An equally good case may be made for placing the Pliocene–Pleistocene boundary at 2.5 Ma, when there was a rapid build-up of ice in the northern hemisphere. Suc and colleagues (1997) have recently argued very persuasively that the Pliocene–Pleistocene boundary should be placed at the Gauss–Matuyama reversal at 2.58 Ma. There is much merit in this proposal, since it fulfils the necessary geological criteria and would be relatively easy to identify both on land and in ocean cores, The choice of Quaternary boundary is very much a matter of personal taste (Vita-Finzi 1973), and has often generated more heat than light. We likewise favour a simple four-fold subdivision of the Quaternary (Fig. 1.7) into Lower Pleistocene (1.8 to 0.75 Ma), Middle Pleistocene (750 to 125 ka), Upper Pleistocene (125 to 10 ka) and Holocene (10 to 0 ka), while noting that none of these somewhat arbitrary divisions or ages is particularly sacrosanct.

RECONSTRUCTING QUATERNARY ENVIRONMENTS

A knowledge of past events and processes can offer useful insights into both present and future environmental changes, but a few preliminary words of caution are necessary here. Earth history is a tale of constantly varying interactions over time between lithosphere, atmosphere, hydrosphere (including cryosphere) and biosphere. Present world landscapes reflect the influence of past as well as present-day

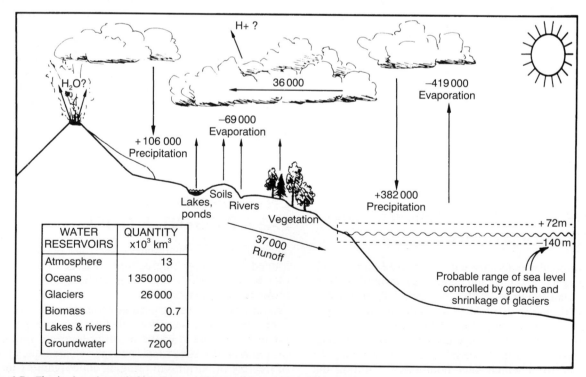

Fig. 1.5 The hydrosphere. (After Bloom 1978; Strahler and Strahler 1987)

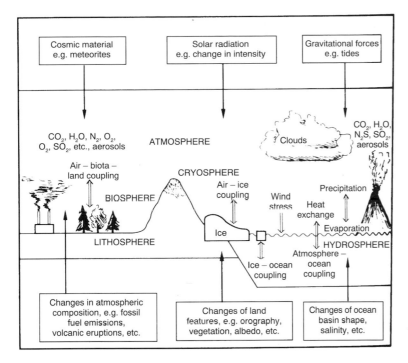

Fig. 1.6 Feedbacks in the global climate system. (After Bloom 1978; Bach 1984; Bradley 1985)

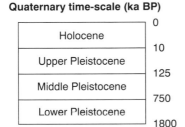

Fig. 1.7 Quaternary time-scale. (Modified from Shackleton and Opdyke 1977; Cowie and Bassett 1989)

processes. Theoretical constructs about the relation between present-day weathering processes and climate (or latitude) are only useful if we are fully aware of their limitations.

Table 1.1 shows some of the types of evidence commonly used to reconstruct Quaternary environments and climates. Each is useful for a specific purpose, and for a particular area or time (Leroy Ladurie 1972; Vita-Finzi 1973; Lowe and Walker 1984; Bradley 1985; Bradley and Jones 1995; Benda 1995; Wadia *et al.* 1995). Difficulties arise immediately when we use Procrustean tactics to force the data to yield palaeo-environmental information at particular scales in space or time for which those data are totally inappropriate. It is essential always to take due note of the time-scales at which the different processes involved in environmental change normally operate (Fig. 1.8). A related issue is the precision available in dating the proxy data or samples used in reconstructing past events (Fig. 1.9). In many cases we are still limited by the imprecision of existing dating methods, which is

often a function of the half-life length of the particular radioactive isotopes involved (see Appendix). Given all of the above caveats, it would seem that the task of Quaternary environmental reconstruction is still more of an art than a science. Such a conclusion is in no way dismissive of some of the excellent progress made in quantifying past fluctuations in temperature and salinity, on land and in the sea, using stable isotopes and trace element composition of the calcareous shells of ostracods and forams. However, we still have a very long way to go to gain the spatial and temporal resolution necessary to test existing models of global atmospheric circulation in the Quaternary (Chapter 12).

QUATERNARY ENVIRONMENTAL ANALOGUES

It is always tempting to use past climatic events as analogues for possible future climatic changes. For example, some workers have suggested that future global warming linked to the greenhouse effect may have an early Holocene climatic analogue. We consider that such claims should be treated with considerable caution, especially since the early Holocene boundary conditions, including sea-level, the extent of the cryosphere, terrestrial albedo, and sea surface temperatures, may have been very different from those used to model future change. Of greater value in understanding possible future change is the geological and biological evidence of past hydrological

Table 1.1 Sources of data used to reconstruct Quaternary environments

Proxy data source	Variable measured
Geology and geomorphology – continental	
Relict soils	Soil types
Closed-basin lakes	Lake level
Lake sediments	Varve thickness
Aeolian deposits – loess, desert dust, dunes, sand plains	
Lacustrine deposits and erosional features	
Evaporites, tufas	Age
Speleothems	Stable isotope composition
Geology and geomorphology – marine	
Ocean sediments	Ash and sand accumulation rates
	Fossil plankton composition
	Isotopic composition of planktonic and benthic fossils
	Mineralogical composition and surface texture
	Geochemistry
Continental dust	
Biogenic dust: pollen, diatoms, phytoliths	
Marine shorelines	Coastal features
	Reef growth
Fluviatile inputs	
Glaciology	
Mountain glaciers, ice sheets	Terminal positions
Glacial deposits and features of glacial erosion	
Periglacial features	
Glacio-eustatic features	Shorelines
Layered ice cores	Oxygen isotope concentration
	Physical properties (e.g. ice fabric)
	Trace element and micro-particle concentrations
Biology and biogeography – continental	
Tree rings	Ring-width anomaly, density
	Isotopic composition
Fossil pollen and spores	Type, relative abundance and/or absolute concentration
Plant macrofossils	Age, distribution
Plant microfossils	
Vertebrate fossils	
Invertebrate fossils: mollusca, ostracods	
Diatoms	
Insects	Type, assemblage abundance
Modern population distributions	Refuges
	Relict populations of plants and animals
Biology and biogeography – marine	
Diatoms	Faunal and floral abundance
Foraminifera	Morphological variations
Coral reefs	
Archaeology	
Written records	
Plant remains	
Animal remains, including hominids	
Rock art	
Hearths, dwellings, workshops	
Artefacts: bone, stone, wood, shell, leather	

Source: After National Academy of Sciences (1975), Bradley (1985) and Williams (1985a)

events. Rather than seeking past climatic analogues to a warmer Earth, it may be more useful for us to focus on how various elements of the biosphere and hydrosphere have responded to former climatic fluctuations. What were the directions and rates of response? What were the thresholds? Was the response synchronous or time-transgressive? How did one set of changes (e.g. deforestation) repercuss upon the rest of the landscape? If we adopt this approach then it is possible to argue that an appreciation of past

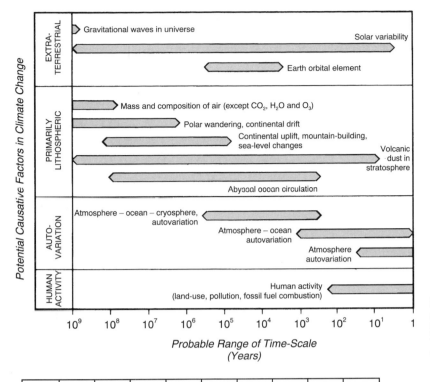

Fig. 1.8 Processes involved in environmental change and their time scales. (After National Academy of Sciences 1975; Bloom 1978; Goudie 1983; Bradley 1985)

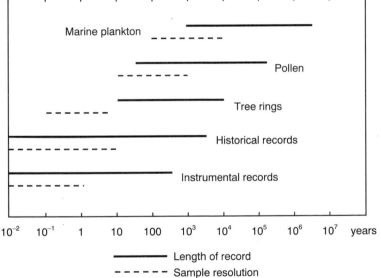

Fig. 1.9 Sample resolution and length of potential paleoclimatic record from various independent lines of evidence. (After Birks 1981, modified from T. Webb, unpublished diagram)

Quaternary events can provide us with insights about possible future events which are unattainable by any other means.

PRACTICAL RELEVANCE OF QUATERNARY RESEARCH

The Quaternary legacy is ubiquitous. Many of our soils formed during the Quaternary, as did many of the depositional features created by moving ice, wind and rivers. Human activities in the last few centuries have served to accelerate many natural processes, including soil erosion by wind and water. Some modern rates of soil erosion are several orders of magnitude greater than the long-term geological rates for those regions. One reason for this discrepancy may simply be the ease with which unconsolidated Quaternary sediments can be mobilised by present-day runoff, but another may be destruction of the vegetation cover, which increases the vulnerability of the soil surface to the erosive impact of rainsplash and runoff.

A knowledge of the rates and magnitudes of past and present environmental change is essential to our understanding of the world we live in. Planners and policy-makers are becoming increasingly attuned to the relevance of Quaternary studies to agricultural and resource management. For instance, Quaternary research can contribute its unique historical perspective to a sensible policy of long-term management of soil erosion, desertification, salinisation, coastal erosion, floods and droughts, and biological conservation. Recent experience shows all too well that to ignore the past is to court future land-use problems. Present rates of plant and animal population changes mean very little unless set in a historical context, in this case Quaternary palaeoecology. The long-term development and preservation of our soil, plant and groundwater resources thus requires a balanced understanding of recent Quaternary environmental changes as well as a thorough knowledge of present-day geomorphic, ecological and hydrological processes.

FURTHER READING

Bowen, D.Q. 1978: *Quaternary Geology. A Stratigraphic Framework for Multi-disciplinary Work*. Pergamon Press, Oxford, 221. (A good clear account with an emphasis on stratigraphy and correlation).

Bradley, R.S. 1985: *Quaternary Paleoclimatology. Methods of Paleoclimatic Reconstruction*. Allen and Unwin, Boston, 472p. (An excellent review of the scope and limitations of many of the methods used to reconstruct former climates).

Butzer, K.W. 1974: *Environment and Archaeology. An Ecological Approach to Prehistory* (3rd edn), Chicago, Aldine. (A masterly and advanced overview of the methods used to reconstruct Quaternary prehistoric environments).

Flint, R.F. 1971: *Glacial and Quaternary Geology*. Wiley, New York. 892 p. (The classic text on glacial geology by one of the great masters. Still well worth consulting).

Goudie, A. 1983: *Environmental Changes* (2nd edn). Clarendon Press, Oxford, 258p. (A clear, concise and eminently readable undergraduate text on Quaternary environments).

Lowe, J.J. and Walker, M.J.C. 1984: *Reconstructing Quaternary Environments*. Longman, London, 389 p. (A useful account, with examples drawn mainly from Europe and North America).

Vita-Finzi, C. 1973: *Recent Earth History*. MacMillan, London, 138p. (A thoughtful, concise and often witty analysis of relative and absolute dating methods used in late Quaternary research).

West, R.G. and Sparks, B.W. 1977: *Pleistocene Geology and Biology, with special reference to the British Isles* (2nd edn). Longman, London, 440 p. (A comprehensive text by two highly experienced practitioners; useful well beyond the British Isles).

CHAPTER

2

PRELUDE TO THE QUATERNARY

Who can tell us whence and how arose this universe?
Anon (c. 1000BC) *Rigveda-Samhita*

In the Quaternary period, the slow and uneven progression toward cooler conditions which had characterised the Earth for tens of millions of years (and about which we shall have more to say shortly) gave way to extraordinary climatic instability. Temperatures changed repeatedly from values like those of the present day to levels many degrees colder, with the growth of enormous ice sheets on land. In the cold phases, sea-level fell dramatically, while on land, mountain treelines shifted to much lower elevations, grasslands replaced forests, ground ice and associated periglacial processes became more common, and dust blew extensively on cold winds. Simultaneously, human cultural development began with the early use of stone tools and of fire, culminating in the Holocene cultivation of crops and the herding of domestic animals. These developments fostered the growth of urban civilisation and the development of writing.

The environments of this remarkable period of 1.8 Ma developed from those of the Tertiary, and are best understood with a knowledge of the global environment as it was prior to the onset of the marked Quaternary instability. We therefore begin our exploration of the Quaternary with a brief investigation of longer-term environmental trends and of the global processes responsible for them. A particular goal of the discussion will be to introduce some of the hypotheses that seek to account for the absence of major ice sheets over western Europe, Scandinavia and the Canadian region until the Quaternary, many millions of years after the Antarctic and Greenland ice sheets developed to modern dimensions.

THE LITHOSPHERIC PLATES AND THEIR MOTION

The environment at the surface of the Earth is controlled by a suite of *external* and *internal* influences. A dominant external influence is the amount of solar radiation received through the atmosphere, which varies in the long term as the sun ages, and more rapidly because of changes in the geometry of the Earth's orbit around it; events associated with the impact of comets provide another.

One of the principal *internal* influences on the environment is the geometry and movement of the lithospheric plates. These plates, fragments of the *lithosphere* (the crust together with the uppermost part of the mantle), move over the Earth's surface because of gradients set up by thermal isostatic processes. The insulating effect caused by large landmasses restricts the escape of geothermal heat and gradually warms the mantle and the lower crust, so that eventually, warping and convective movements begin. Carrying the continents with them, the moving plates determine the form of the ocean basins, and the distribution of land and water. The suggestion has been made that for a major period of global glaciation to develop, land must be located over the polar regions, to provide a site for ice accumulation and to facilitate the reinforcement of cooling by albedo feedback (e.g. see Crowell & Frakes 1970; Burke *et al.* 1990).

The present configuration of the lithospheric plates is shown in Fig. 2.1. This pattern has developed by the disintegration of the supercontinent of Pangaea (and the superocean Tethys), which existed for some hundreds of millions of years before commencing to break up in the Jurassic, around 180 Ma BP. Early rifting opened the proto-Atlantic Ocean, and the development of the other modern oceans followed during the Mesozoic era, concluding with the

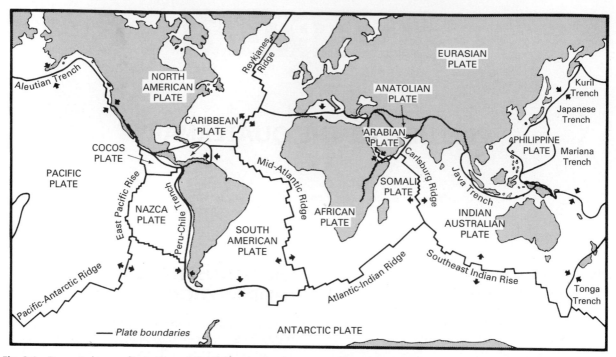

Fig. 2.1 Present-day configuration of the lithospheric plates. Note the relative absence of plate boundaries crossing continental crust, and the remarkable symmetry of the Atlantic Ocean about its axial mid-oceanic ridge system. (After Press and Siever 1986 and Strahler and Strahler 1987)

rapid development of the Southern Ocean after about 50 Ma BP, almost entirely within the Cainozoic era. During this process, the mantle warmth built up under Pangaea was allowed to dissipate, so that the smaller continental fragments cooled and consequently subsided as they moved; simultaneously, the old Tethyan sea floor was replaced with younger, more thermally buoyant oceanic crust, and the average elevation of the sea floor rose. An inescapable consequence of these changes acting in concert was major marine transgression on a global scale.

Associated with the continental fragmentation were other changes, however. The system of oceanic circulation in the Tethys was gradually replaced by the contemporary pattern of smaller, constrained gyres carrying sensible heat across the modern oceans. The transport of heat by ocean currents, especially the global thermohaline circulation system, is a major component of the thermal balancing of the global climate. The Gulf Stream, for example, carries water warmed in the low latitudes poleward where the heat is given off, making regional climates at high latitudes more equable. Not surprisingly, as plate movement changed the geometry and interconnectedness of the oceans, significant regional climatic changes were produced; some of these will be described below. Various seaways opened and closed during the redistribution of the continents, major instances including the final closure of the Panama

Isthmus at 3–4 Ma BP, and the opening of the Drake Passage between South America and Antarctica, and of the major Southern Ocean passage between Australia and Antarctica (see Fig. 2.2). The relatively rapid readjustment of circulation systems that followed these events was one of the factors contributing to the uneven, stepped history of climatic deterioration in the Cainozoic.

Let us consider briefly the effects flowing from the closure of the isthmus of Panama, an event which has been the object of a number of modelling studies. Tectonic processes began this closure at about 12.5 Ma BP, with the present configuration achieved by about 3 Ma BP. By re-opening the oceanic gateway in a numerical model of the ocean and atmosphere, Murdock *et al.* (1997) showed that the northward flow of Gulf Stream water was greatly reduced. Consequently, evaporation in the North Atlantic was depressed, and the cold, saline water that forms there today, and whose sinking provides a major impetus for the global thermohaline circulation, was also curtailed. Indeed, in the model, the production of NADW (North Atlantic Deep Water) ceased altogether. Consequently, heat from low latitudes was not carried poleward by the 'global conveyor belt' of the oceanic circulation. Thus, with closure of the isthmus near the start of the Quaternary, warm water was diverted northward in much greater amounts. This provides one of the key

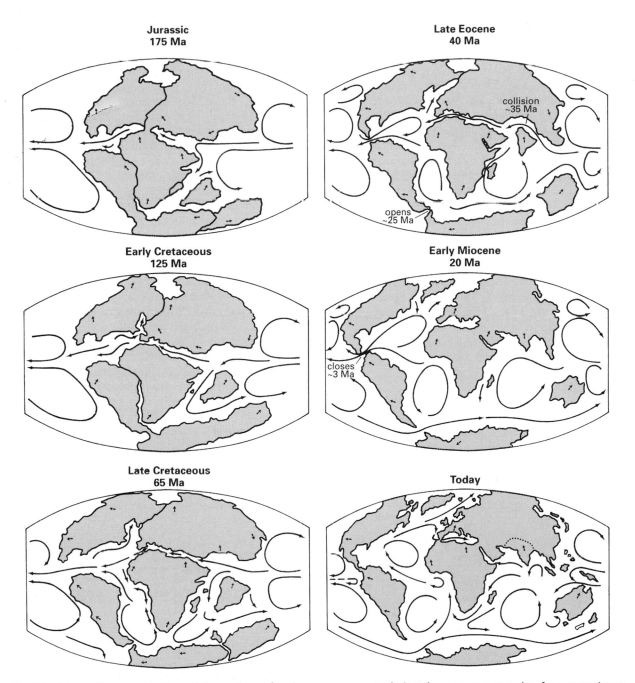

**Jurassic
175 Ma**

**Late Eocene
40 Ma**

collision
~35 Ma

opens
~25 Ma

**Early Cretaceous
125 Ma**

**Early Miocene
20 Ma**

closes
~3 Ma

**Late Cretaceous
65 Ma**

Today

Fig. 2.2 Schematic reconstruction of the pattern of major ocean currents during the most recent cycle of supercontinent disintegration. The timing of certain major events in the development of the oceanic circulation is indicated in the diagram. (After van Andel 1985 and Strahler and Strahler 1987)

requirements for growth of the great ice sheets: a relatively warm oceanic moisture source from which evaporation can draw the precipitation needed for ice accumulation. These inferred changes in the thermohaline circulation of the Atlantic Ocean were confirmed independently from sedimentary evidence by Burton *et al.* (1997). The closure of oceanic gateways like the isthmus of Panama also had effects on

the marine biota, leading to the isolation of subgroups of organisms (e.g. see Vermeij 1993; Webb 1995).

The breakup of Pangaea was associated with elevated levels of igneous activity along the rift systems and the new continental margins. This increase was responsible for a parallel rise in the carbon dioxide concentrations in the atmosphere, both by direct emission, and because the global transgression already referred to reduced the area of exposed land on which chemical weathering could take up CO_2 from acidified rain. Positive feedback of many kinds existed in this environmental change (as in so many environmental processes), acting to reinforce the initial tendency. For example, the greenhouse warming produced by the CO_2 would be likely to warm the surface of the oceans; this would reduce the solubility of the gas and further enhance the warming, as would the increased atmospheric water vapour concentration resulting from evaporation from the warm sea surface. Likewise, the warming would cause thermal expansion of seawater, further supporting global transgression and the reduction in area of exposed land. In turn, the growing global proportion of sea (and the diminishing land fraction), would, because of the lower albedo of the ocean surface, further reinforce a warming trend. Through mechanisms such as this, significant climatic change may be triggered by seemingly relatively minor events.

There are other little-understood effects of the oceanic ridge activity that are associated with plate movement. Seawater cools these sites, and carries with it dissolved silica and other materials. These may exert a significant influence on oceanic productivity, in the same way that materials delivered to the oceans by riverine transport do. At sites of plate collision, tectonic dewatering of sediments undergoing metamorphism may, over the long term, contribute additional fluxes of silica and other dissolved materials (Fyfe 1990a), but such events are difficult to quantify and to date.

TECTONIC CONTROL OF THE OCEANS AND CONTINENTS

In their present-day configuration, the lithospheric plates vary in their geography. Some, like the Pacific plate, are largely covered by ocean. Some (such as the Eurasian) are largely occupied by land. Others, like the Indo-Australian plate, carry major continents as well as oceans.

Certain features stand out in the fundamental structure of this assemblage of plates: the ocean basins contain, in addition to the seawater, the bulk of *divergent* plate boundaries (the mid-oceanic ridge system), and their underlying crust is of basaltic composition.

The continents (those parts of the surface of the Earth not water-covered) are in addition different in composition: they contain assemblages of rocks dominated by aluminosilicate minerals of lower density than those found in the basalts of the oceanic

crust; furthermore, very few plate boundaries cross continental crust.

An explanation for these features of the planet is readily available. The underlying mantle supports, by its own deformation in response to loads applied from above, the mass of the crust. Rocks forming the crust of the Earth sit at two distinct elevations, as a function of their composition, as expressed in the concept of isostasy. The continents have a mean elevation of 840 m, while the ocean floors lie about 4.6 km lower, at an average depth of 3800 m (Gross 1982). This *bimodal* distribution of elevations would exist whether the oceans contained water or not (although the ocean floors, without the weight of water, would rise somewhat). Water condensing at the Earth's surface through time has inevitably accumulated above the lower-lying areas of oceanic crust. Thus, areas of basaltic crust are oceans and areas of continental (granitic) composition are land areas. There are several additional reasons why the continental crust stands high. Water and sediment are continually shed into the oceans as a result of river runoff and erosional processes affecting the land. The considerable weight of terrestrial and marine sediments accumulating in the oceans causes mantle subsidence and corresponding uplift of the land. Furthermore, as a general rule, the larger the continent, the higher its average elevation. This phenomenon is not completely understood; it may reflect accumulation of heat and consequent isostatic rise of larger continents, or it may relate to the thickness of the crust, smaller fragments of continental crust perhaps being thinner in general, and hence lying at lower elevations.

EFFECTS OF PLATE EVOLUTION ON THE ATMOSPHERE AND OCEANS

The episode of Mesozoic and Cainozoic continental dispersal described earlier must have been associated with many changes in the mechanisms which control the global climate.

Marine regression and transgression clearly alter the proportions of land and water at the Earth's surface, and hence the overall planetary albedo. Times of high sea-level would, all else being equal, tend to be associated with enhanced global warming.

The movement of the continents has the potential to alter global heat balance in other ways. Land located at high latitudes can support snow and ice cover if the regional temperature is sufficiently low, reinforcing cooling by the albedo mechanism. Land of sufficient elevation, located at any latitude, and produced perhaps by uplift following continental collision, can act in the same way. It is possible that the movement of land into higher northern latitudes during the episode of Mesozoic and Cainozoic drift acted in just this way, and so contributed to the Cainozoic cooling. A second albedo-based mechanism is possible in view of the fact that continental movement between 100 Ma BP and the present considerably increased the area of land in the northern subtropics. Aridity here, produced by subsiding air, results in high terrestrial albedo, and this effect too may have supported Cainozoic cooling (Kennett 1982).

The distribution of heat *within* the earth–atmosphere–ocean system is also of fundamental importance to global and regional climates. Changes in the configuration of land or oceans which affect the winds or the oceanic currents thus may also be involved in climatic change. The uplift of major mountain belts, particularly if they are *meridionally* aligned, can alter the long-wave structure of the atmospheric circulation, perhaps directing polar air into a more southerly course, and influencing the production and survival of snow and ice.

GLOBAL COOLING AND GROWTH OF THE ANTARCTIC ICE CAP

At the start of the Cainozoic, Antarctica was ice-free. Oxygen isotope palaeotemperature analysis shows that the nearby oceans were quite warm, perhaps 18°C (see Fig. 2.3). Forests grew even at very high latitudes.

Continental movement in key locations may have been pivotal in the subsequent cooling. As Australia moved north after about 50 Ma BP, opening a significant Southern Ocean seaway, the westerly wind circulation was able to establish the major Antarctic Circumpolar Current (ACC). This current completely encircles the Antarctic continent, and acts as a barrier preventing warm currents from lower latitudes reaching the Antarctic coast. The development of this current, therefore, must have substantially reduced the transport of *sensible* heat to the high southern latitudes, and contributed to growing cold there. The tropics may have been less affected by these events, so that the equator–pole temperature gradient perhaps began to increase, so invigorating atmospheric circulation.

In the Cainozoic cooling trend, a series of quite rapid temperature drops can be discerned (see Fig. 2.3). Each may relate to steps in the development of the modern configuration of land and ocean, which we will briefly review.

During the Paleocene and Eocene, it would have been possible for major equatorial ocean currents to encircle the globe, with passages open between the Americas, between India and Europe, and between Australia and Indo-China. The climate must have been generally warm during these periods. For the preceding Cretaceous, with Pangaea only partly fragmented, the temperature difference between the low and high latitudes may have been less than 4°C. As the Eocene progressed, the Southern Ocean began to widen, but the ACC was not able to develop to its full depth because shallow barriers still existed here and in the Drake Passage. However, oceanic cooling proceeded in these latitudes because of the restriction posed by the developing ACC to the poleward transport of sensible heat.

One of the major cooling episodes referred to earlier took place at about 38 Ma BP, at the end of the Eocene. Deep-water temperatures dropped by 4–5°C, and this is taken to indicate the development of freezing conditions at sea-level around Antarctica, with extensive sea ice. Temperatures over land in the high latitudes of each hemisphere may have dropped by 10°C. Development of the ACC in the Oligocene, enhanced by the opening of the Drake Passage at about 30 Ma BP, produced further cooling and the progressive development of ice on Antarctica. The very rapid cooling at the end of the Eocene probably reflected the influence of positive feedback effects relying on the albedo mechanism as land ice grew in Antarctica, with sea ice possibly also extensive. The particular role of the opening of the Drake Passage in Antarctic glaciation was addressed in modelling work by Mikolajewicz *et al.* (1993). This work suggested that the oceanographic changes alone may have been insufficient to account for glaciation, and that perhaps parallel changes in atmospheric CO_2 levels, or some other factor, must also have been involved.

We should note that there are hypotheses that seek to account for Antarctic glaciation by other kinds of oceanographic change. For instance, Prentice and Matthews (1991) have proposed the 'snow gun hypothesis', in which variations in moisture supply feeding precipitation over Antarctica are seen to be a key factor. In this hypothesis, the warming of oceanic deep waters, which upwell at high southern latitudes, is envisaged to have fostered the accumulation of snow in a manner analogous to the process in which water is sprayed over ski resorts to build up snow depth. Variations in deep water temperature could be accounted for by various tectonic and oceanographic changes in the low latitudes.

By the early Oligocene, the approach of India to the Eurasian land mass must effectively have closed off the equatorial currents of the Palaeocene and Eocene (Williams 1986). The widening of the Southern Ocean and Drake Passage oceanic gateways

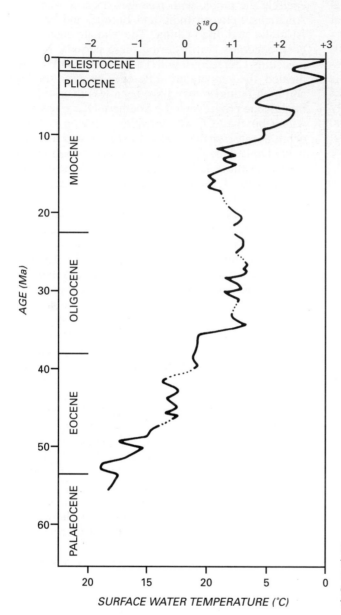

$\delta^{18}O$

Fig. 2.3 Reconstruction of the surface water temperatures of high-latitude southern hemisphere oceans during the Cainozoic era. Early Cainozoic warmth gradually gives way to the cold conditions of the Quaternary and the present day. (From Shackleton and Kennett 1975)

during much the same time permitted full development of the ACC, including bottom-water circulation.

By the middle Miocene, extensive Antarctic ice was present, and significant cooling is shown in the palaeotemperature record (Fig. 2.3). Ice-rafted debris is found in deep-ocean sediments after this time, reflecting the arrival of glacial ice at sea level. It is not clear why it took until the middle Miocene for the ice cap to develop to its present size. A clear consequence of this development, however, was that the system of ocean circulation intensified at this time because of the steepening equator–pole temperature gradient and the resulting strengthened winds. The intensified water circulation was in turn associated with increased upwelling and biological activity in the surface layers.

Further significant global cooling followed in the late Miocene. Cooling and marine regression occurred together at this time, the latter involving a sea-level fall of 40–50 m (Kennett 1982). This regression produced extraordinary effects in the Mediterranean Sea; enhanced perhaps by secondary tectonic effects, it isolated the Mediterranean from the rest of the world oceans except for a small inlet. Evaporation proceeded to lower the level in the closed basin, concentrating salts and eventually laying down enormous thicknesses of evaporite minerals. Because of the destruction of marine biota which this caused, the event is known as the *Messinian salinity crisis* (the Messinian is the name of a local stratigraphic stage); it occurred in the period 6.2–5 Ma BP. Other areas were similarly affected, including the Persian Gulf and the Red Sea (Benson 1984).

The volume of salts laid down represents the evaporation of about 40 times the water volume held by the Mediterranean (Kennett 1982). Water spilling into the basin and being evaporated there concentrated significant amounts of the global salt store into the Mediterranean, so that the salinity in the open sea fell by about 6%. This would have had effects on the global climate because the less saline seawater would freeze at a higher temperature, permitting more sea ice to develop in high latitudes, and reinforcing cooling by yet another positive feedback mechanism.

Re-occupation of the Mediterranean basin by the sea marks the beginning of the Pliocene, at a little before 5 Ma BP.

ONSET OF ICE AGE CONDITIONS IN THE QUATERNARY

After a slightly warmer interlude in the early Pliocene, conditions cooled still further. Finally, in the late Pliocene (around 3 Ma BP) climatic conditions in the northern hemisphere also cooled sufficiently for significant ice growth. Ice-rafted debris is found in sediments of the North Atlantic after this time (Kennett 1982). This marks the beginning of the phase of repeated ice-sheet growth and decay which characterises the Quaternary. The reason for the lateness of ice growth in the northern hemisphere relative to the southern remains unknown. The final closure of the seaway between the Americas may have been involved, possibly through the strengthening of the Gulf Stream, delivering more moisture to feed snow accumulation at high latitudes, noted earlier. It has also been suggested that mountain building may have triggered the ice growth, through the provision of sufficient land at high elevations to permit the survival of summer snow, and the development of albedo feedback reinforcement of the cooling (Raymo 1994).

Moderate-sized ice sheets had grown in the northern hemisphere by 2.4 Ma BP, near the beginning of the Quaternary period (Ruddiman and Raymo 1988). Their effect can be seen in the marine sedimentary record, which shows at this time a dramatic but irregular fall in the proportion of carbonate materials caused by incoming ice-rafted continental siliceous debris. Presence of such debris off Norway and Svalbard is dated to at least 5 Ma BP, so that some ice-sheet development was certainly well underway

prior to the Quaternary. The Greenland ice sheet was present in some form much earlier, at least by 7 Ma BP (in the late Miocene) (Larsen *et al.* 1994).

In the Quaternary, evidence of various kinds shows that there were numerous rhythmic alternations from cold to warmer conditions. These alternations are shown in the oxygen isotope analyses from marine micro-organisms (discussed in Chapter 7), in the record of global sea-level (discussed in Chapter 6), and in various kinds of terrestrial evidence such as the periodic accumulation of wind-blown dust. During the early Quaternary, in the Matuyama *magnetic chron*, the warm-to-cool alternation appears to have had a periodicity of 41 ka (Ruddiman and Raymo 1988) which is the same as the period over which the *obliquity* of the Earth's axis to the plane of the ecliptic cycles from 21.8° to 24.4° (Bradley 1985).

After 0.9 Ma BP, the *amplitude* of environmental fluctuations recorded in marine sediments increased, reflecting increasing size of ice sheets during cold phases. In addition, the fluctuations became less frequent, adopting a cycle time of about 100 ka. Interestingly, this once again correlates with the cycle time of an orbital parameter, but in this case it is the *eccentricity* of the Earth's orbit around the sun, which varies from nearly circular to somewhat elliptical. This frequency became especially dominant in the last 400 ka or so, but there were still 41 ka cycles (and 23 ka cycles, close to that of the cycling or *precession* of the equinoxes in the Earth's orbit) superimposed on the slower variation.

The repeated coincidence between the frequency of environmental change preserved in the Quaternary fossil record and known parameters of the Earth's orbit supports one of the major hypotheses seeking to explain the extreme instability of Quaternary climate: the *astronomical theory* of climatic change (see Chapter 5). We shall consider this theory in more detail in later chapters. First, some important questions remain.

1. If orbital parameters control the glacial–interglacial alternation, perhaps through slight changes in the amount of solar heating received by the Earth, why did they not do so in the same manner during earlier periods? In other words, why was the Quaternary environment so distinctively unstable?
2. We have already seen that during the Cainozoic, climatic change was perhaps more closely related to the progressive development of the tectonic plates, and to the changing configuration of the oceans and of ocean currents. These changes may have been partly responsible for the Cainozoic trend to cooler climates that culminated in the development of the Antarctic ice cap, and subsequently, of ice sheets in the northern hemisphere.

What role, though, was played by *external* processes, such as the orbital variations, in the Cainozoic cooling?

3. Are there other categories of environmental change which need to be considered in a search for fundamental causes? Need we, for example, consider the parallel evolutionary developments occurring in the biological realm, including particularly the plants, which are major users of carbon and thus potential controllers of the concentrations of major global greenhouse gases?

THE NATURE AND POSSIBLE CAUSES OF THE QUATERNARY INSTABILITY

The cyclic alternation of warm and cold phases in the Quaternary, occurring over regular 23, 41 and 100 ka periods, hints at a regular underlying mechanism which it should be possible to discover. The mechanism involved is clearly a relatively complex one, however, probably involving a number of feedback processes. The present day is a Quaternary warm phase, or *interglacial*, but there is still substantial land ice. In the Quaternary *glacial* phases, the volume of land ice grew dramatically, its growth fostered by increasing global albedo and other feedback processes, but never to cover the land completely. The climatic instability of the Quaternary was thus *damped* or limited, such that global temperature only oscillated between values similar to those of the present and 5–9°C cooler. There must therefore be some *negative* feedback mechanisms which act to damp the instability. Some possible mechanisms include the Earth's own radiative cooling as a limit on temperature rise; terrestrial long-wave radiation increases in proportion to the fourth power of the temperature, so that for each additional degree of global warming, much more heat must be trapped in the earth–atmosphere–ocean system. This may prevent warming in the short term of more than, say, 5°C above present values. For a negative feedback process to limit cooling, there is the possibility that the cold oceans of a glacial phase release less moisture to the air, so that the developing land ice sheets are essentially starved of moisture and do not grow beyond the maximum size seen in the Quaternary record. Major uncertainty about these regulatory processes exists, however. It is known that much earlier in the Earth's history, major glaciation affected a larger area, and land nearer the equator, than did the Quaternary ice. Questions of how this more extensive ice formed, and how it eventually retrea-

ted, remain unresolved (e.g. see Evans *et al.* 1997; Kaufman 1997).

A further feature of the Quaternary instability that hints at a mechanism is the *asymmetry* of the changes. The progression into a glacial phase, at least in the upper Quaternary, was relatively slow, but the warming phase or *termination* which ended each was far more rapid. Palaeotemperature curves through the glacial–interglacial cycles of the upper Quaternary thus show a sawtooth pattern. This hints at a positive feedback mechanism which comes into play during deglaciation but which does not act during cooling stages.

Tectonic changes as a precursor to Quaternary instability

One possible explanation for the Quaternary instability is that it was made possible by tectonic uplift generating land at high elevations. Several key locations have been identified which may be critical as sites for snow accumulation: Tibet, the Sierra Nevada and the Colorado Plateau in North America, and the Himalayan–Alpine belt (Raymo *et al.* 1988; Raymo and Ruddiman 1992). One possible role for uplift in these areas is to supply land on which, in response to a decline in solar heating, snow may fall earlier and/or survive later in the year, and so generate an albedo feedback mechanism to promote cooling. Areas of high mountains or plateaux may also act by deflecting major airstreams involved in the global atmospheric circulation. Model experiments indicate that the Tibetan plateau and the ranges of western North America do indeed act to lower atmospheric temperatures by bringing cold polar air southward and reducing summer ice ablation (Ruddiman and Raymo 1988; Ruddiman and Kutzbach 1989; Molnar and England 1990).

It may be that there has been sufficient uplift in the last few million years in these areas to provide snow accumulation sites and sufficient modification to atmospheric circulation that the albedo feedback process has been triggered when orbital geometry reduced solar heating. This hypothesis cannot yet be tested formally, because the exact chronology of uplift remains to be resolved. It may be that the bulk of the uplift preceded the Quaternary, so that it may have contributed to the Cainozoic cooling but does not help to explain the onset of Quaternary climatic instability (Molnar and England 1990).

Other roles for the uplift of mountains are possible. For example, the considerable elevation of the Tibetan Plateau contributes to the strength of the Asian monsoon, by promoting uplift and inflow of warm, moist air from the oceans. Modelling studies by Prell and Kutzbach (1992) showed that strong monsoon circulations only develop when strong solar heating, like that of the present day, interacts with a Himalayan–Tibetan region that is at least half as high as the region today. In view of the recent uplift, this would place the origin of the present monsoonal circulation at some time in the late Miocene. The uplift of this area may therefore have contributed to more rapid physical weathering and erosion processes, with rivers sweeping away rock debris. But in so doing, opportunities for chemical weathering are also enhanced, because of the ready supply of particulates with their enormous reactive surface area. The chemical weathering of silicate rock materials consumes atmospheric CO_2 delivered in rainwater, so that enhanced silicate weathering on a large scale could perhaps draw down atmospheric CO_2 levels sufficiently to contribute to the Quaternary cooling (e.g. Kutzbach *et al.* 1993; Filippelli 1997). Once delivered to the oceans by riverine transport, the increased flux of nutrients could also foster biological productivity there, with further consequences for climate because of the importance of the oceanic carbon balance. Certainly, the timing of the uplift and the onset of Quaternary glaciation do not appear to match perfectly. However, the climate system is complex and changes in the oceans, and in terrestrial ecosystems, may have involved lags that could account for some of the evident delay (e.g. Quade *et al.* 1995).

We have seen that the origin of the Asian monsoon circulation can perhaps be related to the uplift of the Himalayan–Tibetan region. Other plate tectonic processes also contributed to this environmental change, however. At 30 Ma BP, an arm of the Tethys, the former global ocean, still lay over a large area of the Eurasian landmass. Model studies (e.g. Ramstein *et al.* 1997) suggest that the presence of this ocean moderated the climate, and limited the amount of summer heating which is critical in driving the monsoon circulation. Retreat of this part of the Tethys permitted the development of a stronger continental heat low in summer, and the development of the monsoon circulation. The increased wind speeds may have been instrumental in yet more links with the oceans and their biological productivity. Ramstein *et al.* (1997), for example, infer that oceanic upwelling was enhanced by the monsoon circulation, so fostering the growth of marine plankton in sites like the Arabian Sea. The remains of these calcareous organisms provide a sink for carbon, and so periods of more vigorous oceanic productivity can thus draw down atmospheric CO_2 levels (see Chapter 3).

Possible roles for plants in climate change

Vegetation cover globally also stores vast amounts of carbon, extracted from the atmosphere. It has been hypothesised that the rise of vascular plants increased weathering rates by promoting soil microbial activity and hence CO_2 availability in soils, although Keller and Wood (1993) argued that CO_2 production by soil microbes alone may be enough to support weathering. However, ecosystems dominated by angiosperm deciduous plants regrow leaves and flowers annually, taking fresh nutrients from the soil, and hence are associated with higher weathering rates than are conifer/evergreen plant communities. The nature of the soil atmosphere may be important in this difference.

By the early Tertiary, the evolutionary diversification of angiosperm deciduous ecosystems may have resulted in accelerated chemical weathering over large areas, and hence may have produced a trend towards falling atmospheric CO_2 levels and thus global cooling (Volk 1989). The exact contribution that this mechanism made to the Tertiary cooling already described remains uncertain. It is also possible that by binding weathering products, and by isolating these from meteoric water, plants may inhibit chemical weathering (Drever 1994). Much depends on the particular vegetation cover and on the nature of the regolith mantling the land-surface.

Over Quaternary time-scales, a somewhat different role for plants may be envisaged. Glacial phases were associated with lower temperatures, altered land climates, and about 15% greater land area because of sea-level fall. Thus we may speculate that the zonation of vegetation across the globe must have undergone major shifts also. Large areas that were ice-covered would no longer have carried vegetation cover at all, but vast areas on the newly exposed continental shelves would. The net effect of these changes on the amount of carbon stored in plant tissue depends upon the area occupied by particular plant communities (forest, grassland, etc.). If large areas of forest occupied the exposed shelves in the still relatively mild low latitudes and equatorial zone, it is possible that significant additional CO_2 would have been taken from the atmosphere to build the new biomass, enhancing cooling. Estimates of the changed distribution of plant communities that would have existed at the height of the last glacial period, at 18–20 ka BP, do not clearly reveal the magnitude of the effect, but suggest the possibility that it may be a significant one (Prentice and Fung 1990).

Yet more roles for plants may be mentioned. During marine transgression, the vegetation communities of the continental shelves would be inundated. Burial of the vegetation would both store carbon and possibly lead to anoxic conditions in the water column, reducing biological productivity and its associated carbon storage. Changes of this kind might be associated with every glacial termination. The role of anoxia and its possible effects on biological productivity in the oceans remains to be explored.

A proportion of the Quaternary cooling, however it was triggered, can be ascribed to the effects of vegetation change on landscape processes involving hydrology, albedo, and other characteristics. Boreal forests, for example, cannot be blanketed entirely by snow, but retain some leaf exposure. In contrast, tundra landscapes, which would have progressively replaced forest during Quaternary cooling episodes, can be entirely blanketed and adopt the albedo of snow. In this condition, spring-time warming is restricted, permitting snow to remain longer, and so inducing additional cooling in a positive feedback effect. Dutton and Barron (1997) estimated that vegetation-related effects of this kind could by themselves account for up to 1.9°C of the Tertiary cooling. Otto-Bliesner and Upchurch (1997) drew similar conclusions from a climate model of late Cretaceous climates, which suggested that high- and mid-latitude forests present at that time warmed the global climate by about 2.2°C.

Volcanic activity and climate

An association between cooling and volcanic activity has long been considered possible. Eruptions releasing large ash clouds into the atmosphere, for example, might act to reduce the amount of solar radiation reaching the surface, and hence trigger cooling by albedo feedback effects from just a few years of enhanced snow survival. There is indeed an excellent correlation between glacial phases and preceding episodes of volcanic activity (Bray 1977). However, it has been suggested that glacial periods might only be one reflection of wider changes in the Earth's figure and tectonics, induced by orbitally-related changing gravitational stresses (e.g. Mörner 1980a). Changed stress fields within the Earth might thus be linked to both climate change and volcanism. Hence, as is so often the case, a good correlation does not necessarily prove that a causal link exists.

Meltwater and its effects

A final contributor to the complex pattern of environmental instability in the Quaternary that should be mentioned here is glacial meltwater. Land ice is, of course, essentially salt-free; evaporation from the oceans as ice ages developed in the Quaternary removed fresh water and left the oceans more saline than during interglacial times. At the termination of each glacial phase, enormous volumes of fresh water were poured back into the oceans as the land ice retreated. There were dramatic effects on land, because the meltwater was often temporarily dammed by ice

or moraine, and then suddenly released by breaching of the barrier to generate truly catastrophic floods whose impact on the landscape is still evident (e.g. Baker 1988). Enormous amounts of erosional work took place in just a few days during these catastrophic events, and their repetition, quite possibly at the end of each ice age, represents one of the major forces to have moulded the landscapes of areas lying around the Quaternary ice margins.

However, we need here to consider the effects that this fresh water would have had upon reaching the oceans. Being salt-free and less dense than seawater, it is possible that the incoming meltwater produced stratification in parts of the ocean, with fresh water lying above salt, and restricting oxygenation and biological productivity of the lower layers. More importantly, the meltwater might have dramatically lowered the salinity of seawater, and hence weakened one of the major forces which causes poleward surface water currents to sink and strengthen the circulation of bottom water. Normally, continued evaporation from these currents results in a poleward increase in salinity and hence in water density, which results in subsidence and a return flow of bottom water. Dilution by glacial meltwater arriving at high-latitude coastlines may well have weakened or indeed cancelled this effect altogether, altering the strength of the oceanic circulation and hence its ability to transport sensible heat across the Earth's surface. These ideas are considered further in the next chapter.

CONCLUSIONS

The broad trend of global cooling which preceded the Quaternary is well established. However, there seems little hope at present of unravelling the causes of this climatic deterioration from its effects. Undoubtedly, crustal processes and plate tectonic redistributions of land and water are involved; so too may be modification of the global carbon cycle by evolutionary trends in plant communities. The role of atmospheric CO_2 is particularly complex and difficult to identify. Potentially, critical control on atmospheric temperature may be exerted by the biological productivity of the oceans, and the resulting flux of carbon into sedimentary storage. However, the productivity of the oceans is influenced by the strength and configuration of ocean currents, which are influenced in turn by the condition of the atmosphere, so that it is once again difficult to escape from the web of interdependencies. We have also seen that large-scale tectonic processes, such as orogenesis, might influence ocean productivity through the mechanisms of accelerated weathering and riverine transport of nutrients.

The excellent correspondence between the periodicity of climatic change, especially in the late Quaternary, and orbital parameters, leaves little doubt that variations in solar heating exerted the fundamental control on climatic instability during the past few million years. The precise mechanisms by which the small variations in solar heating resulting from orbital characteristics are translated into global glacial and interglacial cycling remain unresolved. So too do the important mechanisms which limit the magnitude of the climatic swings and determine the rates of warming and cooling.

We consider these problems in Chapter 5.

In the following chapters, we will examine in more detail what is known of the environments of the Quaternary period. Only through continued research on the many fronts examined will it be possible to proceed further towards a true understanding of the Earth's environment and its past (and future) variability. We begin in the next chapter by examining the nature of the Quaternary ice sheets, whose rhythmic growth and recession constituted perhaps the most pervasive influence on Quaternary environments worldwide.

CHAPTER 3

QUATERNARY GLACIATIONS: EXTENT AND CHRONOLOGY

The ice was here, the ice was there
The ice was all around:
It cracked and growled, and roared and howled,
Like noises in a swound!
Samuel Taylor Coleridge (1772–1834)
The Ancient Mariner

The Quaternary period involved many environmental changes, but none more dramatic on land (nor more important in their impact on the global climate) than the development of the enormous ice sheets that grew and receded many times, episodically blanketing many land areas in the northern hemisphere and to a minor extent in the southern. In this chapter we introduce some of the evidence for these glaciations, and consider the timing of ice growth and decay in various parts of the world. The many valuable records of Quaternary glacial and interglacial environments derived from ice core records obtained from the remaining ice sheets are also reviewed. The following chapter, Chapter 4, is devoted to an exploration of some of the hypotheses seeking to explain the occurrence of the major glacial and deglacial episodes of the Quaternary.

THE CRYOSPHERE

Those parts of the Earth subject to the effects of ice, including glaciers, ice sheets, ground ice and sea ice, compose the *cryosphere*. The Quaternary period involved considerable expansion of the cryosphere, such that probably 40% of the land area was included, together with vast areas of the oceans. Some of the land and oceanic areas affected were not actually *ice-covered*, but were subject to freezing effects of various kinds, such as permanent or seasonal *permafrost* in which the soil and underlying rock (or the materials of the sea floor) are frozen, often to great depths. Cryospheric effects of this kind are grouped as

periglacial phenomena, a term which identifies the landscape features and processes associated with the cold but *non-glacial* parts of the Earth.

The present-day cryosphere

Because of the behaviour of radiative heat transfer, planets of our solar system relatively near to the sun are hot and those further out are cooler. The coldest planets would require enormous amounts of heat in order to support liquid water; the hot planets would need to be cooled massively. The Earth, however, has an average temperature which is high enough for liquid water to exist, but only marginally, with a mean surface temperature of perhaps 15°C. The surface temperature is not uniform, of course, because of uneven solar heating and the heat transfer effected by the moving atmosphere and ocean currents. Therefore, in places the Earth is even now too cold for liquid water, so that permanent ice cover exists (largely in Antarctica and Greenland), while the low latitudes are warm. Because the Earth's mean temperature is only marginal for the support of liquid water, relatively little cooling is required for the amount of ice to grow substantially; the difference in mean temperatures between the present day and the coldest of glacial times only amounts to 5 or 10°C.

Ice and snow presently cover about 10% of the land area, which amounts to about $15 \times 10^6 \, \text{km}^2$; in addition, sea ice covers about 7.3% of the oceans. The distribution of contemporary glacial and periglacial environments is illustrated in Fig. 3.1.

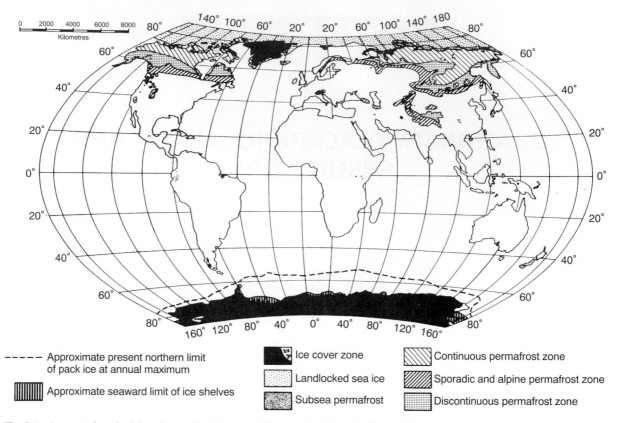

Fig. 3.1 Present-day glacial and periglacial zones of the world. (After Davies 1969; Brown 1970; Péwé 1983a, b; Harris 1985)

The Pleistocene cryosphere

Enormous expansion of the cryosphere is one of the features which characterises *ice ages* like that of the Quaternary. During such ice ages, the vast continental ice sheets come and go repeatedly. Without knowing the future, it is not strictly possible to label the present day (certainly a time of restricted ice extent) as an interglacial, but many lines of evidence suggest that this is indeed the case. It is also possible that the Quaternary ice age, having lasted nearly 2 million years, is now over. Adding uncertainty to this issue is the further modern process of human intervention with its effects on the composition of the atmosphere (notably via release of greenhouse gases) and the albedo of large parts of the Earth's surface, which are altered by deforestation, agriculture, pastoralism and desertification. Hydrological pathways have also been massively altered by dams, irrigation, and river diversion. Advance knowledge of the combined effects of these activities must be sought in climate modelling, the only real alternative being a wait-and-see approach which might prove to be foolhardy. Climate modelling, in turn, must be guided by reconstructions of the ways in which the Earth's environment has responded to previous climatic change,

most importantly the most recent and relatively well-dated glacial period.

During the coldest part of the last glacial period, when temperatures were lowered globally, ice covered nearly *one third* of the land area of the Earth. Since Antarctica and Greenland (which account for about 97% of the area occupied by land ice now) are effectively completely covered by ice, the additional area covered by glacial ice was largely in parts of the globe now ice-free. These included most of Canada, much of northern USA, and large areas in Scandinavia and northern Europe. Figure 3.2 portrays the inferred extent of the affected area. The ice sheets extended from high latitudes to about 36°N (the most southerly point reached by the ice lobes flowing through the basin of present Lake Michigan into Missouri), with alpine glaciers growing even into the tropics where high mountains existed. The additional ice-covered area in the last glacial was almost all in the northern hemisphere, with less than 3% in the southern. It is therefore reasonable to describe the growth of ice in the Quaternary as essentially a northern hemisphere phenomenon, although lowered temperatures and the glacio-eustatically lowered sea-level were global in their occurrence. The small area

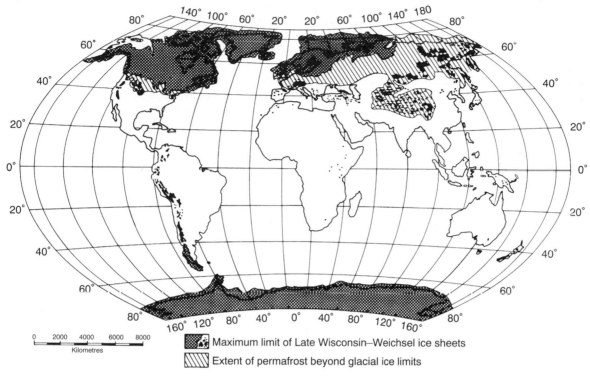

Fig. 3.2 Maximum extent of ice sheets and permafrost at the last glacial maximum. (After Löffler 1972; Washburn 1979; Hollin and Schilling 1981; Denton and Hughes 1981; Nilsson 1983; Péwé 1983a, b)

of ice in the southern hemisphere was largely made up of the Patagonian ice cap in South America, with minor ice caps and valley glaciers elsewhere, including Australia, Papua New Guinea and Irian Jaya, New Zealand and East Africa, together with substantial additional ice on the many sub-Antarctic islands and on Antarctica. In the northern hemisphere, nearly 60% of the additional ice-covered area was in Canada and the USA, and the bulk of the remainder in northern and western Europe and the Alps.

The thickness of the now-vanished ice sheets was remarkable. In places, maximum depths of up to 4 km are inferred, but with more typical depths of 2–3 km. The landscape modification accomplished as the ice ground across the land surface was considerable: Bell and Laine (1985) have estimated that on average bedrock was stripped to a depth of 120 m over the area of the Laurentide ice sheet in Canada and North America. Much of the eroded material now lies as marine sediment offshore, but there are also vast areas blanketed by glacial debris on land. Significant changes were also produced by the massive meltwater lakes and streams that were fed as the ice retreated during climatic recovery. The ice sheets vanished with remarkable speed, being completely gone from large areas only 8000 years after the process of deglaciation had begun. In parallel, recovery

of vegetation proceeded as the environment recovered to the conditions of the present day. The transformation of the landscape of the glaciated areas from what it was at 20 ka BP to its present form (often to forest cover, croplands and pasture) is so dramatic that it is not surprising that early researchers found it difficult to accept the evidence around them and the reality of the massive Quaternary ice sheets.

THE DEVELOPMENT OF IDEAS ABOUT THE QUATERNARY GLACIATIONS

The realisation that a large fraction of the land area had relatively recently formed part of the extended Quaternary *palaeocryosphere* came slowly and fitfully. This 'Glacial Theory', as it used to be known, evolved in the last decades of the 18th century and early in the 19th. It is important to remember that at this time, neither the Greenland nor the Antarctic ice sheets were known at all well, the first crossing of the Greenland ice cap being made in 1888 and the first interior exploration of the Antarctic ice cap not until the first decade of the 20th century. Consequently, the only glacial phenomena familiar to researchers were the smaller valley glaciers of places such as the European Alps and Scandinavia. None the less, geologists and other naturalists correctly identified *glacial erratics*, large blocks of rock found hundreds of kilometres from their nearest bedrock outcrop. The idea that ice had been the agent of transportation slowly supplanted the formerly accepted view that the blocks had been carried in moving water, perhaps the biblical floods, or had been dropped from floating icebergs. Deposits of *till*, the unsorted debris carried and deposited by glaciers, were also common in many areas; in the early 19th century these too were explained as having fallen from floating icebergs, and are still often referred to in consequence as *drift*. Eventually, though, the form of the glacial deposits, the striations left on erratic blocks and on outcropping bedrock, led to the widespread acceptance of the idea that great ice sheets had formerly occupied the landscape.

CHRONOLOGY

As noted in Chapter 2, substantial ice sheets had developed in the northern hemisphere by about 2.4 Ma BP. Ice growth may have begun considerably earlier, perhaps in the Pliocene, in some areas. For example, Clemens and Tiedemann (1997) noted evidence for ice reaching the North Atlantic and North Pacific Oceans by 5.5 Ma BP, while Mangerud *et al.* (1996) inferred that major glaciation in Scandinavia and the Barents Sea–Svalbard area began in the period 2.5–2.8 Ma BP. At this time, there was a marked increase in the amount of ice-rafted debris settling into sediments of the Vøring Plateau. A similarly early onset of ice growth has been demonstrated from ice-rafted detritus found in marine sediment cores taken on the East Greenland shelf, and Larsen *et al.* (1994) suggested that Quaternary glaciation in the North Atlantic region may have begun on Greenland.

The marine isotope record suggests that until about 0.9 Ma BP, the dominant rhythm of growth and recession in these ice sheets was the 41 ka period reflecting the cyclic change in obliquity of the Earth with respect to the ecliptic plane (see Chapter 5). Glacial periods in the late Quaternary, however, occurred with a periodicity of around 100 ka, and displayed a slow and uneven cooling followed by a rapid deglaciation, as noted in Chapter 2. During the slow cooling, lasting perhaps 70–90 ka, there were numbers of slight returns towards warmth (called *interstadials*) as well as unusually cold periods, or *stadials*. Eventually, maximum cooling resulted in the occurrence of a corresponding maximum ice volume, reached at the glacial peak (called *full-glacial conditions*); deglaciation, also punctuated by some

temperature and ice-volume fluctuations, then led rapidly back into *interglacial* conditions with warmth like the present. The period of extensive land ice that ended about 10 ka BP and marked the onset of Holocene interglacial conditions is thus merely the latest *glacial stage* of the many which have occurred during the Quaternary ice age.

There is a vast and confusing array of local names applied to glacial periods, stadials, interstadials and interglacials; often these are locality names chosen from sites providing good exposures of the field evidence. The naming of the last glacial is typical: *Wisconsin* in Canada and the US, *Weichsel* in western Europe, *Devensian* in Britain, *Würm* in the Alps, and so on. Because of their dominance in terms of ice area and volume, the North American and European names are here linked, as in common practice, to identify this glacial period as the Wisconsin–Weichsel.

Periods of marked increase in ice volume have been used to partition this glacial stage into an *early substage, a middle substage,* and a *late substage.* The timing of these divisions is not universally agreed upon, but commonly for purposes of description they are taken as follows:

Late Wisconsin–Weichsel substage: 24–10 ka BP
  (including the glacial maximum at 21–17 ka BP)
Middle substage 74–24 ka BP
Early substage 117–74 ka BP

For some time prior to 117 ka BP, conditions were warm; this is the last interglacial, termed the Eemian interglacial in Europe, and the Sangamon in North America. Full glacial conditions were reached at 21–17 ka BP, after cooling spanning the early and middle substages; this Late Wisconsin–Weichsel substage is best known because the evidence is freshest, and lies within the span of radiocarbon (^{14}C) dating (see Appendix). As we shall see, on the basis of the fragmentary terrestrial record alone, relatively little is known with certainty about events in the early substages, nor about previous glacial stages in the Quaternary. The evidence preserved in deep-sea cores and in the remaining ice masses in Greenland and Antarctica is therefore an invaluable adjunct in reconstructing Quaternary environmental changes, since it is usually more complete than the land record (see Chapter 5).

EVIDENCE OF GLACIATION

Scientific study of the contemporary cryosphere and of the landscape legacy of the Quaternary palaeocryosphere has advanced enormously in the last 150 years. Precise isotope dating techniques, the palynological reconstruction of former vegetation communities (discussed in Chapter 10), and analysis of data from realms other than the cryosphere but which shed light on ice-age events (e.g. the oceans and their sediments, discussed in Chapter 7), have all assisted in this process.

In the modern landscape, including adjacent continental shelves which were crossed by ice during the glacial low sea-levels, many ice-related features have been described and their origins accounted for (e.g. Washburn 1979; M.J. Clark 1988). These are too numerous to recount here, but we can note that the poorly sorted rock debris forming glacial till and the ice-built moraines formed from it, often still standing as barriers marking the limit of ice advance, are vital features enabling the *areal extent* of ice to be deduced. Use is also made of *glacial outwash* deposits, the material carried from the ice margins by meltwater streams and often laid down to great depths further down-valley or in other ice-marginal areas, and *loess*, the fine rock-dust deflated from outwash surfaces and carried by the wind to blanket the landscape beyond the ice margin (see also Chapter 9). Many effects in soil and rock associated with frost shattering are also known. The movement of rock and soil by growing ground ice, and the pro-

duction of the resulting *patterned ground* displaying sorted stone polygons and the like, are very common in areas around the margins of the former ice sheets. Seasonal thawing of the upper layers of the frozen soil permitted the sodden mass to move downslope under the influence of gravity to produce a wide range of distinctive lobed *solifluction* forms. Over vast areas, landscape features such as rounded and polished rock outcrops, bearing crescentic ice chatter marks and linear striations, are abundant (Fig. 3.3). Datable materials are found within glacial deposits, and in the fossil-bearing sediments that accumulated in lakes associated with ice sculpture of the landscape and with meltwater. Indeed, in some lakes a chronology can be built up by counting the annual layers (*varves*) which result from seasonal changes in the nature of the sediment being set down. Pluvial lakes of Quaternary basins show, in some areas, lake floor deposits that connect laterally with the deposits resulting from ice advance from nearby uplands, and the glacial debris can be dated by its association with the lake sediments.

In the case of the major continental ice sheets, quite detailed reconstructions can be made on the basis of the morphology of the glaciated surfaces.

Fig. 3.3 Ice-polished surface of granite in Yosemite National Park, California, USA

These large ice sheets were sufficiently thick that, like the present Greenland and Antarctic ice sheets, they adopted a smoothly domed form which blanketed the underlying topography, with the thickest ice in some central location (see Fig. 3.4). Large ice sheets have complex temperature characteristics, in which heat is generated by friction where movement is fast, and in which the temperature at which the ice would melt is lowered where the ice is thicker and exerts greater hydrostatic pressure at its base. Near the centre of an ice sheet, the ice is nearly stagnant and the ice is frozen to the underlying rock, so that only minimal erosion occurs. As the flowing ice accelerates in its outward movement further away from the centre, more heat is generated and basal meltwater results, permitting the lubricated ice to slide, pluck bedrock, and grind and erode the surface. Further out still, towards the ice margin, ice movement slows as the ice thins in the zone of ablation; freezing once again dominates because of the reduced frictional heating. Between the frozen and thawed zones are zones where the outward-moving ice either becomes progressively warmer (a melting zone) or cooler (a freezing zone). The kinds of deposits left allow these zones to be mapped for the Wisconsin–Weichsel ice sheets, and hence permit reconstruction of the locations of their feeder ice domes, and their patterns of thickness and rates of movement.

As an example, we may consider the Laurentide ice sheet of Canada and North America. The major central ice dome for this sheet was evidently over Hudson Bay, which was consequently isostatically depressed under the weight of ice. Even though this area now displays rapid postglacial rebound, it remains a marine embayment (see Chapter 6). In a zone around the Hudson Bay centre, basal melting would have uncoupled the ice from the bedrock, allowing considerable glacial erosion. This erosion stripped the younger Palaeozoic sedimentary cover and exposed the ancient Canadian Shield rocks in an arc around the presumed ice dome. Bedrock hummocks (around which the pressure melting point is not reached, the overlying ice being thinner) would have generated frozen patches where erosion was restricted; hollows in the subglacial topography (where the ice is thicker and basal melting results) would have favoured erosion and deposition. Thus the deglaciated landscape has the distinctive appearance of numerous lakes (representing thawed patches under the ice sheet) and fields of mounded till deposits (*drumlins*) located in the lee of the bedrock high points. In zones of basal melting near the ice sheet margins, hydrostatic pressure is low because of the thinner ice, and meltwater can gather into subglacial streams. After deglaciation, their sediment load may be left as linear *eskers* (snaking ridges of alluvium) which are distinctive of this zone. Material delivered to the outer freezing zone near the ice margin will be set down during the retreat of the ice sheet to form a *till blanket*, partly modified by meltwater processes taking place during the retreat. In the case of the Laurentide sheet, this area is the zone of water-filled depressions around the margins of the

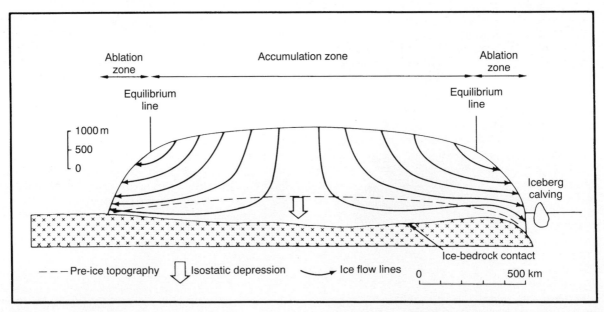

Fig. 3.4 Flow lines and zones of accumulation and ablation in a large ice sheet. (Modified from Sugden and John 1976; West 1977; Reeh 1989)

Canadian Shield, such as the lakes Huron, Michigan, Athabasca, Great Slave, Great Bear, and Coronation Gulf and the Gulf of Boothia.

Finally, along the warmer equatorward margin, a melting zone existed which set the southern limit reached by the ice. Here, lubrication of the ice base by water allowed acceleration of the ice into *ice streams*, which poured out to form lower-lying *ice lobes* around the ice sheet margin, with lobate moraines left after ice retreat. The Des Moines lobe downstream of Lake Winnipeg and the James lobe below Lake Manitoba were major lobes fed from the Laurentide ice sheet in this way.

QUATERNARY CRYOSPHERE RECONSTRUCTION

What, then, do we know about the *geographical extent* of the glaciers, ice sheets, and periglacial phenomena of the Quaternary? Likewise, what is known of the *chronology of the growth and decay* of the vast ice sheets? The answer is a complex one, with great variation from region to region. As we shall see, there are time lags in ice sheet growth and shrinkage following a climatic change which are not constant but depend upon the actual bulk of the particular ice sheet, the extent to which it is confined by ridges or mountain ranges along its margins, and the nature of the climate change which has occurred. These parameters are likely to be different for each of the principal Quaternary ice sheets and for each major climate swing, so that we should expect the record of ice advance and retreat to be far from simple and certainly not in strict agreement as to timing at different sites.

Considering the area affected in a glacial stage on all continents together provides an indication of the size of the palaeocryosphere and the volume of ice involved in glaciated regions. The total ice-covered area at a typical glacial maximum amounts to about $40 \times 10^6 \, \text{km}^2$, compared with the present $15 \times 10^6 \, \text{km}^2$. The volume of water stored as ice at a glacial maximum amounts to about $90 \times 10^6 \, \text{km}^3$, compared with the present $30 \times 10^6 \, \text{km}^3$. Thus, the ice volume is tripled, while the area covered increases by a little more than 2.5 times. In addition, there is a very substantial area affected by periglacial conditions in both glacial and interglacial times; at present this amounts to about $20 \times 10^6 \, \text{km}^2$ (in addition to the ice-covered areas). There is still uncertainty about the reconstruction of conditions at the last glacial maximum, because the field data are not absolutely unambiguous and it is still possible that numbers of individual ice sheets which have been inferred were part of one much larger system.

Resolving this requires considerable understanding of the dynamics of ice sheets that extended much closer to the equator and were larger than anything presently in existence, and is the subject of continuing research. Some uncertainty also remains about the extent of ice on the now-drowned continental shelves; these areas, for obvious reasons, are harder to explore and their deposits are less well mapped and dated.

Timing of ice volume changes: the marine $\delta^{18}O$ record and its interpretation

The most widely employed proxy record for Quaternary ice volume comes from the oxygen isotopes preserved in the tests of marine foraminifera. (A proxy record is composed of data on one variable that can shed light on the value of another that itself leaves no direct record, or else leaves a record that is much harder to evaluate.) Of the two more abundant isotopes of oxygen, ^{16}O is enormously more abundant than ^{18}O. It also forms slightly lighter water molecules, which therefore evaporate more easily than do those of the heavier $H_2^{18}O$. As moisture-bearing air is cooled, water with the heavier oxygen isotope condenses preferentially, leaving the remaining water enriched in ^{16}O. It is well known that at present, the abundance of these two isotopes in snow within an ice sheet reflects the mean annual temperature of the location. Sites further inland (and therefore at higher elevations on the ice dome, and therefore colder) receive snow enriched in the lighter water which has survived the journey inland. The amount by which water or snow deviates from the isotope proportions found in the modern oceans, taken as the standard, is expressed as the $\delta^{18}O$ value (the Greek letter delta, δ, stands for 'difference'). The differences are small, and so are expressed in parts per thousand (per mil) rather than parts per hundred (per cent). Water depleted in ^{18}O, and therefore lighter, receives negative scores; the more negative the value, the lower is the proportion of ^{18}O. Ice sheets are isotopically light (and so display negative values of $\delta^{18}O$, down to $-40\%_o$ or so); the ice-age oceans were left isotopically heavy (and so displayed $\delta^{18}O$ values of up to, say, $+5\%_o$). The calcareous tests of forams reflect the changing isotope abundances in their environment. Originally, these were analysed with a view to using them as a proxy indicator of past water temperatures, since there is a temperature dependence in the isotopic fractionation that takes place as the calcareous tests are set down. Subsequently, however, it was argued (e.g. see Shackleton 1967) that the major influence on the $\delta^{18}O$ record during the Quaternary was instead the periodic removal of large amounts of the light isotope (^{16}O) stored in glacial ice on land. This means that the changes in δ^{18}O values in Quaternary plankton fossils were largely caused by changes in the abundance of ^{16}O and ^{18}O in the oceans. Certainly, the apparent broad similarity of the δ^{18}O record from widely varying ocean sites suggested that they all reflected a common, global influence. On this basis, refined by detailed work, the changing composition of forams is used as a proxy ice-volume indicator.

There is much more to the interpretation of the marine δ^{18}O record than we can discuss here. Issues such as the species dependence of the isotope fractionation, and the blurring effects of burrowing organisms on the resolution and other aspects of the sedimentary record, complicate the separation of an ice-volume record. The benthic species *Uvigerina peregrina*, for example, is known to form its calcareous test so as to reflect the isotopic composition of the surrounding seawater. In some other species, significant 'vital effects', related to metabolic processes, alter isotope abundances and require the use of calibrated correction factors. Typically, isotope work is performed on mono-specific samples of forams hand-picked from the core sediments. Bioturbation confounds the analysis of core records also. There are approaches by which the size of this effect can be estimated, and its influence removed in the process of deconvolution (Bard *et al.* 1987). This essentially attempts to 'undo' the blurring using mathematical techniques. There are also methods by which the data from different cores, perhaps sampled at points representing different past times in each core, and with different sampling intervals, can be cross-correlated and a more representative stratigraphy built up (Pisias *et al.* 1984; Prell *et al.* 1986). These matters are beyond the scope of this chapter. Nevertheless, it will be helpful for us to investigate in outline what this record suggests about the timing of Quaternary ice volume changes. The marine δ^{18}O record is particularly important in Quaternary investigation because Quaternary time has been subdivided on the basis of the fluctuations contained in this record, which are taken to reflect ice volume growth and decay. Many individual events in the record have been identified, and Quaternary time has thus been subdivided into *marine δ^{18}O stages* representing the time between successive events. These oxygen isotope stages are widely employed in the literature, in the context of both marine and terrestrial proxy records. We will examine this chronostratigraphy shortly. For the present, we need only note that the marine oxygen isotope stages are numbered back in time from the present, which is marine oxygen isotope stage (often contracted to MIS) 1. The preceding glacial was stage 2, and so on. In this system, warm periods like the present are always given odd

numbers and cold phases, like the last glacial maximum, even numbers. The stages are subdivided into substages, which are given letter designations. Thus, the warmest part of the last interglacial, which is stage 5, is designated substage 5e. Further details can be found in Pisias *et al.* (1984) and Martinson *et al.* (1987).

For the purposes of glacial reconstruction, one of the main points of contention has been just how to convert the δ^{18}O values from the marine record into equivalent ice volumes. One approach is to take an assumed unvarying δ^{18}O value for glacial ice, and to use this value to work out how much ice must have been in storage at any past time in order to yield the observed δ^{18}O enrichment in the marine record. More sophisticated analyses are based on glaciological estimates of the height of each major ice sheet, and hence of individual δ^{18}O values.

For example, Mix (1987) has presented estimated data on the δ^{18}O (ice) values for the major ice masses of the last glacial maximum. These range from -25‰ for British ice to -60‰ for East Antarctic ice. A mean value of around -40‰ is indicated. Taking the observed glacial/interglacial marine δ^{18}O change of about 1.7‰, this permits an approximate estimation of the volume of ice required in order to generate the observed marine isotope signal. This kind of calculation can be made as follows:

Estimate of additional Quaternary ice volume: $\sim 50 \times 10^6 \, \text{km}^3$.
This is about 3.5% of the total present ocean volume of $1.37 \times 10^9 \, \text{km}^3$. Take the mean δ^{18}O of the glacial-period ice to be -40‰.
Thus, δ^{18}O change in the remaining ocean volume, which is 97.5% of the initial volume, is:

$$\delta^{18}\text{O shift in glacial ocean} = \frac{3.5 \times 40}{97.5} \cong 1.45‰$$

Since the observed shift in the foram record is about 1.7‰, an ice volume of $50 \times 10^6 \, \text{km}^3$ would thus leave about 0.25‰ of the overall shift as the product of oceanic temperature change.

A major difficulty with this approach as a basis for decoding a temporal record of ice volume from the foram archives is that it is certain that the isotopic composition of glacial ice was not constant (Shackleton 1987; Mix 1987). The increasing elevation of a growing ice sheet results in progressive changes in the isotopic fractionation of water forming the ice. Likewise, during melting, water of different composition would return to the oceans as a function of the elevation, age, and geographical location of the ice that was melting. Thus, while a mean δ^{18}O value for the last glacial maximum (stage 2) ice is perhaps adequate for a simple dilution calculation of the kind made above, it is not really sufficient to turn a continuous record of marine $\delta^{18}O$ values into a record of ice volume or of sea-level through time.

A way around this that has been suggested is to employ only the $\delta^{18}O$ record from benthic forams, from the deep ocean. These should experience much smaller temperature variations, because the deep ocean is such an enormous mass of cold water, and thus provide a more reliable ice volume record than near-surface (planktonic) forams (Streeter and Shackleton 1979; Duplessy and Labeyrie 1994).

There has been uncertainty, however, about the validity of the presumption that temperatures have changed little in the deep ocean. Evident disparity between the sea-level record inferred from $\delta^{18}O$ and that derived from marine terraces led Chappell and Shackleton (1986) to propose that glacial deep-ocean temperatures had indeed been up to 2.5°C colder than interglacial values. Shackleton (1987) indicated that sea-levels at 82 ka BP and 104 ka BP (MIS 5a and 5c) inferred from marine terraces were only about 20 m lower than at the peak of the previous interglacial (stage 5e, 124 ka BP). According to the $\delta^{18}O$ record, however, the stage 5a and 5c sea-levels were 70 m below their interglacial position. These differences support the idea of a significant temperature effect on the foram record. The same conclusion was reached by Birchfield (1987), who concluded that the deep oceans were 1.4°C colder during the last glacial.

Recently, this issue has been appraised using a different proxy record as a means of resolving the remaining uncertainty. Dwyer et al. (1995) used the ratio of Mg/Ca in ostracods from the North Atlantic sediment record to do this. The ratio has a strong temperature dependence, but since the residence time and abundance of Ca and Mg in the oceans prohibit much change in these over Quaternary time, it has almost no ice volume dependence.

Analyses of modern ostracods of the genus *Krithe* collected from waters of varying temperature permitted the establishment of the relationship:

$$\text{deep water temperature (°C)} = (0.854 \times Mg/Ca) - 5.75 \qquad (3.1)$$

This was then employed to estimate water temperature for the Quaternary fossil ostracods. A good general correspondence to the well-known $\delta^{18}O$ pattern was found, but with variation in detail, including a phase lead in the ostracod record of about 3.5 ka. Glacial–interglacial shifts in bottom-water temperature were estimated to be 3–5°C. The Mg/Ca palaeotemperature data in principle permit the separation of temperature and ice-volume effects in the marine $\delta^{18}O$ record. Using data from the Late Pliocene and running through the Quaternary, Dwyer et al. (1995) inferred that bottom waters have cooled progressively. They interpret this to signify diminishing deep

water formation in the North Atlantic, and a more widespread influence of colder Antarctic bottom water. The rate of such a process would undoubtedly undergo significant fluctuations, and this must contribute some of the variability in Quaternary climates, and some residual uncertainty in ice volume estimates.

It is generally accepted that of the 1.7‰ change in foram $\delta^{18}O$ values between glacial and interglacial periods, about 1.3‰ was caused by ice volume changes and the remaining 0.4‰ by ocean temperature change. Working on this basis, a global ice-volume curve for the Quaternary can be calculated (see Fig. 3.5). This provides a valuable global tally of ice volume but does not reveal the geographical location of the various ice masses. More can be done to address this by using geophysical models to interpret the pattern of ice loads that is required to be removed to account for the displacement of dated Quaternary shorelines (e.g. see Tushingham and Peltier 1991; Peltier 1994).

We will now review separately the extent of Quaternary ice, especially in the Late Wisconsin–Weichsel glacial, in several key areas of the world.

ALPINE GLACIATIONS

It is difficult to reconstruct glacial history in valley glacier areas such as the Alps because major ice growth can erase the sedimentary record left by earlier ice, especially in a rugged environment where high erosion rates exist. Loss of evidence is particularly likely in the case of the minor advances associated with stadial phases and also with short interstadials. Minor fragments of the glacial moraines or interstadial lake sediments from these times may exist, but be exceedingly hard to find or to date.

Despite its relatively small extent (less than 0.2% of the additional ice-covered area at glacial maximum), the area of Quaternary glaciation in the European Alps is the starting point for our discussion of ice extent for three reasons:

1. The Alps, an area rich in erratics, striated bedrock, and well-preserved moraines, provided the field area which led to the early advancement of the 'glacial theory' already mentioned.
2. Research in the Alps led to the early suggestion, now known to be incorrect, that four major glacial episodes had taken place.
3. Refinement of knowledge of the field evidence in this area makes it clear that there are problems with attempting to reconstruct ice age chronologies on the basis of evidence from valley glaciers, a conclusion pertinent to analysis of glacial records from other parts of the world.

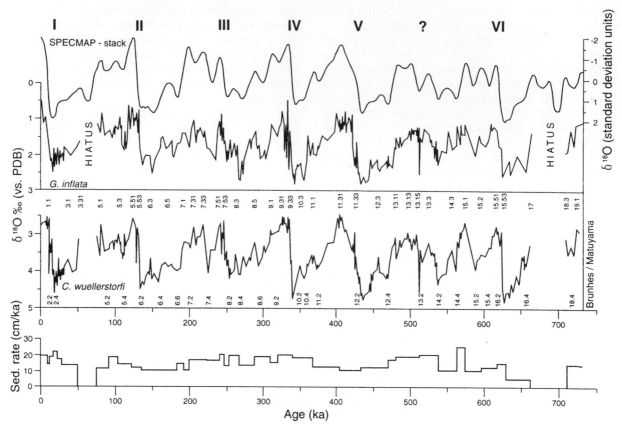

Fig. 3.5 Marine oxygen isotope record covering the last 700 ka. The uppermost curve is the smoothed SPECMAP record (Astronomically calibrated time-scale, notably using the oxygen isotope record of the oceans); below this is a record based on the foram *G. inflata*. The third curve is a record based on *C. wuellerstorfi*, and below this is the inferred record of sedimentation rate in the source core (Ocean Drilling Program Site 658). (From Sarnthein and Tiedemann 1990)

The small area of Alpine glaciation can be seen in Fig. 3.6, which also shows the extent of glacial and periglacial phenomena affecting Britain and Europe during the last glacial maximum.

In the Alps, some earlier Quaternary glaciations were evidently more extensive than the most recent, their terminal moraines lying further downvalley than those of the Late Würm. It is important to remember, therefore, that reference to the glacial maximum of the last glacial stage does not imply that this represents the maximum *Quaternary* extent of ice.

Early this century, Penck and Brückner (1909) identified a sequence of four sets of putative glacial sediments in the Alpine foreland or piedmont zone. These were represented by the older and younger *Deckenschotter* (cover gravels) and the *Hochterrassenschotter* and *Niederterrassenschotter* (high and low terrace gravels). They were taken to represent four sequential cold periods, which were named after four small tributaries of the Danube west of Munich, the Günz, Mindel, Riss, and Würm. The

older cover gravels are high-level remains found on plateau surfaces; the materials are strongly weathered glaciofluvial sediments, and have been linked to deeply weathered tills ascribed to the Günz and Mindel glacial stages. The terrace gravels are set within broad valleys cut lower in the landscape. The high terrace gravels have been connected with Riss moraines, and the low terrace with Würm moraines. The materials of the terrace gravels are substantially fresher than those of the *Deckenschotter*. The fact that the terrace gravels are set into broad valleys suggested to the early workers that a substantial period of incision had followed the cold phase which resulted in the formation of the younger cover gravels, since these lie above the valley sides on the upland surfaces. The period of incision was taken to represent a long interglacial.

Mapping in the Alps has shown that in the Late Würm, the snowline was about 1200 m lower than at present. During the older of the glacial events allegedly responsible for the *Deckenschotter*, the

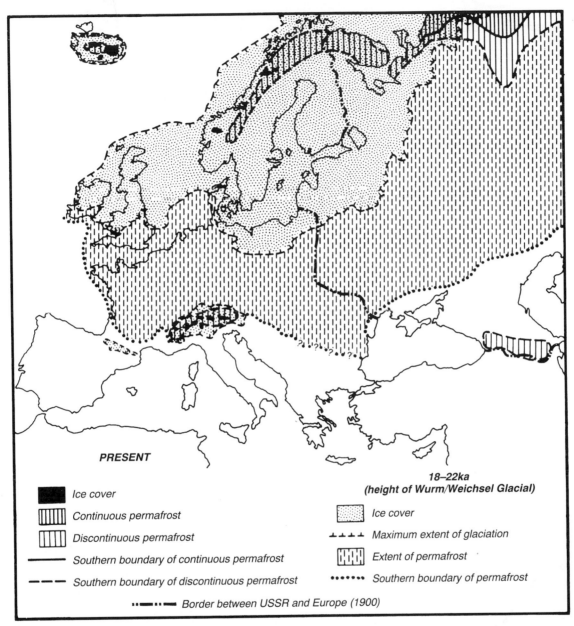

PRESENT

- ■ Ice cover
- ▥ Continuous permafrost
- ▯ Discontinuous permafrost
- ── Southern boundary of continuous permafrost
- ─ ─ Southern boundary of discontinuous permafrost
- ▪▬▪▬ Border between USSR and Europe (1900)

18–22ka
(height of Wurm/Weichsel Glacial)

- ░ Ice cover
- ⊥⊥⊥⊥ Maximum extent of glaciation
- ▦ Extent of permafrost
- •••••• Southern boundary of permafrost

Fig. 3.6 Glacial and periglacial zones of Europe and Scandinavia at the present and at the last glacial maximum. (From Maarleveld 1976; Anderson 1981; Baulin and Danilova 1984; Harris 1985)

snowline was lower by an additional 100–200 m (Nilsson 1983). During the glaciations, ice occurred largely in the form of separate valley glaciers, not an integrated ice sheet. Some glaciers descended into the plains to the north and south to form large ice lobes. The location of these is very clearly preserved by lobate terminal moraines, such as those on the Po plain of northern Italy; the scoured rock basins lying inside the moraines now hold major lakes, such as Lago Maggiore and Lago di Como north of Milan in Italy, and lakes Lucerne, Zurich and Geneva on the northern flank of the Alps. These basins would have been sculpted to a degree by each glacial, with their final form being a result of the most recent action of the Würm ice. Downvalley from the ice lobes, streams were fed with sediment-laden meltwater. These overloaded streams proceeded to aggrade, and to build up floodplains. During interglacial times, as the ice retreated up-valley, and vegetation re-established itself, sediment delivery was reduced, and the undersupplied streams proceeded to incise, creating the terraces. The upper surface of each terrace therefore dates from the start of an interglacial stage. Uplift of the Alps has ensured that the chronological sequence of terraces from old to young is also a height sequence, old remnants being at the greatest elevations, with fresher materials nearer present river level. The outwash trains, as the alluviated valleys are called, supplied extensive dusts which blanket the landscape as loess. Soils developed slowly in these loess deposits, and now provide one of the kinds of field evidence used to recognise and define interglacial conditions.

The effects of four major glacial phases do indeed dominate landscapes around the Alps. However, this is not to say that we still accept, as Penck and Brückner did, that this means that there were *only* four glacials. Indeed, revision of the system of four glaciations has proceeded slowly ever since it was proposed. Detailed mapping has suggested to subsequent workers that each of the glacial stages was indeed a compound event of two to four separate ice advances delimited by interstadials. The interstadials and interglacials in turn were recognised not to be simple single events, on the evidence of multiple loess layers separated by palaeosols in the wind-blown deposits which cap the moraines. Sediments taken to date from the last interglacial (i.e. post-Riss and pre-Würm) show a pollen record consistent with warmer conditions than present, with *Quercus* (oak) forest subsequently replaced by *Picea* (spruce) and cold-tolerant species as the next glacial developed (Husen 1989). The fossil fauna of the interglacials is also revealing, with elephant, rhinoceros and beaver. In cooler, early Würm interstadials which followed, the fauna includes reindeer, giant deer, and woolly mammoth. The disappearance of these animals, now extinct, is considered further in Chapter 10.

In addition to subdividing the glacial and interglacial phases, modern work in the Alps has suggested the existence of glacials older than the Günz (they have been named the Donau and the Biber). The chronology of these older events is not well established. Normal magnetic polarity is displayed by the older *Deckenschotter* sediments of the Günz glacial, as well as by Günz tills, on which basis it has been argued that the younger four main glacials at least occurred within the Brunhes magnetic chron, i.e. within about the last 730 ka (Kohl 1986).

Recession of the Würm glaciers in the Alps was underway by 15 ka BP, and it has been suggested that the retreat was very rapid, with glaciers contracting to half their full-glacial length in only 1–2 ka (Husen 1989). The retreat is dated on the basis of evidence such as ^{14}C-dated tree stumps, in growth position, standing within the Würm moraines. In the French and Italian piedmonts, deglaciation was evidently interrupted by two major readvances. By 10 ka BP the main valleys were ice-free, and indeed ice became less extensive than today by 8.4 ka BP (Billard and Orombelli 1986). There were minor Holocene readvances, mentioned later in this chapter.

NORTH AMERICA AND GREENLAND

The largest ice accumulations of the Late Wisconsin–Weichsel glaciation occurred as a series of domes and ice sheets which, at their maximum areal extent, covered essentially all of Greenland, all of Canada and offshore islands, and many of the northern states of the US (see Fig. 3.7). This amounts to about $16 \times 10^6 \, \text{km}^2$, about twice the area affected in Europe ($6.7 \times 10^6 \, \text{km}^2$) and slightly more than the present Antarctic ice sheet (nearly $14 \times 10^6 \, \text{km}^2$). In the east, major ice lobes extended equatorward through the low-lying parts of the Mississippi basin, reaching to just past St Louis at 36°N. For most of the southern margin of this ice, the Late Wisconsin–Weichsel advance reached about the same point as earlier Quaternary glacial stages. Sea ice was also more extensive during the Wisconsin–Weichsel, as shown in Fig. 3.8. As we shall see later, the greater extent of sea ice has an important bearing on the availability of the moisture required to sustain the terrestrial ice sheets.

The land ice cover is envisaged by many to have consisted of multiple ice domes and accumulation centres. In the west, a Cordilleran ice sheet was supported by the Rocky Mountains, and flowed north and south to produce major ice lobes spreading on to lower elevations in Alaska and Washington State, as well as generating offshore ice shelves along the Pacific coast and glaciers draining to the east to link with the Laurentide ice sheet.

In the Queen Elizabeth Islands, an Innuitian ice sheet seems probable. This, however, is one of the geographical regions where the true extent of Late Wisconsin ice remains uncertain. In terms of annual precipitation, much of the area rates as very dry desert today. Thus, it seems unlikely that ice could accumulate rapidly now. The greater cold of the last glacial, surprisingly, might not have helped; it would have permanently frozen many of the surrounding oceanic moisture sources, and it is therefore reasonable to imagine that the region would then have been even drier. As a result, the chronology of glaciation may be unusual in areas like this, with most ice growth occurring early on in the glacial, perhaps even with retreat occurring in the coldest phases, and possibly renewed growth as conditions warmed somewhat and precipitation increased.

The largest of the ice sheets was the Laurentide, which was probably composed of multiple ice domes.

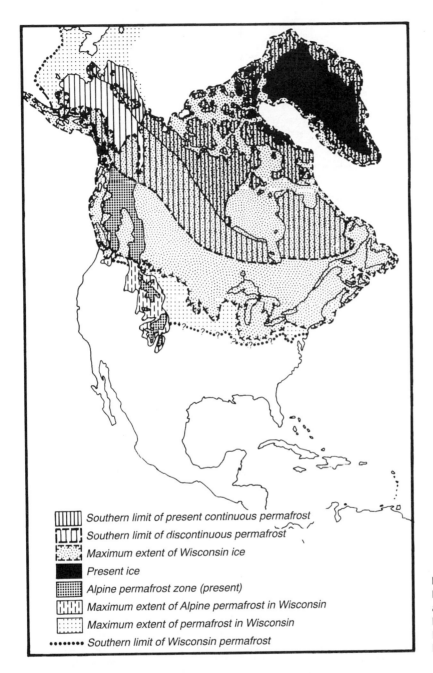

Southern limit of present continuous permafrost

Southern limit of discontinuous permafrost

Maximum extent of Wisconsin ice

Present ice

Alpine permafrost zone (present)

Maximum extent of Alpine permafrost in Wisconsin

Maximum extent of permafrost in Wisconsin

Southern limit of Wisconsin permafrost

Fig. 3.7 Glacial and periglacial zones of North America and Greenland at present and at the last glacial maximum. (From Davies 1969; Prest 1969; Brown 1970; Hamilton and Thorson 1983; Péwé 1983a, b; Harris 1985)

These may have included a Keewatin ice sheet covering much of western Canada (and adjoining the Cordilleran ice sheet), a Labrador ice sheet in the southeast, and a Foxe ice sheet lying over Baffin Island and the Foxe Basin (between Baffin Island and Hudson Bay). A single dominant ice dome located over Hudson Bay is viewed by some as better describing the state of the ice sheet. The Laurentide ice may have reached maximum thicknesses of nearly 4 km; others favour lesser values of 2–3 km (e.g. using their geophysical model, Tushingham and

Peltier (1991) deduced an ice thickness of 3.2 km over central Hudson Bay), with quite thin margins composed of ice lobes perhaps 500 m thick. Many such ice lobes have been inferred from the pattern of moraines along the southern margin of the ice, including the major Des Moines lobe of Iowa already referred to, and lobes produced by ice moving through the low-lying basins of the Great Lakes. Some of the ice movement toward these marginal lobes must have involved flow up the topographic gradient from the most isostatically depressed area

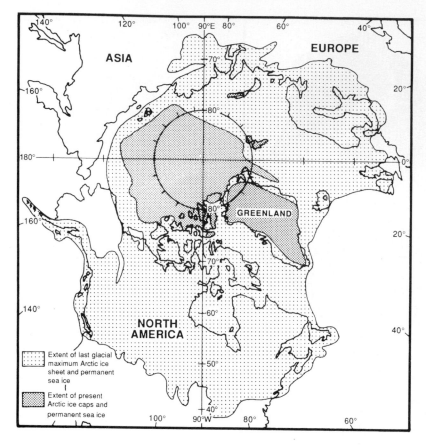

Fig. 3.8 Distribution of Arctic sea ice and adjacent land ice at present and at the last glacial maximum. (Adapted from Bartholomew *et al.* 1980; Denton and Hughes 1981; COHMAP 1988)

around Hudson Bay. Along the southern margin of the ice, a major zone of periglacial conditions existed. Casts of *ice wedges*, ice-filled frost cracks resulting from repeated freeze–thaw alternation of water moving in from the surface soil, and filled by inwashed sediments as the ice melted, are abundant indicators of this zone; their distribution along the former ice margin is as distinctive as that of ice-margin moraines. The presence of ice wedges is taken to mean that the deeper soil, below a surface *active layer* which may have thawed each summer, was permafrost.

In Greenland, the present extensive ice sheet was enlarged somewhat at the glacial maximum. Uncertainty remains about the extent of ice movement on to the continental shelf. In eastern Greenland, there is evidence for restricted ice growth, possibly because of limited moisture availability, as in the case of the Queen Elizabeth Islands.

The chronology of ice growth and retreat over this vast area is not straightforward (e.g. see Andrews 1987). The areal extent of ice was greatest in the early Wisconsin–Weichsel, and the middle substage was a time of extensive deglaciation in some areas. Ice readvanced in the late substage, generally reaching its maximum for that advance in the period 21–17 ka BP. The southern margin of the Laurentide ice sheet certainly reached its maximum extent in this period, and readvanced close to this again at about 15–14 ka BP (Denton and Hughes 1981). The Des Moines lobe, which flowed into Iowa, did not reach its maximum extent until about 14 ka BP, a date similar to that of 15 ka BP for the time of maximum southward extent of the Cordilleran ice. Mann and Peteet (1994) placed the last glacial maximum in the Alaskan Peninsula–Kodiak island area in the interval 23–14.7 ka BP. Relatively little is known of the detailed chronology of ice retreat, except that it was proceeding by 14 ka BP, and was fastest along the southern ice margins. Ice margin recession here reached speeds of a few hundred metres per year (i.e. a few hundred km per ka). By 11 ka BP the ice had contracted to the margins of the Canadian Shield, and the Canadian Plains and Cordillera were completely deglaciated by 10 ka BP (Fulton 1989). By 8 ka BP the sea invaded Hudson Bay and the Laurentide ice sheet ceased to exist as one continuous ice mass. Ice lasted in the Foxe Basin until about 7 ka BP, and remnants of Labrador Ice until about 6.5 ka BP (Fulton 1989). Complex sequences of readvances are documented during the overall retreat of the Laurentide ice sheet (e.g. see Dyke and Prest 1987). These events seem to have been related to changes in the ice bed, to changes in ice sheet elevations and elevation gradients, and other kinds of dynamic flow-related processes within the ice mass. This record of ice retreat really only tells us about the diminution in the *area* covered by ice, and does not shed light on whether the ice sheet also became thinner as it retreated. Simultaneous thinning would result in a much greater rate of ice *volume* loss than would simple retreat of the ice margin. This distinction is an important one to which we will return later in this chapter.

EUROPE

In northern Europe, the Late Wisconsin–Weichsel ice did not cover quite so extensive an area as did some of the earlier Quaternary glaciations, older moraines lying several hundred kilometres further south in some areas. The largest ice sheet was the Scandinavian, the terrestrial extent of which is well known but which also extended on to the Norwegian continental shelf. Smaller ice sheets occurred in other areas, including the Siberian Plateau, Svalbard, Franz Joseph Land, Novaya Zemlya, and Severnaya Zemlya (see Fig. 3.9). Uncertainty about the extent of ice over the area of Svalbard and the Barents Sea remains. Siegert and Dowdeswell (1995) employed a model of ice behaviour in this area to infer the growth of ice over Svalbard at about 28 ka BP, culminating at about 20 ka BP in an ice thickness of about 1.3 km. After a two-step deglaciation (Elverhøi *et al.* 1995; Polyak *et al.* 1995), with a recession from the outer continental shelf at about 14.8 ^{14}C ka BP and a second phase at 13–12 ^{14}C ka BP, this ice was gone by 10 ka BP.

The Late Wisconsin–Weichsel ice did not reach Holland, permitting the accumulation there of one of the best sedimentary records of glacial times. There

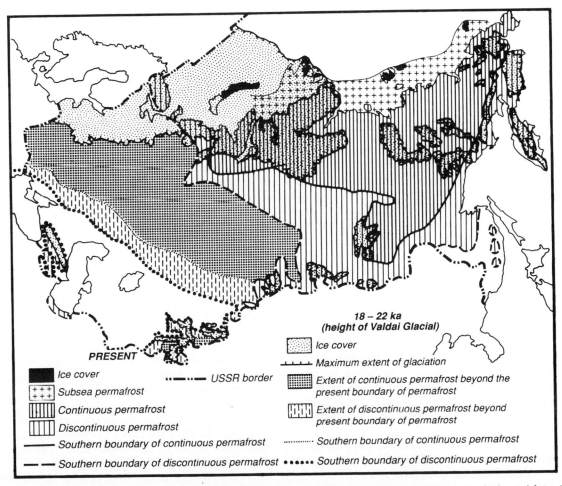

Fig. 3.9 Glacial and periglacial zones of the former USSR at present and at the last glacial maximum. (Adapted from Davies 1969; Washburn 1973; Anderson 1981; Péwé 1983a, b; Baulin and Danilova 1984; Harris 1985)

is also a good record in Denmark, which lay near the southern ice limit during the Wisconsin–Weichsel. The whole of Denmark was, however, ice-covered in earlier glacial stages; four episodes of ice growth are recognised. In the last glacial, ice did not reach Denmark until after 25 ka BP, reaching its maximum extent in the period 20–18 ka BP (Lundqvist 1986).

In Norway and Sweden, the record of earlier glaciations is very fragmentary, as both areas were completely glaciated in the Late Wisconsin–Weichsel. Mangerud et al. (1979) used pollen analysis at a site in western Norway to show that sea-level decline had begun during the last interglacial (Eem) before any large ice masses had developed in Scandinavia. The first signs of ice in this area seem to have been the product of local mountain glaciers, rather than a large inland ice source. The Swedish and Norwegian west coasts show evidence of several phases of ice retreat during the last glacial, whereas the central areas show only one early Wisconsin–Weichsel retreat. Regional variation in the glacial record of this kind must be expected, as a consequence of Quaternary gradients in temperature, elevation and moisture availability. The southern ice margin was in retreat by 15 ka BP, and a significant cooling of the North Atlantic is inferred to have resulted at that time from the influx of ice and meltwater (Lehman et al. 1991).

In the British Isles, the area affected by Late Wisconsin–Weichsel ice was less extensive than that covered by some earlier glacial episodes. In the Devensian, as the last glacial stage is known in Britain, ice spread from various upland centres, most importantly the Scottish Highlands. At the glacial maximum, ice covered most of Scotland, the northern parts of Ireland and Wales, as well as a large part of England, notably the Pennine uplands (see Fig. 3.10). The ice extended south to about the latitude of the Thames Valley, but the northern ice margin was sinuous, running northward in a broad loop to skirt the lower-lying Vale of York, and thus leaving much of eastern England ice-free but affected by permafrost. An ice stream moved southward through the Irish Sea, while in the east, movement of the ice was deflected into northern and southern branches by the adjacent Scandinavian ice sheet. The exact location of the ice front linking the ice in the British Isles with that of the Scandinavian ice sheet remains uncertain. Using a geophysical model to interpret postglacial sea-level records, Tushingham and Peltier (1991) have rejected the possibility of the North Sea having been ice-covered.

There is no firm evidence of glaciation in the early or middle substages of the Devensian, although it is likely that ice existed in upland areas in the west and north. Major glaciation began in the Dimlington stadial, early in the late Devensian. Ice growth is

inferred to have begun after 26 ka BP, reaching its maximum extent in the period 18–17 ka BP, and retreating from most areas by about 14.5 ka BP (Bowen et al. 1986). After the Windermere interstadial, during which ice completely disappeared, a Loch Lomond stadial has been identified. This renewed ice growth was largely restricted to the Scottish uplands, where many valley and corrie glaciers advanced. This stadial is correlated with a widespread cooling, termed the Younger Dryas event, which has been identified in both hemispheres. A discussion of the environments of this period may be found in Troelstra et al. (1995). The name of this stadial comes from a group of Arctic and alpine plants belonging to the genus Dryas. The Younger Dryas (YD) stadial was a brief reversion to glacial temperatures that occurred midway through the period of deglaciation. The age of this event has permitted good dating of deposits from this period in many environments, both marine and terrestrial. The interpretation of this evidence has provided vital clues about the processes that act during deglaciation, a topic that we will return to later in this chapter.

SOUTHERN HEMISPHERE

We will only briefly consider the evidence from the southern hemisphere, which is more fragmentary and less well dated. To the extent that it is known, the chronology of the last glacial stage here is very similar to that of the northern hemisphere already described, and there is no doubt that the southern hemisphere ice growth reflected control by the same instability of global climate.

The largest area of ice growth that is well known was in South America, where there were locally enlarged glaciers on high peaks of the Andes even in the tropical north. However, the largest fraction of the ice developed in the far south of the cordillera, in Chile and Argentina. In the more northerly parts of this zone, the ice mostly took the form of valley glaciers which poured down towards the Pacific continental shelf, where they calved into the sea. Further south, over Tierra del Fuego, ice accumulated over about 480 000 km^2 to form the Patagonian ice cap; this reached a thickness of about 1.2 km. A model of this ice cap by Hulton et al. (1994) has supported a probable volume of about 440 000 km^3, and indicated that temperature changes in the region might have been modest (3°C). Tushingham and Peltier (1991) inferred a thinner ice sheet in this area, adopting a maximum depth of only 400 m. In northern Peru, deglaciation began by 13.5 ka BP, 50% of the glacial maximum ice extent had been lost by 12.1 ka BP, and cirques were ice-free by about 10 ka BP (Rodbell 1993).

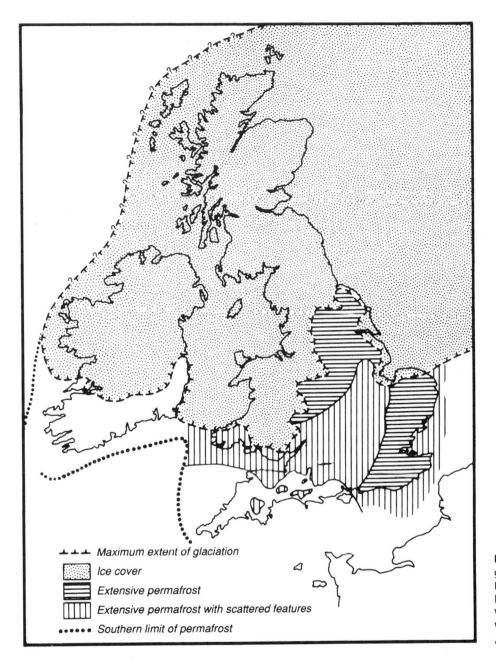

Fig. 3.10 Glacial and periglacial features of the British Isles at present and at the last glacial maximum. (From Williams 1969; Sparks and West 1972; Maarleveld 1976; Anderson 1981)

Legend:

＿▲＿▲＿▲＿ Maximum extent of glaciation

Ice cover

Extensive permafrost

Extensive permafrost with scattered features

••••• Southern limit of permafrost

On the Australian mainland and on the island of New Guinea, snowlines were lowered and small cirque and valley glaciers developed, together with more extensive periglacial phenomena. The snowline in tropical New Guinea was lowered by about 1000 m, an amount comparable to that recorded for northern hemisphere sites. In Tasmania, a small ice cap of about 8000 km^2 reached its maximum extent at about 19–20 ka BP, and had receded by 10 ka BP.

The largest ice-covered area in the southwest Pacific region occurred in the Southern Alps in the South Island of New Zealand. Here the most recent glacial stage is called the Otiran; it is equated to the Wisconsin–Weichsel of the northern hemisphere. At its maximum, which occurred just before 18 ka BP, ice covered about 40 000 km^2. The ice was relatively thin, however, possibly no more than 100–200 m on average.

The extent of ice growth in Antarctica during Wisconsin–Weichsel time remains uncertain. Denton and Hughes (1981) argued for a significant growth of the West Antarctic ice sheet outward to the edge of the continental shelf, together with a lesser increase in the volume of the main East Antarctic ice. Their estimate of the maximum probable growth in ice volume amounts to more than a 50% increase over the volume of the present ice sheet, but the field evidence is not yet resolved and this may prove to be a substantial overestimate. The timing of ice retreat around Antarctica also remains unresolved; Peltier (1988) and Tushingham and Peltier (1991) have argued that a late onset of deglaciation (perhaps beginning 7 ka later than in the northern hemisphere) is required in order to explain the global pattern of Holocene sea-level rise (see Chapter 6). No good explanation for the lag is available, although it is known that the conditions of the present day do result in different circumstances of sea ice melting in the Arctic, where melt ponds develop and lower the albedo, rather than in the Antarctic, where a reflective snow cover persists on the sea ice until the air temperature increases above freezing in summer (Andreas and Ackley 1982). Different behaviour might therefore have been displayed during Quaternary episodes of deglaciation. It appears probable that most melting of the excess Antarctic ice occurred during the Holocene, when most of the additional glacial ice had disappeared from the northern hemisphere (Nakada and Lambeck 1988; Tushingham and Peltier 1991).

As was the case in the Arctic Ocean and North Atlantic, sea ice around Antarctica during the last glacial maximum was considerably more extensive than at present. Significant seasonal expansion and contraction would have occurred during glacial times, as at present, with contraction of the sea ice during the southern ('Austral') summer and renewed expansion in winter (see Figs 3.11 and 3.12).

A summary of the ice sheet volumes inferred in the geophysical model of Tushingham and Peltier (1991) is presented in Table 3.1.

DATING CRYOSPHERE GROWTH AND RETREAT

One of the major benefits of the documentation of the extent of the Quaternary cold climates, through the use of the kinds of field evidence just referred to, is the ability to use this information to make inferences about past temperatures, snowlines, and so on. The use of evidence from the cryosphere to shed light on past climates is most useful if the *timing* of events can be established. For building and testing hypotheses about causation, we need to know, as precisely as possible, *when* the climatic cooling that triggered a particular advance of ice sheets itself began, and when climatic warming reversed the trend. A major problem here is that the enormous Quaternary ice sheets, by virtue of their bulk, could not respond

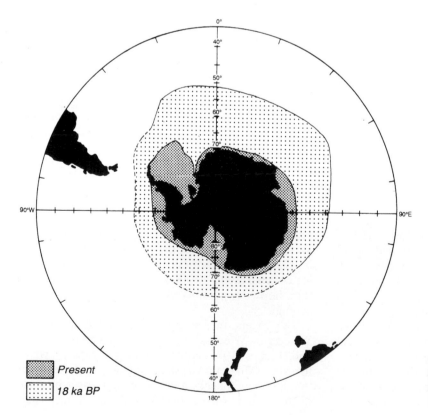

Present

18 ka BP

Fig. 3.11 Estimated extent of Antarctic summer sea ice at present and at the last glacial maximum. (From Hays 1978; Bowler 1978)

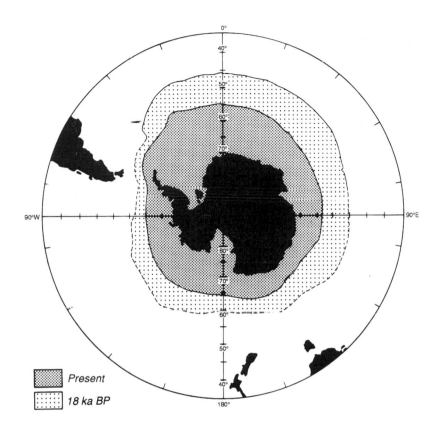

Present

18 ka BP

Fig. 3.12 Estimated extent of Antarctic winter sea ice at present and at the last glacial maximum. (From Hays 1978; US Navy Hydrographic Office 1961)

Table 3.1 The estimated volume of the last glacial maximum ice sheets as employed in the modelling study of Tushingham and Peltier (1991), compared with present-day ice volumes

Ice sheet	Maximum volume $(10^6\ km^3)$	Present volume $(10^6\ km^3)$
Innuitian	0.9	0.034
Laurentide	21	0.035
Greenland	5.5	3.0
Fennoscandian	2.9	0
Barents Sea	2.2	0.009
Kara Sea	2.6	0
East Siberian Sea	1.1	0
Iceland	0.2	0
Scotland	0.01	0
Patagonian	0.2	0
Antarctic	35	26
TOTAL	72	29

rapidly to changes in global or regional climate. The movement of great masses of ice is a response to the mechanical weakness of crystalline ice and the resulting gravity flow processes. Even if there was a sudden warming, such that no further snow fell on a major ice sheet like that of Greenland, its outlet glaciers would continue to flow for considerable periods as the kilometres-thick accumulated ice slowly dissipated by outward flow. Thus, we cannot take a date identifying the onset of glacier retreat also to identify the onset of climatic warming: this may well have been considerably earlier. Similarly, the time elapsing during the accumulation of sufficient snow and ice to establish flow, and the further time taken for the advance of the ice margin to locations hundreds or thousands of kilometres distant, means that we cannot take the date of ice advance recorded at a site to represent the date of climatic deterioration. Small valley glaciers can respond much more quickly, and in principle could provide useful data on the timing of climate swings. The problem with such small valley glaciers, however, is that in repeatedly advancing and retreating along their valleys, former deposits are destroyed so that the record of earlier fluctuations becomes incomplete or absent altogether. This is the case with the glacial record of the Alps, already described. Additional problems arise in attempting to date cryosphere instability on the basis of terrestrial evidence. Because of the ^{14}C half-life (see Appendix), the timing of the growth and recession of the continental ice sheets is only adequately known from terrestrial evidence for the Late Wisconsin–Weichsel, since only materials from this period lie within the range of ^{14}C dating. However, it must be remembered that not only is the chronology blurred by the time lags in ice response mentioned above; in addition the radiometric dates really only *bracket* the time of ice advance or retreat, yielding maximum or minimum ages. For example, the age of interstadial plant remains overlain by till only sets a *maximum* age for the ice advance; materials from meltwater lakes only set a *minimum* age for ice retreat. Trees in growth position upvalley of a moraine simply indicate that ice retreat had occurred at some earlier time; how much earlier remains unclear. In a similar way, dated raised shorelines, used to infer the onset of isostatic rebound following deglaciation, and hence the timing of the deglaciation itself, really only indicate a *minimum* date for the ice retreat, which might have begun gradually at an earlier time.

Therefore for the most reliable *chronology* of climatic fluctuation, we must in general rely on environments with shorter *lag times*. Living organisms, which may in a single lifetime reflect the characteristics of their environment (say, the salinity of the water in which they grew), are vital here, especially the micro-organisms of the oceans, whose remains have accumulated in marine oozes throughout the Quaternary. Proxy data derived from these remains form the basis for the MIS division of Quaternary time described earlier. The microfossil remains of these organisms now provide one of the most sensitive *chronometers* of Quaternary environmental change (see Chapter 5), and a palaeoclimatic record of much finer time resolution is obtained from them than from the terrestrial record of the Quaternary cryosphere. The use of oceanic evidence, however, raises the additional issue of whether the two environments, terrestrial and marine, experienced the same environmental fluctuations. Certainly, the controls on conditions in the oceans (including factors such as salinity, speed and depth of currents, and degree of oxygenation of the water) are not the same as those (largely temperature, windiness and precipitation) which dominate conditions on land, and we must interpret the separate records with care.

It is none the less the case that terrestrial evidence of ice sheets or ground ice is definitive of cold climatic conditions. Knowledge of the areal extent and general chronology of these features remains vital in our developing picture of the timing of the Quaternary glacial and interglacial periods. The extent of glacial conditions at the Wisconsin–Weichsel peak in particular is employed in climate modelling on the assumption that this represented a few thousand years of steady-state conditions, in which the ice was neither advancing nor retreating. In such steady-state conditions, it is reasonable to take the extent of the ice to be a reflection of approximately stable environmental conditions during the full-glacial state.

Useful evidence on the nature and timing of glacial fluctuations has been derived from well-dated records of events in the Holocene. We will examine the Holocene record briefly before turning our attention to high-resolution records spanning larger fractions of late Quaternary time, especially the records derived from ice cores and marine sediments.

HOLOCENE GLACIER RECORDS

In addition to shedding light on the glacial and interglacial stages of the Quaternary, study of valley glaciers has the potential to reveal the chronology of climatic fluctuations over shorter time periods. The use of the pattern of advance and retreat of valley glaciers has mostly been applied within the Holocene with the aid of ^{14}C dating of moraines and through the use of historical records, sketches and photographs (see Leroy Ladurie 1972; Grove 1988). Glacial episodes of the Holocene are generally termed *neoglaciations*.

The reconstruction of Holocene environments from glacial evidence is based upon the fact that the downvalley extent of glacial ice in a particular glacial valley reflects to some degree the *mass balance* of the glacier, which controls the physical form that is adopted to balance snow accumulation in the snowfields above and ablation losses at the melting snout of the glacier. A climatic change resulting in greater snowfall or lessened ablation can result in thickening of the glacier, and its advancement downvalley. Changes in the opposite direction result in thinning or recession upvalley. Stable environmental conditions result in essentially constant glacier form, once sufficient time for adjustment has elapsed.

The evidence of glacier fluctuations is mostly in the form of moraines, and analysis of these is beset by the problems already described for the Würm glacial in the Alps. Additionally, there is again the problem of lags in glacier response, such that downvalley advance may not begin until some decades after the climatic change has triggered greater snow accumulation; indeed, during rapid climatic changes, insufficient time may elapse for the equilibrium glacier form to be established at all.

The pattern of Holocene glacier advance and retreat therefore shows an unsurprising variability. In part this reflects the differing lag times of large and small glaciers, those on steep versus gentle slopes, and variation in ice temperature, which enable some glaciers to be advancing while others are in recession. The numerous records of Holocene glacier fluctuation indicate that ice growth was particularly marked in the second half of the Holocene, earlier millennia presumably being somewhat warmer.

One major period of ice advance is known in the 'Little Ice Age' of the 16th to mid-19th centuries (i.e. 0.1–0.4 ka BP). Much support for the climatic cooling interpreted from glacial evidence is provided by indirect historical records dating from this time. Records of grain and wine harvests show substantial reductions during this period, and stock losses in some areas were considerable; increasing winter sea ice made navigation difficult or impossible in places. Interestingly, the Little Ice Age is reflected in records extracted from the Quelccaya ice cap in the mountains of tropical Peru, at latitude 13° S, indicating that this period of cooling must have been essentially global in its effect. (Some of the records extracted from the Quelccaya ice are presented in Fig. 3.13.) The global temperature drop in this cold period, however, is inferred to have been less than 2°C (e.g. see Grove 1988). Denton and Karlén (1973) identified an earlier period of Holocene glacial advance dating from 3.3–2.4 ka BP, and a later, less significant one, at 1.25–1.05 ka BP, relating these to variation in solar output. They suggested that there may indeed be a regular periodicity of about 2.5 ka in the occurrence of ice growth. A major period of glacier recession is also known, dating from the last half of the 19th century and running up to the present day. The extent to which this may reflect anthropogenically-produced climatic warming, following the Industrial Revolution, remains to be resolved.

ICE CORE RECORDS

We now turn our attention to changes occurring over longer periods of late Quaternary time, since these will provide evidence that is useful in unravelling the nature and causes of the major glacial–interglacial climate instability, which is the focus of the next chapter.

In recent times, new techniques have made it possible to extract quite sensitive records from the large ice sheets still in existence in Greenland and Antarctica, despite their relatively sluggish response to climatic change. This is done by examining the layers of ice, corresponding to each year's precipitation, that are preserved in the cold accumulation zones of the ice sheets. Major ice cores have been recovered by drilling programmes from various sites in Greenland, the Canadian Arctic and Antarctica, as well as from sites in Peru and Tibet. Cores of more than 2 km have been collected, with the ice at the base ranging in age to beyond 200 ka, but typically only to 100–130 ka (the age of the last interglacial). In addition to ice, the annual snow accumulation layers contain a wealth of other materials including dust, sea salts, pollen, volcanic debris, cosmic particles, and, in modern times, isotopes from nuclear weapons testing. These materials, set down year-

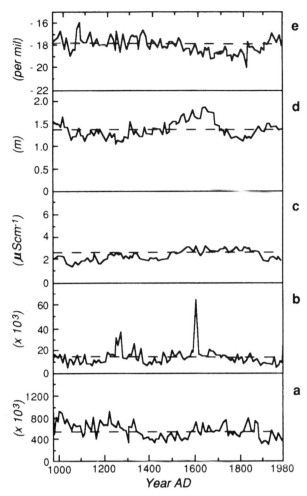

Fig. 3.13 Records from the Peruvian Quelccaya ice core (from Thompson *et al*. 1986). Note the depression in the ^{18}O values (curve e) in the last few centuries, corresponding to the Little Ice Age
(a) Particulate abundance per ml of sample
(b) Abundance of large particles (over 1.59µm in diameter) per ml of sample
(c) Electrical conductivity
(d) Accumulation rate (m per year)
(e) δ^{18}O

by-year, form what has been referred to as 'a continuous diary' of the environment around the ice sheet (Stuiver *et al*. 1995). This can provide data with good time-resolution despite the sluggish overall response of the ice sheet itself.

The layering representing the snow from each year is progressively compacted as it is buried, and undergoes the transition through firn to dense ice. As this happens, air is trapped as small bubbles within the ice, and this can be extracted to provide samples of air roughly as old as the ice itself. Analyses of the air can provide indications of the gas composition of the atmosphere at the time the firn was finally compacted to dense ice. A number of effects arise to complicate this kind of analysis of entrapped air bubbles. Today, bubbles in firn are typically squeezed shut once they are about 80–100 m below the surface (Sowers *et al.* 1991). Until that time, they can have continued contact with the atmosphere. In the light of snow accumulation rates, this means that the air in a newly sealed bubble will be surrounded by ice that is about 2.5 ka old. In other words, the air sample is younger than the ice in which it is contained. The difference, which must be subtracted in order to 'date' the air sample, can vary systematically with climatic conditions (Barnola *et al.* 1991). For example, in the glacial periods, accumulation rates were lower, so that it would take longer to accumulate the required depth of snow to close off bubbles. Consequently, the air from glacial periods is up to 6 ka younger than the ice in which it is contained. There are also processes that can influence the composition of the air trapped in bubbles. Before they are pinched shut, CO_2, which is denser than air, can move downwards through the firn, increasing the apparent CO_2 concentration found later when air samples are analysed. This is a potentially important effect, because glacial–interglacial changes in CO_2 concentration in the atmosphere must be known to understand how much of the temperature change was produced by the altered greenhouse action of this gas. However, Barnola *et al.* (1991) estimated that in the case of the Antarctic Vostok core, the gravitational increase in CO_2 concentration amounted to no more than 2 ppmv. Chemical reactions taking place if the snow melts during summer, such as those between water and particulates contained in the snow, can also affect the gas composition of trapped air (Staffelbach *et al.* 1991).

Layers in the ice sheet can be dated by counting downwards from the surface, the annual layers being recognised by seasonal variations in dust content, or by patterns of acidity or isotope content. This can be done with good precision over at least 10 ka (i.e. through the span of Holocene time), although layer counting has been carried out in ice well beyond this age. Deeper, older ice is more difficult to date precisely because compaction leaves the layers increasingly thin and ice movement deforms them. Annual layering then becomes more difficult to recognise with certainty, and the ages are inferred from glaciological models of ice compaction and deformation.

One of the principal records extracted from the long ice cores is that of the isotopic composition of the oxygen in the water molecules. Data on temperature and precipitation $\delta^{18}O$ from multiple sites at present indicate that an isotope ratio deviation of 0.62‰ is produced by a temperature change of 1°C at the site of condensation of water vapour. Analysis of this ratio on samples of ice from within the cores reveals the temperature of the snow accumulation site at the time the snow fell. There are some effects which could act to interfere with this analysis, however:

1. if the samples represent a growing ice cap, whose height will therefore be increasing, the one ground point will experience increasingly cold conditions, and the temperature record derived from it will show cooling. This, however, is not external global climate cooling, but merely reflects the normal decline of temperature with elevation above the Earth's surface.
2. if the warmth of the moisture source changes, or the ice cap is fed from winds blowing from a different direction, then the isotopic composition of the snow may change. Such a change could be produced as a glacial stage progressed by the development of sea ice, for example, or by shifts in the general circulation of the atmosphere.

There are other complications relating to such factors as the role and significance of seasonal variation, and to additional isotope fractionation taking place between cloud level and the ground. Such factors are revealed in the mixed success achieved in studies seeking correlations between observed contemporary temperatures and $\delta^{18}O$ values (e.g. Siegenthaler and Oeschger 1980; Yao *et al.* 1996).

Various calibrations of the ice core palaeothermometer are available. For instance, Kapsner *et al.* (1995), working with the Greenland Ice Sheet Project 2 (GISP2) core, employed the relationship:

$$T = [(\delta^{18}O + 18.2)/0.53] + 273 \qquad (3.2)$$

where T is the temperature in kelvin. Cuffey *et al.* (1995) have used a calibration method that does not rely on the assumption that present-day relationships between temperature and $\delta^{18}O$ values applied throughout the Quaternary. Indeed, their borehole calibration technique uses information on the present temperature profile of the ice sheet to infer the general form of the temperature time-series that must have been experienced at the site in order to impress the observed temperature profile into the ice. Using a best-fit model, they have shown that the coefficient that relates temperature to $\delta^{18}O$ actually varies with the external climate. During the transition from the LGM to the Holocene, the data suggest that a lower value of about 0.33‰/°C is appropriate. This increases the suggested temperature change from the LGM to the Holocene (Peel 1995; MacAyeal 1995).

Deuterium (D), the heavier isotope of hydrogen, is employed in a manner similar to that described for the oxygen isotopes. Water molecules containing deuterium are heavier, and display a lowering of vapour pressure similar to that resulting from the presence of heavier oxygen isotopes. Thus, water containing deuterium is preferentially lost as moist air proceeds inland. Concentration differences with respect to modern seawater are in this case written as δD, and show a good relationship to mean air temperature at the point where the water containing deuterium falls as snow. Typical δD values in ice from the last glacial are around −480‰; the value rises to −420 to −440‰ for interglacial ice.

Greenland ice core records

Despite the possible complications noted above, ice cores provide extremely valuable records of temperature reflecting external global climate change. The core from Camp Century at 77.2° N in northern Greenland, for example, shows the transition from the last glacial stage to interglacial conditions at a depth of 1150 m; here the $\delta^{18}O$ values shift from the very light values of −40‰ which characterised the glacial to −29‰ or so. By counting the layers, the date of the warming at the end of the Younger Dryas stadial, which may be taken as the transition from glacial to interglacial stages, can be dated to 10.75 ka BP ± 150 years (see Fig. 3.14). The final warming marking this upper boundary of the glacial stage appears to have been extremely rapid; in the ice cores, the transition is reflected in only 2 m of ice, which corresponds to about 100 years! Of course, the form of the ice sheets themselves could not possibly respond to such rapid climatic change; lag times of thousands of years are involved.

Employing the conversion to temperature mentioned earlier, the isotope record indicates that at Camp Century, the glacial maximum was about 11°C colder than the early part of the Holocene. In contrast, the borehole calibration derived by Cuffey *et al.* (1995) for the GISP2 record, mentioned earlier, suggests that central Greenland averaged 14–16°C cooler than present during the Wisconsin–Weichsel, and in extreme periods was at least 21°C colder. In the Greenland Dye-3 core, the glacial maximum (identified by the $\delta^{18}O$ minimum in the ice) can be placed by layer-counting at 17.2 ka BP (Paterson and Hammer 1987). There appears to have been a linear warming trend from the glacial maximum until about 14 ka BP, during which time about 35% of the full glacial–Holocene warming occurred. More rapid

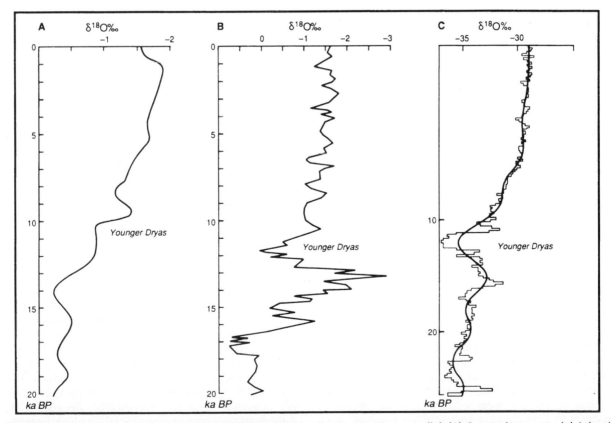

Fig. 3.14 Oxygen isotope records for the last 20 ka indicating the Younger Dryas stadial. (A) Composite equatorial Atlantic Ocean record, based on an assumed constant sedimentation rate (from Berger *et al.* 1985). (B) Record from marine sediments in the Gulf of Mexico dated by ^{14}C (from Leventer *et al.* 1982). (C) Record from the Greenland Dye 3 ice core dated by counting annual accumulation layers (from Dansgaard *et al.* 1985, 1989)

warming is indicated beginning at 13 ka BP, but this gave way to the Younger Dryas stadial at 11 ka BP; in this stadial, the $\delta^{18}O$ values fell back to levels comparable to those of the glacial maximum. The rapid warming referred to above then followed at 10.75 ka BP, marking the end of the Late Wisconsin–Weichsel and the transition to the Holocene. Warming continued for 1–1.5 ka after the boundary, reaching levels typical of the Holocene by about 9 ka BP.

Former precipitation is also estimated from ice core records. This technique is based upon measurements of the isotope ^{10}Be, which is produced in the atmosphere by cosmic radiation, and then washed out attached to aerosol particles. If the intensity of the cosmic radiation remains constant, so too will the abundance of ^{10}Be. The abundance of the isotope in ice will then be inversely proportional to the precipitation. Years of abundant snowfall, which produce a thick snow layer, will be associated with low average concentrations of the isotope in that snow; years which only produce a thin snow layer will result in higher ^{10}Be concentrations in the smaller volume of ice that would result.

The pattern revealed in this way for the last glacial stage over Greenland shows a value of precipitation of only about 30% of the present; thus, as would be expected, the extremely cold glacial maximum was also dry. This result is matched by a similar thinning of the annual accumulation layers during the glacial stage. Precipitation reached the typical Holocene value by about 13 ka BP. The record for earlier parts of the Quaternary shows relatively high precipitation in the intervals 125–115 ka BP, 80–60 ka BP and 40–30 ka BP. These time periods match those deduced from the oceanic isotope record to be times of continental ice growth (see Chapter 5).

Dust concentration in ice cores generally shows levels up to 70 times higher during the last glacial stage than during the postglacial. The additional dust is thought to represent two conditions: deflation from dry continental interiors (such as the Australian deserts), from glacial outwash deposits and from exposed continental shelves and increased wind strength, which may have assisted in the deflation of material from these sites.

One of the most interesting records obtained from ice cores is provided by analysis of the gas samples preserved in air bubbles, especially the abundance of greenhouse gases (reviewed by Raynaud et al. 1993). Because of the delayed pinch-off of air bubbles described earlier, the actual age of the enclosed air is younger than the surrounding ice, by an amount that depends on the ice accumulation rate and the temperature. In Greenland, where pinch-off occurs at a depth of about 90 m (Barnola et al. 1991), the air is 100–400 years younger; in Antarctica, the difference in the Vostok core is 3–4 ka (Barnola et al. 1987;

Paterson and Hammer 1987). Nor is the gas at any level in an ice core all of the same age, since different pores are pinched shut at different times. The analyses which are obtained thus represent averages spanning perhaps a few centuries, and as a result of these uncertainties, the gas composition records derived from bubbles are always published in the form of 'envelope curves' reflecting the analytical uncertainty, rather than as exact values. Carbon dioxide concentrations at the glacial maximum turn out to be about 190–200 ppmv (parts per million by volume); in the Holocene these values rise to 260–280 ppmv. Thus, greenhouse warming because of higher CO_2 levels must have been a significant process fostering deglaciation. The ice cores also show rapid fluctuations in CO_2 content during the Wisconsin–Weichsel, some of up to 60 ppmv in 100 years. Some of these apparent jumps may reflect the melting of snow while at the surface, and resulting CO_2 concentration, but others are considered to be real. The mechanisms that control these variations in atmospheric CO_2 abundance, and which accounted for the major postglacial increase, are not known. It is likely that they relate to processes in the oceans, which contain vastly more CO_2 than the atmosphere. We will return to these ideas and to the role played by the oceans shortly.

Methane (CH_4) concentrations have also been analysed from air bubbles in the Greenland ice cores. This has revealed significant concentration fluctuations amounting to a doubling of CH_4 concentrations from glacial to interglacial time (350 ppbv (parts per billion by volume) to 700 ppbv). High-resolution studies, like the 10 m sampling interval employed by Chappellaz et al. (1993) on a core from the Summit site at the top of the Greenland ice sheet (at 72°34' N and 3230 m elevation), have revealed that the CH_4 concentrations are quite dynamic. A marked decline corresponding to the Younger Dryas has been recorded, together with six oscillations in the period 40 ka to 28 ka BP. These appear to correspond to previously recognised interstadial events seen in the $\delta^{18}O$ record from Greenland. This correspondence has been verified and some further detail provided by detailed methane analyses on the GISP2 core (Brook et al. 1996). These analyses have shown a sawtooth pattern of CH_4 changes in which concentration changes parallel temperature shifts. Methane peaks correspond to interstadials, and are followed by slow declines in concentration that take 7–20 ka before another peak occurs. This distinctive pattern has been identified in other records, where it is called a Bond cycle. We will consider these records shortly. Because CH_4 concentrations reached interglacial levels during the Bølling interstadial (just prior to the Younger Dryas), when the great ice sheets were still areally extensive, Chappellaz et al. (1993) and Brook et al. (1996)

inferred that wetland sources of the methane must have lain in lower latitudes, so establishing that the climate changes recorded in the Greenland $\delta^{18}O$ values must have much wider climatic significance. Times of lower methane abundance would correspond to periods of drying-out in the wetland sources, with less decay of organic matter to yield methane. If wetland expansion is the source of the additional methane represented in the air-bubble peaks, then the moisture supply at lower latitudes seems likely to be affected in some way by the same mechanism that accounts for the rapid $\delta^{18}O$ fluctuations seen in the ice itself (which may reflect temperatures in the Greenland region). A possible mechanism, discussed in Chapter 4, involves changes in the major ocean currents, especially in the North Atlantic region. Whatever the explanation, the methane record emphasises the strong coupling that can be manifested between the biosphere and climate over quite short time-scales. Although the climatic greenhouse effect of the variations in CH_4 concentrations is small (estimated to be only about $0.1°C$; Chappellaz *et al.* 1993), tracing the mechanisms responsible for the methane fluctuations provides further evidence of the kinds of environmental changes that have occurred during glacial and interglacial periods, and helps to tie down their locations. Furthermore, mechanisms through which short-lived but very large emissions of methane could arise are known (e.g. Thorpe *et al.* 1996), and it is possible that these may have been missed in ice core sampling. In association with feedback effects, it is possible that such brief methane 'spikes' could have had climatic consequences. We will return to these ideas in our discussion of mechanisms responsible for deglaciation towards the end of this chapter.

While discussing the ice core record of greenhouse gases, we should note that the concentration record for nitrous oxide, N_2O, also a greenhouse gas, is known from Antarctic ice core records. The levels fell to about 190 ppbv at the LGM, compared with Holocene values of 270 ppbv (Leuenberger and Siegenthaler 1992). The radiative forcing arising from N_2O is larger than that from methane, and amounts to about 15% of that arising from CO_2. Changes in N_2O parallel those of CH_4, suggesting a common source in biomass or soil processes. The temporal behaviour of CO_2 is different, emphasising that other processes control the concentration of this gas (Leuenberger and Siegenthaler 1992). Fluctuations in CO_2 will be discussed more fully later in this chapter.

Antarctic ice core records

Important records have been obtained from some long cores collected at Vostok, located at 78.5° S in East Antarctica. The mean annual temperature here is about −55°C (Jouzel *et al.* 1993, 1994). The deepest core was drilled to more than 2.5 km, and is about 220 ka old at the base, but more than 1 km of the ice (extending to below sea-level) remains to be penetrated.

The Vostok core shows changes in CO_2 concentration similar to those described from Greenland (see Fig. 3.15). Analysis of ^{10}Be also shows that precipitation there during glacial conditions was only about 50% of that received during the last interglacial or the present (Yiou *et al.* 1985).

Ice core analyses have also provided data on former atmospheric methane (CH_4) concentrations. Atmospheric concentrations of methane are much lower than those of carbon dioxide, amounting to about 1700 ppbv at present. The Vostok core shows that much lower values, near 350 ppbv, occur under glacial conditions, with interglacial values reaching 650 ppbv (Chappellaz *et al.* 1990). The record from the last glacial termination shows a sharp drop in concentrations at 11 ka BP, corresponding to the Younger Dryas stadial recognised in the terrestrial and marine records of the northern hemisphere. Increases this century, presumably anthropogenic, have generated higher concentrations than occurred at any time in the last 160 ka, and also the highest *rate* of increase in concentration recorded.

The deuterium temperature record deduced from the Vostok ice is based on sampling at 1 m intervals throughout the 2083 m long core. It shows that at this site, the last glacial maximum was about 9°C colder than the average for the Holocene (Jouzel *et al.* 1987). The transition to Holocene warmth is revealed as a two-step process, beginning at about 15 ka BP and interrupted by about 1 ka of cold conditions at 12–11 ka BP (probably reflecting the same events as the Younger Dryas stadial of Europe). The Holocene is shown to have been warmest early on (around 9 ka BP) and to have cooled subsequently. Temperatures in the last interglacial were about 2°C warmer than present for a period of about 5 ka. Cooling towards the last glacial maximum was interrupted by major interstadials at 106–73 ka BP, when temperatures peaked at about 6°C above those of the glacial maximum, and at 58–30 ka BP, when warmest temperatures were only about 4°C warmer than the glacial maximum.

Many complications arise in the interpretation of the ice core records, especially those of temperatures derived from the oxygen isotope analyses. One of the major problems to be evaluated is the degree to which the inferred postglacial warming described above might be due to effects other than temperature rise. The obvious effect to be considered is that during deglaciation, the ice cap might become thinner; the snow would then be falling at lower (and warmer) elevations, so that the temperature record subsequently

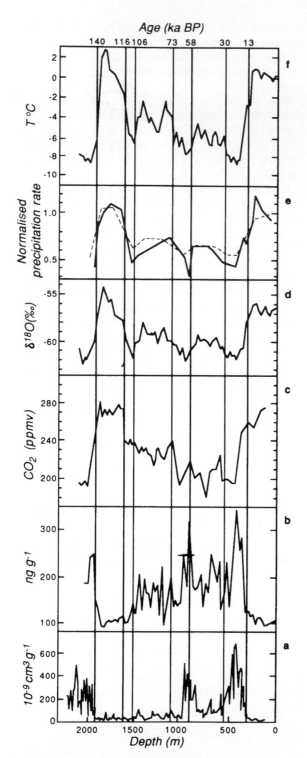

Fig. 3.15 Records from the Antarctic Vostok ice core
(a) Volumetric abundance of particulates (from Petit *et al.* 1990)
(b) Non-seasalt sulphate content (from Legrand *et al.* 1988)
(c) CO_2 content (from Barnola *et al.* 1987)
(d) $\delta^{18}O$ (from Lorius *et al.* 1985)
(e) Estimated precipitation rate (from Yiou *et al.* 1985)
(f) Inferred palaeotemperatures (from Jouzel *et al.* 1987)

extracted would partly be a record of elevation change, and not simply temperature. The unravelling of this effect requires that the amount of lowering be known separately, but this is rarely the case. A similar effect arises, unless the sampled site has always been located exactly on an ice divide, because the ice will have been moving down the slope of the ice dome since its accumulation. This means that the ice sampled down a vertical core will not all have accumulated at that point; the basal material will be from snow which fell higher up the ice cap. Thus the indicated temperatures at the base are likely to be colder than present, whether or not there was an external climate change. Correcting for this effect again requires that the rate and direction of ice movement be known for the whole period represented in the core, so that a correction can be deduced and applied to the isotope record. There is often insufficient information to allow this to be done with complete confidence; ice cores drilled at sites with little ice movement (e.g. the Greenland Summit site) minimise these difficulties.

Other records derived from snow and ice cores

Samples of snow and ice derived from cores can provide various additional kinds of information about Quaternary environments. The warmth of summer may generate meltwater within a snow pack that can percolate deeper into the snow and then refreeze. This ice can be recognised by a distinctive 'bubble' texture.

Koerner and Fisher (1990) used records of such melt layers from ice cores taken on Ellesmere Island to examine the record of Holocene summer warmth, assessed by the abundance of melt ice present. They reported that, just as at Vostok, the warmest summers were in the early Holocene, when there was complete summer melting and possibly some meltwater runoff. The record then suggests progressive summer cooling, but up to 40% of the apparent change could be the result of isostatic recovery of the crust, which has experienced about 100 m of postglacial uplift in the study area. The trend in summer temperatures broadly follows the decline in solar radiation calculated from Milankovitch orbital perturbations (see Chapter 5). Interestingly, the summers of the past 100 years in this record appear to have been the warmest for more than 1000 years.

We should note here that periodicities corresponding to the Milankovitch orbital terms have been clearly demonstrated from ice core data using spectral analysis methods. This is an important piece of information which relies on the continuous diary that ice cores yield. Yiou *et al.* (1991) made extensive analyses of the periodicities that could be detected in ice core records from Vostok (Antarctica), and showed periods of 41 ka, 23 ka and 19 ka (orbital obliquity and precession periods), as well as numerous shorter ones ranging from 11.1 to 2.4 ka. All of the shorter periods were shown to be probable combination tones produced by interactions among the fundamental Milankovitch periods. This effect is explained in Chapter 5. This is a very interesting and important demonstration that various combinations of environmental processes are capable of extending the influence of slow orbital changes to yield climate changes having periods of just a few thousand years.

There are many other kinds of records derived from ice cores which we cannot discuss at length here. Examples include the use of entrapped carboxylic acids (Legrand and De Angelis 1996) and ammonium (Fuhrer *et al*. 1996) to establish a chronological record of biomass burning. Ice cores also provide detailed records of gas composition through the historical period, from which human effects on the gas composition of the atmosphere can be evaluated (e.g. see Etheridge *et al*. 1996). These show, for example, the rapid rise in CO_2 levels that has accompanied the Industrial Revolution.

High-frequency climate changes seen in ice core records

One of the most interesting aspects of the records derived from ice cores is that they provide very fine temporal resolution. In the case of marine sediments, from which many vitally important proxy environmental records are obtained, time resolution is less good because of the activity of burrowing organisms at the sea floor. This bioturbation smears the sediment layers and homogenises the upper few centimetres. Because the sediment often accumulates at a rate of a few cm per ka, bioturbation amounts to a kind of averaging over a period of several thousand years. No finer resolution of the proxy data obtained from, say, foram remains in the sediment is thus possible. Often, neither sediment deposits nor ice cores are analysed at the maximum possible time resolution, because of the cost of studying great numbers of closely spaced samples taken along a core. Only parts of the km-long ice cores, for instance, have been sampled in great detail at centimetre spacings as needed to resolve the timing of key events.

Such very detailed sampling has revealed some remarkable features of environmental change in the Quaternary. Let us consider a few examples. Greenland ice cores, like the Camp Century core, reveal persistent oscillations in dust content and $\delta^{18}O$ values. Through most of the last glacial period, these swings occur with a period of about 2.5 ka. Possible non-climatic causes for these patterns, such as ice surging and consequent episodic lowering and warming of the ice surface, have been carefully evaluated and rejected (e.g. see Dansgaard *et al*. 1984).

A typical detailed ice core record is shown in Fig. 3.16. The high-frequency swings in $\delta^{18}O$ are clearly evident. The amplitude of the $\delta^{18}O$ fluctuations is about 3–5‰ from lightest to heaviest, but the absolute values vary between cores taken at different elevations on the ice cap (Summit, Dye-3, Camp Century, etc.). Typically, during warm (interstadial) phases, the $\delta^{18}O$ record moves back to levels typical of the Allerød-Bølling interstadial period. Intervening stadial periods have values like that of the glacial maximum, which appears in the core at between 25 ka and 20 ka BP. In other words, the fluctuations in the $\delta^{18}O$ record revealed in these ice cores represent dramatic oscillations in environmental conditions over quite short periods of time. Johnsen *et al*. (1992) reported at least 11 short warm interstadials in the last 40 ka, individual interstades lasting from as little as 500 years to around 2 ka. Dansgaard *et al*. (1993) documented a total of 24 interstades in the last glacial period from the GRIP core (see Fig. 3.16). Their records from the 2321 m Summit core, which has a mean ice accumulation rate of 23 cm/ka, confirmed that each interstade began abruptly, with a very rapid temperature rise of up to 7°C over just a few decades. Following this, conditions deteriorated rather more slowly, reaching 12–13° colder than the present, before the next abrupt jump to an interstade. The Younger Dryas cold phase had many of the characteristics of the earlier stadials, and appears to have ended with similar rapidity. Indeed, Hughen *et al*. (1996) have suggested that the Younger Dryas terminated in less than 10 years, while Dansgaard *et al*. (1989) reported from the Dye-3 core a significant amelioration in 20 years, with a 7°C warming in south Greenland over only 50 years. Precipitation increased by 50% across this transition. The rapid temperature oscillations recorded in the ice cores are now known as *Dansgaard–Oeschger* (D–O) events. They remain unexplained, although some hypotheses to account for them do exist. We will return to these later.

THE PATTERN OF RAPID CLIMATE CHANGES IN THE LATE QUATERNARY

High-frequency shifts in the upper Quaternary isotope record of the Greenland ice cores are present not only during the period of deglaciation. Rather, they are known from the last glacial period and from the previous interglacial as well (the Eemian of Europe).

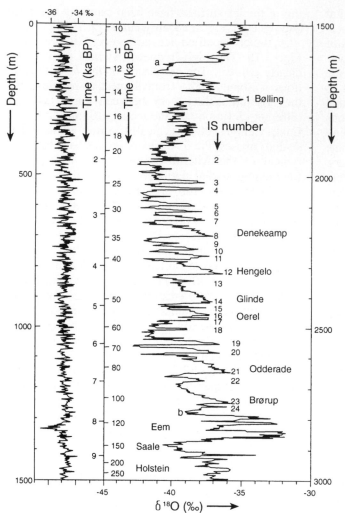

Fig. 3.16 Ice-core oxygen isotope data from the GRIP Summit core. On the left is the relatively steady Holocene section of the record. The right-hand section of the diagram demonstrates the dramatic contrast provided by the preceding 200 ka or so. The 24 glacial interstades are numbered from the most recent (1) at the top of the diagram. (From Dansgaard *et al.* 1993)

For example, the GRIP (Greenland Ice-Core Project) Members (1993) have also documented a series of remarkable isotope jumps in the 80 m or so of ice from the Greenland Summit core that corresponds to the Eem interglacial. (This is named after the Eem river in the Netherlands, where materials of this age have been studied.) During this time (around 125 ka BP, in MIS 5e), the isotope values suggest that average conditions may have been 2°C warmer than during the present (Holocene) interglacial. But unlike the Holocene, which shows a remarkably steady isotopic record, the Eem shows three clear warm intervals during the time of MIS 5e, separated by intervening cold events. The warm periods may have been 4–5°C warmer than present, and the cold periods 4–5°C cooler than present (Johnsen *et al.* 1995). Some of the temperature jumps suggested in this record are remarkable. One involves an apparent jump of 14°C in as little as 14 years, another 10°C in

30 years. Moreover, the period for which the climate appears to have remained in one or other mode is quite variable. The GRIP Members indicated (1993) that the events may have been quite transient, lasting only decades or centuries, or may show a kind of 'latching' in which the temperature change persists for up to 5 ka. These results paint a picture of the Eem interglacial that is quite different from what we know of the Holocene, which has displayed only more gradual climate changes, such as the broad period of warmth (the 'Holocene climatic optimum') which is seen to peak at about 7 ka BP in the Renland ice core from eastern Greenland (Larsen *et al.* 1995). This is of relevance to the analysis of the future of our contemporary climatic environment. The major evident difference between the Eem and the Holocene is that the Eem was a little warmer. Consequently, the idea that contemporary anthropogenic warming of the climate through the emission of greenhouse gases

may only cause a slow temperature rise has to be questioned. Perhaps warming of a few degrees above Holocene levels triggers the kind of climatic instability apparently seen in the Eem isotope record. Human activities may thus be worryingly close to the threshold of such a change.

The reality of these remarkable climate changes remains to be completely established, however (see Zahn 1994; McManus *et al.* 1994). There is support for them from pollen data derived from Lac du Bouchet in France, where five climate oscillations have been recognised during Eem time. Similarly, An and Porter (1997) have documented millennial-scale fluctuations of texture in the Chinese loess record from this period, which reflect fluctuations in monsoon vigour. Some investigators nevertheless suspect that the ice core records have been corrupted by deformation within the ice. Moreover, the ice core $\delta^{18}O$ swings could be generated by other influences upon the isotopic composition of the snow, rather than by temperature changes. We can identify a number of such factors. A change in the temperature of the source water bodies could act in this way, as could a change in the trajectory of the air streams delivering the water to the ice cap, since a different path could allow altered isotope fractionation along the route. Also, air masses from different sources could mix in varying proportions to yield changes in the isotope composition. Modelling studies have suggested that some of these options may provide plausible alternative explanations. Charles *et al.* (1994) used the GISS (Goddard Institute for Space Studies) climate model to track isotope fractionation and moisture supply for the Greenland ice. When Laurentide ice was present over North America, they found that that area was capable of supplying less moisture, so that the Greenland ice had to be fed from North Pacific and North Atlantic sources. Water vapour from the North Pacific followed a long, cold path to Greenland and arrived 15‰ heavier than North Atlantic water. Under these conditions, a change from a local North Atlantic moisture source to one half derived from the Pacific could result in a 7‰ isotope change without any temperature change being required (although one would almost certainly be involved). Changes in these airstreams could reflect physical changes in the bulk and elevation of the Laurentide ice sheet, so that the $\delta^{18}O$ swings could reflect aspects of ice sheet behaviour. If this is so, then the risk that contemporary global warming may set off such instability (referred to above) seems much reduced.

Another explanation for the apparent Eemian instability involves the ocean currents in the North Atlantic. If the East Greenland Current weakened, then the oceanic front separating this cold water from the warmer water of the Atlantic (e.g. the Norwegian Current) might retreat westward, bringing warmer water to the Greenland coast. Resulting altered isotope fractionation might then account for the $\delta^{18}O$ shifts.

Among the other mechanisms that might be active is one that involves the altered atmospheric dust loading seen in the ice cores. The additional dust particles might act as condensation nuclei and affect the degree of cooling required for crystal growth in clouds feeding the ice sheet. If so, the $\delta^{18}O$ swings might reflect changes in the isotopic composition of the snow delivered to the ice, rather than temperature changes. Evidence concerning the real role of temperature change has been presented by Kapsner *et al.* (1995). They have confirmed that snow accumulation in Greenland today is indeed controlled primarily by atmospheric dynamics rather than temperature. The increased moisture-holding capacity of warmer air suggests that accumulation rates on the ice should increase by about 4% per °C warming. But according to the ice core data, accumulation rates in the Holocene have only varied by about 0.9% per °C, with greater influence being exerted by varying storminess and storm tracks in the region. During glacial times, storms were likely to have been more frequent, with storm tracks displaced southward to the boundary of the open ocean and the much-expanded sea ice. These ideas have been supported by statistical analyses of variability in a Greenland core by Ditlevsen *et al.* (1996), who showed greater climatic instability in the last glacial period. More variable storm tracks, relating to movement of the sea ice margins and to more vigorous atmospheric circulation generally, seem to be required to account for this. This debate about the reality and possible causes of Eemian climatic instability serves to illustrate the care required in the interpretation of proxy records, especially when they are derived from environments unlike any that can be observed directly today.

High-frequency climate change in oceanic cores and correlations with ice core records

Subsequent work has revealed further structures within the ice core records from Greenland, and shown strong associations with $\delta^{18}O$ fluctuations recorded in ocean sediment cores displaying low bioturbation, and sampled at resolutions of 300–500 years. This correlation of ice core and marine records provides confirmation that both archives reflect climatic oscillations affecting at least a broad area of the North Atlantic. Bond *et al.* (1993) demonstrated such correlations using the $\delta^{18}O$ record from the Summit ice core, together with $\delta^{18}O$ data from several oceanic sediment cores including DSDP-609 and VEMA 23–081. The correlations which they established suggest that temperature changes in the ocean must have been akin to those over the ice cap: several degrees within a period of some decades. Bond *et al.* (1993) confirmed that these fluctuations occurred throughout the last 90 ka, therefore taking place both as ice volumes increased before the LGM, and decreased after it. Marine sedimentary records from the Santa Barbara basin off the Californian coast in the northeast Pacific have shown high-frequency fluctuations which also correspond to D–O events. Sedimentation rates here were high (> 120 cm/ka), facilitating detailed analyses. Furthermore, the bottom waters are very depleted in oxygen, which restricts the possibility of significant bioturbation. Changes in the nature of sediment being laid down are thus preserved clearly in fine laminations. Behl and Kennett (1996) used detailed measurements of sediment texture to infer when bioturbation took place in the core, reflecting better oxygenation. They found a good correlation with the Greenland ice core D–O events, in which stadial phases seen in the ice corresponded to periods of oxygenation and benthic foram growth in the Santa Barbara Basin. The Younger Dryas was clearly revealed as a phase of oxygenation. Behl and Kennett inferred that changes in ocean ventilation and the mode of thermohaline circulation must have altered the characteristics of the water carried into the basin from intermediate depths. The strong temporal correlation with events in the North Atlantic requires that ocean circulation changes in that region be rapidly transmitted to the Pacific through the global thermohaline circulation system. Similar findings from the Cariaco Basin in the Atlantic off the coast of Brazil were reported by Hughen *et al.* (1996). Here, century-scale fluctuations in marine productivity were demonstrated, related to fluctuations in the strength of upwelling presumed in turn to be related to changes in the strength of the trade winds. These wind fluctuations in turn are a function of variations in the pattern of sea-surface temperatures (SSTs) influenced by the thermohaline system.

The marine record has revealed a distinctive pattern in the incidence of D–O events: they are 'bundled' into cycles of progressive cooling, spanning 10–15 ka, followed by a very rapid warming episode (Bond *et al.* 1993). About seven of these 'bundles' have been identified from the last 90 ka. These sawtooth patterns of climatic variation (a characteristic form already noted in the ice core δ swings; see Lorius 1989), which have been the subject of intense study, are called Bond cycles. They are clearly linked to the D–O events in some way, and, like the D–O events, remain unexplained. However, as we shall shortly see, there are further events recorded in ocean sediments that provide clues to the explanation of both phenomena.

CONCLUSION

We have seen that land-based evidence in the form of tills, moraines and ice-sculpted features provided the initial evidence of the great Quaternary ice sheets. Although detail is lacking in some areas, the broad geographical distribution of the most recent ice sheets is reasonably well known, and some significant regional differences in the chronology of ice growth and decay are known. However, oceanic sediment records and ice core data provide a rich source of additional information on the environments of the glacial and interglacial periods. These include estimates of temperature, precipitation, and gas and dust composition of the global atmosphere. Oceanographic events are also suggested by the data. Evidently, very rapid environmental changes may take place under both glacial and interglacial conditions, superimposed on the longer-term pattern of ice growth and decay. In the next chapter, we will concentrate largely on the interpretation of this store of evidence, seeking to reveal some of the mechanisms responsible for the complex glacial record of the Quaternary.

QUATERNARY GLACIATIONS: CAUSES AND FEEDBACK MECHANISMS

Wir müssen wissen. Wir werden wissen.
(We must know. We shall know)
David Hilbert (1862–1943)
Epitaph on his tombstone in Göttingen

Having examined something of what is known about the areal extent of the cryosphere during the last glacial stage, and the broad patterns in the timing of the ice growth and recession, upon which are superimposed various patterns of high-frequency instability (D–O events, Bond cycles, etc.) we can now turn to the complex issues involved in identifying the *causes* of the glacial and deglacial episodes. We will concentrate on some of the mechanisms involved in ice sheet growth and decay, especially those involved in ice sheet collapse and deglaciation. The broad pattern of glacial and interglacial cycles, alternating at about 100 ka intervals in the late Quaternary, seems likely to relate to external forcing provided by solar radiation changes. Because of the complex issues involved in this forcing, and in terrestrial responses to it, a fuller discussion is provided in a separate chapter (Chapter 5) which focuses on the Milankovitch mechanisms.

TRIGGERS FOR ICE SHEET GROWTH AND DECAY

Let us begin by considering further the nature of the rapid climate swings reflected in the Dansgaard–Oeschger events, and the Bond cycles formed from bundles of D–O events, introduced in Chapter 3. The clear patterns of the Bond cycles in particular evidently reflect a mechanism that has operated over a considerable period of the late Quaternary. In the search for causes, such patterns cannot be overlooked.

The study of sediment cores from the North Atlantic has revealed another phenomenon which is clearly linked with the mechanism generating the Bond cycles. Heinrich (1988) reported periodic sediment layers, rich in limestone and dolomite detritus and poor in foraminifera, which had evidently accumulated quite rapidly. The $\delta^{18}O$ values derived from the forams contained in these layers indicated very cold water, as did the presence of the particular foram *Neogloboquadrina pachyderma* (left-coiled, or sinistral 's-form'), which is a polar species. Bond *et al.* (1992) dated six such layers within the past 70 ka in the North Atlantic, while Andrews and Tedesco (1992) and Andrews *et al.* (1994) recorded two in sediments of the northwest Labrador Sea. The layers are thin, and forams are rare within them. Using detailed core records (see Fig. 4.1), Bond *et al.* (1992) made the following observations:

1. Each layer reflected an episode of ocean cooling; this was associated with a reduced flux of planktonic foraminifera.
2. During an interval occupying part of the cold period, the abundant detrital materials were laid down.

The exciting interpretation made of this (Bond *et al.* 1992; Broecker *et al.* 1992) was that each layer of detritus had been deposited by a massive armada of icebergs released from eastern Canada and possibly elsewhere. The cold ocean temperatures slowed the melting of the icebergs, and cores taken at various sites across the Atlantic show that on two occasions, the icebergs were able to cross the Atlantic, depositing material progressively as they melted. The detritus (IRD, or ice-rafted debris) is found over a wide area spanning the Atlantic Ocean between 45°N and 60°N. Records of these massive iceberg flotillas (termed 'Heinrich events') were related to the timing

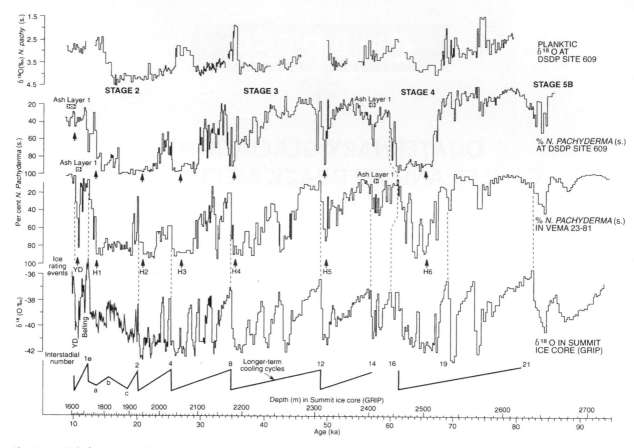

Fig. 4.1 High-frequency climate change events (Dansgaard–Oeschger events, Heinrich events, and Bond cycles) in marine sedimentary records and the GRIP Summit ice core (Greenland). The lowermost curve shows the generalised sawtooth-structure of the Bond cycles. (From Bond *et al.* 1993)

of the Bond cycles by Bond *et al.* (1993). It became clear that the Heinrich events had occurred in the coldest stadials, immediately prior to the rapid warming marking the boundary of a Bond cycle. The ages of Heinrich events from the last glacial period are set out in Table 4.1.

Other environmental changes are known to be synchronous with the Heinrich events and so with the phasing of the Bond cycles. In the polar North Atlantic, sea ice broke up to yield open water conditions that are reflected in an increase in foram abundance in sediment cores (Dokken and Hald 1996). This has been interpreted to signify that Atlantic water moved northward into the polar North Atlantic six times during MIS 4, 3 and 2, remaining abundant for over half the total time. This certainly suggests that sea ice cover was not continuous during the glacial period, and implies that moisture starvation of the regional ice sheets may have been episodic at best during the last glacial.

What does the correspondence between the Bond cycles, recognised from ice core records of tempera-

Table 4.1 The dates of the six most recent Heinrich events from the North Atlantic, together with the Younger Dryas stadial

Heinrich event	Age (^{14}C years BP)
Younger Dryas (?)	11–10 ka
H1	14.5–13.5 ka
H2	22–19 ka
H3	27 ka
H4	35.5 ka
H5	52 ka
H6	69 ka

Source: Bond *et al.* (1993).

tures over the Greenland ice sheet and from the marine ^{18}O record, and Heinrich events, which show oceanic cooling and evidence of ice sheet collapse and sea ice breakup, tell us? The forceful implication is that over the last glacial period, the atmosphere and the oceans behaved as a linked system. This leads us to ask whether all of the correlated events were controlled externally, say by a Milankovitch-type mechanism, or whether they reflect an internal control such as a periodic thermal or mechanical instability in the great ice sheets. Finding the answer to this question is clearly of great importance in the task of understanding the nature and causes of the Quaternary glacial cycles. The Heinrich events evidently do not occur at regular intervals, and they seem to recur too frequently to be directly related to even the shortest of the Milankovitch orbital periodicities, that of the precession (19–21 ka). Let us consider some of what has been discovered.

There is important evidence for similar events occurring at even higher frequencies than the Heinrich events. Deep-sea core data employed by Bond and Lotti (1995), for instance, show that there were multiple episodes of deposition of IRD in the North Atlantic during the last glacial period. Including the Heinrich events, Bond and Lotti recorded a total of 13 ice-rafting events in the period 10–38 ^{14}C ka BP. The duration of these cycles of instability averaged 2–3 ka, and Bond and Lotti interpreted them to reflect increases in the discharge of ice from the Icelandic ice cap and perhaps from ice in the Gulf of St Lawrence. Remarkably, all but one of these ice rafting events could be correlated with a known D–O event from the Greenland ice core records. Bond and Lotti observed that ice rafting episodes of the kind evident in their record appeared to have arisen when the air temperatures in the North Atlantic region dropped below a critical threshold value. This suggests that both the D–O events and the episodes of ice rafting might be related to a separate prior adjustment in the external climate.

The rapidity of the temperature changes involved in the Dansgaard–Oeschger events suggests that the cause cannot be wholesale advance and retreat of the Quaternary ice sheets themselves. These were too massive, and their response times too long, to account for decadal climate swings. The more widely spaced Heinrich events, however, involving enormous iceberg flotillas, clearly do reflect some kind of catastrophic collapse of at least a part of one or more of the major ice sheets. Oceanic cooling, however, preceded the iceberg release and so was not directly caused by this event.

The major heat source for large areas of the higher latitudes in the northern hemisphere is the oceanic circulation system. This delivers sensible heat from the tropical and subtropical latitudes, and the sensible heat is then carried into the atmosphere by convection. A possible cause for the temperature swings of the D–O events is thus a change in the rate at which the oceanic currents carried heat polewards. The oceanic circulation system in the North Atlantic ocean is known to play a key role in contemporary global climate, and changes in this system have long

been inferred (e.g. Smythe *et al.* 1985). The suspicion that changes in this region might be involved in the short-term variability of Quaternary climates is reinforced by the fact that much of the evidence for this variability is strongest in the North Atlantic region. Many of the changes, though smaller in magnitude, have nevertheless been detected at places widely spread around the globe. The D–O cycles, for instance, are present (although in a much attenuated form) in Antarctic ice core records, and have been correlated with events recorded in non-bioturbated sediments from the Pacific coast of the US (Behl and Kennett 1996). Similarly, the Younger Dryas stadial has been reported from as far away as the South Island of New Zealand (Denton and Hendy 1994), from the Sulu Sea (Linsley and Thunell 1990), and from the equatorial Andes, where a small ice cap formed at this time in Ecuador (Clapperton *et al.* 1997). It has also been recorded in tufa deposits at Pyramid Lake in Nevada, part of the former Lake Lahontan system, a site where shifts in atmospheric jet stream paths are inferred to control water balance (Benson *et al.* 1995). Finally, events synchronous with the marine Heinrich events have been identified in records of land vegetation in Florida. In a study of the pollen record of Lake Tulane, Grimm *et al.* (1993) showed that peaks in the abundance of *Pinus* pollen were coincident with the Heinrich events, suggesting a significant climatic impact. All of this demonstrates clearly that Quaternary climatic events lasting only 1 ka or so have indeed had global incidence.

Given that it is a prime suspect in all of this, let us consider briefly how the thermohaline ('temperature–salinity') circulation works and how it might have been modified during the Quaternary. The major oceanic current systems (termed gyres) are steered by wind drag at the sea surface and by the Coriolis effect. But the currents themselves arise primarily from temperature-driven variations in the density of seawater that arise at certain locations. Key among these are the polar and subpolar latitudes in each hemisphere. In the south, around Antarctica, winter extension of sea ice creates a cold, dense brine because the salts dissolved in seawater are excluded during the freezing process. The brines then sink to abyssal depths and create upper inflows as required in order to form a closed circulation system. Similarly, in the northern hemisphere, winter heat loss to the atmosphere chills seawater so that it becomes dense and sinks in a similar way. In the Greenland, Iceland and Norwegian seas, this manufacture of dense cold water feeds a bottom-flow back into the Atlantic. This draws in a compensating upper-level flow from the subtropical Atlantic where the water is warmed by the greater solar heating experienced there. The loss of this heat to the atmosphere as the water chills, elevates atmospheric temperatures in high northern latitudes. The cold outflow generates deep water in the Atlantic Ocean that has a temperature of only a few degrees, and which circulates globally in the abyssal parts of the ocean basins. In the production of these cold, dense circulation systems, remarkably small areas of sea seem to be critical. Around Antarctica, for example, 80% of the water that sinks to become Antarctic Bottom Water (AABW) is generated in the Weddell Sea next to the Antarctic Peninsula (Dickson and Brown 1994). Here, relatively shallow shelf waters evidently permit seasonal cooling of the water mass without convective mixing being able to dissipate the cool water and so restrict the temperature lowering that can be achieved. Likewise, small areas of the Greenland, Iceland and Norwegian seas generate enormous amounts of bottom water. Fast, thin currents of water that are steered by the ocean bottom topography then become involved in what has been termed the 'global conveyor belt' of the ocean circulation system (Broecker 1992). The flows of water involved are very large. They are conventionally expressed in units called sverdrups (symbol Sv), after a famous oceanographer. One sverdrup is equal to a flow of 10^6 m^3/s. Many major thermohaline currents involve flow in the range 5–20 Sv. According to Hay (1993), the flux of AABW is about 38 Sv. Because the oceans are a closed and interconnected system, the outflow of cold bottom water from the North Atlantic Ocean has to be matched by an inflow of water at shallower depths. The warm Gulf Stream, originating at low latitudes, is part of this compensating flow.

The rate at which heat is carried northwards can be reduced if the strength of these ocean circulation systems is diminished. Indeed, W. Broecker (e.g. Broecker and Denton 1989) and others have speculated that these circulations might actually 'turn off' or 'turn on', essentially like a switch, if the supporting conditions are perturbed. Attention has consequently been paid to the kinds of events that might have arisen during the Quaternary to do this. One immediate possibility is that fresh water delivered to the polar seas might dilute the seawater there, making it insufficiently dense to sink even when fully cooled in winter. If this were the case in the North Atlantic, for example, then the reduced production of NADW (North Atlantic Deep Water) would immediately mean a reduced compensating inflow of warmer subtropical surface water which normally takes its place. Among the possible sources of additional fresh water that has been widely investigated is meltwater supplied from waning ice sheets. This idea was proposed by Rooth (1982), who realised that regional influences on the thermohaline circulation system could have consequences for the global climate, since the currents indeed transport

heat throughout the oceans. Rooth also raised the possibility that Quaternary changes in regional precipitation patterns and amounts might affect the thermohaline system. Additionally, the blocking of drainage systems by advancing ice, and the re-establishment of drainage during ice retreat, were seen by Rooth as potentially significant influences on the strength of the circulation. During glacial onset, it is possible that reduced fresh-water discharge may have strengthened the thermohaline circulation, providing an increased flux of warm water to feed evaporation and hence nourish ice sheet growth. This would be particularly effective during the early phase of ice build-up, before ice had reached the coast and iceberg calving had once again increased the delivery of fresh water to surrounding seas.

A particular case addressed by Rooth (1982) involved the drainage from Lake Agassiz, a large meltwater lake that developed south of the ice front during deglacial retreat of the Laurentide ice sheet. Flow from this lake principally followed the Mississippi River system into the Gulf of Mexico. But when the path to the St Lawrence River was opened by the retreat of an ice lobe, the massive discharge was carried off in that direction to the Atlantic Ocean. Rooth speculated that this might have delivered sufficient fresh water to the Atlantic to retard the thermohaline circulation and so trigger the Younger Dryas cooling, which was roughly coincident with the drainage diversion. Teller (1990) tabulated estimates of the rate of flow in the Mississippi and the St Lawrence river systems at 500 ka intervals through the period in question, and inferred discharges along the St Lawrence system of at least six times modern values. He also inferred that there had been several abrupt doublings of the flow rate near the start of the Holocene. In addition, Miller and Kaufman (1991) have suggested that there was a simultaneous increase in the iceberg flux from the Labradorean part of the Laurentide ice sheet, which may have assisted in causing the YD cooling through mechanisms like direct cooling, albedo increase, and reduced wind-mixing of surface waters.

Rooth's suggestion was taken up by subsequent researchers, who looked for evidence of ocean circulation changes in the marine sedimentary record. Let us examine some of the evidence that has been amassed concerning possible Quaternary fluctuations in the thermohaline system of the kind that Rooth proposed, and the environmental effects that have been postulated to arise as consequences of the oceanic events. The possibility of major changes in this system has attracted enormous attention, because they provide a plausible mechanism *internal* to the Earth's own systems that could generate rapid climate changes otherwise hard to account for because of the excessive response time of the great ice sheets or of the slowly-changing orbit of the Earth.

On the basis of isotopic composition changes in benthic forams, Berger and Vincent (1986) suggested that the production of NADW had indeed been shut down episodically. Boyle and Keigwin (1987) employed similar data to support the idea that ocean circulations in the North Atlantic had undergone re-arrangements in the period following the glacial maximum, including conspicuous changes coincident with the Younger Dryas (YD) stadial.

Broecker *et al.* (1989) presented somewhat equivocal $\delta^{18}O$ evidence that the estimated $30\,000\,m^3/s$ flow from the Lake Agassiz basin had indeed affected the ocean composition, following Rooth's suggestion. They used data from a core taken off the SE coast of the US, in a basin where anoxic conditions restricted bioturbation. Here *Globigerinoides ruber* showed more positive $\delta^{18}O$ in the period 11–10 ka BP. There was a sharp decline at 10 ka BP which Broecker *et al.* (1989) termed the 'cessation event'. However, Broecker *et al.* were unable to show how much of this signal was related to a changed water flux, and how much might have been related to a possible global temperature change consequent upon the Younger Dryas itself, and caused by some other mechanism.

Not surprisingly, these ideas were accepted by some (e.g. Street-Perrott and Perrott 1990) but rejected by others. Jansen and Veum (1990) suggested a contrary view that the production of NADW had actually *increased* during the YD as a result of reduced overall meltwater flow as expected in a cold period. This was a possibility noted by Broecker *et al.* (1989) who observed that reduced meltwater output from elsewhere (e.g. the Fennoscandian ice masses) might have been capable of offsetting the flow of Laurentide meltwater from Lake Agassiz. The increased thermohaline circulation in the scheme of Jansen and Veum (1990) would provide a greater delivery of heat, and thus provide a suitable mechanism for bringing cold events like the YD to an end. Broecker envisaged other ways of achieving this.

Data on the rate of sea-level rise provided by Fairbanks (1989) from reefs in Barbados seemed to conflict with the Rooth–Broecker meltwater hypothesis. Fairbanks showed that in fact rates of sea-level rise were lower in the YD than before or after, hardly consistent with the shutdown of NADW production because of lowered salinity arising from a major influx of meltwater. Broecker (1990) countered that the rapid rise in sea-level prior to the YD might have left the thermohaline system operating but vulnerable to an additional push like that of the Lake Agassiz diversion.

In a continuing investigation of these issues, Veum *et al.* (1992) used isotope data to confirm that overturning (ventilation) in the North Atlantic–Norwegian Sea continued during the YD, and rejected Broecker's 'on/off' model of the periodic shutdown of NADW production. But the ability of the NADW system to undergo high-frequency shifts in operation has been verified using $\delta^{13}C$ data (Keigwin *et al.* 1994). Modelling efforts have also shed light on this topic. Fichefet *et al.* (1994) used a global ocean circulation model to infer that NADW production *was* altered during the LGM, although its production did not switch off. Rather, they found that Antarctic Bottom Water (AABW) flowed further north and invaded the Atlantic, displacing the deep water circulation there to a shallower, intermediate depth. Fichefet *et al.* estimated that this reduced northwards heat transport by the thermohaline system from the present value of $\sim 1 \times 10^{15}\,W$ to only $0.7–0.8 \times 10^{15}\,W$ at the LGM. Similar conclusions were reached by Yu *et al.* (1996) in a study of the abundance of protactinium (Pa) and thorium (Th) in oceanic sediments. Pa bonds to particulate matter and its abundance in the North Atlantic sediments depends partly on how much was swept from the basin by ocean currents. Yu *et al.* (1996) found that just as much Pa was exported to the southern oceans at the LGM as during the Holocene, and confirmed that the thermohaline circulation had not 'turned off' in this period. Rather, their data support the modified pattern of circulation, in which NADW was underlain by water of southern origin at the LGM. Similar modifications to the system were modelled by Rahmstorf (1994). A coupled ocean-ice sheet model employed by Paillard and Labeyrie (1994) has nevertheless suggested that on/off cycles of the thermohaline system can occur. These workers found that during 'off' periods, heat accumulated in lower latitudes and was rapidly delivered polewards once the system turned 'on' again. This switching would provide a ready means of accounting for the very rapid warm jumps at the termination of the D–O events and Bond cycles. Recently, using fluctuations in the abundance of ^{14}C independently timed with reference to an absolute chronology based on counting of annual tree rings, Björck *et al.* (1996) have added support to the idea that fresh water inflow does inhibit NADW production. Fluctuations in the levels of ^{14}C reflect ocean circulations and ventilation because oceanic upwelling can deliver old carbon to the atmosphere, and uptake of new ^{14}C can also be affected. Björck *et al.* (1996) inferred that a period of reduced ocean ventilation may have led the global climate into the YD stadial period. They suggested that the strength of the North Atlantic part of the oceanic conveyor decreased substantially at the boundary between the Allerød interstadial and the YD stadial. After the YD, abrupt warming in their data suggested that the onset of increased northwards heat advection was almost instantaneous. Similar findings were reported by Goslar *et al.* (1995). This suggests that there may be at least a component of the NADW system that has 'switch-like' behaviour. In addition, there seems to be good evidence that glacial variation in the production and movement of NADW involved a range of possible adjustments, not just an alternation of 'on' and 'off' modes. This has been likened to NADW production being controlled in a way more analogous to an *adjustable* valve, capable of variable regulation of heat delivery, than by a simple switch (e.g. Boyle and Rosener 1990; Taylor *et al.* 1993).

Models of the oceanic thermohaline circulation system have been used to shed light on the general phenomenon of rapid climate variations in the Quaternary. Sakai and Peltier (1996) have performed modelling experiments with an ocean model containing a representation of the three major oceans. They show that the circulation system that arises in the model can display oscillatory behaviour on 100–1000 year time-scales, and that fluctuations of this kind may indeed be intrinsic in the deep ocean circulation. This clearly supports the suggestion that changes in the thermohaline system, by affecting the warmth of the atmospheric environment, are responsible for the D–O events. Furthermore, Sakai and Peltier have used their model results to confirm that the rapid discharge of meltwater pulses into the ocean may indeed weaken this system and so cause events like the Younger Dryas.

The importance of the oceanic circulation system in regulating global climate and in influencing Quaternary climate change thus appears established. There is much still to be learned, however. In particular, the role of the southern hemisphere cannot be overlooked, even though events in the North Atlantic region seem most prominent. Crowley (1992) has shown that with NADW production 'off' (or perhaps weakened), the South Atlantic 'banks' the heat that would otherwise be transported northwards. He estimates this to be equivalent to a southern hemisphere gain of 2–4 W/m^2. With NADW production 'on', there is thus a significant net cooling of the southern hemisphere as the heat is exported northwards. Crowley showed that the glacial temperature lowering seen in the Vostok ice core could be best accounted for by a 1:3:1 proportion of effects arising from NADW production, CO_2 decline, and orbital (Milankovitch) insolation forcing. This scenario suggests that NADW production may simultaneously warm the high northern latitudes while cooling the southern. If this is so, then interglacial NADW production might provide a trigger for glacial onset around Antarctica at a time before any direct insolation cooling was experienced in the northern hemi-

sphere. Southern hemisphere oceanic cooling could then lead to significant ice volume growth in the north. Variation in the rate at which the South Atlantic could 'bank' héat with NADW production turned off would vary in accordance with the dominant precessional influence on insolation in the low latitudes. Because of this, the Milankovitch forcing of the low southern latitudes could become significant in glacial onset (see McIntyre *et al.* 1989). This possibility is in direct opposition to the traditional view that heating at high northern latitudes, in the areas actually occupied by the Quaternary ice sheets, must provide the key to glacial cycles. Here we see a clear demonstration of the importance of understanding global teleconnections and interlinkages between events in the two hemispheres, matters which we will explore further below.

The Heinrich events provide additional clues to the nature of short-period climate variability in the Quaternary. Fronval *et al.* (1995) reported cooling events and iceberg discharges analogous to and contemporaneous with Heinrich events, but seen in a sediment core from the Norwegian Sea. This suggests that both the Laurentide ice sheet, known to be the source of the Heinrich icebergs from the distinctive composition of the rock detritus, and the Fennoscandian ice sheet, the source of the icebergs from the Norwegian Sea, underwent fluctuations in phase with one another and at a frequency higher than that of the Milankovitch periods, which are thought to control the broad pattern of Quaternary ice ages. Events that correlate with the Heinrich events have also been recorded from more distant sites. For example, the great deposits of loess in China preserve a record of past wind strengths (see Fig. 4.2). Porter and An (1995) used the median diameter of quartz grains in the loess as a proxy for monsoon wind strength, inferring that stronger winds would entrain and carry larger particles, primarily from Mongolian source areas to the northwest. They were able to identify six grain-size maxima in a loess sequence that evidently corresponded in timing to the last six Heinrich events. Furthermore, the grain-size fluctuations also showed good correlation with the changing abundance of the foram *Neogloboquadrina pachyderma* (sinistral) seen in North Atlantic core V28–32 (the Bond cycles), and also with the D–O events seen in the GRIP $\delta^{18}O$ record. These striking correlations clearly suggest that events in the North Atlantic had significant consequences in areas greatly removed from the local oceanographic and atmospheric changes. Variability in the ocean record extends beyond the North Atlantic region, too, adding support to this idea. Thunell and Mortyn (1995), for example, showed glacial variability to exist in the NE Pacific Ocean. They also used fluctuations in the abundance of *N. pachyderma* (sin.) to deduce when

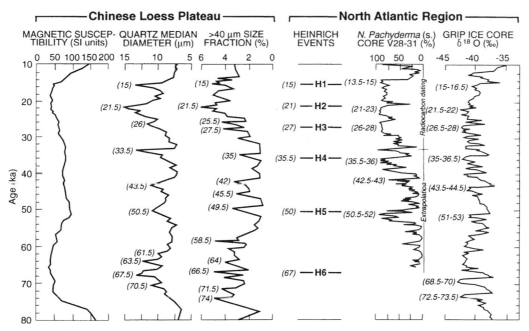

Fig. 4.2 Record spanning the bulk of the last glacial, derived from Chinese loess. The loess data, including grain-size fluctuations taken to represent wind strength, are on the left. The right-hand side of the diagram shows records from the North Atlantic region, including the marine record of the timing of the Heinrich events, and the GRIP ice core $\delta^{18}O$ record. Note the good correspondence between the times of loess grain-size maxima and Heinrich events. (From Porter and An 1995)

subpolar waters had extended to their core site at 33° N, and found major changes in the period 35–15 ka BP. These changes in ocean circulation are presumably driven by shifts in the atmospheric wind systems. These in turn may be related to changes in the elevation of the Laurentide ice sheet and consequent changes in the merging or splitting of the atmospheric jet stream (see Chapter 12). Such a mechanism has also been invoked to account for the ice core δ shifts. Stuiver *et al.* (1995) suggested that cold periods were associated with a split jet stream, bringing polar air southwards. This they termed 'atmospheric circulation mode A'. Warmer periods in contrast related to a polewards retreat of polar air, in what was termed 'atmospheric circulation mode B'. Rapid oceanic changes have also been recorded in the Arabian Sea, where Sirocko *et al.* (1996) employed dust abundance in a sediment core as a proxy for monsoon strength. They found several episodes of monsoon intensification, and a period of low monsoon activity coincident with Heinrich event 1. Spectral analysis of the monsoon strength proxy showed peaks at periods of 1.15, 1.78 and 2.8 ka, periods also found in analyses of the Greenland ice cores. The widespread occurrence of climate changes at these frequencies is suggestive of a global cause, rather than one originating in the North Atlantic. Indeed, Sirocko *et al.* (1996) argued that the cause may rather lie in the low latitudes, related perhaps to

changes in the amount of water vapour exported poleward, with the ultimate cause possibly being a non-linear response to solar insolation changes. The nature of the apparent global teleconnections between low and high latitudes during the Quaternary is an area that requires additional study. Mountain glaciers in the western US are known to have advanced simultaneously with Heinrich events (Clark and Bartlein 1995), and this requires a means for transmitting climate changes over significant distances. Since the Heinrich events were of necessity associated with a major collapse (lowering) of Laurentide ice, it would appear that this collapse affected climate over a sufficiently wide area for the glacier response to be elicited, perhaps through altered global wind systems. Clark and Bartlein (1995) envisaged that the major glacial cycles represented 'first-order variations' in ice abundance, with the collapses associated with Heinrich events forming smaller 'second-order' variations induced by non-astronomical causes. The possibility that both events might relate to global cooling triggered in some other way has also been investigated. For example, using an ice sheet model, Oerlemans (1993) imposed 3000 years of cooling prior to the date of the last three Heinrich events. The imposed cooling actually decreased the calving rate for H3 and H2, evidently because of lower accumulation rates on the ice sheet. Increased calving did result in the case of H1, but this was an unusual Heinrich event, occurring during the deglaciation process when the Laurentide ice sheet was undergoing disintegration. Oerlemans (1993a) concluded that external climatic cooling did not cause the ice sheets to release the iceberg armadas. A cause internal to the ice sheets themselves thus appeared likely. Let us consider some of the ideas that have been put forward in support of this internal mechanism.

INTERNAL ICE SHEET PROCESSES AND RAPID CLIMATE CHANGES

This idea has been developed in a series of models that relate the iceberg release represented by the Heinrich events to instability in part of the Laurentide ice sheet. MacAyeal (1993a, b) and Alley and MacAyeal (1994) have presented models of this kind. These 'binge/purge' models incorporate the idea that large ice sheets may inherently undergo alternate periods of slow growth and rapid disintegration, and that this behaviour may explain the quasi-periodic Heinrich events that recur every 7–12 ka. The Heinrich events are also, as we have seen, coordinated with the D–O events in slow cooling trends that last about 7 ka and end in abrupt warming events. In the models developed by MacAyeal, the 7 ka of cooling in these Bond cycles corresponds to a time of ice accumulation over the Laurentide ice sheet (which has been picturesquely termed a 'binge'). During this time, geothermal heat progressively warms the bed of the ice, providing liquid water to lubricate till materials there. Ice motion then accelerates, further promoted by frictional heat. The rate of heat production is taken to exceed greatly the rate at which heat is conducted upward through the ice. However, as the rapid ice flow event (the 'purge') continues, the ice becomes thinner, basal pressure is lowered, and less frictional heat is generated. Finally, the rate of upward heat loss through the thin ice exceeds the rate of heat generation by friction, and the temperature begins to fall. This finally allows the ice to refreeze to the bed, and the purge comes to an end. In the numerical models, driven by realistic ice properties and geothermal heat flux, the binge phase of heat build-up lasts about 7 ka while the purge phase lasts for about 750 years, during which time an estimated 10^{15} kg of rock material is exported. Overall, height loss modelled for the ice sheet in a purge amounts to 1200–1500 m, releasing $\sim 10^{14}$ m^3 of water into the Atlantic Ocean. The mean flow rate amounts to about 0.16 Sv. This volume of water is sufficient to cause a sea-level rise of about 3.5 m. In the view of MacAyeal, this quasi-cyclic drawdown of part of the ice episodically left a two-lobed Laurentide ice sheet,

with the interior portion remaining at its full elevation. This leads to a possible explanation of the warm jumps that follow each Heinrich event at the end of a Bond cycle. The lower elevation of the remaining ice sheet would permit the re-arrangement of wind systems around the ice sheet noted earlier. Such a shift may have directed more airflow over the North Atlantic, favouring increased evaporation there. This may have been sufficient to increase salinity and re-start the thermohaline circulation. This would then provide a supply of warm water to nourish renewed ice build-up and start the 'binge' process again. MacAyeal has speculated that the rapid D–O cycles that follow the warming jump may reflect 'ringing' or persistent oscillation in this unstable system (Broecker has envisaged such behaviour in his 'salt oscillator' model of the thermohaline system). If this is the case, then the D–O cycles seen in the ice core record owe their development to the same cause as the iceberg release events. MacAyeal's models imply that this is entirely internal to the behaviour of the ice sheets, and is governed by thermal and mechanical considerations governing ice sheet stability.

The Hudson lowlands area seems to have been the site where these events took place because of its underlying geology. The interior part of Canada lies on the hard crystalline basement of the Canadian Shield, whereas the Hudson lowlands lie on softer sedimentary materials. These latter rocks appear to have been more suitable for the development of a saturated till layer beneath the ice which could facilitate rapid flowage. Indeed, Clark (1994) has argued that the Heinrich events could only occur while Laurentide ice was located on the sedimentary rocks. The cessation of Heinrich events at about 14 ka BP can then be related to retreat of the remaining ice on to the harder crystalline rocks of the Canadian Shield.

Many aspects of this model await verification and testing against field data. Much also remains to be learned about the configuration of basal till materials beneath large ice sheets and the way in which frictional drag might be distributed spatially. Nevertheless, these ice flow models have also been applied to analysing the stability of the contemporary West Antarctic ice sheet (MacAyeal 1992). Under various kinds of temperature forcing, ice sheet instability has been suggested, with short periods of iceberg discharge (akin to Heinrich events) delivering ice volumes of up to 500 km^3 per year into the ocean. Significantly, MacAyeal found that the distribution of basal till was the prime determinant of whether the ice sheet was stable or not. External temperature variations were less significant. This suggests that global warming may not be a sufficient condition for the destabilisation of the contemporary ice sheet, unless the till distribution and the amount of till accumulated are appropriate for failure to proceed.

ICE SHEET GROWTH AND GLACIAL INCEPTION

Let us turn now to consider the processes responsible for the periodic slow and unsteady increase in ice volume that marked the onset of each Quaternary glacial period.

The exact growth *mechanisms* for the major ice sheets remain unclear. Insolation changes during summer at high northern latitudes have often been considered the 'forcing function' for climate change, as described in the Milankovitch or *astronomical theory* (see Chapter 5). It is envisaged that reduced solar heating resulting from the orbital characteristics (primarily obliquity and precession) periodically allows summer snow to survive without ablation, with albedo feedback from the growing snow cover, and other positive feedback mechanisms, reinforcing atmospheric cooling. Under this conception, we would expect slow accumulation of snow and eventually glacier growth from initial accumulation areas on uplands in the high northern latitudes, with a gradual extension of the ice-covered area. It is also possible that ice growth accelerated quickly once the regional snowlines began to fall in response to climatic cooling, because of the albedo feedback mechanism, so that snow persisted through summer and accumulated annually and simultaneously over large areas. This has been termed the 'instant glacicrisation' hypothesis. Even in this case, ice accumulation would have begun first in upland regions such as those on Baffin Island, Labrador, and the Rocky Mountains, the Alps, and the mountains of Scandinavia.

In addition to an astronomically-caused drop in high latitude summer insolation and suitable land area with uplands in the high latitudes, we can identify two further necessary precursors to the establishment of a glacial stage. These are:

1. an adequate supply of moisture, which implies a sufficiently warm ocean located somewhere upwind;
2. minimal loss of accumulated snow and ice.

In particular, an inland area of uplands, with no glacier connection to the sea that could result in iceberg calving and reduced accumulation, would favour ice-cap development. Lower abundance of ice-rafted debris during phases of glacial build-up is indeed seen in the marine record (Ruddiman and McIntyre 1981), and this is consistent with net build-up of land ice as required for the albedo feedback effect. Ice sheets partly grounded below sea-level would also be more prone to disruption by sea-level changes. Thus, an ice cap located away from the coastline is immune to the disturbing effect of meltwater floods that might be created by temporary retreats during a slow growth phase.

According to the oceanic oxygen isotope record, which as we have seen is taken to be dominantly a reflection of aggregate global ice volume, there were rapid periods of ice volume growth (revealed by rapid increases in the foram $\delta^{18}O$ values) centred on 115 ka, 90 ka, 75 ka and 25 ka BP. The two most important of these periods were at 115 ka BP and 75 ka BP; the phase of ice growth in each case lasted for about 10 ka. According to estimates by Ruddiman *et al.* (1980), these periods each contributed nearly half of the ice volume growth of the last glacial stage. The net rate of ice volume increase in these short growth phases was thus very rapid, amounting to 5% per ka.

According to the temperatures indicated by foram assemblages, the first period of ice growth at 115 ka BP occurred before the sea surface at 40–45° N in the Atlantic had begun to cool significantly. Thus we can conclude that ice growth *precedes* oceanic cooling, in this case by about 4–4.5 ka. The mechanism envisaged to explain this is that the ice growth was in an area not connected to the sea, so that despite major ice accumulation, no iceberg calving which might have chilled the oceans (or weakened the thermohaline circulation and the polewards transport of heat in the Atlantic Ocean) took place. This appears to confirm the idea of inland ice accumulation as envisaged in the list of preconditions mentioned above. Interestingly, 115 ka and 70 ka BP were the times of lowest summer insolation at 70° N, according to astronomical calculations, which gives us further confidence that the ice growth was driven by the Milankovitch mechanism. A likely candidate for the location of this ice was in the eastern sector of the developing Laurentide ice sheet of southeast Canada, adjacent to the warm ocean.

Ruddiman *et al.* (1980) reported a major increase in the abundance of ice-rafted sands in sediments of the North Atlantic after 75 ka BP, and concluded that iceberg calving began at about that time. Subsequently, the North Atlantic cooled rapidly. If the scenario outlined here is typical of the late Quaternary, we may conclude that northern hemisphere ice growth in glacials is out-of-phase with oceanic cooling, preceding it by 4–5 ka; this is consistent with the suggestion made earlier that a sufficiently warm moisture source must be available to sustain ice growth. This immediately raises the question of how the ice sheets fared in terms of moisture nourishment after the oceans had cooled. We shall return to consider this shortly.

Ruddiman and McIntyre (1981) explored further aspects of the ice build-up phase. In particular, they observed that lingering oceanic warmth during the early period of ice build-up had consequences beyond the nourishing of ice growth. For instance, the sustained evaporation would have kept the ocean salinity high. Because ice growth is presumed to occur during warm summers, the winters would be correspondingly mild (see Chapter 5), restricting sea ice growth and maintaining an open water surface. Ruddiman and McIntyre also envisaged that the temperature gradient between the warm open ocean adjacent to land ice would constitute a corridor favouring northward storm motion, with storms following this track delivering moisture to the growing ice sheets.

The exact pattern of events in the area of individual Quaternary ice sheets is difficult to resolve, because of both dating problems (ice growth having begun in MIS 5, at 130–116 ka BP) and a lack of unambiguous field evidence. The timing and geographical pattern of glacial onset of the Laurentide and Cordilleran ice sheets has been evaluated by Clark *et al.* (1993). This work, using extensive land-based evidence, reveals major episodes of both growth and decay, with quite dynamic fluctuations in ice volume. Detail of this kind cannot be obtained from the aggregate ice volume record preserved in the marine sedimentary archive.

ICE SHEET RETREAT AND DEGLACIATION

In order to be able to document the decline in ice volume, it is once again necessary to refer to the marine isotope record, because the moraines and other features of the terrestrial record only indicate the decline in ice area; volumes can only be estimated by making assumptions about the thickness of the waning ice sheets. The rapidity of deglaciation is, however, clear in all the records; it was completed in only about 8 ka, which contrasts with the 90 ka or so of the Wisconsin–Weichsel that elapsed before full-glacial conditions were reached.

The isotope record of deglacial times is unfortunately a complex encoding of many related processes, and the history it contains is still not resolved with any finality. Let us consider some of the processes that may confuse analysis of the deglacial record in marine isotopes.

The elevation of a growing ice sheet changes continually, as the ice thickens to its final 2–4 km. Thus, the snow which feeds a developed ice sheet is deposited from air which has had to rise higher than would have been the case earlier in the period of ice growth. As a result, the isotopic make-up of the ice in an ice cap will not be uniform through its depth, as noted earlier in this chapter, but is likely to be more negative in younger ice. This in turn implies that, depending on the age of the ice that is melting at any

time during the deglaciation, meltwater of different isotopic character will be returned to the oceans, and the marine record will partly reflect the compositional layering of the ice sheets, not just their remaining volume. The actual effect will depend, for example, on whether most meltwater is being returned from ablation along the equatorward ice margin which lies at low elevation, or is mainly by iceberg calving from fast ice streams draining the poleward flanks of the ice sheet. The only way that these issues can really be approached is by glaciological modelling. Additional complication stems from the fact that the isotopic make-up of land ice varies with latitude, in the same way as temperatures vary, so that ice formed in lower latitudes is less isotopically negative. Thus, the effect of meltwater return on the isotope record of the oceans also depends upon the latitudinal distribution of the ice retreat. In all likelihood this would have involved a systematic tendency for most retreat in low latitudes first, followed by recession at more poleward sites.

Both the terrestrial and marine records indicate that ice retreat was episodic. Major moraine deposits located along the margins of the retreating Laurentide ice sheet, for example, are testimony to halts or reductions in recession rate following major ice loss (Andrews 1987). The marine isotope record from the North Atlantic confirms two or possibly three separate major episodes of ice volume loss, one at 14–12 ka BP, another at 10–9 ka BP, and possibly a third at 8–6 ka BP (Mix 1987; Jansen and Veum 1990). A very large fraction of the northern hemisphere ice volume, perhaps a third, was lost during the first two episodes of retreat. Only relatively small areal contractions of the ice were coincident with the first episode, according to the moraine evidence; the northern hemisphere ice sheets still occupied 75–80% of their full-glacial area at 13 ka BP (Ruddiman and McIntyre 1981) so that much ice *thinning* must have occurred. Interestingly, the same timing of ice volume loss is inferred from the sea-level record preserved in corals at Barbados. Here, Fairbanks (1989) identified two intervals of rapid sea-level rise. The first involved a very rapid rise of 24 m in less than 1 ka, centred on 12 ka BP; the second, lasting slightly longer and involving a 28 m rise, was centred on 9 ka BP. The period of slower sea-level rise separating these two intervals presumably reflects the Younger Dryas stadial. The dates for the periods of rapid sea-level rise are very similar to those inferred from the marine oxygen isotope record to have been times of rapid ice volume loss. Fairbanks has termed the two episodes of rapid sea-level rise Meltwater Pulse IA and Meltwater Pulse IB. Modelling and stratigraphic studies have suggested that IA might have been fed from the Eurasian ice (including the marine-based ice sheets which could be readily destabilised by some initial sea-level rise) and IB from the Laurentide ice (Lindstrom and MacAyeal 1993; Stein *et al.* 1994). This remains unresolved, however, and Clark *et al.* (1996) favour the Antarctic ice as the source of the first meltwater pulse. Very rapid rates of sea-level rise are known from these events. Blanchon and Shaw (1995) observed that these can outstrip the growth rates possible for corals such as *Acropora palmata*, and from analyses of coral stratigraphy in Barbados, infer rates of sea-level rise of up to 70 mm per year. (See Chapter 6 for a more detailed discussion of these events.) The identification of meltwater pulses in the marine record does not require that ice retreat was primarily by wholesale ice melting. It is more likely that ice sheet collapse, and the transfer of large masses of ice to the oceans by iceberg calving, was a dominant mechanism in deglaciation. As Hughes (1987) observed, ice streams drain up to 90% of the modern Greenland and Antarctic ice sheets, and it seems reasonable to infer a similar scenario for the great Quaternary ice masses.

The major periods of ice retreat during the late Quaternary seemed so rapid to early workers analysing the records that the events were termed *terminations*. It was thought that these reflected very abrupt climate swings. As the evidence presented in this chapter has shown, we now have indications of a much more complex sequence of events during deglaciation, including some periods of change that were extremely rapid indeed, and periods in which, part-way through the deglaciation, the climate swung back towards a glacial character (like the YD). These instabilities suggest that one or a number of processes internal to the climate system exert a major influence during these times. This is suggested clearly by the contrast between the relatively smooth and gradual changes in seasonal insolation arising from the Milankovitch mechanisms, and the rapid jumps during deglaciation. Furthermore, while some aspects of the insolation changes have opposite effects in the two hemispheres, glacial fluctuations are in essence global and synchronous.

The nature of a number of the late Quaternary 'terminations' has been examined in high-resolution ocean cores, in order to verify that these instabilities are more general. This was done by Sarnthein and Tiedemann (1990) using a core from ODP Site 658 on NW Africa. Here the mean sedimentation rate was about 15 cm per ka, giving a core length of about 100 m for the whole Brunhes chron. All of the last six 'terminations' examined in detail from this core did indeed display multiple steps with reversion to cold conditions akin to the Younger Dryas cold phase of the last deglaciation (Termination I) (see Fig. 3.5). Sarnthein and Tiedemann were able to show that the durations of complete terminations varied consider-

ably, termination V being slow and III being rapid (see Table 4.2). Termination I, for example, involved two periods of rapid warming, called termination IA ($\delta^{18}O$ change from 4.55–3.15‰) and IB (3.5–2.85‰). These correspond to the episodes of very rapid sea-level rise documented by Fairbanks from Barbados corals. Event IA is further subdivided into two subevents, IA1 and IA2. Termination IA1 lasted about 710 years, IA2 1370 years, and IB 2600 years. Periods of less rapid change during the overall termination bring its aggregate length up to the 8.5 ka listed in Table 4.2.

The fact that all of the studied terminations show instability during the progression from glacial to interglacial conditions indicates that some kind of systematic instability arises at such times. The hypothesis examined earlier concerning the role of meltwater in modulating heat delivery to the North Atlantic region seems very applicable in this context. Early deglaciation perhaps generates sufficient meltwater to turn off or modify the production of NADW; this in turn refrigerates the North Atlantic region leading to readvance of the ice and a reduction in the meltwater input. NADW production then begins again, triggering the next phase of ice retreat. Changes in the $\delta^{13}C$ record are consistent with the hypothesis of reduced and then enhanced ocean ventilation following this sequence (Boyle 1990).

It appears that the terminations have a consistent relationship to the Milankovitch insolation forcing. Raymo (1997) has shown that there is a good correlation between deglaciations and times of increased northern hemisphere summer insolation. Sarnthein and Tiedemann (1990) showed that all deglaciations that they investigated occurred at times of high obliquity. If the precessional index is also high (see Chapter 5) a rapid termination occurs, taking perhaps 0.25–0.5 of the precessional cycle time, while if the precession index is lower, a slower termination results (taking longer than half a precessional

Table 4.2 The $\delta^{18}O$ amplitude, number of steps, and duration of the last six glacial 'terminations'.

Termination	Total $d^{18}O$ amplitude (‰)	No. of steps in the deglaciation	Duration (ka)
I	1.85	3	8.50
II	1.73	2	8.34
III	1.06	2	5.81
IV	2.15	3	10.70
V	2.20	2	29.00
VI	1.90	3	7.52

Source: Sarnthein and Tiedemann (1990).

period). The slowest termination, termination V, indeed occurred during a relatively lengthy period of low precession index. Rapid mid-deglaciation fluctuations like those documented from these six terminations were at one stage thought to be restricted to periods of *transition between glacial and interglacial regimes* (e.g. Berger *et al.* 1987) but, as we saw earlier, the episodic instability of the Bond cycles persists through the *glacial* period itself, and we have seen ice core evidence for climatic instability in the Eem *interglacial* also. Indeed, the conclusion suggested by all of this evidence is that episodic instability is more the norm than the exception in late Quaternary environments. The distinctly mild and slow climatic fluctuations that led the GRIP Members (1993) to refer to 'the strangely stable Holocene' now appear to be a real anomaly that remains to be accounted for, although there were some jumps in temperature even in the Holocene (Meese *et al.* 1994).

From studies of the timing of events during deglaciation, much has been learned. An example is provided by the work of Ruddiman and McIntyre (1981). Using analyses from a North Atlantic core, these workers showed that through the last 185 ka, calculated northern hemisphere high-latitude summer radiation values fluctuate first, followed by a response in ice volume that lags by about 3 ka, and by a response in sea-surface temperatures (SSTs) that lags by a further 6 ka. This means that SSTs lagged behind insolation changes by an aggregate of 9000 years. Ruddiman and McIntyre envisaged that the periodic increase in summer heating generated by orbital changes boosted melting and iceberg calving in the eastern parts of the Laurentide ice sheet, with major discharges passing into the Atlantic Ocean. Some of the effects there have already been noted, including ocean cooling and extensive winter sea ice growth, with resultant starvation of the ice sheet. This is a feedback effect that reinforces the tendency to ice sheet decay triggered by a period of strong summer insolation over the latitudes where the ice sheets developed. Ruddiman and McIntyre (1981) envisaged that for these effects to develop strongly, ice sheets had first to grow to sufficiently large size. This is one of many ideas that have been proposed to account for the fact, not yet adequately explained, that major glaciations have displayed about 100 ka periodicity through the late Quaternary. This is the so-called '100 ka problem' discussed further in Chapter 5.

WHAT ACCOUNTS FOR THE RAPIDITY OF DEGLACIATION?

It is necessary now to consider the mechanisms behind deglaciation. Summer insolation over the high northern latitudes was a little higher than that of today at 17 ka BP, but did not reach its maximum until 11 ka BP. While the last deglaciation is *centred* on 11 ka BP, it began considerably earlier, being well underway by 14 ka BP. Now, the additional solar heating even at its maximum at 11 ka BP was only slight (just a few per cent), so that the fact that deglaciation began well before even this slight push requires explanation. The most attractive idea is that once a slight deglaciation had begun in response to the early minor increase in solar heating, one or several positive feedback mechanisms came into operation to reinforce the warming. The intervention of feedback processes introduces what is termed *non-linearity* into the cause-and-effect sequence. Many ideas about the nature of the feedback processes have been put forward, and we should consider the most important ones.

The sea-level rise caused by ice melting has the potential to lift grounded marine ice sheets, allowing them to break up and drift offshore. A most important consequence of this is that the buttressing support provided by the grounded ice to the land ice lying upslope of it is removed; massive accelerations of the ice flow from the inland ice domes might then result. Fast-moving or *surging* ice streams would then deliver ice to the coast very rapidly; by draining the interior parts of the terrestrial ice sheets they could cause just the kind of ice sheet thinning referred to above as being required by the otherwise inconsistent record of terrestrial moraines and marine oxygen isotopes. This process of ice sheet collapse has been termed *downdraw*. The most important part of this mechanism for reinforcement of deglaciation is that it is not the slight increase in solar heating which melts the ice, but rather heat taken from seawater as the calved ice floats away to melt. Chilling of the oceans should result; it has been estimated that this effect could cool the upper 100 m of the North Atlantic at a rate of 1°C per annum. This indeed appears to be confirmed by the isotope record of the mid-latitude North Atlantic, where lowest water temperatures appear to have been reached at about 13–9 ka BP, well after the terrestrial glacial maximum, and during the time of maximum solar heating at high latitudes: this is taken to reflect the chilling produced by the influx of icebergs and meltwater (Ruddiman and McIntyre 1981). A mechanism related to this has been envisaged along the terrestrial margins of the ice sheets, but involving proglacial meltwater lakes rather than the sea. Accelerated ice loss into the darker waters of the proglacial lakes, where the icebergs would melt more readily, could accelerate frontal ice melting in a similar way to that just described at the oceanic margins.

Another feedback mechanism that has been considered is moisture starvation. Early melting would return fresh water to the North Atlantic. If this water formed a stratified layer overlying the salt, at least two effects might follow. First, since fresh water freezes at $0°C$ (rather than the $-1.9°C$ of seawater), sea ice cover might become more extensive, cutting off a source for the evaporation of moisture to sustain the ice sheets. Second, since the layer of fresh water would not readily mix with the underlying seawater, it would store most of the heat from summer, leaving the water below cooler than normal, and again restricting evaporation in the autumn and winter seasons. The more the ice melted, the more the remaining ice sheets would be starved of moisture, ensuring their rapid collapse. These conditions seem probable given that deglaciation is driven by high summer insolation, according to the Milankovitch mechanism. Correspondingly, winters must have low radiation, since varying seasonal insolation must sum to a constant annual total, and this would promote extensive development of sea ice in the colder part of the year.

Moisture starvation is considered by some to have become significant rather earlier, once the North Atlantic began to cool rapidly after 75 ka BP. First, let us consider the cause of this oceanic cooling. The best explanation in fact comes from global climate models seeking to reconstruct atmospheric conditions as they were during the last glacial (e.g. Manabe and Broccoli 1985). These models in turn rely on the reconstructions of the Quaternary ice sheets based upon the field evidence of tills, moraines, and periglacial features reviewed earlier. The models show that the huge Laurentide ice sheet, and to a lesser extent the Scandinavian ice sheet, acted as physical barriers to the atmospheric circulation. The Laurentide ice caused the westerly jet stream of the middle troposphere to split into two substreams, one of which ran around the south of the ice and the other across its northern margin. Subsiding air over the ice resulted in a northward surface flow which also adopted a course parallel to the northern ice margin. Air following such a route would be cooled substantially by the chilled surface. The cold air mass then flowed between the Laurentide and Greenland ice sheets, across the Labrador Sea, and out over the North Atlantic. The movement of this air over the warmer sea surface would have extracted heat from it, resulting in the oceanic cooling revealed by foram $\delta^{18}O$ analyses, and producing more extensive sea ice. According to the models, sea ice may have existed seasonally south to $46°N$ or so. This sea ice in turn

would have restricted evaporation and moisture levels in the atmosphere. Hence, according to the models, the ice sheets at glacial maximum were suffering net ablation rather than accumulation. The presence of meltwater ponds on the ice surface would have lowered its albedo, and further enhanced ablation. Thus, the small additional trigger provided by the Milankovitch maximum in summer heating may have been all that was required to trigger a small further ice retreat and then catastrophic deglaciation through the intervention of one or more of the many possible feedback processes. The rapid 'terminations' seen in the marine isotope record would then reflect the re-establishment of a single jet stream, and rapid warming of the North Atlantic.

A role for isostatic rebound behaviour of the crust under the ice sheets has also been envisaged (Peltier 1987). Initial ice loss would lower the ice cap surface, with the weight on the underlying crust thus being reduced. Because the mantle is viscous, and only responds slowly, isostatic recovery would be delayed. Thus, instead of the ice cap rebounding, and so re-elevating the surface to colder heights, the melting would be accelerated by the lower elevation of the ice surface. Further melting would simply lower the surface even more, which in turn would bring it down to yet warmer elevations, and so forth.

Potentially the most important feedback mechanism to promote deglaciation involves the atmosphere, and particularly its greenhouse gas concentrations. The ice core evidence presented earlier shows that in glacial times, levels of nitrous oxide, methane and carbon dioxide are reduced, and some possible mechanisms for this were outlined in Chapter 2. Associated with glacial cooling there was a decline in atmospheric water vapour concentrations, so that this greenhouse gas was also less abundant. Whatever mechanism in fact controls the glacial–interglacial changes in greenhouse gas concentrations, initial deglaciation would evidently be associated with an increase in concentrations of these gases. This would result in additional warming beyond that provided by the Milankovitch mechanism, and hence reinforce the warming.

A particularly interesting aspect of some of these feedback mechanisms is the way in which they relate to deglaciation in the southern hemisphere. Because of the largely oceanic nature of the high latitudes in this region, and the high heat capacity (i.e. thermal inertia) of water, the Milankovitch mechanism must have had a much reduced impact on temperatures in the southern hemisphere. None the less, deglaciation occurred just as rapidly there, and the broad Quaternary glacial cycling was synchronous globally. It might be imagined that the loss of the northern hemisphere ice sheets and their albedo-fostered cooling would simply warm the global atmosphere and so induce worldwide deglaciation. However, climatic modelling (e.g. Manabe and Broccoli 1985) suggests that ice loss only resulted in a small increase in the net amount of warmth available: rather too little to result in sufficient inter-hemisphere heat transport through the atmosphere to promote deglaciation in the south. The explanation is as follows: although the high albedo of the ice sheets did reduce solar heating, the colder surface also emitted correspondingly less terrestrial long-wave radiation, so that the change in net radiation was much less than might be expected. It has thus been argued that one of the feedback processes needs to be involved to transmit the deglaciation trigger into the southern hemisphere. This might have involved the greenhouse gases in the atmosphere (which mix globally quite rapidly), heat transport by ocean currents, or destabilisation of grounded ice sheets around Antarctica by the global sea-level rise triggered by ice loss in the northern hemisphere. These matters remain to be resolved.

THE MECHANISMS OF ATMOSPHERIC CO_2 CHANGE

Because of its evident importance in feedback processes favouring deglaciation (and glacial onset), and in synchronising responses in the northern and southern hemispheres, we will consider in a little more detail the mechanisms that might account for the observed variation in atmospheric CO_2 levels from interglacial to glacial climates. Lower ocean temperatures during a glacial period would of themselves draw down atmospheric CO_2 levels, because this gas is more soluble in colder water. Broecker (1992) estimated that the ocean temperature decline alone would account for about 33 ppmv of the observed decline in the atmosphere. However, a coeval increase in ocean salinity, arising from the removal of large volumes of water, would act in the opposite sense and compensate for about two-thirds of the decline arising from the colder water. The net result is that these effects balance each other, so that very little change in CO_2 concentration would have arisen without some other aspect of ocean chemistry changing.

The oceans, which contain about 60 times as much carbon as resides in the atmosphere (Broecker 1992), control the availability of CO_2 to the climate system. The level of CO_2 in the atmosphere is consequently controlled to a large extent by uptake or release from the vast ocean reservoir, and so it is clear that the oceans must provide the dominant mechanism behind the glacial–interglacial changes in atmospheric CO_2 concentrations (Adams et al. 1990a). However, the exchanges involved are complex, and

involve both water chemistry and the activity of marine organisms (Archer and Maier-Reimer 1994).

One major cause of CO_2 fluctuations that has long attracted attention is the amount taken up by calcareous organisms in the ocean. The death of forams that make their tests from calcium carbonate, for example, generates a continual rain of remains to the sea floor, where the materials are partly redissolved in the colder unsaturated water and partly buried in the ocean sediments. The sequestering of some of the carbon used by these organisms permits the entry of more from the atmosphere. The oceans also release CO_2 to the atmosphere. Present patterns of ocean upwelling provide water that is rich in nutrients at certain key locations, such as the eastern equatorial Pacific. Here, biological productivity is fostered. But these upwelling waters, which warm at the surface, also provide the largest natural source of CO_2 to the atmosphere (Murray et al. 1994; Paytan et al. 1996).

An appealing idea is that the biological productivity of parts of the ocean increased during glacial periods, so drawing more atmospheric CO_2 into the oceans. Various hypotheses have been put forward to account for the kind of shift in oceanic productivity required.

Orbital changes have been cited as a possible cause. In high latitudes, the amount of light available for photosynthesis can limit productivity. During glacial periods, the growth of sea ice as a result of lowered insolation (because of changes in the obliquity of the Earth) may have displaced major plankton activity southwards away from the ice. At lower latitudes, with greater annual total insolation, productivity may have been enhanced and more CO_2 taken up (Sarmiento and Toggweiler 1984).

In addition to lack of light, lack of certain nutrients is known to suppress oceanic productivity in some areas, and so changes in nutrient availability provide another means of fostering greater oceanic productivity in glacials.

Ganeshram et al. (1995) for example have highlighted the role of nitrogen. This is lost from the oceans to the atmosphere by bacterial denitrification. When this takes place, an isotopic fractionation occurs so that the remaining nitrogen is enriched in the heavier form, ^{15}N. Ganeshram et al. were thus able to use changes in the ratio $^{15}N/^{14}N$ (written as $\delta^{15}N$) in organic matter as a proxy for the intensity of denitrification during the Quaternary. They showed that the rate was lowered during glacial periods, leaving more available as a nutrient. Increased biological productivity could then have removed CO_2 from the atmosphere. The abundance of nitrogen was further improved by the reduction in the area of submerged continental shelves, which are important sites of denitrification in the modern ocean. Also, the erosion of shelf sediments may have provided an additional source of nitrogen. Similar arguments were raised by Altabet *et al.* (1995) from an analysis of $^{15}N/^{14}N$ ratio changes down three sediment cores from the Arabian Sea, which is one of the principal areas of denitrification in the modern ocean. Spectral analysis of the variations showed the periods 100, 41 and 23 ka, dominant orbital periods in the Milankovitch mechanism. Altabet *et al.* therefore concluded that the $^{15}N/^{14}N$ changes are likely to parallel other climate swings, and to have perturbed marine biogeochemical cycles during the late Quaternary.

Such complex environmental changes offer considerable scope for interpretation. Alternatives are often equally plausible, and resolution of differences often has to await better data. In the case of the Quaternary nutrient changes, Farrell *et al.* (1995) have offered a contrary view. They employed $\delta^{15}N$ analyses from organic matter at sites in the east Pacific Ocean. Using these data, they showed that there was increased biological productivity during the last glacial maximum in this region, supported by a greatly increased availability of nutrients. This suggests that there must have been greater delivery of nutrients by stronger upwelling at that time. Farrell *et al.* concluded that the eastern Pacific therefore was not a site of net CO_2 uptake during the glacial period, but merely one of greater turnover. The increased productivity of the eastern Pacific during the last glacial was confirmed by Paytan *et al.* (1996) using analyses of precipitated barite. These authors estimated that productivity levels at the LGM were about twice present values. This area may thus have been a source of CO_2 to the global atmosphere. Similar arguments in favour of increased nutrient supply to support increased biological activity during glacial periods were presented by Broecker and Peng (1989). These authors argued for a different mechanism to cause CO_2 absorption into the glacial oceans, involving a change in the level of alkalinity. Alkalinity changes were also envisaged by Archer and Maier-Reimer (1994), but there are other approaches based on ocean chemistry that seek to explain the changing CO_2 levels. Indeed, there is a wide diversity of models seeking to explain CO_2 changes in terms of control by oceanic chemistry. Mangini *et al.* (1991) pointed to the record contained in manganese (Mn) abundance in the sediment cores, and envisaged a mechanism in which the deposition of Mn carbonates influenced the alkalinity and through this the atmospheric CO_2 levels. Leuenberger *et al.* (1992) attempted to resolve the effects of ocean alkalinity changes and biological productivity by analysing the $\delta^{13}C$ record from ice core air-bubbles, and concluded that biological productivity changes alone are insufficient to account for observed glacial–interglacial changes in isotope abundance.

The details of these various chemical and isotope analyses lie beyond the scope of this chapter.

Iron has been highlighted in several studies as another key nutrient that is presently undersupplied in parts of the ocean (Martin and Fitzwater 1988; Watson 1997). Experiments in which soluble iron was delivered to the ocean have been conducted by Kolber *et al.* (1994) and Cooper *et al.* (1996). Results of these trials show that significant increases in biological productivity follow. The interesting possibility suggested by these studies is that iron may have been supplied to the glacial ocean in increased quantities because of wind transport from the enlarged areas of bare, arid continental surfaces, and so fostered increased biological productivity. Increased wind speeds during glacial periods have been suggested on the basis of changes in continental dust abundance in Antarctic ice (Petit *et al.* 1981). Dust concentrations in ice at the Antarctic Dome C site suggest dust delivery of about 20 times present rates at the end of the last glacial period, while the Vostok core suggests values about 15 times greater (Petit *et al.* 1990). Petit *et al.* (1981) estimated that wind speeds would need to be increased by a factor of 1.3–1.6 in order to account for the additional dust entrainment required. (Yung *et al.* 1996 have suggested that higher dust concentrations in part reflect lower rates of loss during atmospheric transport because of the less vigorous hydrological cycle and reduced washout rates.)

Effects on the continents have also been invoked to account for the glacial–interglacial fluctuations in atmospheric CO_2 levels. The role of chemical weathering was noted briefly in Chapter 2. The enlarged land area, and exposed continental shelf sediments, suggest that more carbonic acid weathering might have contributed to the decline, although Adams (1995) opposed this idea. In contrast, a model study by Munhoven and François (1996) has confirmed that this is a significant effect. Using a multi-box model of the ocean carbon cycle, these authors have estimated that CO_2 consumption in glacial-period silicate rock weathering might have increased by a factor of 2–3.5 compared with the present rate. Corresponding estimates of the decline in atmospheric CO_2 levels lie in the range 50–110 ppmv. The cyclic growth and exposure of coral reefs caused by sea-level changes was also shown in this model to contribute to fluctuations in the atmospheric CO_2 concentration.

Marine nutrient supply changes related to iceberg abundance during glacial periods have been proposed as another mechanism to contribute to CO_2 concentration decline during glacials. Icebergs foster mixing in the upper layers of the ocean, perhaps bringing up more nutrient-rich subsurface water. But they may additionally deliver nutrients such as nitrate, ammonia and silica from contained terrestrial detritus (Sancetta 1992). Iron might be among the materials supplied. Normally, iceberg abundance could well be too low for significant effects to arise, but periods of ice surging (for example, Heinrich events) might deliver large pulses of nutrients.

A quite different mechanism for CO_2 drawdown which also involves silica abundance has also been proposed. Pollock (1997) has suggested that progressive interglacial warming and sea-level rise may periodically destabilise the West Antarctic ice sheet, which is grounded below sea-level. The resulting massive ice discharge would deliver to the oceans a large amount of finely-ground silicate rock flour from the basal till sheets. This additional silica could then be supplied to upwelling zones through the global ocean conveyor system. At these sites, elevated silica use by diatoms could promote the more intense use and storage of carbon, fostering absorption of additional CO_2 from the atmosphere. In this scenario, the drawdown of CO_2 follows as a *consequence* of the ice sheet collapse, rather than being a cause, and siliceous organisms provide the key, not calcareous ones.

A final illustration of the diversity of processes involved in the glacial climate instability is provided by studies of possible changes in cloud processes. The role of elevated levels of glacial dust in altering cloud processes was referred to earlier in the discussion of possible causes for the dramatic $\delta^{18}O$ swings seen in the Greenland ice core records. An additional mechanism of glacial climate change which is related to this involves sulphate particles generated in the atmosphere by the oxidation of dimethylsulphide (which is produced by biological activity in the upper ocean). The sulphate particles may act as condensation nuclei in clouds, and are additional to the sulphate particles supplied by the evaporation of spray droplets released from breaking waves and bubbles. Analyses of the abundance of the non-seasalt sulphate particles in the Vostok core has shown that their abundance was increased by 20–46% during the last glacial period, and that the changes correlated strongly with those of atmospheric CO_2. Both changes might therefore be related to altered biological processes in the glacial ocean. Calculations by LeGrand *et al.* (1988) suggested that the elevated particle abundance might trigger increased cloudiness and so raise the global albedo by 0.03, sufficient to cause about a 1°C cooling. This then is another mechanism capable of contributing to the lowered temperatures of glacial periods. Once again, however, differing interpretations are possible, and the combined effects of the dimethylsulphide albedo feedback and other active feedback responses must be analysed (e.g. see Lovelock and Kump 1994).

METHANE AND ITS ROLE IN GLACIAL CYCLES

We have already seen that methane concentrations fluctuate in a similar way to those of CO_2, falling by 50% in glacial periods. Indeed, as the GISP 2 ice core analyses demonstrated, methane fluctuations broadly follow orbitally-driven insolation changes, suggesting a climatic control at least for the slower fluctuations (Brook *et al.* 1996). The amount of methane in the atmosphere is considerably smaller than the amount of CO_2, so that the radiative climate forcing by methane is much smaller. However, there are significant reservoirs of methane that have the potential to undergo exchange with the atmosphere quite rapidly. This must clearly be so, in view of the parallel fluctuations in methane concentrations and temperature seen in the six brief stadial periods in the interval 40–28 ka BP recorded in ice cores, and noted earlier in the chapter. We will examine briefly some of the ideas about the involvement of methane in the Quaternary glacial cycles, and in particular whether it is possible that methane fluctuations could account for major deglaciations as well as for the brief stadial periods.

A methane source that has the potential to release enormous volumes of gas exists in cold sediment bodies, such as in permafrost areas and in accumulations of marine sediments. In such environments, methane is held within the crystal lattice of ice. The ice adopts an altered crystal form to accommodate the gas, and the resulting materials are called *clathrates*. Thorpe *et al.* (1996) have estimated that Arctic permafrost may hold 800 Gt of carbon in the form of methane clathrates, with a further 24 000 Gt of carbon in clathrate deposits in oceanic continental slope deposits. These figures may be compared with the estimated glacial to Holocene increase in atmospheric CO_2 levels, which involve only about 200 Gt of carbon.

The clathrate deposits can be destabilised if they are warmed, or if the confining pressure of overlying sediments (or seawater) is reduced. Details of the conditions required were reviewed by MacDonald (1990). Massive amounts of active greenhouse gas could then be released to the atmosphere. Confining pressures could be relieved by falling sea-levels as the great ice sheets developed; the 120 m decline at the LGM represents a 12 atmosphere lowering of pressure. One possibility is that the collapse of the clathrates generated enormous gas pressure within the marine sediments, causing massive sediment slumps that released the gas. Very large submarine sediment slumps were relatively common in the Quaternary, and involved areas of up to 34 000 km^2 (Nisbet 1992; Haq 1995; Thorpe *et al.* 1996).

Modelling results show that although methane only has an atmospheric residence time of about 10 years, a major methane release could perturb levels for 50–80 years, and yield an initial warming of nearly 1°C, with a warming of 0.3°C lasting for several decades (Thorpe *et al.* 1996). Various feedback effects could amplify the direct greenhouse action of the methane, including changes in stratospheric ozone and water vapour levels, and increased cloudiness which might lead to increased trapping of terrestrial radiation. Despite the fact that a methane 'spike' has not been observed in the ice core records, it is still possible that such a brief event has been missed.

An alternative methane source which is commonly considered is that of wetlands, where fluctuations in the areal extent of peat deposits in particular provide a changing source for methane or a sink for CO_2. Franzén (1994) outlined a hypothesis for glacial cycles based on the progressive sequestration of carbon in high-latitude peat deposits, which is envisaged to reduce steadily the atmospheric greenhouse gas concentrations. An alternative, noted earlier in this chapter, is that fermentation of the biomass in tropical swamps provides the most important source of methane variation (Chappellaz *et al.* 1993; Street-Perrott 1993). The drying out of tropical swamps could be related to changes in monsoon strength and to hemispheric changes involving the production of NADW (see Petit-Maire *et al.* 1991). This idea is supported by good correlations between periods of low lake levels in monsoonal Africa and low atmospheric methane concentrations seen in the ice cores. The low lake-levels can in turn be related to altered conditions in the oceanic moisture sources, perhaps modulated by the changing vigour of NADW production (Street-Perrott 1993).

THE ROLE OF THE TROPICS AND TROPICAL CLIMATE CHANGE

A second issue which we should now consider is whether it is possible to identify a part of the Earth that provides the lead in Quaternary glacial cycles. From what has been said earlier in this chapter, it can be concluded that climate change in the area actually occupied by the great ice sheets at high northern latitudes was only one component of a complex array of globally interlinked processes. In particular, we have seen that the major oceanic heat sources for the North Atlantic are the tropical Atlantic and the oceans of the southern hemisphere. These heat sources provide a path by which the dominant insolation variations of the low latitudes, which relate to the orbital eccentricity cycle, could be transmitted to high latitudes. Furthermore, changes in oceanographic variables can be transmitted, through changes in SSTs related to altered upwelling, to variations in rainfall in the tropics and to the strength of the Hadley circulation system of the atmosphere. These changes in turn, especially the monsoon circulations, have the potential to affect the hydrology of the continents, the flux of nutrients like iron blown to the oceans, and the flux of methane released from tropical swamps.

All of these ideas suggest that the tropics may have been far more important in the glacial cycles than early workers may have imagined. They also suggest the possibility that although the growth of ice sheets was dominantly a high-latitude phenomenon, other kinds of Quaternary environmental change might have been just as significant within tropical environments (Charles 1997).

There is a growing body of evidence that suggests that this is so, and that the once-favoured notion of relative temperature stability in the tropical oceans may be an oversimplification. Tropical land areas are known to have experienced significant glacial cooling. For example, in Java, van der Kaars and Dam (1995) used terrestrial pollen records to infer glacial temperatures 4–7°C cooler than now. It seems unlikely, in view of this, that tropical oceans escaped significant glacial temperature change. A study of noble gas concentrations in ^{14}C-dated groundwater from lowland tropical Brazil led Stute *et al.* (1995) to conclude that the last glacial maximum temperatures there were about 5.4°C colder than today, and Colinvaux *et al.* (1996a) used pollen data to confirm a glacial cooling of 5–6°C. This suggests that the nearby equatorial Atlantic Ocean must also have been cooler. Detailed studies of the SST palaeotemperature record of Barbados coral led Guilderson *et al.* (1994) to the conclusion that SSTs in the Barbados area were about 5°C colder at 19 ka BP, while from a study of the [18]O record of the ice core record from the Peruvian ice cap at Huascarán, Thompson *et al.* (1995) estimated that the tropical Atlantic was up to 6°C cooler during the late glacial. From the tropical Indian Ocean, Bard *et al.* (1997) were able to demonstrate, using alkenone palaeothermometry, that temperature fluctuations of up to 3°C had occurred. These may have been synchronous with some of the millennial-scale instabilities of the northern hemisphere discussed earlier in this chapter. Finally, Beck *et al.* (1997) have shown similar results from the SW Pacific, in the region of Vanuatu. Here, using a technique of palaeothermometry based on the Sr/Ca ratios of closely dated coral sections, they have shown that SSTs were as much as 6.5°C cooler as recently as 10.3 ka BP, and rose abruptly over the following 1.5 ka.

These and other studies suggest that although

glacial cooling was less in the tropics, it was none the less very significant, amounting to perhaps half of that experienced at higher latitudes. Furthermore, it appears that the oceans were a dynamic part of this system, displaying surprisingly rapid temperature fluctuations. Given the importance of the warm tropical seas as moisture sources for monsoon rain, consequences for the delivery of monsoon rains to the land, and hence of river runoff to the oceans, suggest that major readjustments of many aspects of the global climate would follow from cooling of the tropical oceans. The full interpretation of the results just cited has not yet been completed. However, they clearly suggest that the tropical oceans were not immune to glacial climate change, and the possibility exists that the tropical oceans were intimately involved in the mechanisms of global glacial cooling.

A GCM (general circulation model) study by Webb *et al.* (1997) explored this idea. In this study, it was assumed that, in accordance with evidence presented earlier in this chapter, some form of NADW production continued in glacial times, probably as cold water outflow at a shallower depth (constituting Glacial North Atlantic Intermediate Water, or GNAIW; see also Boyle and Weaver 1994). When the model was driven by other glacial conditions (lowered atmospheric CO_2, lowered sea-level, major ice sheets), significant cooling of the tropical oceans was found. A major decrease in atmospheric water vapour contributed to the cooling, as well as changes in albedo and cloud cover. Because of heat export northwards, relatively limited North Atlantic sea ice cover was modelled. Overall, the model suggested that tropical SSTs under the modelled scenario would be 5.5°C cooler than modern temperatures, in good accord with the coral palaeotemperature data noted above.

The issue of just where the Quaternary glacial cycles originate clearly is a complex one. It is evident that resolving this issue depends heavily on the correct dating of the sequence of events at various locations. With reliable dating, it is possible to examine the relative timing of events in different geographical regions, and in this way to attempt to identify those with a response that leads that seen in other areas. Cortijo *et al.* (1994) have performed this kind of analysis using records of $\delta^{18}O$, together with SST estimates based on micropalaeontological transfer functions derived for sites in the Norwegian Sea and the North Atlantic. They inferred that insolation increases during interglacial times caused an increase in freshwater flux into the Norwegian Sea, shutting down thermohaline circulation there. The increased freshwater flux was related in part to the warm interglacial SSTs, feeding more precipitation into the area. This diminished the salinity of the Norwegian Sea, turning off deep water formation and

causing a relocation of the main site of deep water formation into the North Atlantic. Climatic consequences then follow in a cascade because of the key role of the NADW production in oceanic heat transport, and hence in tropical conditions like the monsoon circulations already described.

Similar conclusions have been indicated in comparisons of the Greenland and Antarctic ice core records. The GISP 2 core shows 22 warm intervals in the period 20–105 ka BP, while the Vostok record reveals only nine. Bender *et al.* (1994) have shown that it was the longer (> 2 ka duration) events that apparently affected Antarctica (see also Jouzel 1994). Furthermore, in addition to being more numerous in Greenland, the interglacial events are also sharper there, and display more rapid warming and cooling than in the Vostok record. All this led Bender *et al.* (1994) to favour the explanation that the events began in the northern hemisphere and only spread to the southern if they lasted sufficiently long.

This kind of scenario building has been employed with quite large data-sets, and generally points to the high northern latitudes as being a site with a very early response to insolation changes. We will consider these ideas a little further in Chapter 5.

CONCLUSION

Our discussion of the glacial phenomena of the Quaternary so far has been biased, of necessity, toward the last glacial stage. This has meant that we have neglected one of the major issues which must be tackled if we are to understand fully the controls on Quaternary environments, namely, the longer-term rhythm of the glacial cycles.

As has been indicated in earlier chapters, the most prominent features of the Quaternary environmental instability after 0.9 Ma BP were the dramatic glacial 'terminations', in which a volume of ice which took perhaps 90 ka to accumulate was returned to the oceans in only 8 ka. As we have seen, some of the environmental changes taking place during the terminations were remarkably rapid; at the end of the Younger Dryas stadial, southern Greenland warmed by 7°C in 50 years, according to the records from the Dye-3 ice core. Similarly, in only a few decades, the climate of the North Atlantic became milder and less stormy, following the retreat of sea ice there (Dansgaard *et al.* 1989). The rapidity of these events confirms the intervention of one or more feedback processes in reinforcing an initial tendency toward ice ablation produced by a very slight increase in summer solar heating in the high northern latitudes.

The oceanic record makes it clear that at least the last eight or nine such terminations have occurred about 100 ka apart, a frequency which matches that of the change in orbital eccentricity. This is another indication that Milankovitch rhythms underlie the Quaternary glacial cycles. However, it is problematical because there is negligible direct forcing resulting from changes in eccentricity; these only have a significant effect by modulating the amplitude of the 23 ka precessional variation. That is, solar heating of the Earth does not vary significantly with a period of 100 ka. Therefore it is necessary to look for a separate mechanism which could be responsible for the striking 100 ka periodicity of late Quaternary glacial terminations. This is done in Chapter 5.

Presently, no completely acceptable explanation for the 100 ka periodicity exists. Additionally, we have seen abundant evidence in this chapter for very rapid climate change. Clearly, therefore, the global climate system experienced change at a wide range of frequencies during Quaternary time. Some of these, like the dominant 100 ka rhythm, are at periods that match orbital periods outlined in the Milankovitch theory, or are what Yiou *et al.* (1991) called superorbital periods, formed by combinations of shorter orbital periods (see Chapter 5 for explanation, and also Pestiaux *et al.* 1988). The high-frequency changes (e.g. the D–O events and the Bond cycles) may correspondingly reflect suborbital frequencies, again formed as interaction tones. On the other hand, the climate system possesses many feedback processes and some of these (like the partial ice sheet disintegrations involved in the Heinrich events) may have their own independent periods. Resolving actual causes in this complex system remains one of the major puzzles in the continuing investigation of the Quaternary cryosphere. In the view of some theoreticians, the search for individual causes and mechanisms in this situation is intractable. There are cascades of positive feedback effects, and isolating a cause in circumstances of strong positive feedback can be extremely difficult because the 'initial conditions' must be known with great precision. Small perturbations in some parameter, when amplified highly in one of the feedback loops of the climate system, can be responsible for important climate outcomes. Conditions of strong positive feedback also involve each 'effect', in essence creating a further 'cause' (this is the essence of regenerative change), so that there are indeed significant difficulties in attempting to identify a fundamental trigger for any environmental change. One approach that has proven fruitful in the face of these obstacles is to supplement information on the timing and sequence of global events like deglaciation with information on the *geographical pattern* of occurrence over the Earth's environments, in a search for sites where change seems perhaps to begin first. The analysis of timing (phase) relations among the various climate and other parameters involved also has the potential to help to separate cause from response (e.g. see Saltzman and Verbitsky 1994).

Whatever their cause, clear evidence of the impact of the rapid 'terminations' is to be found along the present coastlines. These have existed in their present configuration only since the most recent marine transgression, which resulted from the return of meltwater to the oceans. At full glacial times, sea-levels were about 120 m below their present position. This lowering, and the very rapid *Holocene marine transgression* that resulted during termination I, are among the effects of continental glaciation and deglaciation felt in the oceans. We consider these events in more detail in Chapter 6, following a review of the important Milankovitch hypothesis of Quaternary environmental change.

CHAPTER 5

THE MILANKOVITCH HYPOTHESIS AND QUATERNARY ENVIRONMENTS

Time present and time past
Are both perhaps present in time future,
And time future contained in time past.
T.S. Eliot (1888–1965)
(*Four Quartets* (1944), 'Burnt Norton')

The previous chapters reviewed some of the evidence for the Quaternary glacial–interglacial cycles, and some of the feedback and other processes that were associated with these great environmental changes. We observed that external factors, related to the geometry of the Earth's orbit around the sun, might be pivotal in influencing the chronology of these events. In this chapter, we pursue this important hypothesis in more detail, considering both the mechanisms involved and the evidence for them.

Variation in the amount of radiant energy received from the sun seems a plausible cause of Quaternary climatic instability (Berger 1981b; Imbrie 1981). The long-term evolution of the sun as a star, its luminosity increasing at about 6% per Ga (Hunten *et al.* 1991), has exerted a major although incompletely understood influence on the global environment throughout Earth history. Otto-Bliesner (1996), for example, employed a solar irradiance 3% less than the present value in developing a model of glacial onset in the Carboniferous (360 Ma BP). The first proposal, that shorter-term quasi-periodic orbital variations might cause oscillatory changes in solar heating, was made in 1842 by the mathematician Joseph Adhémar in France, and later developed by the Scotsman James Croll. Accounts of the fascinating early development of these ideas are provided by Imbrie and Imbrie (1979), Imbrie (1982), Berger (1980, 1988) and Hartmann (1994).

However, the 'astronomical theory' of the ice ages, as this idea has come to be known, is most associated with the name of Milutin Milankovitch (1879–1958). This Serbian mathematician, a professor at the University of Belgrade, devoted many years of his life to the complex task of calculating how the orbit of the

Earth had changed in the Quaternary, and relating the orbital changes that he found to the planetary climate. Some of this was done during a period of captivity during the First World War. Milankovitch (1941) developed his own version of the astronomical theory of Quaternary glaciation, as others have done since, in which he saw particular orbital changes and their presumed effects on key parts of the Earth as providing the key to the glacial cycle. Some of these ideas have had to be abandoned subsequently, as better dating of Quaternary events has become available.

Although much of Milankovitch's work has been refined and improved, its significance in the building of modern knowledge is enormous. He is remembered for developing the first comprehensive analysis of the relevant aspects of the Earth's orbital variation in the Quaternary, and for systematically exploring links between orbital characteristics and global climate. In this chapter, we will review technical aspects of these matters in outline only, because of the complex mathematics required in a formal analysis. Our goal is rather to understand the ways in which solar radiation delivered to the Earth varied during the Quaternary, and to explore some of the diverse ways in which planetary systems like the climate, the biosphere and the cryosphere responded to these imposed changes and generated further perturbations in the operation of many global environmental processes.

Following Imbrie *et al.* (1993a) we can take as our expression of the Milankovitch theory of palaeoclimates the idea that *temporal changes in insolation that are astronomically driven cause significant changes in the global climate, and in particular, that they account for much of the climatic variability of the Quaternary glacial–interglacial cycles.*

We can logically break down our discussion of the Milankovitch mechanisms by exploring a series of questions:

1. What are the aspects of the Earth's orbit, from the perspective of celestial mechanics, that result in changed patterns of solar radiation at the top of the Earth's atmosphere? Over what periods do the orbital changes occur, and how big are the changes in solar insolation that result? (Note: we must consider the incident radiation at the top of the atmosphere, because the amount of radiation actually absorbed in the Earth–atmosphere–ocean system depends upon the prevailing planetary environment, including the absolute amounts of land and ocean surface, the cloudiness of the atmosphere, the amount of land covered by ice and snow, and so on. This cannot be calculated for past times, whereas the external insolation values can. This is an issue that will be discussed later in this chapter.)

2. What sensitivity does the climate system display to these externally-imposed insolation changes? What kinds of feedback processes arise that are capable of further altering the radiation balance? How much effect do these feedback processes have, and over what time-scales do they have effect? Feedback processes are of critical importance because they have the potential to amplify or attenuate the '*insolation forcing*' at the top of the atmosphere. Amplification of an orbit-related decline in solar radiation could follow from increased reflection by a growing cover of snow and ice, for example.

3. What resulting changes occur in the global climate? That is, how do air temperatures, atmospheric and oceanic circulations, the great ice sheets, and other systems respond when the orbital forcing is modified by internal feedbacks? There are two ways in which this question may be answered. We may use *palaeoenvironmental indicators* such as plant fossils, which provide proxy climate records that may show how the system actually responded in the past. Berger (1991) called such sources 'natural archives'. Alternatively, past climates can be explored through computer models of the system. The use of models has the potential advantage over the archives that it permits forecasts to be made and hypotheses tested in 'what if' studies, where modelled solar radiation can be deliberately changed by set amounts. The realisation of this potential is hindered by the vast complexity of the planetary climate system and the difficult task of embodying its operation in computer models.

4. Given that it takes a considerable time for large ice sheets like those of the Quaternary to accumulate, and later to melt, and for forests to retreat or spread across large areas, how does the *timing* or chronology of Quaternary environmental changes (the global 'response') correspond to the orbital forcing? In some cases we will see important *time-lags* in the climatic response, for example as the great ice sheets slowly grow. In more general terms, we have to anticipate various kinds of *phase shifts* in the natural archives which might obscure links with the orbital forcing. In some of these, peak environmental responses might *precede* the maximum signal in the insolation forcing, while in others, a lagged response may be shown. For instance, all the ice in a particular region might melt *before* the peak of solar heating was reached in a particular past era, if the ice were thin, or *after* the peak of heating, if it was massive and melted only slowly. Thus a record of past sea-level change might not show good correspondence with the timing of orbital changes. In many cases like this, establishing a *causal link* between insolation forcing and any response found in the natural archives will require careful and detailed work.

SEARCHING FOR SIGNS OF THE MILANKOVITCH MECHANISM

Because of the possible phase shifts just referred to, the search for evidence that the Milankovitch orbital mechanisms have indeed influenced past climates (a suggestion that was long rejected) has been conducted in two broad *domains*, where separate kinds of evidence have been sought.

The frequency domain: The orbital changes which wc will examine below occur at well-known rates, and there are reasons for believing that the environment might respond by changing at similar rates. Evidence for changes at these frequencies may be sought in the sedimentary record. This is a search for evidence in what is termed the *frequency domain*. Simple phase shifts do not cause problems in this kind of study, because although the environmental response might lag somewhat, it can still exhibit or reflect the underlying frequency of the controlling orbital changes.

The time domain: We also require knowledge of how the environmental changes have corresponded in their timing with the orbital fluctuations. For example, did the last glacial maximum at about 18–20 ka BP correspond to a period of unusually low insolation? Alternatively, did it perhaps follow such a period after a significant lag time? The checking of correspondence here is in the *time domain*, and requires a careful unravelling of the phase shifts just referred to, which can otherwise generate apparent mismatches of orbital triggers and environmental responses. A significant mismatch might lead us to reject (perhaps erroneously) the Milankovitch mechanism as the principal cause of a particular environmental change. Time domain work thus requires that we have both reliable calculations of the past insolation forcings, from celestial mechanics, and good ground-based work on the natural archives of data, including sound representative sampling of the Earth's environments and good age determinations on the materials being studied.

We will now consider separately these four sequential questions about the Milankovitch mechanism. This will equip us to understand the results of searches for evidence of the Milankovitch mechanism in both the frequency and time domains, which we will subsequently review.

THE NATURE OF THE EARTH'S ORBIT AND ITS CHANGING CHARACTERISTICS

First, we will see how it is that the changing orbit of the Earth influences the amount of solar radiation received at the top of the atmosphere. Some background material is provided to help in the task of understanding the astronomical discussion.

At present, the Earth makes a revolution around the sun in 365.25 days, following a very slightly elliptical track (see Fig. 5.1). As the Earth moves toward the sun in this orbit, it accelerates under the strengthening gravitational attraction, and as it moves away from the sun, it slows down. The *average* speed of travel is about 30 km/s (over 100 000 km/h). The other planets follow similar orbits, but move at varying speeds (e.g. Mercury, 48 km/s, and Neptune, ~5 km/s) (Wyatt 1977). The movement of each planet in the solar system is affected by the complex gravitational field created by all the others, which may be gathered together or widely dispersed around their various orbits. These shifting gravitational fields slowly perturb such features as the shapes of the orbits, their orientations in space, and the orbital speeds of the planets. Some of the original orbital calculations made by Milankovitch showing these changes have been improved upon, partly because there are now better estimates for the mass of each planet, which enable a more correct analysis of the perturbing gravitational fields. We also have the power of computers to help solve the difficult equations required in order to work out the mechanics of the solar system. This task requires the evaluation of tens of thousands of terms, work which took the early investigators years of hand calculation (see Milankovitch 1941).

The planets all revolve around the sun in orbits close to the plane of the Earth's own path (mostly within 5°). This is a legacy from the revolving nebula that was the common origin for the planets of our solar system. The plane of the Earth's revolution around the sun is called the *ecliptic*, because when the moon comes between the Earth and the sun in

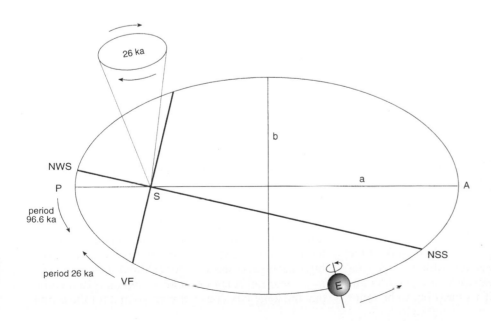

Fig. 5.1 Schematic diagram of the major elements of the Earth's orbit. The sun is at one focus of the ellipse, marked S. The major and minor axes are labelled a and b. E indicates the earth, A the position of aphelion, P the position of perihelion, and VE the location of the vernal equinox. The precessional cone is indicated. NWS and NSS refer to the northern winter solstice and northern summer solstice respectively

this plane, it can obscure the sun, yielding a solar eclipse. Likewise, when the Earth is between the sun and the moon in the ecliptic plane, the shadow of the Earth can fall across the moon and cause a lunar eclipse. (The origin of these words is the Greek *leipo*, meaning 'leave' or 'be without'. With the same derivation, the Greek word *ekleipsis* means 'forsaking'.)

When the Earth in its elliptical orbit approaches the sun most closely, it is said to be at *perihelion*. At the other end of the ellipse, when the Earth is furthest from the sun, it is said to be at *aphelion* (both terms come from the Greek *helios*, meaning 'sun'). We will employ these terms shortly.

Changes in the orbit

Three principal aspects of the Earth's orbit around the sun are subject to variation over time periods relevant to Quaternary climates. These are:

- the shape of the orbit (expressed by its *eccentricity*);
- the tilt of the Earth's spin axis (termed *obliquity*);
- the season of the year when the Earth is nearest the sun (this changes because of a process termed *precession*).

The changes in these three orbital parameters *do not cause any significant changes in the total amount of energy received at the top of the atmosphere in the course of a year*. Thus, the Quaternary climatic instability was not the result of changes in the total amount of solar radiation reaching the Earth. Changes in the three orbital parameters do, however, cause two key variations in the pattern of insolation in time and space:

- *Geographical distribution.* The distribution of solar radiation across the geographical zones of the planet can be affected by orbital changes. For example, increasing the obliquity widens the tropics and permits more energy to reach polar latitudes, which in summer are tilted more towards the sun.
- *Seasonality.* The seasonal timing and amounts of insolation received can also be varied. Thus, if summer in one of the hemispheres is reached when the Earth is near perihelion, insolation is greater than for a summer occurring when the Earth is at aphelion. Similarly, because of the variation in speed around the elliptical orbit, changing eccentricity can alter the relative lengths of the various seasons.

Both kinds of change are potentially important for the Quaternary, and may cause variations in the amount of solar radiation *actually absorbed by the Earth–atmosphere system*. This turns out to be quite a complex subject, and to understand it we need to consider the nature and variation of eccentricity, obliquity and precession in more detail.

SHAPE OF THE ORBIT (ECCENTRICITY)

If the orbit becomes increasingly elliptical, the planet accelerates more toward the sun and slows down more passing away from it. This shortens the fraction of the year spent near the sun, where heating is slightly stronger, and increases the amount of time spent far from it, where solar heating is less intense. This affects the contrast between summer and winter seasons, as well as changing by some days their relative durations. Through the last 1 Ma, the four seasons have in fact varied in duration from 82.5 days to 100 days (Berger and Loutre 1994a, b), departing significantly from the mean of about 91 days.

Elliptical orbits are conventionally described by their *semi-major axis*, of length *a*, and *semi-minor axis*, of length *b*. The long axis of the Earth's orbit is known to remain essentially constant. The short axis is the one whose length varies through time, becoming almost the same as the long axis when the orbit becomes circular, and shortening when the orbit become more elliptical. The Earth actually revolves around the *focus* of the ellipse, which is located on the major axis but away from the *centre* of the ellipse (where the long and short axes cross) (see Fig. 5.1). The usual measure of out-of-roundness for the orbit is the *eccentricity*, *e*, defined as the dimensionless ratio of distance between the centre and the focus and the length of the semi-major axis, *a*. Elementary geometry shows that

$$e = \frac{\sqrt{a^2 - b^2}}{a} \tag{5.1}$$

At present, the value of *e* for the Earth's orbit is 0.016. Accordingly, at aphelion the Earth is about 1.67% further from the sun than would be the case in a circular orbit, and 1.67% nearer the sun at perihelion. The calculated range of variation in *e* through the Quaternary shows that the short axis never declines further than 3.5% below the long axis, so that the orbit does not depart far from being circular. Diagrams of the orbit typically show greatly exaggerated eccentricity for the purposes of illustration.

The effect of eccentricity on solar radiation

Changes in orbital eccentricity can actually alter very slightly the total amount of insolation received at the top of the atmosphere during a year. This is because with a more eccentric orbit, the planet is actually a little nearer the sun on average, and thus receives a more intense beam of solar radiation. (Remember that the long axis does not vary, so that for the orbital eccentricity to increase, the length of the short axis must decline.) When most elliptical, 0.27% more solar energy would be received than in the present orbit, while for a circular orbit, annual total radiation would be 0.01% less than today's value (Berger and Loutre 1994a, b). These changes in total annual radiation receipts are minute, and even though they would undoubtedly be heightened somewhat by changes in the planetary albedo, they seem far too small to be the dominant control on climate change in the upper Quaternary. We know that the present climatic sensitivity for the Earth is about 1°C for a 1% change in absorbed solar radiation (Ramanathan *et al.* 1992). To cool the Earth to a glacial maximum of, say, 5°C lower than now thus requires a far larger decline in absorbed radiation than eccentricity changes alone could possibly cause. Nevertheless, various studies have suggested that in the upper Quaternary, low volumes of ice correspond to times of high orbital eccentricity. (In contrast, Milankovitch originally speculated that glacial times would be associated with certain highly eccentric orbits, thinking that cooler summers far from the sun would favour snow survival and build-up from year to year.) This major climatic response to what seems a negligibly small insolation forcing raises a problem (the '100 ka problem') which we will return to later.

The eccentricity of the Earth's orbit is a factor of enormous importance for the global environment. We need to consider the role of eccentricity in two separate time frames. We will examine the significance of *e* for world climate with the present orbit, and then the effects of the varying value of *e* for insolation change over Quaternary time-scales.

The present-day effect of *e* on solar insolation

A note on solar insolation: Solar radiation is often described in terms of a hypothetical 'solar constant'. This is defined as the rate of receipt of solar radiation on a surface at right-angles to the solar beam, and located at the mean Earth–sun distance. For our considerations of Quaternary time-scales, it is more appropriate to focus attention on the *varying* solar insolation received as the distance of the Earth from the sun varies. Although the solar irradiance on a surface permanently exposed normal to the solar beam, S_o, averages about $1360 \, W/m^2$, the *mean irradiance* at the top of the atmosphere is far less. This is because any point on the Earth's surface spends about half the time in darkness as the Earth rotates daily. Moreover, the curving surface of the Earth is not normal to the solar beam. The solar energy is in fact distributed across the larger surface area of the spherical Earth, not a flat disk having the diameter of the Earth. Thus, the *mean irradiance*, S, is the energy received on a disk of Earth size divided by the surface area of the spherical Earth. Using R_e for the

radius of the Earth and A for the planetary albedo, we can write

$$e = \frac{S_o(1-A) \, \pi R_e^2}{4\pi R_e^2}$$

so that

$$S = \frac{S_o(1-A)}{4} \tag{5.2}$$

or a quarter of the mean solar irradiance. Taking the mean planetary albedo to be 0.3, this amounts to about $1360 \times (1–0.3)/4$ or about $238 \, W/m^2$. The local value varies across the Earth's surface, reaching an annual mean of about $400 \, W/m^2$ at the equator but less than $200 \, W/m^2$ at the poles (e.g. see Peixoto and Oort 1992). In summer, maxima reach more than $450 \, W/m^2$, while in winter the polar areas receive almost no direct solar radiation.

Now let us return to the impact of the present elliptical orbit on insolation. We can consider the size of the effect quite readily. The mean distance of the Earth from the sun is about $1.496 \times 10^{11} \, m$. This is called one *astronomical unit* (AU). At present-day perihelion, the Earth–sun distance is about 1.67% less than this mean, at 0.983 AU, while at aphelion, it is about the same amount greater, at 1.017 AU. (Note that these two values add up to exactly 2 AU, which is the unchanging length of the major axis of the orbit.) The solar irradiance at the top of the atmosphere obeys the inverse-square law, declining with increasing Earth–sun distance. Thus, at perihelion the irradiance is increased a little, and we have

$$S = 238 / (0.983)^2 = 246 \, W/m^2 \tag{5.3}$$

while at aphelion the irradiance is diminished so that we have

$$S = 238 / (1.0167)^2 = 230 \, W/m^2 \tag{5.4}$$

The annual range of S at the extreme ends of the orbit thus spans about $16 \, W/m^2$, or nearly a 7% change from the mean value of S during the course of one year.

Actual differences in the energy received at the Earth's surface during the year are even larger than this, because the changes in S produce changes in snow cover and in the amount of cloud, as well as affecting plant growth. Consequent seasonal changes in surface reflectivity enlarge the annual variation in insolation by a few W/m^2 (e.g. see Peixoto and Oort 1992). These irradiance changes are sizeable, so that the nature of the seasons that we presently experience is in part a function of the present value of *e*.

The influence of varying values of *e* during the Quaternary

Once again, a simple calculation allows us to see the size of insolation changes related to the variation of eccentricity during the Quaternary.

The range of eccentricity for the whole Quaternary is roughly from $e = 0.0$ (circular orbit) to $e = 0.067$ (most elliptical). (Berger (1978a) lists the absolute maximum value of e that is reached as 0.0728.) With $e = 0$, the Earth maintains a steady distance from the sun and so receives an unchanging solar insolation. With $e = 0.067$, the same calculations as made in the previous section show that the difference in insolation between perihelion and aphelion would grow to reach about 27% of the mean value (with values 15% greater at perihelion and 12% less at aphelion). This would result in an annual variation in insolation about four times greater than we have now. The more eccentric orbit would thus have stronger seasonal contrasts, and seasons that were also slightly different in length than those which arise in a more circular orbit.

These great Quaternary changes in the seasons would balance each other, with almost no net effect on the *external* forcing *summed over an entire year*. In other words, higher solar insolation in summer when the Earth is near the sun is counterbalanced by a lower insolation in winter about six months later when the Earth is further from the sun. The annual total insolation at the top of the atmosphere remains essentially unchanged (although the net radiation at any actual place on the surface of the Earth will probably not do so, because of the seasonal changes in snow cover, plant communities, and so on, referred to earlier).

Let us summarise the material covered so far. We have seen that variations in eccentricity over Quaternary time can only cause minute changes in the total amount of energy delivered to the top of the atmosphere in a year. Large *seasonal* variations in insolation do occur with large values of *e*, but balance over the course of a full year. Variations in *e* can thus only become important in longer-term climate change if there is some process capable of converting a seasonal deficit or surplus of radiation, lasting perhaps a few months, into a more permanent one. This is where the climate system itself intervenes. Let us take a very simple example of such a process. If insolation during a summer were cut by say 10%, it might be possible, in the right location, for winter snow to survive the summer without melting. This snow would then keep the albedo of the surface higher than it would have been had the snow vanished completely, to be replaced by summer plants. In this way, more of the summer radiation would be reflected away from the surface, leaving it cooler still. This effect would diminish the amount of radiation actually absorbed by the Earth by *more than* the initial 10% drop. More importantly, the area of snow might be able to increase year by year, reflecting more radiation as it did so. This is a simple instance of a regenerative change or a *positive feedback mechanism*. Many of these mechanisms are known, and they offer an important means by which a small external 'trigger' change in insolation produced astronomically might grow into a major global climate change. We return to this intriguing subject after some more detail of the Earth's orbit.

TILT OF THE SPIN AXIS (OBLIQUITY)

While orbiting the sun, the Earth of course also spins on its axis, once per day. The axis of rotation is tilted at an angle to the ecliptic plane (almost 23.5° at present) which is termed the *obliquity* of the Earth.

The obliquity exerts a primary control on the geographical distribution of insolation. One polar region, being tilted away from the sun in winter, experiences long periods of darkness, while the other, tilted towards the sun, receives continuous sunlight (Fig. 5.2). In contrast to this extreme variability, the low latitudes receive sunlight for about half of every day of the year. Consequently, the largest yearly totals of radiation are recorded at the low latitudes and lowest totals at the poles. This generates the relative warmth of the tropics and the familiar temperature gradient between the equator and the poles.

This pattern of received radiation varies when orbital changes affect the obliquity, which has varied through a range of about 2.5° during the Quaternary. The difference in solar heating between the equator and the poles, and hence in surface and air temperatures, is a major driving-force for both atmospheric and ocean circulations globally. Therefore it is clear that changes in obliquity have the potential to generate major changes in mechanisms regulating the global climate.

The amount of energy received in a unit area exposed to the sun depends on the angle at which the solar rays strike. A low angle of incidence, such as that typical of high latitudes where the Earth's surface curves away from the sun, spreads the solar beam over a larger area and thus lessens its intensity. Reflection of solar radiation can also be enhanced if the rays strike the surface at a low angle. The effective intensity of the solar beam indeed varies with the cosine of the *zenith angle*, which is the angle of incidence measured from the vertical (Iqbal 1983). We can write an expression for the irradiance on a horizontal surface at the top of the atmosphere readily. Writing z for the zenith angle, d for the distance of the Earth from the sun at a particular time, and d_m for the mean distance:

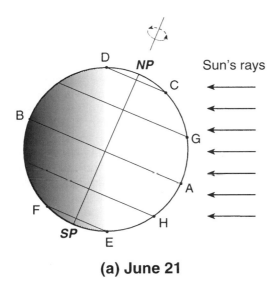

(a) June 21

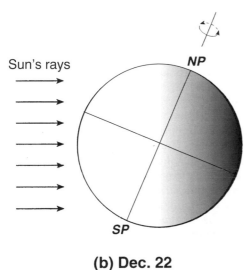

(b) Dec. 22

Fig. 5.2 Day length variation between the solstices.
(a) The northern summer solstice. Points lying on or above the latitude CD experience 24 hours of daylight; points on or south of EF experience 24 hours of darkness. The equator (AB) has 12 hours of daylight and 12 hours of darkness. Points in the summer hemisphere (e.g. point G) spend more than half of the day in daylight, while points in the winter hemisphere (e.g. point H) spend more than half of the day in darkness.
(b) The northern winter solstice. This situation is clearly the reverse of that shown in (a), so that all points in the northern hemisphere spend more than half of each day in darkness. (After Wyatt 1977)

$$\text{irradiance} = s \left(\frac{d_{\mathrm{m}}}{d}\right)^2 \cos z \qquad (5.5)$$

This irradiance changes rapidly for large zenith angles, and less and less as the angle approaches 0° (i.e. rays striking the surface from directly above).

Thus, the biggest effect of changing obliquity during the Quaternary would be found where the sun is low in the sky, that is, in the high latitudes. For example, if changing obliquity decreased the zenith angle at a high latitude location from 65° to 63°, a 3% change, the change in the cosine of the zenith angle would result in an increase in the irradiance of more than 7%. A similar 2° change, but at a latitude of 10°, yields less than a 1% variation. Thus we can see that the effect of obliquity is greatest at high latitudes. The amount of solar radiation received here can be greatly increased with increased obliquity, and diminished with lower obliquity. Milankovitch calculated that the greatest winter half-year effect arose at 66° N, where a 1° increase in obliquity results in a 5.9% reduction in insolation. (Note that this figure is for the top of the atmosphere, and does not incorporate feedback effects like that of snow albedo.) The effect is small at low latitudes, where neither zenith angle nor day length vary greatly with obliquity.

Thus we can see that increasing obliquity means greater receipt of solar radiation at high latitude sites, and thus means *diminished contrast* in annual total insolation between low and high latitudes. On the other hand, because an area inclined *towards* the sun in summer is tilted *away* from it in winter, increased obliquity also means *greater seasonal contrast* during the year. Both hemispheres would experience the same effects with greater obliquity, each receiving more radiation in their respective summers and less in winter. Since the total received insolation for the year is almost constant, warmer summers are necessarily accompanied by cooler winters, and vice versa. The annual total amount of heat received *at the top of the atmosphere* is thus unaffected by changes in the obliquity, although the amount received at any particular high latitude can be affected greatly.

THE SEASON REACHED AT PERIHELION (PRECESSION)

Finally we must consider the third orbital change in our list, that of the changing season of year when the Earth is nearest to the sun. Clearly, if the Earth is near perihelion in summer, insolation will be larger than if summer coincides with aphelion.

How do changes in the positions of the seasons around the orbit arise? The seasonal contrasts with which we are familiar arise from changes in the amount of solar radiation received in different parts of the year, to which both orbital eccentricity and obliquity contribute.

The orientation of the Earth's axis involves more than an angle of obliquity: the *direction* in which the spin axis points in space is also important. Let us see what governs this direction of tilt.

The Earth rotates every 24 hours, so that the rotational speed of a point on the equator is about 1700 km/h. This high rotational speed causes the planet to deform: the equatorial zone bulges outward a little, and the equatorial diameter of the Earth is about 20 km larger than the polar diameter. The planet is said to be *oblate*.

The gravitational attraction of the moon, sun, and other planets pulls on the equatorial bulge (see Fig. 5.3). Most of the mass of the bulge lies outside the ecliptic plane because of the tilted rotation axis, so that the force tends to cause a change in the tilt of this axis. More than two-thirds of the effect arises from the attraction of the moon, and most of the remainder from the much more distant but far more massive sun. Because of its spin, the Earth responds to this pull not by tilting but instead by pivoting and so shifting the *orientation* of its spin axis. The spin axis remains tilted at a fixed angle. The angle of tilt actually does also change a small amount as the precession takes place (the changing obliquity), but the main effect is that the spin axis moves in space, and the Earth revolves around the sun with its spin axis remaining tilted, but changing so as to point to various points in distant space. The re-alignment of the direction in which the spin axis points is called *precession* (from the Latin *prae*, 'before', and *cessus*, 'be in motion'), or more properly, 'lunisolar precession'. The process is *precession* because, as shown in Fig. 5.1, the rotation axis shifts in a retrograde direction around the orbit (i.e. opposite to the direction of planetary orbit). The spin axis moves in such a way that it progressively traces a cone, rather like that of a spinning top. This causes an effect called *precession of the equinoxes*.

To visualise the effects of these precessional changes, consider the Earth near perihelion (closest to the sun). As the rotation axis is oriented now, at perihelion the southern hemisphere is tilted toward the sun (i.e. is in summer) and the spin axis through the N pole points away from the sun into space (winter in the northern hemisphere). This means that southern hemisphere summers have slightly stronger solar heating than northern hemisphere summers which take place at aphelion. (Southern hemisphere summers are also slightly shorter, because of the higher revolution speed around the orbit near perihelion.) But if we could suddenly jump ahead to a time when the axis of spin had slowly precessed so as to be inclined in the opposite direction, having moved 180° around the precessional cone (an effect which would ordinarily take about 9000 years), the northern hemisphere would instead be inclined *towards* the sun, and thus have *summer*, at perihelion. Thus, it would then be the northern hemisphere that had the higher summer solar radiation, and the shorter summer. This is important, because of its influence on the climate of the land in the high northern latitudes where the Quaternary ice sheets developed. Cool northern hemisphere summers like those of the present (i.e. occurring near aphelion) would allow snow to survive more readily than the warmer summers happening near perihelion. Furthermore, the correspondingly mild northern winters, occurring at perihelion, would promote evaporation and transport of water vapour from the oceans, thus possibly nourishing the ice sheets. In summary, just which season is reached near perihelion (as a result of the slow precessional changes in spin axis orientation in space) has the potential to alter the climates of the hemispheres. It is important to emphasise that the effects of precessional change are opposite in the two hemispheres. A hemisphere having summer at perihelion and winter at aphelion experiences a great annual range in insolation and extreme seasonal contrasts. But the other hemisphere would necessarily have

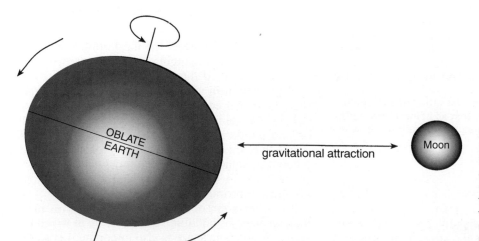

Fig. 5.3 The forces causing lunar precession. The rotating Earth is oblate, so that the gravitational attraction of the moon (together with that of the sun) would by itself tend to draw the Earth into an upright position, with the rotation axis normal to the ecliptic plane

winter near perihelion (mild winter) and summer at aphelion (cool summer) so that the seasonal range there would be depressed. Consequently, the two hemispheres may be expected to display contrasting climatic responses to any precessional change in the orbit. As precession proceeds, the boundaries of all the seasons, and the timing of the equinoxes and solstices, are shifted. Indeed, every day of the year and every season coincides with perihelion in a slow, ongoing sequence. The expression 'precession of the equinoxes' should bring to mind all of these related changes.

To understand the significance of solstices and equinoxes, let us consider the orbital processes involved here a little further. Because of the Earth's axial tilt, the noon sun in summer appears briefly to be overhead at places as far north (or south) as 23.5°, as the Earth makes its annual revolution around the sun. The zone of the Earth lying between these limits we call the *tropics*, and this zone experiences zenith angles that never exceed the obliquity angle. The day on which the apparent poleward travel of the sun reaches its limit, and the movement of the planet further around the elliptical orbit begins to move the sun back toward the equator, is a *solstice* (when the sun appears to pause or stand in its progression: this comes from the Latin *sol* 'sun', and *stare*, 'stand'). At the solstices, the tropics receive their maximum duration of solar radiation. As the Earth moves from the solstices, it approaches, after about four months, the intervening *equinoxes* ('equal nights', from the Latin *aequus* 'equal' and *nox* 'night'). At these locations in the orbit, the day is evenly divided into halves of sunlight and darkness, since the rotation axis is inclined neither toward or away from the sun. For each hemisphere there is a summer solstice, when the sun reaches its furthest into that hemisphere (around June 21 for the northern summer solstice) and a winter solstice when the sun is furthest into the other hemisphere (around December 21 for the northern winter). The intervening equinoxes are near March 20 and September 22.

Exactly when the solstices occur during the Earth's annual orbit is significant. Currently, the northern hemisphere summer solstice indeed happens when the Earth is near the far end of its elliptical orbit (i.e. near aphelion). But 11 000 years ago, around the time of the Younger Dryas cool phase, the axis was shifted such that the summer solstice, although still near June 21, happened when the Earth was near perihelion. Thus, northern summers then would have been warmer than now, although shorter (because of the higher orbital speed near perihelion), while northern winters would have been cooler and longer than now. Precession thus influences the seasonal contrast between whatever seasons are reached at perihelion and aphelion. The greatest annual varia-

tion arises where it is summer at perihelion and winter at aphelion.

As we shall see, the effects of precession are conventionally described in terms of the angle (or time) that separates the vernal (spring) equinox from perihelion.

Time variation of orbital parameters during the Quaternary

We have outlined the features of the orbit that are known to undergo continual change, and can now turn to consider the rates at which the changes take place. The changes in the orbital parameters do not display strictly regular periodicities, but rather they have what are termed 'quasi-periods', or periods that vary somewhat through time. The calculated orbital parameters, as well as proxy records from the natural archives, are therefore often analysed by spectral analysis methods, which describe how much of the fluctuation in a record involves changes at each frequency present in the record, from the slowest change to the most rapid. Then the 'quasi-periods' are those that account for the greatest proportion of the variation displayed by the parameter being examined.

Details of the procedures for calculating the Quaternary variability of the orbital parameters are set out in the works of Berger (1978a, b, c, 1984 and other papers) and Vernekar (1971).

ECCENTRICITY

The eccentricity changes quite slowly, and in an irregular manner (see Fig. 5.4). Eccentricity has a lower limit of zero, and from time to time a larger value is displayed. Eccentricity reached a peak of more than 0.02 at about 13.5 ka BP, and is currently declining. For the last 1 Ma, the largest value of *e* tabulated by Milankovitch (1941) was 0.063, and the smallest 0.0018. He recorded large variations in both the amplitude and period of individual cyclic variations of the value of *e*. The mean period determined over the last 1 Ma was 92 ka, with a range from 77 ka to 103 ka, which spans 28% of the mean value.

The spectral power of the calculated eccentricity variation is concentrated in a number of peaks at about 100 ka and a major one at 412 ka, although there are actually many periods associated with eccentricity variation, including long periods at 1.3, 2.0, and 3.5 Ma (Berger 1977). Over the next 1 Ma, according to Berger (1977), the mean period will fall to 91.4 ka.

OBLIQUITY

The periodicity of the changes in obliquity presently averages about 41 ka (Fig. 5.5). The last maximum of

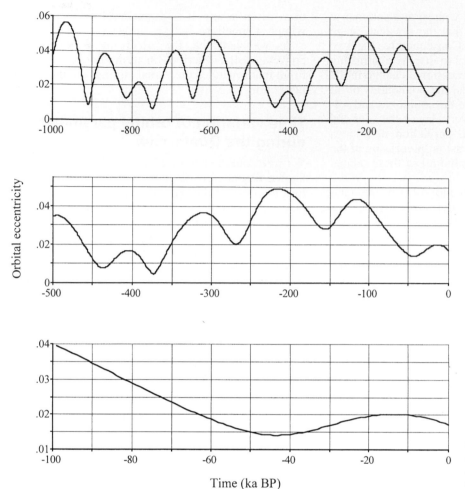

Fig. 5.4 Eccentricity variation over 0–1 Ma BP; 0–500 ka BP; and 0–100 ka BP. (Plotted from data in Berger and Loutre 1991)

obliquity was about 10 ka BP, and obliquity is now decreasing at 0.5″ per annum. For the last 1 Ma, the largest value of obliquity tabulated by Milankovitch (1941) was 24°28′, and the smallest 22°0′, a range of almost 2.5°. He further reported that the period between successive obliquity maxima ranged from a maximum of 45 ka to a minimum of 38 ka, a range which represents about 17.5% of the mean value. Berger (1977) noted that the relative constancy of the 41 ka variation in obliquity reflects the fact that this is not only the dominant period, but is near to the period of five of the following ten terms in the relevant trigonometric series expansion.

An interesting historical indication of the ongoing shift in the Earth's obliquity was provided by Chao (1996), who gave an account of a monument in Jia-Yi County in Taiwan which was erected in 1908 to mark the position of the Tropic of Cancer. But since the obliquity is currently decreasing, the old monument is now 1.27 km north of the correct current latitude of the Tropic of Cancer. Worse still, according

to Chao, the correct location will continue to move south for about a further 9300 years, when it will have shifted over 90 km. Following that, the tropic will of course begin to move north again, as the obliquity cycle continues.

It has been suggested that orbital processes like the value of obliquity (as well as the precessional process) might have been affected by the additional mass of the Quaternary ice sheets and related crustal depression (e.g. Sabadini et al. 1982), leading to true polar wander. Recent work suggests that the effect was probably negligible (Jiang and Peltier 1996). A discussion of the longer-term mechanisms controlling obliquity is provided in Ward (1982) and Williams (1993).

PRECESSION OF THE EQUINOXES

A little additional background is required to understand how precession is analysed and its variation in time described.

Positions such as the location of the equinoxes and

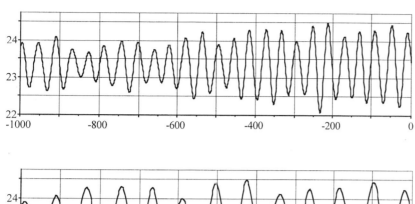

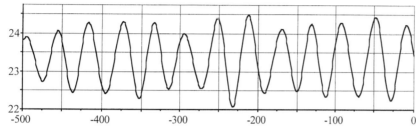

Orbital obliquity (degrees)

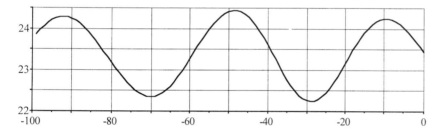

Fig. 5.5 Obliquity variation over 0–1 Ma BP; 0–500 ka BP; and 0–100 ka BP. (Plotted from data in Berger and Loutre 1991)

the boundaries of the seasons have to be judged with reference to some standard location. This is done conveniently by using a measure called the *longitude of perihelion*. This refers to the angle, measured as an arc around the Earth's orbit, from the location of the vernal equinox to the position where perihelion occurs. There are two ways in which this could be done. The reference location could be the position of the vernal equinox at some fixed date (like 1950), or the actual location of the equinox in any particular year. Generally, the longitude of perihelion is measured from the current equinox location.

The vernal (northern spring) equinox is the equinox at which the path of the sun crosses the ecliptic into the northern hemisphere. As the location of this equinox undergoes precession, the longitude varies slowly from 0° to 360°. Through this process, the day of the year when the Earth reaches perihelion gets about one day later every 70 years (Lamb 1972). This is equivalent to about 50″ of arc per year.

Because of this effect, the nature of seasonal insolation (e.g. summer heating) can vary depending upon how near or far the Earth is from perihelion. But clearly, this is only so if the orbit is eccentric, so

that the solar distance is less at perihelion than at aphelion. If the orbit were circular, precession would have no effect on the seasonal insolation. This means that the precession effect can only exert an influence when the Earth's orbit is eccentric – and the more eccentric the orbit, the larger the precessional effect.

It is because of this that reference is often made to the *eccentricity-modulated precession effect*. This is expressed in terms of a *climate precessional parameter*, $e \sin \omega$, where e is the eccentricity and ω (omega) is the longitude of perihelion measured from the vernal equinox.

This climate precessional parameter has a physical meaning. It is a measure of the Earth–sun distance, r, at the northern summer solstice. This distance is given approximately by the expression

$$r = a\,(1 - e \sin \omega) \qquad (5.6)$$

where a is the length of the semi-major axis of the Earth's orbit (Berger and Loutre 1994).

The period of the precessional cycle is actually set by two orbital periodicities. The first, termed *general precession*, is the steady change in the orientation of the long axis of the Earth's orbit in space. The orientation completes a 360° revolution in 96.6 ka. Simultaneously, the Earth's spin axis rotates around the precessional cone in 26 ka. But since the general precession is prograde and the axial precession retrograde (see Fig. 5.1), the location of perihelion moves one-quarter of a revolution in the time of a single precessional cycle. Thus, the spin axis only has to migrate three-quarters of the way around the precessional cone to once again coincide with perihelion. Consequently, the period of the precessional cycle is effectively reduced from 26 ka to (0.75 × 26) ka or about 19.5 ka.

This means that precession has the highest frequency of variation, with peaks of spectral power presently at 19 ka and 23 ka. Over the last 1 Ma, Milankovitch (1941) determined a mean period of 21 ka, but noted that a full precessional cycle had been completed in as little as 16.3 ka or as much as 25.8 ka. This range spans about 45% of the mean value. The value of $e \sin \omega$ swings from about -0.05 to $+0.05$ as $\sin \omega$ goes from -1 to $+1$ (see Fig. 5.6). The value of ($e \sin \omega$) is presently about $+0.017$ and

declining from a recent peak that was reached at about 2 ka BP. The amplitude of the variation is modulated to large values with a quasi-period of about 100 ka by the fluctuations of eccentricity. Table 5.1 summarises the various orbital periods discussed above.

In particular intervals of Quaternary time, different orbital periodicities are known to dominate the sedimentary proxy records. The 100 ka period appears strongly in the last 1 Ma or so, and especially the last 500 ka, when major glaciations appear to have waxed and waned over just such periods. Before about 1 Ma BP, the 100 ka periodicity was not dominant. On the basis of such changes, Pisias and Moore (1981) subdivided the Quaternary as follows:

Late Quaternary: 0–900 ka BP (major increase in amplitude of climate variability near 100 ka and a lesser increase in variability at 41 ka).

Middle Quaternary: 0.9–1.45 Ma BP (increase in amplitude of the 21 ka and 41 ka periods).

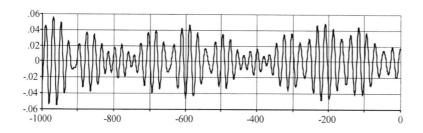

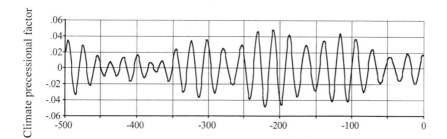

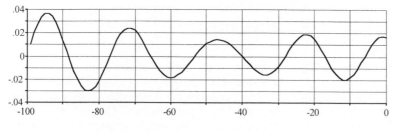

Time (ka BP)

Fig. 5.6 Climate precessional parameter (*e* sin ω) variations over 0–1 Ma BP; 0–500 ka BP; and 0–100 ka BP. (Plotted from data in Berger and Loutre 1991)

Table 5.1 Late Quaternary ranges and present-day values of orbital eccentricity, obliquity, and the longitude of perihelion

Orbital parameter	Range during last 800 ka	Dominant periodicities in insolation forcing	Value at present day	Value at 9 ka BP	Value at 18 ka BP (last glacial maximum)
Eccentricity (dimensionless)	0.0005–0.0607	Average quasi-period 95 ka; spectral components at 410, 120 and 100 ka	0.0167	0.0193	0.0195
Obliquity (degrees)	22°00′–24°30′	Stable quasi-period 41 ka; spectral components at 54 and 29 ka	23°27′	24°14′	23°30′
Longitude of perihelion from vernal equinox (degrees)	0–360°	Average quasi-period 21.7 ka; spectral components at 23 and 19 ka	102°30′	311°12	164°

Early Quaternary: 1.45–2.0 Ma BP (generally lower amplitudes of climate variability, with low amplitude at 41 ka).

The nature of the change from middle to late Quaternary has not yet been established. Analysis by Park (1992) and others suggests that the transition may have been a gradual one. Likewise, Ruddiman *et al.* (1989) suggested that the transition required several hundred thousand years, with the most rapid shift in the period 0.7–0.6 Ma BP. An important anomaly pointed out by Imbrie *et al.* (1993b) is that the greatest spectral power in the forcing function throughout the period of 100 ka dominance of the global environment has actually been at the 412 ka eccentricity period. According to these authors, this argues against external insolation forcing as a cause of the 100 ka periodicity in the Quaternary record, and lends support to those theories that relate these long-period responses to aspects of the internal behaviour of the climate system (e.g. tectonic effects, ice-sheet growth/decay behaviour, and other mechanisms mentioned in Chapter 4). Resolving this issue remains a major challenge in the understanding of Quaternary climate change; we will return to these issues later.

THE MAGNITUDE OF INSOLATION CHANGES DURING THE LATE QUATERNARY

So far we have discussed the key orbital parameters and their time variation. Before turning to consider the absolute *magnitudes* of the insolation forcings that have been determined for the Quaternary, measured in W/m^2, we must emphasise again that analysis of the Milankovitch mechanisms involves two

different kinds of data. The first are the direct insolation forcings arising from orbital processes. These are calculated without making any allowance for the internal feedback processes that might arise within the climate system (such as the ice albedo effect mentioned earlier). The data on insolation at the top of the atmosphere are taken to be the primary insolation forcings which could drive Quaternary climate change. The net radiation at the surface may, however, change by a greater or lesser amount, depending upon the complex mechanisms of the climate system. The growth of increased cloud cover in a warmed environment could act to limit the increase in ground-level radiation arising from direct solar forcing, for example, or the albedo effect of growing ice cover could act to enlarge the deficit arising from a minimum in the solar forcing. In a similar way, dust entrained from arid surfaces could act to backscatter incident solar radiation, but the amount of dust formerly present in the atmosphere cannot be calculated exactly for points on the Earth's surface, while the external insolation can.

Data reflecting the final climate-system outcome of the insolation forcing can be obtained from a number of sources. Fossil indicators from the natural archives, such as pollen (see Chapter 10) or marine microfossils (Chapter 7) provide one source. Comprehensive models of the Earth–atmosphere–ocean–cryosphere–biosphere system provide another possibility. Although the fossil record provides, in principle, evidence of the actual response of the climate system, it is not easily possible to resolve causes from such data. Sophisticated global environment models, though, do have the capability of revealing the relative importance of the geography of the planet, and of the effects of such factors as atmospheric clouds and water vapour, for no direct fossil records exist. Appropriate kinds of global models are widely used, but are still far from being sufficiently complete in their representation of global processes. This is partly because many planetary processes, such as changes in ocean circulation, and resulting alterations in the pattern of marine biological activity, are themselves not yet well enough known to be incorporated adequately into global climate models. Thus the natural archives, for the present, remain the final arbiter of the debate about the fluctuating global environments of the Quaternary.

Insolation forcing at the top of the atmosphere

There are many sources of data dealing with both the direct insolation forcing and the climate system amplification or attenuation of this signal. Let us consider the external forcing first. Here researchers have always encountered the difficulty of deciding what insolation record should be analysed. Milankovitch and others have stressed the key position of land in the high latitudes, where much of the Quaternary ice indeed accumulated. A common choice is to study the pattern of insolation at latitude 65° N. This parallel runs through the Scandinavian countries, northern Europe, Alaska, Canada and Greenland, over sites occupied by the great ice sheets. (In the southern hemisphere, 65° S runs almost entirely through the present sea ice zone surrounding the Antarctic continent.) The choice of a key latitude like 65° N is a difficult one, but simplifies the task of sifting through the complex patterns of insolation which are calculated for different parts of the Earth's surface. A second choice involves the kind of tallying of insolation that is undertaken. The radiation received in particular astronomical seasons cannot readily be compared from one era to another, because the lengths of the seasons vary. Milankovitch had calculated that the lengths of the summer and winter half-years had varied by ±31 days through the last 1 Ma. He therefore devised the idea of using the total energy received during what he termed *caloric half-years*. These partitions of the year were based on daily insolation, the summer caloric half-year being made up of all those days receiving more insolation than days in the winter half-year. The caloric half-years always remain exactly 182 days 14 h 54 min in length (Milankovitch 1941). Initially on the advice of the climatologist Köppen, Milankovitch focused his attention on the search for cool summers that were thought to be the key to ice accumulation. But others have analysed insolation in terms of monthly values or daily values, often of days around the northern mid-summer (mid-July). The difference that is observed between the calculated insolation for some time in the Quaternary

and modern insolation for the same site depends upon which of these values is considered. The caloric half-year values only depart from modern totals by 3–4% (Berger 1991). In general, annual values would show the least difference, seasons a larger difference, and daily values the largest departures. Insolation values have even been determined for different parts of the day (e.g. near noon) or when the sun is low in the sky (Cerveny 1991; Berger *et al.* 1993a). The idea here is that the albedo of some surfaces varies significantly with the angle at which the solar radiation strikes, so that a climatic response might be elicited by changes in the amount of energy received during some part of the day.

Insolation data are certainly required at greater time resolution than half-yearly. As shown by Berger *et al.* (1993a), for instance, the monthly insolation for March at 60° N began to decrease at 133 ka BP, while that for September did not begin to decline until 121 ka BP. Overall, summer insolation consequently peaked near 123 ka BP, a value which is clearly important when interpreting records that relate to the initiation of glaciation. The commonly tabulated mid-month insolation values typically show departures of more than 10% from modern values. Mechanisms have then to be sought linking these various measures to the climatic environment.

Let us take a few examples. Berger and Loutre (1991) have provided a thorough analysis of the external insolation forcing arising at 65° N· during the last 800 ka, with departures compared with the 1950 AD value for that latitude (427 W/m²). During the last 200 ka, for example, and taking the mid-month July insolation, the deviations listed in Table 5.2 are found.

As can be seen from Table 5.2, the fluctuations in insolation are sizeable, the amplitude of the swing

Table 5.2 Extreme insolation deviations from the value of S for 1950.0 AD (427 W/m²) occurring in mid-July at 65° N during the past 200 ka

Negative deviations		Positive deviations	
Date (ka BP)	Deficit (W/m²)	Date (ka BP)	Surplus (W/m²)
185	28	197	46
160	9	173	54
137	11	148	28
114	35	126	60
93	6	104	48
70	19	82	40
41	10	56	34
22	9	33	15
		10	43

Source: Berger and Loutre (1991).

from biggest deficit to largest surplus being 95 W/m^2, or about 22% of the 1950 value of 427 W/m^2. Many of the departures listed are about 10% or less of the 1950 insolation. However, these are instantaneous values at mid-summer, not time-averages, which would yield lesser values.

These values are typical of those revealed by the calculation of external insolation. The data can be broken down into separate components that arise from the three orbital parameters discussed earlier. This has been done by Imbrie *et al.* (1993a, b), once again for the mid-summer insolation at 65°N. They used band-pass filters tuned to the 23, 41 and 100 ka periods of the orbit to isolate the components of the calculated insolation changes arising from each parameter (see Fig. 5.7). Their results show that the largest component in the time variation of insolation comes from the precession band (19–23 ka), followed by the obliquity band. In the eccentricity band (100 ka), the signal amplitude was only about 2 W/m^2, an order of magnitude smaller than the variation in the other two bands.

The data for 65° N provide a key illustration of the size and timing of the insolation forcing. But values for a single latitude provide no evidence of latitudinal gradients; neither do they tell us anything about the other hemisphere, where the precessional effects are reversed. Often, therefore, calculated values of the insolation are portrayed graphically, in a way that covers the planet from pole to pole, as well as showing the changing values through time. (Some workers have searched for 'signature' patterns of insolation that might correspond to interglacial conditions, or to glacial onset. The work of Kukla *et al.* (1981) and Kukla and Gavin (1992) provides exam-

ples.) An example of a global time–space insolation plot is shown in Fig. 5.8. Further examples of such presentations may be found in Berger (1978a, 1979), and Berger *et al.* (1993a). A graphical presentation of the insolation for a whole hemisphere, plotted on a coloured projection of a sphere, was introduced by Wolters *et al.* (1996). This too provides a compact representation of the large amount of data generated in analyses of the geographical pattern of forcing.

These external insolation forcings are not felt at the Earth's surface. The varying cloudiness and opacity of the atmosphere, arising from aerosol particles, intervene. The fate of the radiation entering the Earth–atmosphere system is harder to resolve. This task was attempted by Tricot and Berger (1988), who calculated the amount of radiation reaching the surface, as well as the amount actually absorbed there. These values carry perhaps more significance for the global climate than do the external insolation forcings, as stressed in early modelling work (e.g. Schneider and Thompson 1979). In calculating these values, Tricot and Berger (1988) made reasonable allowance for clouds and for scattering and absorption within the atmosphere. They found that the amplitude of the fluctuations in insolation declined after transmission through the atmosphere and after any reflection at the surface. The radiation reaching the surface was diminished by a factor of 2–4 compared with the external forcing. Moreover, the amount of radiation finally absorbed by the surface did not show a maximum at high latitudes, as the external forcing did, but in tropical and middle latitudes. The explanation for this is that areas at the high latitudes also have high albedo values. Less radiation is consequently absorbed there, and so the fluctuations in

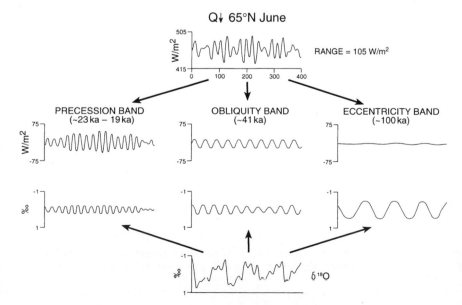

Fig. 5.7 Band-pass filter analyses of the insolation record for 65° N (top) and a marine ^{18}O record (bottom). Note the small amplitude of the insolation forcing in the eccentricity band and the surprisingly large response in this same band in the oxygen isotope record. (After Imbrie *et al.* 1992)

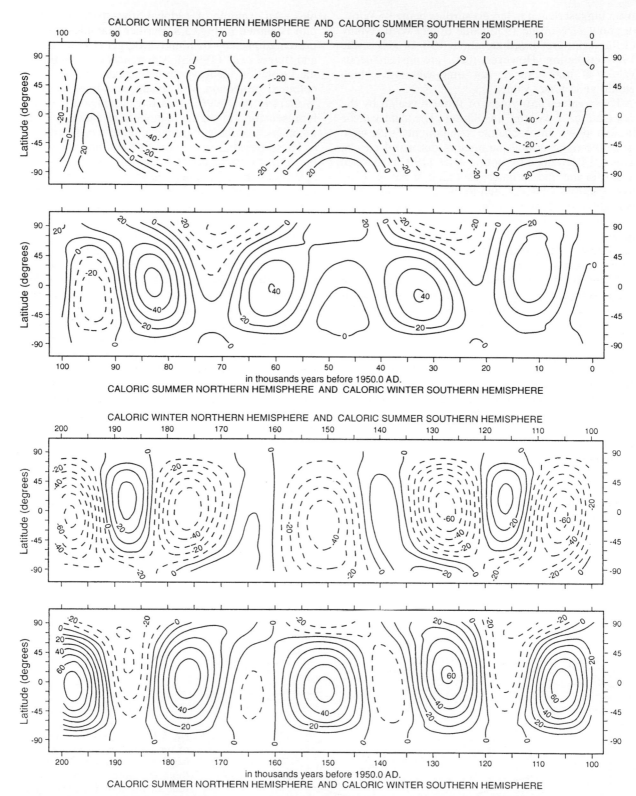

Fig. 5.8 The pattern of caloric insolation across the latitudes 90° N to 90° S for the last 100 ka. Upper diagram: data for the period present to 100 ka BP; lower diagram: data for the period 100 ka BP to 200 ka BP. (After Berger 1978b)

the amount of radiation absorbed are also diminished. Larger deviations from present values were consequently calculated for $30°\,\mathrm{N}$ than for $70°\,\mathrm{N}$, and the latitudinal gradient in received insolation (taken as the difference between the values absorbed at $30°$ and $70°$) was dominated by the values found at $30°\,\mathrm{N}$. Because the major changes at $30°$ relate to the precessional cycle, and not the obliquity cycle which is dominant at high latitudes, we see here a mechanism through which the largest changes in external forcing might not be those which in the end most affect the planetary climate. Further details of the role of the hemispheric insolation gradients were explored by Young and Bradley (1984). They stressed the importance of these gradients in influencing the position of the subtropical high pressure belts, and in the polewards transport of moisture. In their interpretation, glacial periods were related to times when the insolation gradient was stronger than normal, while interglacial periods displayed weak insolation gradients. Further exploration of these ideas is unfortunately beyond the scope of this chapter.

DO THESE INSOLATION CHANGES ACTUALLY AFFECT THE GLOBAL CLIMATE?

If the global climate is responsive to the changes in insolation just outlined, then evidence for this should be found in the natural archives of oceanic sediments, thick ice sheets, or the deposits in large continental lake basins. Because the changes in the Milankovitch trigger are short-lived and quite small, when summed over a whole year, many who considered the hypothesis became convinced that they were inadequate to account for glacial periods (e.g. see van Woerkom 1953; Sellers 1970; Saltzman and Vernekar 1971; North and Coakley 1979). There were also workers who, in parallel with the work of Milankovitch, presented supporting time-domain evidence from the presumed chronology of glacial advances or sea-level fluctuations. Examples can be found in the work of Zeuner (1950). Mason (1976) gave renewed support to the Milankovitch hypothesis, on the basis of heat balance calculations (see Gribbin 1976). But the intricate feedback processes of the global environment mean that a simple consideration of the size of the initial radiation forcing is not a sound basis for estimating the final response of the climate system. A surer way to discover whether the climate system responds significantly is to look for evidence that it has done so during the Quaternary or earlier times. In any case, a simple accounting of the global energy budget shows that the amounts of energy needed to melt the great ice

sheets, or which must be taken away for them to form, are not great. A short calculation, after Oerlemans (1991), will show the annual flux of energy through the top of the atmosphere. Taking the solar constant as $1360\ \mathrm{W/m^2}$ and the Earth's radius to be 6371 km, and allowing for a planetary albedo of 0.3, we obtain the following estimate:

$$(1360\ \mathrm{W/m^2}) \times \pi \times (6.371 \times 10^6\,\mathrm{m})^2 \times (3.1536 \times 10^7\,\mathrm{s}) \times 0.7 \approx 3.8 \times 10^{24}\ \mathrm{J}$$

To melt ice, latent heat of fusion amounting to 0.33 MJ/kg is required. Therefore, taking the density of ice to be 0.9 t/m3, we find that to melt all of the additional ice present at the last glacial maximum (about $49 \times 10^6\ \mathrm{km^3}$) in a period of 5 ka, the total energy input required would be

$$(49 \times 10^6\,\mathrm{km^3}) \times (0.9 \times 10^{12}\,\mathrm{kg/km^3}) \times (0.33 \times 10^6\,\mathrm{J/kg}) \approx 1.5 \times 10^{25}\ \mathrm{J}$$

This amounts to only $\approx 3 \times 10^{21}$ J/a, or about 0.08% of the annual incoming solar energy. Similar calculations made by Saltzman and Sutera (1984) led them to conclude that not even the small insolation change associated with eccentricity variation can be rejected outright as being too small to achieve significant effects on ice sheet growth and decay, given the long time periods involved. These authors estimated the energy flux required to supply the latent heat of fusion during deglaciation to amount to only $0.1\ \mathrm{W/m^2}$, which can be compared to the few $\mathrm{W/m^2}$ range yielded by eccentricity variation. In other words, finding sufficient heat energy is not a major difficulty; the trick is to get the energy to the right place at the right time. In this context, it is interesting to draw to mind a related subject. To form the great ice sheets, large amounts of water had to be transferred to the solid state, so releasing a correspondingly large amount of latent heat, called by Adam (1975) the 'heat of glaciation'. The climate system had to remove this heat so that it did not offset the cooling that was a response to the Milankovitch decline in insolation. Mason (1976) calculated that in the development of the last great ice sheets, in the period 83–18 ka BP, the Milankovitch mechanisms created a radiation deficit of 1.05×10^{26} J in the region poleward of $45°\,\mathrm{N}$, indeed sufficient to compensate for the latent heat that was liberated. A calculation of the amount of latent heat released in the condensation of vapour and subsequent freezing of the liquid water can be made as follows, taking the total of latent heat of condensation and fusion to amount to 2.83 MJ/kg:

$$(49 \times 10^6\,\mathrm{km^3}) \times (0.9 \times 10^{12}\,\mathrm{kg/km^3}) \times (2.83 \times 10^6\,\mathrm{J/kg}) \approx 1.25 \times 10^{26}\ \mathrm{J}$$

This shows an approximate equality with the insolation deficit. Likewise, according to Mason (1976), during the phase of ice melting between 18 ka BP and 6 ka BP, the radiation excess once again was sufficient to supply the required latent heat of fusion taken up as the ice melted.

In seeking confirmation of his ideas, Milankovitch made comparisons between his calculated insolation curves and what was known about Quaternary glacial chronology at that time, noting good agreement. The insolation history determined by Milankovitch for the past 600 ka is shown in Fig. 5.9, together with his suggestions for the insolation minima corresponding to the Quaternary glacial periods as they were then envisaged. The speculative chronology used by Milankovitch is now known to have been quite erroneous, and so his difficult time-domain tests were not conclusive (although his calculated insolation curves indeed match modern radiometrically dated proxy records rather well). The first soundly-based test of the Milankovitch hypothesis was reported in a famous paper by Hays, Imbrie and Shackleton (1976a). In this work, Hays *et al.* used analyses of oxygen isotopes from long cores of oceanic sediment, with limited radiometric dating control. The ^{18}O record that they analysed largely reflects global ice volume, since the locking-up of water in the great ice sheets left growing concentrations of ^{18}O in the oceans. (In fact, the ^{18}O record can be roughly converted to a sea-level record by using the relationship

that a 1‰ change in δ^{18}O corresponds to about a 100 m change in global sea-level: Imbrie *et al.* 1984; Shackleton 1987). Hays *et al.* used spectral analysis methods to show that the variance in the ^{18}O record had a red-noise profile, with increasing variance at lower frequencies (see Fig. 5.10). But superimposed on the red-noise profile were spectral peaks which they identified as corresponding to periods of 106, 43, 24 and 19 ka. These periods are very close to those calculated by Milankovitch in his work on the external insolation. In particular, the detection of the peaks at 24 and 19 ka, close to the two expected peaks in the precession band, constituted a clear demonstration, in the frequency domain, that Milankovitch rhythms really were present in the natural archives. Berger (1991) described this finding as '...one of the most delicate and impressive tests for Milankovitch's theory'.

Many subsequent studies have confirmed conclusively that abundant natural archives show variations in chemical composition, or sediment texture, or some other feature, at rhythms corresponding to those calculated for the external solar insolation. For example, Kashiwaya *et al.* (1991) recorded 20 ka and 40 ka periods, matching those of precession and obliquity, in the patterns of speleothem growth from British caves. Indeed, spectral analysis revealed the presence of distinct 19 ka and 23 ka periods in their

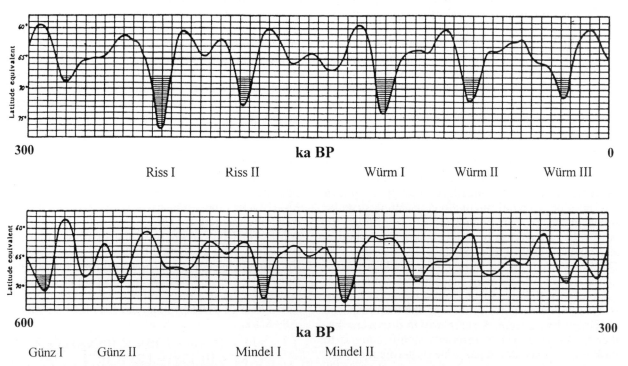

Fig. 5.9 Insolation history for the past 600 ka, as calculated by Milankovitch. The insolation minima corresponding to supposed phases of the Würm, Riss, Mindel and Günz glaciations that Milankovitch identified are shown shaded. (After Milankovitch 1941)

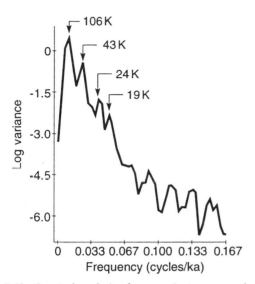

Fig. 5.10 Spectral analysis of oxygen isotope records from the famous paper by Hays *et al.* (1976a). This graph reveals the presence of eccentricity, obliquity, and the two main precession frequencies within this record, and was the first solid demonstration that Milankovitch periodicities were indeed evident in the marine record

data. The fluctuations in speleothem growth presumably reflected environmental characteristics such as wetness, temperature, and the surface vegetation. All these things can affect the solubility of carbonates in the water, and the processes within the cave by which secondary carbonates are deposited.

Interestingly, Olsen (1986) reported that fluctuations in bedding thickness in ancient lake sediments in eastern North America showed Milankovitch periodicities. The lacustrine sediments he studied were from the early Mesozoic, which was globally warm and ice-free. The changing sediment accumulation again presumably reflects the influence of climate on vegetation and erosional processes in the catchment area of the lake. If caused by the Milankovitch mechanism, these deposits show that the presence of snow and ice, to generate an albedo feedback, is not a precondition for the Milankovitch processes to become significant in global environments. The recent reporting of precessional rhythms from tropical Pangaea during Triassic times raises similar ideas (Olsen and Kent 1996). Indeed, there are many studies that demonstrate that Milankovitch rhythms are very probably represented in sediments of Palaeozoic as well as of Mesozoic age. Ancient desert dune sands in Scotland have revealed such evidence (Clemmensen *et al.* 1994). Here, the bedding thickness in the sandstones averages 22 m, in a deposit of some 700 m thickness, with the intervals between beds interpreted to signify the reversion to wetter conditions. Frequency analysis of these climate swings strongly

suggests the presence of Milankovitch rhythms. In the marine record, Boyd *et al.* (1994) have shown that repeated fluctuations in sediment composition at a site on the Exmouth Plateau off NW Australia probably reflect Milankovitch rhythms also. At this site, the marine sediments show changes in character about every 20 cm, with fossil-rich layers alternating with more clay-rich layers. These changes are thought to relate to changes in the input of terrestrial materials and fresh water, with resulting changes in the planktonic productivity of the marine communities. In this case, the sediments suggesting the Milankovitch rhythm are of Cretaceous age (late Mesozoic, about 74 Ma BP).

Similar periodicities have been reported from deposits elsewhere. Kerr (1987) noted that the 40 ka obliquity period is evident in the marine sediments exposed in cliffs along the English and French coastlines, for example. Orbital forcing of the Asian monsoon has been suggested (e.g. Morley and Heusser 1997). The eccentricity periods have been revealed in Pliocene marine [18]O records (Clemens and Tiedemann 1997), and variation at the obliquity period, measured in terms of benthic ostracod species diversity in Pliocene sediments of the North Atlantic, has also been demonstrated (Cronin and Raymo 1997). Orbital periods have also been demonstrated from the Oligocene/Miocene (Zachos *et al.* 1997). Other examples may be found in the work of Boyd *et al.* (1984) in the Arctic Ocean, Herbert and Fischer (1986) on Cretaceous shales in Italy, Alaskan loess stratigraphy (Begét and Hawkins 1989), equatorial Alantic ocean temperatures (McIntyre *et al.* 1989), and in Cretaceous sediments of the US (Sageman *et al.* 1997). Many of the long, late Quaternary ice core records discussed in Chapter 3 have also shown Milankovitch periods in data on such parameters as dust concentrations, oxygen isotope values, and sulphate particle abundance.

Clear evidence thus exists that global and regional environmental changes have taken place at rhythms that strongly correlate with the Milankovitch orbital periods, over vast periods of geological time, and without the ice albedo feedback being a prerequisite. Berger (1989b) presented evidence of Milankovitch rhythms detected from the mid-Palaeozoic record (400 Ma BP). The actual periodicities of the orbital changes differed from their Quaternary values in the more remote geological past. The gradual retreat of the moon (at a present rate of > 1 cm/a), for example, exerts a significant effect. Berger and Loutre (1994b) calculated values of the obliquity and precession periods back to 2500 Ma BP. At 500 Ma BP, for instance, the 23 ka precession period becomes 19 ka, and the 41 ka obliquity period becomes about 30 ka. Such shorter periods have been confirmed by field data. De Cisneros and Vera (1993), for example, reported a 39 ka periodicity, taken to be the shortened obliquity period, in peritidal limestones from the lower Cretaceous of Spain.

The recognition of these Milankovitch rhythms in the archives has resulted in various developments in stratigraphic analysis. Perlmutter and Matthews (1990) have erected a model of the rhythmic sediment bodies that result, which they term *cyclostratigraphy*. They have hypothesised that global environments may pass through a regular series of environmental changes as the Milankovitch cycles proceed, with some areas oscillating from humid, at an insolation peak, to arid, at a time of insolation deficit. Through this sequence, changes in weathering processes, and in the dominant particle size of eroded material, are envisaged. These are just the kinds of changes required to produce the sedimentary evidence referred to above, and seem to be revealed in a 24 ka precessional signal detected in early Quaternary sediments in Death Valley reported by Hsieh and Murray (1996). Perlmutter and Matthews (1990) have related the environmental changes to shifts in the location and strength of the atmospheric circulation systems, such as the Hadley cells, in response to the insolation forcing. Schwarzacher (1993) provides a fuller treatment of cyclostratigraphic ideas, and includes a number of photographs of cliff sections through bedded deposits that have been shown to reflect Milankovitch rhythms.

Evidence for the Milankovitch mechanisms has also been amassed in the time domain. The sea-level history derived from the Huon Peninsula marine terrace sequence shown in Fig. 6.6 is a well-known example, corresponding well with the timing of the calculated insolation forcing (Broecker *et al.* 1968; Mesolella *et al.* 1969; Chappell 1973, 1974; Veeh and Chappell 1974a; Aharon 1984). We can also note the good general temporal correspondence between late Quaternary climates and the graphical insolation plot shown in Fig. 5.8. The LGM at 18–20 ka BP, for instance, clearly followed a period of summer insolation deficit in the high northern latitudes, while the previous interglacial at 125 ka BP corresponded closely with a period of significant insolation surplus there. Further evaluation of the Milankovitch chronology from time-domain data will undoubtedly proceed as dating methods improve. An instance of this was provided by Slowey *et al.* (1996), who reported U–Th dating of calcareous marine sediments belonging to the last two interglacial periods. The dates obtained (120–127 ka BP and 189–190 ka BP) both correspond well with predictions made on the basis of calculated insolation forcing at those periods in the past. Lourens *et al.* (1996) have provided an evaluation of the astronomical time-scale in a comparison of sedimentary records and insolation forcing for periods extending into the Pliocene. Time-domain

checking beyond this time is difficult, because of the problem of extending the orbital calculations reliably. As Berger (1980) emphasised, even an error of 1% in calculating an orbital period would put the signal a whole period out after 100 periods, and the time-based data hence become less and less reliable into the distant past. Berger and Pestiaux (1984) expressed confidence in frequency-domain checking of insolation forcing back to 5 Ma BP, but only to about 1.5 Ma BP in the time domain without further improvement in the data from celestial mechanics.

EXPLORING THE MECHANISMS LINKING MILANKOVITCH FORCING TO CLIMATE

To understand how the environmental responses which have been documented come about, we need to consider the mechanisms that elicit a response in global environments from the transient Milankovitch forcings.

The climate system involves many processes capable of amplifying or attenuating the insolation changes expressed at the Earth's surface. This involves a change in the *magnitude* of the effective insolation change. Furthermore, the albedo feedback or other mechanisms are capable of extending a seasonal effect into a year-long or semi-permanent one, which is a change in *duration* of the effect. It is not surprising, therefore, that the intervention of the climate system (involving also the oceans, the cryosphere and the biosphere) has generated a rich pattern of spatial and temporal variations in global Quaternary environments.

Without resorting to the analysis and interpretation of palaeoclimate proxy data, which is a specialised matter discussed in other chapters, we can explore some of the climate system responses from suitable global models, either energy balance models or larger models involving the interaction and spatial extent of atmosphere, land, vegetation, ice and oceans. Seeking explanation of the observed Quaternary environmental changes provides an opportunity to refine and test our understanding of natural environmental processes. Such an understanding could be of great assistance in the management of future human disturbance of these systems.

Let us first consider the magnitude of climate changes that have been estimated on theoretical grounds, employing a general understanding of the way in which terrestrial environments of the present (tropical, polar, monsoonal, etc.) vary in relation to their insolation conditions. Milankovitch (1941) provided early calculations of these effects, and estimated that seasonal impacts of the insolation changes would have been quite large in the Quaternary. For example, he calculated that the mean temperature of the warmer half of the year had dropped by up to 14°C in the northern hemisphere during times of insolation minimum. These changes are larger than those of the southern hemisphere, more of which is covered by ocean, whose larger heat capacity damps temperature changes. Even so, Milankovitch calculated that temperature drops of up to 8°C had occurred in the warmer half-year in the southern hemisphere. These hemispheric averages are a useful indication of the kinds of climatic responses that are possible, but they provide no indication of how the effects would vary geographically. To see something of these patterns, we will examine first a relatively simple, two-dimensional energy balance model. Such a model, using a simplified map of the planetary land and ocean areas, was developed by Short *et al.* (1991), who ran it to analyse conditions during the past 800 ka. The Berger (1978a) formulae were used to derive the values of the orbital parameters for this time period. The model contained no changes in cloudiness nor in the snow and ice cover, and hence no albedo or other feedbacks. Thus, many of the changes in temperature determined through the model may be substantially smaller than the final changes that would arise through the fuller operation of the climate system. None the less, they provide much useful information on the effects of orbitally-induced insolation forcing over the oceans and land areas.

The climatic parameter selected for analysis by Short *et al.* (1991) was the maximum temperature reached in summer, which they termed T_{max}, a parameter clearly relevant to the survival of snow and ice through the summer season.

Orbital parameters were shifted singly in the model. At the extreme, steepened obliquity resulted in about a 4°C increase in T_{max} in the high latitudes. At the same time, T_{max} was depressed about 0.5°C over the equatorial oceans. Shifting summer from perihelion to aphelion with eccentricity held at 0.06 and obliquity at 23.25° resulted in changes of T_{max} of 10°C over large continents. In the southern hemisphere, the effects are opposite, and cooling results when northern summers are at perihelion.

Using, as extreme cases, maximum obliquity occurring at perihelion and minimum at aphelion increased the seasonal changes in temperature. The combined effect yielded summers up to 14°C warmer over central Eurasia. This is a large temperature change given that the predicted value does not incorporate any amplifying feedback effects.

Short *et al.* (1991) provided an interesting analysis of the timing of the insolation forcing experienced at various points on the Earth's surface, and of the tem-

perature response predicted by the energy balance model. Over north central Eurasia, for example, their data showed that the precessional forcing had the greatest magnitude, and fluctuations in temperature over the past 800 ka took place dominantly at the 23 ka and 19 ka periods. In contrast, over the northern Atlantic ocean, although again the precessional terms dominated the forcing, the dominant temperature fluctuations were at the 41 ka obliquity periodicity, the insolation forcing for which was much smaller than that of precession. The temperature fluctuations here were small (less than 1°C) because of the high heat capacity of the ocean. The obliquity effect seems to dominate the temperature response because this forcing changes the annual total of radiation at these latitudes, whereas the precessional effect is strictly seasonal, and is strongly damped by the thermal inertia of the ocean.

Over equatorial Africa a most significant result was found. The insolation forcing had most of its power at the precessional periods, but the temperature response, while being dominated by the 23 ka and 19 ka periods, also showed major responses at 100 ka and 400 ka, related to the eccentricity variation (see Fig. 5.11).

The eccentricity-modulated precession forcing is apparently demodulated in the low latitudes, where T_{max} occurs close in time to an overhead passage of the sun that happens near perihelion. Thus, the alignment of equinoxes with perihelion is more significant here than is solstice–perihelion alignment, which has greater significance at higher latitudes. Lows in the insolation forcing are thus relatively unimportant, but the climate system shows a response to peaks which occur with the eccentricity periodicity. The individual temperature fluctuations corresponding to the eccentricity-driven forcings were only small, about 0.35°C at the 100 ka period, for example. But the various periods near to 100 ka sum to a total effect of at least 1°C, perhaps sufficiently large to have climatic consequences. It is significant that the 100 ka and 400 ka periods are evident in the output of the energy balance model, because most investigators have appealed to other mechanisms to generate this (discussed later), in view of the small size of the external eccentricity forcing.

A second example of the use of models to identify the global response to Milankovitch forcing will be useful. In this case we will look briefly at a more sophisticated model, consisting of a climate model linked to a model of ice sheet behaviour. Gallée *et al.* (1992) ran their model first with ice extent fixed at the modern value, to exclude the feedback effects that arise when ice sheets grow. This revealed that surface temperatures averaged over a year and over the northern hemisphere changed by less than 1°C. When ice sheets were allowed to develop in the

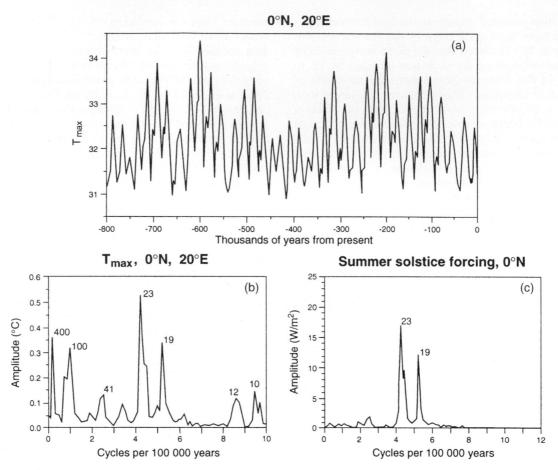

Fig. 5.11 Results of model calculations of temperature variation over an equatorial site at 20° E, for the past 800 ka:
(a) Times series of modelled maximum temperatures
(b) Spectral analysis of the T_{max} time series. Note the presence of peaks at 400 ka and 100 ka in the eccentricity bands
(c) Spectral analysis of the summer solstice insolation forcing at the equator.
(After Short *et al.* 1991)

model, larger changes were found. In the region 65–70°N, annual mean temperatures over the last 120 ka fluctuated over a 9°C range. As noted earlier, annual mean values conceal larger seasonal effects. Seasonal perturbations of temperature and precipitation were observed in an atmospheric GCM coupled to a simple mixed-layer ocean (without any heat transport) by Phillipps and Held (1994). In a series of numerical experiments, Phillipps and Held simulated global climate under conditions of high eccentricity ($e = 0.04$), low and high obliquity (22° and 24°), and with northern summer solstice happening at perihelion and aphelion. Certain parameters were neglected in the model, such as any variation in cloud cover.

With northern summer solstice at perihelion, for example, mean annual temperatures in high northern latitudes were found to be almost 4°C warmer than for aphelion alignment. In contrast, under these conditions land areas near 20° N were cooler by more

than 2°C. This arose because increased monsoonal rainfall had generated moister soils, with correspondingly greater evaporation. But seasonal effects were much larger than annual mean values. For instance, summer warming with the solstice at perihelion was in the range 9–12°C in central Asia, and >15°C in central North America. This exceeds the 6°C warming modelled by Mitchell *et al.* (1988), but corresponds well with the values modelled by Lautenschlager and Herterich (1990) using the European T21 GCM. Drier soils in North America contributed significantly to the warming modelled by Phillipps and Held (1994), in a positive feedback effect also noted by Mitchell *et al.* (1988) in model studies of the climate at 9 ka BP. Further comments on climate sensitivity are provided by Suarez and Held (1979).

An important mechanism noted in these model experiments was the influence of the warm high-latitude summers on sea ice. The area covered by sea ice was dramatically reduced, and the ice became significantly thinner. We will consider this effect later.

Despite various attempts to encourage ice survival through the summer, the model used by Phillipps and Held could not be made to generate stable ice sheets, and so clearly fails to incorporate one or more key climate processes that reinforce the effects of the insolation changes. This same finding was reported in earlier studies, like that of Pollard et al. (1980), in which modelled ice sheet behaviour displayed the higher frequencies of the obliquity and precessional forcing, but not the 100 ka period of the eccentricity. The ability to model the development of ice sheets in response to insolation forcing also eluded Rind et al. (1989). A problem that commonly arises in this kind of investigation is that, for a given ice sheet size, two different mass balances must apply: one slightly positive, during the phase of slow ice sheet growth, and the other very negative, during rapid deglaciation. This immediately poses the task of finding the group of parameters that control the change in mass balance.

The models mentioned briefly here, and others like them, thus indicate that the response of the climate system to the Milankovitch forcing is indeed a substantial one. All aspects of climate are affected, including surface and air temperatures, precipitation, evaporation, and land and sea ice. No model yet incorporates all of the additional effects that may have arisen in the terrestrial and oceanic parts of the biosphere, or in chemical weathering, or in oceanic thermohaline circulation systems and heat transport, atmospheric cloudiness or aerosol loading. The model results are thus indications of the kinds of climate outcomes that are plausible, rather than prescriptions of likely or actual responses. None the less, we have seen that summer temperatures over large land areas may be altered by more than 10–15°C in response to the Milankovitch trigger, and that key parameters like sea ice extent are greatly modified. (Actual measurements confirm values of this magnitude. From Greenland borehole temperature data, Cuffey et al. (1995) reported that the coldest phases of the last glacial were about 21°C colder than now at the ice surface, with a shift of about 15°C from average glacial conditions into the Holocene.) We can therefore safely dispense with any suggestion that the Milankovitch trigger represents an insignificant influence on the global environment. It clearly has the potential to produce major climatic perturbations. Additional work will be required to clarify exactly how these translate into the waxing and waning of the great Quaternary ice sheets. At present, the natural archives provide what may be considered more reliable evidence of the kinds of processes that are

involved. Let us highlight just a little of what has been discovered.

A single example will suffice, the literature providing a rich source of additional information. Imbrie et al. (1989, 1992) studied a large set of proxy palaeoenvironment data assembled from published work. The data included calculations of solar insolation (the Milankovitch forcing), ^{18}O data from benthic forams, plankton assemblages to indicate surface water temperatures, δ^{13}C data from benthic forams to indicate oceanic carbon flows, indicators of dust abundance, and so on. The data came from a range of oceanic sites. Time series records of all of the variables, taken from long sediment cores, were employed to reveal the phase relationships among them as the Milankovitch cycles of the past 400 ka proceeded. The environmental variables were found to fall into two categories, which Imbrie et al. (1992) termed the 'early response group' and the 'late response group'. The sequence of responses to the insolation forcing was found to reflect a regular progression of geographical locations, some of which responded rapidly and others only later. This information was used to generate a conceptual model that stressed the importance of initial responses seen in the Arctic, where changes in the flux of fresh water modulate oceanic processes involved in heat export to the deep ocean. Terrestrial changes follow later. Details of this work cannot be included here, and the reader is referred to the paper cited above, as well as to the companion paper (Imbrie et al. 1993b), where the '100 ka' problem is addressed further. Clearly, however, this kind of detailed analysis of the store of information in the archives provides a vitally important way to unravel some of the complex behaviour of the global environment which lies beyond the capability of contemporary numerical models.

In the following sections we return to consider further some of the processes that translate the Milankovitch forcing into these global responses, to understand better how it is that the relatively small forcings create the large environmental changes. We will also pay attention to the kinds of mechanisms that create phase shifts between the insolation forcing and the response of the environment (the fourth of the questions posed at the start of this chapter).

FEEDBACK PROCESSES IN THE CLIMATE SYSTEM

Among the processes that arise in the complex response of the climate system to insolation change, are some that magnify the change itself. These processes are *feedback* processes, since they produce a response that is bigger (or smaller) than would arise without their involvement. Thus, it is as though some of the initial climate change is fed back to the factors causing the change, increasing their magnitude. Such processes seem to be essential in the chain of events that links an external insolation forcing to a final response in the global environment. An analysis of some of the many feedbacks that are known will assist us in understanding the magnitude of environmental change related to the Milankovitch mechanisms.

A most obvious potential feedback process in the Quaternary is the *ice albedo feedback*. If Milankovitch forcing results in sufficient cooling to cause a small increase in the area covered by snow, then the higher reflectivity of the white surface alters the radiation balance. The effect is to reflect a larger proportion of the incident radiation back to the atmosphere without any surface warming. The regional climate must then become cooler, so perhaps promoting a further increase in the snow-covered area, more reflection of radiation, and so on. Through this albedo feedback, the radiation absorbed by the surface declines far more than the initial amount resulting from the Milankovitch forcing. Thus, we cannot judge whether the size of the insolation changes produced directly by orbital changes, and tabulated earlier, is sufficient to cause glaciation, without knowing how much amplification of the change is generated by the response elicited in the climate system.

The albedo feedback is itself not as straightforward as the above account might suggest. Indeed, there are always multiple responses, linked in a cascade, to any disturbance of the global environment (Cess and Wronka 1979). In the case of albedo feedback, the temperature lowering produced by the albedo change may cause, in its turn, changes in the moisture content of the atmosphere, in vertical convective mixing, and in cloud cover. According to Cess (1992), the direct albedo effect itself might account for less than half of the overall feedback. The bulk of the effect, rather, arises from changes in the atmospheric lapse rate, caused by altered convective mixing of water vapour into the troposphere, which influences the greenhouse processes acting there. Further details of these mechanisms were provided by Raval and Ramanathan (1989), who showed that changes in the moisture content of the atmosphere cause major variations in the greenhouse effect which are strongly correlated with atmospheric temperature. There may even be thresholds in this effect, such that deep convection over the tropics might feed sufficient water to contribute to ice-dominated cirrus clouds, which act to reduce terrestrial radiation loss from the Earth–atmosphere system. (Warm clouds lower in the atmosphere are effectively opaque to terrestrial radiation, so that increasing their depth largely acts by increasing short-wave reflectivity, so

tending to cool the Earth.) But the effects of growing snow and ice create further changes. For example, a growing ice sheet becomes thicker as annual snow layers are added to it. Thus, the height of the ice surface above sea-level increases, and this is associated with the normal cooling associated with elevation (about 6°C per km). In this way, the temperature over the ice sheet falls even more than would be expected from the albedo effect alone. This, and the simple albedo effect, involve long response times – up to several thousand years or more – during which the ice accumulates progressively. Thus, these various feedback processes develop slowly, and the response of the climate system cannot be expected to keep pace with the Milankovitch forcing function.

But now things become even more intriguing. The substantial weight of a large ice sheet is sufficient to cause crustal subsidence. The mantle below the crust has to deform to permit this subsidence, and the material of the mantle is enormously viscous, responding to loading very slowly. This introduces a further internal process (although hardly part of the climate system) that will be completely out of phase with the insolation forcing. These effects also arise during global warming, when the melting of ice sheets unloads the crust below. The crust then consequently experiences isostatic uplift, with mantle material slowing flowing inwards beneath it. The diminishing ice cap is thus slowly lifted in elevation, which might tend to retard melting. However, the crustal response is very slow, and so meltwater might form lakes in the crustal depression, favouring more rapid ice sheet collapse. We will consider these ideas a little further later. However, a complete analysis of these subjects is beyond the scope of this book; fuller analyses can be found elsewhere (e.g. Oerlemans and van der Veen 1984).

A positive feedback related to albedo, and which is active during melting, was highlighted by Gallée *et al.* (1992). They pointed out that unless fresh snow falls, the albedo of a snow surface progressively declines because of various changes in the crystal structure. In their model, snow albedo decline because of this process provided the key to deglaciation, which did not occur if the albedo decline was excluded from the model. Birchfield and Weertman (1982) also showed that such albedo variations greatly increased the sensitivity of a palaeoclimate model to orbital insolation forcing.

As a final example of the compounding effects that arise, we can mention the ideas of Mörner (1994). He has pointed to the various links between orbital effects and the solid Earth. Orbital changes, for example, will be associated with some tendency for the Earth itself to deform. The changing distribution of mass arising as ice sheets grow on land, and weight is correspondingly removed from the ocean floors, might also be associated with changes in the figure of the Earth and hence in its rotation. These changes within the solid Earth might influence volcanic activity and hence feed back to the climate system directly, through the emission of particulates to the atmosphere. Tying all of this behaviour down is no simple task, and highlights the need for Quaternarists to engage in work that links what might appear to be disparate parts of the Earth and environmental sciences

It is worth considering briefly some of the many other feedback processes that may come into play in linking insolation changes to environmental responses. There are, for example, other albedo feedbacks. If thickly vegetated regions experience vegetation loss because of drier or colder conditions, the bare surface may display a higher albedo than that of the green foliage, so generating an effect like that of the ice mechanism. Also, exposed continental shelves provide an additional area, as sea-level falls, whose albedo may provide a feedback effect. A significant role is seen by many workers for a feedback arising in the interconversion of boreal forest and tundra vegetation that accompanies climate change in high latitudes. Gallée et al. (1992) employed this in their climate modelling. They pointed out that snow rarely covers tree tops, so that forest cover lowers the surface albedo in comparison with values that can be achieved in the case of snow-covered tundra. In another modelling exercise, Gallimore and Kutzbach (1996) further emphasised the potential importance of this feedback. In a study employing the NCAR GCM, they showed that the increased surface albedo of expanding tundra was sufficient to permit glacial onset over northeastern Canada, given the impetus of the 8% decrease in summer insolation there at 115 ka BP.

The gas composition of the atmosphere provides a different mechanism of feedback (Pisias and Shackleton 1984). Cooling of the oceans increases the solubility of CO_2 from the atmosphere. If this gas is lost into solution, the atmospheric greenhouse effect is diminished and further cooling may result. But the biology of the oceans is vital here, since calcareous organisms, whose abundance depends upon patterns of ocean circulation and nutrient supply, may also sequester varying amounts of CO_2. Thus biological processes must also be evaluated in order to understand the planetary response to Milankovitch forcing. Further complications ensue, because the biological productivity of the oceans depends upon factors like nutrient supply from river runoff, and the salinity of the ocean surface, which can be altered by changing volumes of fresh water draining from the land. Nutrients derived from dust blown from the continents could also foster biological activity in the oceans, and

thus the sequestering of CO_2 from the atmosphere. This has been demonstrated in large-scale experiments in which an area of the Pacific Ocean was fertilised with iron. Major phytoplankton growth resulted (Behrenfeld et al. 1996). Feedback effects may arise in various ways through the marine biological productivity effect. For example, global cooling might cause extension of unvegetated soil surfaces, which might yield more dust. If this fostered biological productivity in oceans downwind, CO_2 loss could cause further cooling, and an increase in aridity, in a positive feedback (Cooper et al. 1996). Other feedbacks may be created by the varying rate at which CO_2 is consumed in the weathering of rock (Munhoven and François 1996). The increasing land area exposed during a cooling phase has the potential to increase the volume of rock material being attacked by chemical weathering involving carbonic acid derived from atmospheric processes, so enhancing the diminution of the greenhouse effect in a positive feedback. The reverse process would be active during a period of rising sea-levels.

Using a model that was successful in reproducing the timing and amplitude of climate changes over the past 200 ka or so, Berger et al. (1993b) explored the relative importance of the various feedbacks incorporated in the model. This was done by excluding individual feedback effects and noting the consequences for the modelled climate variation. The estimates derived from this work suggest that about 67% of the cooling at the last glacial maximum arose from astronomical forcing and its associated albedo feedback, while the remaining 33% was associated with diminished atmospheric CO_2 levels. Within either of these feedbacks, about 40% of the effect was found to be attributable to the further feedback effects of changing atmospheric water vapour content. It is also clear from this work that phase differences exist between the insolation and CO_2 forcings, providing scope for the climatic response to depart in phase as well as amplitude from that expected from the insolation forcing alone. Model experiments addressing this point led Berger et al. (1993b) to conclude that the phase of ice volume changes was more strongly linked to the phase of the insolation change than to that of the CO_2 forcing.

In the case of planetary warming, other effects arise. Rising sea-level might inundate vegetation growing on the continental shelves, whose decay might then release greenhouse gases to the atmosphere, promoting yet more warming. These effects are quite difficult to pin down, since they depend upon a detailed knowledge of what vegetation grew on the continental shelves, what its biomass was, and whether decay products were buried in accumulating sediments or were released to the atmosphere.

Warming oceans may release CO_2 also, in a positive feedback related to planetary warming triggered by the Milankovitch forcing. Physical feedbacks may also operate, some of which were mentioned in Chapter 4. For example, rising sea-level might act to destabilise ice shelves grounded below sea-level in the Antarctic, giving rise to catastrophic collapse of the remainder of the ice sheets that were buttressed by the ice shelves. Albedo loss might then promote further warming. These feedbacks, too, are difficult to evaluate for past periods, depending as they do on patterns of ice shelf anchorage, ice thickness, and land-ice stability, as well as on the rate and magnitude of any sea-level rise. Once again, sophisticated models offer the best chance of understanding how these processes operate and what effects they have on global climate. Some further discussion of these feedback processes can be found in DeBlonde and Peltier (1993) and in Brubaker and Entekhabi (1996).

FREQUENCY RESPONSE AND PHASE SHIFTS IN THE CLIMATIC RESPONSE TO INSOLATION FORCING

In much of the stratigraphic data referred to earlier, the evidence for the Milankovitch mechanism was derived from spectral analysis, and thus lies in the frequency domain mentioned at the beginning of this chapter. This leads us to consider aspects of time-domain checking. Here we will consider issues such as the phase shifts which may alter the timing of environmental changes to either precede or follow the putative insolation trigger. We will examine the magnitudes of some of the significant phase shifts that have been discovered in palaeoenvironmental data.

We have already noted some reasons why the response of the climate system, including the cryosphere, cannot be expected to be in phase with the Milankovitch solar forcing. Simple time-lags as oceans warm or cool, or ice sheets grow or decay, will unavoidably be involved. However, there are more complex issues than simple time-lags. A key issue is why, at least in the last 0.5–1 Ma, the palaeoclimate record has been dominated by fluctuations with the 100 ka periodicity of the orbital eccentricity, which causes a much smaller insolation change than does obliquity or precession. Indeed, the change in the dominant period of climatic instability has been highlighted by Raymo (1992), who referred to the earlier '41 ka world' of the early Quaternary which was supplanted by the '100 ka world' of the last 700 ka or so. Given that the relative importance of these periodicities in the Earth's orbit only changes in subtle ways,

especially over the short Quaternary time-scales, it is tempting to infer that the sizeable change in frequency response seen in the climate system reflects *internal* adjustments. What factors could be involved in such a change in internal system behaviour? Slow growth in the sizes of the Quaternary ice sheets, related to global cooling, might be involved (Shackleton 1993). Larger ice sheets display longer time constants in growth and decay. Pisias and Moore (1981) have suggested that glacial erosion, whose scouring and remodelling effects on the Earth's surface build up as successive glacial periods take place, might be involved. Massive glacial erosion can strip the land-surface, creating basins such as Hudson Bay in Canada, and may also modify links between various seaways (e.g. the Atlantic and Arctic Oceans). Through the course of Quaternary glaciation, lowered bedrock elevations may have allowed the growth of ice sheets buttressed by sea ice, with the associated instability that can be caused by sea-level change. We can envisage that glacial erosion itself may in due course be capable of bringing about an end to conditions that could support the onset of further glacials. It also seems possible, given the importance of topography for atmospheric processes, and oceanic circulation patterns for heat transport as well as biological productivity and the carbonate system, that progressive glacial erosion might also change some important rhythms in the climate system as seen in the natural archives.

Harmonic and tone effects in the climate system

Various researchers have drawn attention to the fact that interactions among various frequencies present in the Milankovitch forcing may be significant in the production of harmonic terms of both lower and higher frequency.

Indeed, as shown by Stothers (1987) and by Berger and Loutre (1994a), just such combination tones arise in the production of various of the orbital periodicities. Thus, the periods of variation in eccentricity arise from combination tones developed from terms expressing the longitude of perihelion, $e \sin \omega$ and $e \cos \omega$, in the equation

$$e = \sqrt{(e \sin \omega)^2 + (e \cos \omega)^2} \qquad (5.7)$$

The series expansion of the relationship generates a series of periods, including 308.04 ka, 176.4 ka, 72.5 ka and 75.3 ka. The combination tones P_{comb} may be calculated from relations of the form

$$\frac{1}{P_{comb}} = \frac{1}{P_1} \pm \frac{1}{P_2} \qquad (5.8)$$

where P_1 and P_2 are the primary periods. Using the first two periods (308.04 ka and 176.4 ka) and taking a difference tone yields a period of

$$\frac{1}{308.4} - \frac{1}{176.4} = 412 \, \text{ka}$$

which is one of the major eccentricity periods that has been detected in marine proxy records (e.g. Moore *et al.* 1982).

Using these same principles, Wigley pointed out that a difference tone created between the two major precessional periods, 23.1 and 18.8 ka, has a period of about 100 ka, so that a sufficiently large response in the climate system to this combination tone might account for the 100 ka period seen in the natural archives. The calculation is

$$\frac{1}{P_{comb}} = \frac{1}{18.8} - \frac{1}{23.1}$$

$$P_{comb} \approx 101 \, \text{ka}$$

Using the same principles, Berger (1991) showed that combinations of the 41, 23 and 19 ka orbital periods can generate combination tones with short periods such as 10.3, 4.7 and 2.5 ka.

The modulation of the precession effect by orbital eccentricity is in some senses analogous to the modulation of a carrier frequency by a lower frequency signal in amplitude modulation (as employed in radio transmission). In this context the precession signal is the carrier and the eccentricity is expressed as the signal.

Two kinds of effects may now be considered to arise. The first involves the generation of harmonic frequencies in the modulation system, and the second involves the effective demodulation, with enhancement of the modulating signal (the eccentricity effect) by the terrestrial climate system.

Wigley (1976) explored the factors likely to be involved in generating climate response at the 100 ka period. He showed that a non-linear response of the climate system is required. This means that the sensitivity of the climate system must vary as a function of the value of the forcing signal. Various mechanisms have been envisaged for this.

Differential sensitivity of ice sheets to warming and cooling can yield a non-linear response. Thus, ice sheets may develop slowly, owing to such factors as increasing dryness of the air above them as cold intensifies. The coldness of an ice-sheet surface reduces the moisture-holding capacity of the air, so restricting the delivery of snow to promote further ice sheet development. The thickening of the ice sheet also acts in this way by increasing the elevation of the surface that must be reached by moist air (recall that the large ice sheets are up to several kilometres

in height). This has been termed the 'elevation desert' effect.

However, when insolation changes trigger ice sheet recession, quite different processes come into play. A widely cited one is the rising sea-level produced by some initial ice melting. This could act to destabilise coastal ice margins, since ice floats, and the contact between water and ice may deliver sensible heat to promote collapse of the ice margin. In response to such effects, ice sheet decay may be much more rapid than the growth phase. In other words, the phase relation between Milankovitch cooling and ice sheet growth may be quite different from that between Milankovitch warming and ice sheet decay.

Wigley (1976) also considered it likely that significant non-linearity may exist in the albedo feedback processes related to ice sheet growth and decay. Thus, the aggregate behaviour of the climate system may be the result of multiple non-linear responses acting in concert.

An additional important point also made by Wigley (1976) is that various components of the climate system may display individual response functions with differing degrees of non-linearity. Consequently, as different responses arise in the system and interact, new interaction frequencies may also arise. Once more, therefore, it appears necessary to expect more in the palaeoclimatic record than a simple synchronicity of warm periods with insolation maxima and cool periods with insolation minima.

Tectonic processes including crustal isostasy

Pollard (1982), Peltier and Hyde (1984), Hyde and Peltier (1985), Birchfield and Grumbine (1985), Birchfield and Ghil (1993) and others have developed models capable of simulating several aspects of the Quaternary glacial–deglacial chronology. Within a reasonably broad range of conditions, the model of Hyde and Peltier yielded glacial–deglacial cycles with periods around 100 ka even when the radiation forcing used to control it had shorter periods (say 40 ka, similar to the obliquity period). The response was governed by the slow, oscillatory growth (incorporating the effects of assumed meteorological feedback) but resulted in rapid decay of large northern hemisphere ice sheets. The rapid decay in the model is related to crustal subsidence that slowly develops under the burden of the ice sheet. When solar forcing causes ice sheet decay to begin, this of course happens earliest and fastest along the southern margin where the ice sheet is ablating. In the model, the ice retreats faster than the isostatically depressed crust can rise. Thus, the retreating ice sheet margin leaves a crustal depression of some hundreds of metres in depth. Rapid ice flow then takes place southwards, into this depression. Catastrophic disintegration of the ice sheet follows soon after, and the depression finally reaches equilibrium. Isostatic uplift during ice sheet collapse also tends to maintain the elevation of the ice surface, helping to sustain the 'elevation desert' effect in which precipitation on a high ice surface is reduced because of the limited moisture-holding capacity of the cold air, so starving the ice sheet. The rapid 'terminations' noted in the Quaternary record might thus arise from an isostatic depression of the crust, a mechanism internal to the Earth's response. Interestingly, in experimenting with various periodicities of the model's forcing function, Hyde and Peltier (1985) observed that maximum predicted ice sheet volumes occurred with long periods such as the 100 ka eccentricity period. Thus, although of small amplitude, it may be that the long eccentricity period may be important in the growth of major ice sheets, especially if the forcing is in phase with the isostatic response. Here we see suggestions of why glacial periods have not always been associated with Milankovitch cooling episodes; it may be necessary to have land of the correct elevation, and lying on continental crust of the right dimensions. Some continental fragments might be too small or too large to have a tectonic response that can phase-lock to the 100 ka eccentricity period.

Differing phase response at different latitudes

Because of the Earth's obliquity, the precessional process can generate warm summers and cool winters, or the reverse, at high latitudes in either hemisphere. Attention has always been focused on the northern latitudes, because in contrast to the ocean-dominated southern hemisphere, land is present there to support growing snow and ice cover.

But the obliquity effect is minimal at equatorial latitudes, which always remain a belt of strong insolation. During the year, the sun passes overhead at the equator and adjacent latitudes twice, moving toward the solstice position in first one hemisphere and then the other. Because of this, whatever the configuration of the orbit at any particular epoch, the maximum temperature in the low latitudes will be experienced in the season whose equinox happens nearest to perihelion. Consequently, the warmest part of the year does not necessarily correspond to a single season, as with summer in higher latitudes. Therefore, since the obliquity effect is small and the timing of the warmest season is not fixed in the year, equatorial latitudes do not experience a cool mini-

mum temperature effect as the higher latitudes do when they have winter at aphelion. Rather, the equatorial latitudes experience warming when one or the other equinox happens near perihelion.

This contrast in latitudinal sensitivity is significant. It means that the timing of summer with perihelion or winter with perihelion is most significant for creating warm excursions at high latitudes, while at equatorial latitudes, it is *equinox* alignment with perihelion that is important. These two effects are 90° (or several ka) out of phase with each other.

The climatic consequences of these differences were pursued in the important modelling experiments by Short *et al.* (1991) cited previously. Using an energy balance model, they showed that the absence of strong insolation minima at equatorial latitudes significantly alters the response to Milankovitch forcing that is exhibited there. For a site at 0° N 20° E over Africa, the frequency response of modelled annual maximum temperature over the last 800 ka shows a strong precessional signal (19 ka and 23 ka). However, the 400 ka and 100 ka periodicities of the eccentricity are also strongly evident. The obliquity signal is the smallest of the orbital terms. Over equatorial oceans, the eccentricity effect is muted, because of the enormous heat capacity of water, and Short *et al.* (1991) determined small temperature effects there.

Harmonic effects arise here also. Because the sun is overhead at the equator at an equinox twice during each full 19–23 ka precessional cycle, the maximum temperature so induced occurs about every 11.5 or 9.5 ka. These higher frequencies are *precessional harmonics*. According to Short *et al.* (1991), some palaeoenvironmental records appear to show these harmonic effects.

The model used by Short *et al.* (1991) indicates that the 100 ka frequency effect in equatorial latitudes is associated with a 1.6–1.8°C range from warmest maximum temperatures to coolest maximums. Thus, the response of the Earth to the very small eccentricity forcing is a potentially significant effect on low-latitudes temperatures. The effects could in principle be transmitted from these latitudes to other parts of the globe via effects on the amount of evaporation feeding polewards water vapour transport, and on resulting water vapour feedbacks in the planetary climate system. In this context it is necessary to remember that the major energy source driving the planetary Hadley circulation (and indeed the zonal Walker circulations) is latent heat released in the troposphere. Atmospheric warming by latent heat release exceeds that from direct radiation. Thus, the evaporation of water exerts a vital control on the planetary climate system, and changes in the moisture flux have great potential significance.

Phase lag effects arising from differential thermal response times of land and ocean

Effects arising from the differing thermal response of land and ocean to insolation forcing have been identified as possible sources of significant lag effects in climate response. Because of its higher heat capacity, ocean water displays a smaller temperature rise for a given change in solar radiation than does the surface of the land. This is true despite the lower albedo typical of the ocean surface. In addition, the resulting temperature rise develops more slowly over the oceans where sensible heat is mixed through a much greater depth than happens in soil and rock, which have no mixing and which are poor thermal conductors. Thus, in response to a given increase in insolation, land warms more, and more quickly, than ocean. Thus, the land may already be experiencing cooling, as insolation falls, while ocean temperatures continue to rise or at least remain relatively steady.

These effects are potentially significant in a range of climatic processes. Kutzbach (1981) and Kutzbach and Street-Perrott (1985), for example, consider the effects likely to be significant controls on the levels of tropical lakes. They examined the enhanced seasonality of insolation in the period 15–6 ka BP, when northern hemisphere summers happened nearer perihelion and when the obliquity was about 1° larger than now. At 11–10 ka BP, when the forcing was greatest, the change was about $+30\,W/m^2$ (+7%) in July, compared with today's value, and about $-25\,W/m^2$ (−7%) in January. Given that these are only seasonal changes, not mean annual values, the rapidity of possible thermal response in the climate system is of key importance. Kutzbach and Street-Perrott (1985) anticipated a large thermal response over land (up to ±5°C) but a heavily damped response over the ocean, where the large thermal inertia would limit the swing to perhaps ±0.5°C. This differential warming (and cooling) would then reduce atmospheric pressure over large land areas in summer, and increase them there in winter. Consequently, the forcing should have had the effect of increasing the strength of the seasonal monsoon circulation, with consequent effects on the water balance over land. As Kutzbach and Street-Perrott (1985) observed, the opposite effect, a weakening of seasonality, would have arisen in the southern hemisphere where the solar forcing was reversed, perihelion being in winter at that period.

Effects arising from differential warming of land and oceans were again examined in the context of tropical climates by Short and Mengel (1986) using an energy balance model. These researchers examined conditions during the last 30 ka, a period that covers a full precessional cycle. In this period,

solstice–perihelion alignment occurred at about 8.5 ka BP, while in contrast, modelled temperatures over the tropics did not peak until 8–9 ka BP, i.e. until several ka after the perihelion. At the time of the modelled temperature peak, the Earth was reaching perihelion about six weeks after the summer solstice. Thus, at a point at 15° N, the sun would have passed overhead twice, once about six weeks before perihelion was reached and again about six weeks after perihelion. The energy balance model results thus suggested that warmest summers are generated when perihelion is reached as the sun is making its second overhead passage. We can infer from this that the initial overhead passage of the sun begins the warming of the oceans, and 12 weeks later, the second overhead passage occurs with slightly greater heating, since the Earth was then at perihelion. This provides a further boost to temperature rise.

A conclusion of some importance arises from this work: that in order to maximise the thermal response of the climate system, the phases of the forcing function need to be set appropriately in relation to the properties of the climate system.

The model results of Short and Mengel (1986) suggest that when this is achieved, the warmest summers over Eurasia can be up to 12°C warmer than the coolest summers. Over NE Africa, the value is about 6°C. The time-lag between perihelion and the maximum temperature rise reaches a maximum, according to the model, of about 3.5 ka, over tropical Africa. The modelled lags over most of Eurasia and North America were < 1 ka. Here again, we see the complex nature of the climatic response, with phase lags that vary with the geography of the Earth's surface.

Short and Mengel (1986) observed that the response of the southern hemisphere might well be phase-locked to that of the northern, in view of the larger extent of tropical Africa north of the equator than south of it. Thus, the strengthened monsoon of the northern hemisphere may dominate over the weakened circulation of the southern hemisphere. It is true at the present day that the monsoon systems of the northern hemisphere are stronger than those of the southern. This is aided to some degree by the presence of the high land of the Tibetan Plateau, which strengthens the advection of warm air from the oceans in the flow across the Indian subcontinent.

The natural archives provide ample evidence of the response of the water level in African and other lakes to the orbital forcing of monsoons (e.g. see Kutzbach and Street-Perrott 1985). The seemingly ever-present feedback effects appear to have been important in further modulating the environmental outcome, however. Kutzbach *et al.* (1996) have analysed these effects by using one of the NCAR (National Center for Atmospheric Research) climate models. They showed that increased insolation in the mid-Holocene could have increased rain amounts by 12% between 15° N and 22° N, but by incorporating biosphere and soil responses, larger changes could be generated. For example, allowing the extra rain to foster grassland in place of desert, and allowing more loamy soils to form, increased the rainfall further by amounts of 6% and 10% respectively. This brings the aggregate effect arising from the orbital insolation forcing to a precipitation increase of 28%. This is a very substantial change. There have been changes in the phase relations of the monsoon systems and the global ice volume which suggest that there is a non-stationary component in this system (e.g. Clemens *et al.* 1996). This topic lies beyond the scope of this chapter.

Phase lead effects arising from the transient response of terrestrial environments

All of the phase shifts so far discussed have involved a phase lag, in which the environmental response reaches its peak some time after the maximum change in the insolation forcing. Instances of an apparent global response to Milankovitch forcing which *precede* or show a *phase lead* are also known, although these at first seem less likely than the phase lag effects just discussed.

An interesting example is provided by the work of Pokras and Mix (1987), who analysed the abundance of a fresh-water diatom in deep-sea sediments of the equatorial Atlantic Ocean. The diatoms were inferred to be delivered to the ocean by aeolian processes following deflation of dry lake beds in tropical Africa. Thus, the relative abundance of the diatoms in the marine sedimentary record may serve as a record of environmental changes affecting tropical lakes. Pokras and Mix performed spectral analyses on the record of diatom abundance, and found strong evidence for the presence of a 23 ka periodicity, corresponding to the precessional cycle.

However, interestingly, they found that the diatom abundance changes *led* the precessional index by about 2–3 ka. This can be explained in terms of a limitation on the supply of diatoms at the lake sources. Either the diatomaceous sediments were depleted rapidly, or supply was restricted by the development of some kind of resistant crust on the lake floor. The model envisaged by Pokras and Mix (1987) is shown in Fig. 5.12. This diagram indicates how exhaustion of the diatom supply leads to a peak in the palaeoenvironmental indicator (diatom abundance) prior to the peak in insolation forcing.

In view of the precessional signal present in the diatom record, Pokras and Mix concluded that mon-

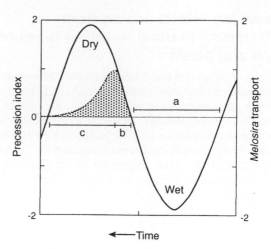

Fig. 5.12 Phase lead in palaeoenvironmental data. The shaded region suggests the rate of deflation of fresh-water diatoms into marine sediments; supply limitation results in peak flux preceding the peak in the precessional astronomical forcing (solid line). See text for details. (After Pokras and Mix 1987)

soonal climates in tropical Africa are modulated in intensity by the changes in low-latitude insolation produced by orbital effects. They further concluded that the apparent lack of any 41 ka obliquity signal in their record suggests that insolation forcing at higher latitudes has little significance for the changes in tropical climates over Africa.

The '100 ka problem'

Various hypotheses have been put forward to account for the unexpectedly strong presence of the 100 ka periodicity in late Quaternary climate records: the '100 ka problem' referred to earlier. Let us consider this issue briefly.

In the spectral analyses of ^{18}O records from marine cores, such as that of Hays *et al.* (1976), about half the power resides in the frequencies corresponding to the 100 ka periodicity of orbital eccentricity. Thus, the major temperature changes associated with glaciation in the last 700 ka or so have occurred at this period, and only much smaller temperature changes can be ascribed to the larger insolation forcing occurring at higher frequencies.

Many workers have consequently been attracted to the idea that the response of the climate system to insolation forcing displays non-linearity at the lower frequencies. That is, the presumption is made that the gain in the feedback processes is not fixed (Imbrie 1994 discusses gain variation), and may be able to increase the response from the small 100 ka forcing. The problem that immediately arises is just how this can be done without the environmental response being dominated not by the 100 ka period, but by the stronger peak in the eccentricity variation at 412 ka. This dilemma (which has in turn been called the '400 ka problem') suggests the alternative view that the dominance of the 100 ka period reflects some internal control. The role of the slow growth of the major ice sheets, and associated crustal depression, has already been mentioned.

A final problem with the idea that the 100 ka periodicity is a response to insolation forcing is that, at least in the marine oxygen isotope record, the phase lag between the forcing and the response in the proxy variable is too small to be consistent with the anticipated 10–15 ka lag as ice sheets grow (Imbrie *et al.* 1993b).

Many models attempting to resolve these issues have been created, and have been categorised and reviewed by Imbrie *et al.* (1993). Some involve systems that oscillate freely, while others invoke various forms of rectification, in which power is derived from the precession envelope. Inasmuch as these problems remain unresolved, it will serve our purpose here to highlight one or two illuminating aspects.

The existence of self-sustained oscillations within the climate system can be explained simply, following Ghil (1991). The ice albedo feedback already described provides a basis for one kind of oscillation. On the basis of this feedback, the temperature T goes down as the albedo A goes up. Thus, writing for the first derivative of $\dot{T}$ in time, and using the symbol ~ to mean 'varies directly with' we have

$$\dot{T} \sim -A \qquad (5.9)$$

The albedo must increase as the ice volume V increases, so that

$$A \sim V \qquad (5.10)$$

Eliminating the albedo term between these two relations yields

$$\dot{T} \sim -V \qquad (5.11)$$

For a growing ice sheet, the net precipitation P must govern volume growth so that

$$\dot{V} \sim P \qquad (5.12)$$

and since P will be greater in warmer periods when evaporation supplies vapour to the atmosphere,

$$P \sim T \qquad (5.13)$$

Eliminating P from these last two equations we can write

$$\dot{V} \sim T \qquad (5.14)$$

Equations (5.11) and (5.14) clearly provide the basis for a system that could oscillate in a self-sustaining manner. Other models of this kind can be developed (see Ghil 1981, 1989, 1991).

Models that involve the systematic isostatic sinking of a growing ice sheet can be exemplified by the work of Oerlemans (1980). His calculations show that, once ice sheet growth is triggered by an insolation change, its growth can be self-sustaining because of the kinds of feedback effects described earlier. However, over about 100 ka, the burden causes crustal depression which lowers the elevation of the ice sheet surface and so triggers rapid decay of the ice. In a model like this, the system is phase-locked to the insolation forcing, but the 100 ka periodicity is derived from internal constants to do with the speed of isostatic adjustment. A similar situation was modelled by Le Treut and Ghil (1983), who developed a model that behaved as a self-sustaining oscillator. This generated a form of red-noise spectrum like that reported from marine ^{18}O records, as well as phase-locking to the frequencies of the insolation forcing. Periods near 100 ka arose in the model as combination tones from higher frequencies in the precessional band.

Other components of the cryosphere have been invoked to control long period responses like this. In particular, Saltzman and co-workers have drawn attention to the potential significance of interactions in the system involving the ice sheets, sea ice, and oceanic temperatures (Saltzman 1977; Saltzman and Moritz 1980; Saltzman *et al.* 1982, 1984; Saltzman and Sutera 1984). These ideas were developed partly from the work of Newell (1974). In essence, the suggestion is that increasingly extensive sea ice during glacial development progressively insulates the northern oceans against heat loss. On the other hand, increasing ocean temperatures seem likely to be linked to diminishing extent of sea ice. These two tendencies form the basis for an unstable system capable of free, damped oscillations, analogous to that of ice sheet extent and temperature outlined previously. (Warm oceans have the potential to provide a supply of moisture to nourish ice sheet growth, so that early in glacial onset, there may be an 'interglacial' ocean adjacent to a 'glacial' landmass – see Ruddiman and McIntyre 1979.) Long time-constants arise in this system from the thermal inertia of the oceans, but the outcome of the models depends significantly on how the thermohaline circulation system behaves during glacial onset and deglaciation (see for example Broecker 1992; Dwyer *et al.* 1995). These ideas were made slightly more general by Saltzman and Sutera (1987), who explored the behaviour of a model with three parameters, namely ocean water mass, ice mass, and atmospheric CO_2 abundance. They showed that it is possible for interactions among these parameters to generate a temporal record with many of the characteristics of the Quaternary δ^{18}O series, including a major change in system behaviour at about 900 ka BP, when a 100 ka period emerges. Having shown this, Saltzman and Sutera concluded that the Milankovitch mechanisms are not those that drive the 100 ka cycles, although accepting that they may exert an influence on the precise form of the cycles and upon their phase. Finally, the model of Saltzman and Sutera (1987) requires some elaboration, since it fails to produce the classical 'saw-tooth' cycle shape, with rapid deglaciations. These seem critically important, since the $\delta^{18}O$ records show steady ice growth right up to the point of an evident mode switch, where deglaciation sets in rapidly. There is no sign of an exponential approach to a new equilibrium, for example, prior to this mode switch. Therefore, the issue of what sets the amplitude of the ice volume variations, and controls the rapid mode switch, must be answered by any model designed to provide a full explanation of the 100 ka cycles.

An alternative explanation, strongly linked to orbital processes, has been put forward by Liu (1992). In an analysis of the time variation of obliquity, Liu showed that the cycle is indeed irregular, and sometimes asymmetric, with the axis remaining at higher tilt for less time than at lower tilt. The difference was approximately 3.5 ka. Furthermore, it was shown that the variation in frequency of the obliquity itself has a period of 100 ka. Liu proposed that the onset of deglaciation related to times of unusually fast obliquity variation, while the initiation or prolonging of glacial phases related to periods of slow variation of the obliquity period.

In another study of the mid-Quaternary onset of the 100 ka periodicity in the proxy records, Watts and Hayder (1983) employed a model resembling that of Weertman (1976). They showed that the growth of an ice sheet depends on the duration of the climatic cooling, represented numerically as the time for which snow accumulation persists over northern land areas, rather than retreating into the cooler polar seas. Once the cooling lasts past a critical minimum duration, the glaciation persists. In this model too, no mechanism existed to terminate a glacial phase, although this could presumably be accomplished by invoking crustal subsidence and consequent lowering of the ice surface. Interestingly, variation in the climatic conditions appeared to be able to sustain higher-frequency glacial cycles, resembling those of the Quaternary prior to the onset of the 100 ka periodicity. Watts and Hayder (1983) thus speculated that the 100 ka period may have become dominant through the progressive cooling of the global climate. At the change-over point, the output of their model is very sensitive, and very small changes in the environmental conditions are capable of generating a rapid change from high to low frequency glacial cycles.

This brief account of the '100 ka problem' leads one to agree with Saltzman (1983) that '...we are dealing with an extremely complex, forced, dissipative system that undoubtedly contains a rich assortment of linear and nonlinear, positive and negative, feedbacks.'

CONCLUSIONS

In the light of the results presented in this chapter, it is hard to imagine attempting to understand Quaternary environmental change without reference to the Milankovitch processes. A great deal of research effort has been expended in the calculation of insolation values for various points or regions on the Earth's surface, and using various tallying schemes for the insolation (caloric half-year, monthly, daily). As emphasised by Berger *et al.* (1993a), a record from the natural archives that displays a precessional or obliquity period can very possibly be matched to the results obtained from one of these forms of tallying. A task that awaits full completion is the evaluation of how the various measures of insolation actually have effect on and through the climate system. It is here that better models of the planetary environment will have much to contribute. However, from the data reviewed here we can conclude that the Milankovitch mechanism can provide the framework to account for much of the Quaternary climatic instability, in concert with the rich set of feedbacks provided by terrestrial processes. We would be incautious, however, to discard other ideas (e.g. see review in Berger 1981b). For instance, although they are assumed to be relatively minor in conventional Milankovitch analyses, variations of solar output certainly occur, and might be the explanation for brief excursions like the Little Ice Age. Evidence for this possibility has been widely discussed (Wigley and Raper 1990; Magny 1993; Schönwiese *et al.* 1994; Hanna 1996; Karlén and Kuylenstierna 1996). The Little Ice Age corresponds in time to the Maunder minimum in solar output, when solar luminosity may have been 0.25% less than now (Board on Global Change 1994). Also, volcanic eruptions were relatively frequent in the Quaternary, and eruptions of the right kind throw dust particles into the atmosphere. These may cause global cooling because they scatter radiation back to space. Rampino and Self (1992) noted that the eruption of Toba in Sumatra at 73.5 ka BP happened at a time of summer insolation minimum, and may have been involved in the strengthening of global cooling. They estimated a subsequent annual hemispheric lowering of surface temperature in the range 3–5°C (Ramaswamy 1992). Similarly, early Holocene climatic cooling produced by volcanic ash is proposed by Zielinski *et al.* (1994). The effects of an eruption can be prolonged by the raised surface albedo created by volcanic ash. This was proposed for the pale rhyolitic ash produced by explosive eruptions in western North America by Bray (1979), who estimated that in the case of these Quaternary deposits, the hemispheric temperature might have been depressed by 0.07–0.41°C for decades. Such events might provide explanations for short-lived climatic swings such as that of the Younger Dryas cool phase, especially when coincident with a time of lower solar luminosity. Plumes of smoke arising from major fires also have the potential to attenuate the solar beam reaching the Earth's surface. Christopher *et al.* (1996) recently estimated that a heavy loading of aerosol particles can cause a decline in S of $36 \, W/m^2$, a change as large as many of those related to the Milankovitch mechanism listed earlier in this chapter. Aerosol dusts, derived from sites like the great deserts and their margins, may have acted in the same way (Tegen *et al.* 1996). Harvey (1988) estimated that in total, marine and terrestrial aerosols could have accounted for 2–3°C of global cooling at 18 ka BP. The climatic effects of the higher aerosol loading are not uniform, and Harvey suggests a net tendency towards warming in high latitudes but towards cooling elsewhere. Overpeck *et al.* (1996) have suggested that warming may indeed be the critical effect, with peaks in dust loading implicated in the rapid glacial terminations. Finally, we have seen that the climate–cryosphere–solid Earth system involves many intricate responses and interlinked feedbacks of its own. Some of these involve significant time-lags and phase shifts. Overall, then, it is unsurprising that the palaeoenvironmental record is complex and that unravelling of cause and effect remains a challenge for Quaternarists. The Milankovitch mechanisms, since they are deterministic, are readily incorporated into global climate models. The effects of volcanic eruptions and events of this kind are much harder to establish, and many models that now incorporate Milankovitch mechanisms still ignore these other influences on the global environment. As progress in the modelling of Quaternary climates proceeds, it seems certain that some mixture of deterministic and stochastic–chaotic processes will have to be developed (e.g. see Kominz and Pisias 1979; Hasselman 1981). Many such background processes within the climate system are after all required to account for the overall 'red-noise' profile of the Quaternary climate record reported earlier in this chapter from the work of Hays *et al.* (1976a).

The '100 ka problem', which space has prohibited us from discussing at any length, provides an instance of the many other areas of uncertainty that remain. The direct insolation forcing at this frequency appears, as we have seen, too small to account for the dominance of the 100 ka period in the last 700 ka or so. Many attempts have been made to identify non-linearities in the global response that could account for the strength of the 100 ka signal. This approach thus attempts to derive more power from the Milankovitch mechanisms. But an alternative that has recently been highlighted is that the 100 ka period may indeed reflect an orbital process, but not one operating directly on the pattern and timing of insolation as the classical Milankovitch mechanisms do. Muller and MacDonald (1995, 1997a, b; see also Kerr 1997) drew attention to a previously overlooked orbital process that has a 100 ka period: the variation of the orbital inclination. This refers to a periodic change in the tilt of the Earth's orbital plane (the ecliptic). Muller and MacDonald speculated that this parameter might affect the planetary climate through some effect like a control on the abundance of extraterrestrial dust through which the Earth's orbit carries it. Subsequently, Farley and Patterson (1995) reported, that using [3]He as an indicator of the presence of such dust, a 100 ka periodicity could indeed be detected in sediments from the Atlantic Ocean. These authors stress that it is as yet unclear whether the 100 ka periodicity of the global climate is a consequence of changing dust abundance, or whether the dust abundance in the sediments reflects the 100 ka period of some climatic process. They also observe that the two phenomena might be totally unrelated! Clearly, there is still much to be learned in this area. It seems quite likely that short-term fluctuations are driven by the obliquity and precession changes in the Earth's orbit through the classical Milankovitch mechanisms, while the 100 ka variability that dominated the upper Quaternary was driven by the newly proposed attenuation process, or by one or more of the endogenous processes outlined earlier. The long-term crustal processes, invoked by many to explain the 100 ka periodicity, undoubtedly do occur, and so we find here another circumstance where careful weighing of data, and detailed modelling experiments, will be required in order to sift out the truth from the growing multitude of plausible hypotheses that have been set out.

It is also well to recall, in reviewing what has been learned from this chapter, the great complexity and diversity of the global environment and its processes. Researchers have paid much attention to temperatures, patterns of monsoon rainfall, ocean circulations, and so on. Less well known are the changes that have taken place in local and regional water balances, weathering rates, and related terrestrial processes. Terrestrial processes are not only of great importance in human endeavour; they also link to the oceanic realm, through the agency of fresh water, sediment and nutrient fluxes. At present, attempts to simulate the operation of the hydrological cycle under conditions of altered Milankovitch orbital parameters are only partially successful (e.g. Coe

1995). There is also a need for a more wide-ranging analysis of the possible effects of the Milankovitch forcings. An illustration of this was provided by McIntyre and Molfino (1996), who analysed a possible effect of variation in wind strength, related to the precessional cycle, on a number of important oceanographic processes. Effects on winds and storms are significant, since they are closely involved in the polewards transport of heat, momentum and water vapour (Saltzman and Vernekar 1975).

Another clear conclusion that can be drawn from what has been said is that the global climate appears incapable of achieving equilibrium. The time constants involved in the response of the great ice sheets prohibit this. As Held (1982) pointed out, this conclusion strikes one forcefully when it is realised that the present values of eccentricity and obliquity are quite similar to those that applied at the last glacial maximum. The difference in the two climatic regimes reflects the climatic disequilibrium. (The similar orbital values arise because the precessional cycle is around 20 ka – the period since LGM – and the present obliquity is near the average value, which is reached about every 20 ka through the 40 ka obliquity cycle.)

With an eye to the future, we can observe that today's orbital configuration is quite like that which typically leads into a glacial period; the Earth is near perihelion during the northern hemisphere winter, and eccentricity and obliquity are moderately large. Whether glaciation follows depends upon many factors, not least human modification of planetary albedo, atmospheric greenhouse gas concentrations, and deliberate interference with the hydrological cycle. Human activity has the potential to overshadow the Milankovitch mechanism as a control on the future of the planetary environment, which may therefore be derailed before we have even come to a full understanding of it. Assuming that climatic changes occur in the manner of the past, orbital calculations allow estimation of likely future trends. On this basis, Berger *et al.* (1981) and Berger (1981a) foreshadowed a gradual cooling over the next 6 ka followed by descent into glacial cold between about 54 ka and 60 ka after present (AP). Even though the orbital changes that we have studied here take place only slowly, the time periods involved are certainly appropriate when considering issues such as the long-term integrity of radioactive waste storage sites that might be affected by changing lake or ocean levels, or by intensified monsoon rains (Berger 1991). Some of these issues are considered in Chapter 13.

The enormous complexity of the global environment still stands in the way of a final unravelling of the causes of Quaternary environmental instability, despite the view of Milankovitch (1941) that:

> ...the march of insolation in the remote past, taking into account the varying reflective power of the Earth, is absolutely sufficient to explain the full extent of even the greatest climatic events of the Quaternary, and to clearly show their causes.

But this challenging field consequently offers the opportunity for researchers today to continue the great task, the solid foundations of which he set out.

CHAPTER
6

QUATERNARY SEA-LEVEL CHANGES

I know not what I may appear to the world, but to myself I seem to have been only like a boy playing on the sea-shore, and diverting myself in now and then finding a smoother pebble or a prettier shell than ordinary, whilst the great ocean of truth lay all undiscovered before me.

D. Brewster
Memoirs of Newton, Vol. 2 (1855)

NATURE AND DESCRIPTION OF SEA-LEVEL CHANGES

In this chapter we will consider one of the few environmental parameters that might in principle be expected to show some uniformity in its Quaternary variation: sea-level. The oceans are interconnected, so that any fluctuation in sea-level must be transmitted throughout the oceans of the world and be felt everywhere. While this is indeed the case as a general rule, some rather complex effects influence the behaviour of sea-level and make global fluctuations less regular than we might expect. Because the water occupying the ocean basins can deform easily, the surface of the sea is a *potentiometric surface*, which always lies horizontally with respect to the local gravity field except where it is temporarily distorted by winds and currents. The local gravity field at any point consists principally of the field produced by the Earth itself, but modified by the particular water depth and by the properties of nearby rock masses or ice sheets. Consequently, the sea surface actually lies in a rather complex and peculiar configuration of highs (e.g. over the North Atlantic) and lows (e.g. over the equatorial Indian Ocean), the maximum elevation difference among which amounts to about 200 m. The three-dimensional form of this uneven surface is termed the *geoid*. It is very similar to an *ellipsoid*, which is the regular geometric figure often used to represent the shape of the solid Earth. The terrestrial ellipsoid is a sphere flattened by about 1/300, or about 40 km out-of-round. Global sea-level fluctuations are thus actually changes in the detailed form of the geoid, and the magnitude of the change experienced at any point depends partly on whether it lies relatively high or low on the geoid surface, as we shall see below. It is necessary to understand that the depth of the ocean at any place (and hence the height of sea-level) is the thickness of the water column lying between the ocean floor and the ocean surface (the geoid). A sea-level change could thus come about without any change in the actual amount of water present in the oceans. A change in the elevation of the ocean floor, brought about by tectonic processes, or a perturbation of the geoid caused by a change in the local distribution of mass (rock, ice or water), or both simultaneously, could do this (J.A. Clark 1980), and the bulk of the sea-level change thus produced could be relatively localised and not worldwide as might at first be expected.

The realisation that geoid perturbation could cause significant sea-level variation, and a quantitative analysis of the size of effects that are possible, is relatively recent. Much of the terminology employed in Quaternary sea-level studies comes from the older notion that the main cause of sea-level variation was a change in the amount of water contained within the ocean basins. Changing this volume, it was once thought, really would displace sea-level up or down by a more or less uniform amount worldwide (although local crustal movements would produce local variations from the norm).

Water added to the oceans by melting land ice or taken from them in phases of global cooling supplies a mechanism for such major notional *worldwide* sea-level fluctuation, which, referred to the land, locally produces either *transgression* or *regression*. When a sea-level change results from changing water volume in the ocean basins in this way, it has long been described as *eustatic*. The amount of notional eustatic change is generally calculated from the following relationship:

$$\text{Eustatic sea-level change} = \frac{\text{Change in ocean water volume}}{\text{Surface area of the oceans}} \quad (6.1)$$

Using the total excess ice volume at the last glacial maximum of about $50 \times 10^6 \, \text{km}^3$ (equivalent to $55.6 \times 10^6 \, \text{km}^3$ of water if the density of the ice was 0.9), and taking the area of the oceans to be $361 \times 10^6 \, \text{km}^2$, this simple relationship suggests that the eustatic sea-level change resulting from the last deglaciation should have been about

$$55.6 \times 10^6 \, \text{km}^3 / 361 \times 10^6 \, \text{km}^2 = \text{approx. } 154 \, \text{m}$$

This estimate could be refined by allowing for the changing area of the oceans as their depth varies (see Marsiat and Berger 1990), and by various other effects such as sagging of the ocean floor under the increased load, which would lessen the sea-level rise. It must therefore be expected to be somewhat different from field estimates of the actual glacial lowering of sea-level derived from suitable stable sites, such as the 118 m from Barbados (Bard *et al.* 1990a) or the <130 m for southeastern Australia (Ferland *et al.* 1995).

Even after making such corrections, we now understand that the amount of rise or fall would not be *exactly* equal everywhere because the changing distribution of ice and water mass alters the local gravity field simultaneously, so that the geoid readjusts. This means that in reality the glacial lowering of sea-level with respect to its present level really was by different amounts at different locations on the Earth's surface. Consequently, true eustatic changes, in which sea-levels go up or down by a set amount worldwide, cannot exist. Nevertheless, sea-level changes resulting primarily from changes in the water volume contained in the oceans, or by a change in the holding capacity of the oceans themselves, are still often referred to as eustatic changes. They can correctly be referred to in terms of an 'equivalent sea-level rise' (e.g. Lambeck and Nakada 1990), that is, by referring to the sea-level rise that would result if the effect of a change were indeed spread uniformly over the oceans. In this way, the sea-level equivalent of a given volume of Quaternary ice can be referred to without the inference being drawn that a sea-level change of this exact amount would necessarily be observed.

The record left by past sea-levels around the margins of the continents is made even more variable by the tectonic movement of the land. Sea-level as we ordinarily measure it is, of course, just the elevation of the line along which water (the geoid surface) and land meet; an apparent fall of sea-level interpreted, say, from old beach deposits or sea caves, can thus in reality be caused by the land rising, and vice-versa. A common cause of changed elevation of the land is *isostatic readjustment*: this is produced by a change in the load borne by the crust, and is a consequence of the fact that the crust in turn is supported by the highly viscous but deformable mantle below. An increase in the load on the crust results in slow subsidence; removal of the load results in a similarly slow *isostatic rebound* or recovery (Fig. 6.1). This problem is particularly pertinent to the study of Quaternary sea-level changes, which were caused by the growth and melting of land ice; the changing ice mass borne by the continents resulted in isostatic subsidence or uplift and hence the evidence of old shorelines and other coastal markers has inevitably been moved, especially near glaciated areas. The changing water mass in the oceans causes a similar effect, varying the load that is borne by the rocks of the sea floor. Considerable ingenuity often needs to be used to unravel the resulting distorted record of sea-level change. It is worth noting that the present coastline lies in a position that we could term a hinge zone, separating the sinking ocean basins, loaded with water, and the rising continents, recently freed of their Quaternary ice load.

In working through the complex record of sea-level, it is helpful to distinguish *gauge* and *relative* sea-levels (Chappell 1983). *Gauge* sea-level is the absolute level of the water surface which might be measured against some stable datum, unaffected by tectonics or isostasy. The centre of the Earth would provide such a datum. There are relatively few Quaternary environments that record gauge sea-level, because any such recording site has to be completely stable tectonically, but the concept is a helpful one. The second way to consider sea-level is in terms of *relative sea-level* (RSL), the position as we record it against the land. Evidence from the field (ancient shorelines, coral reefs, wave-cut notches, mangroves, etc.) generally only indicates RSL, and in the discussion of Quaternary sea-levels that follows, this convention is adopted, so that generally, wherever sea-level is referred to, RSL is what is meant.

The study of sea-level history has been a major theme in Quaternary studies. Sea-level controls or influences many environmental processes that are of climatic and environmental significance, as well as being one of the main determinants of coastal zone geomorphology and shoreline evolution (Bird 1993). The evolution of the coastal zone has exerted a significant evolution on patterns of human settlement and cultural development. It has been suggested that the slowing of postglacial sea-level rise at about 6000 BC permitted the beginnings of silt accumulation in the Nile delta region and hence fostered agriculture and the development of the Predynastic settlements (Stanley and Warne 1993a, b). Major fluxes of nutrients to and within the oceans are also modulated by sea-level. Sea-level controls the areal extent of dry land, and through the agency of carbonic acid weath-

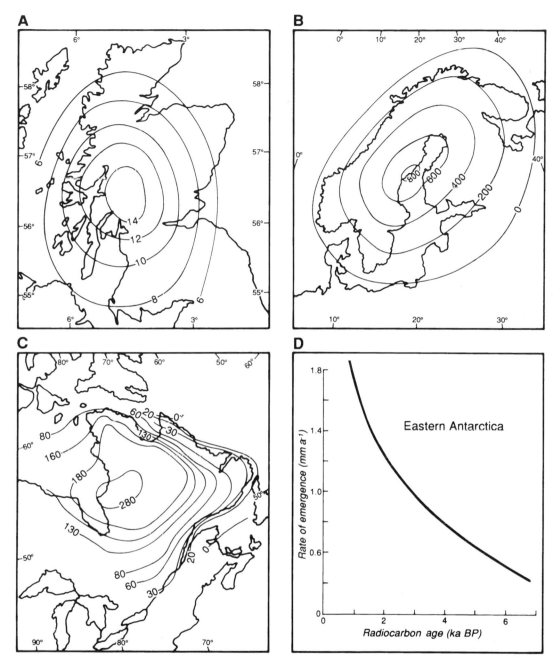

Fig. 6.1 Patterns of glacio-isostatic rebound resulting from the diappearance of Wisconsin–Weichsel ice. Emergence in metres for: (A) Scotland during the past 6.8 ka; (B) Fennoscandia during the past 13 ka; (C) northeastern North America over the last 7.5 ka; and (D) Antarctica over the last 7 ka. (After Sissons 1983; Mörner 1980b; Hillaire-Marcel and Occhietti 1980; Adamson and Pickard 1986)

ering, variations in exposed land area may exert a major control on the global carbon cycle and on the climatic greenhouse effect. Sea-level is also of significance for its relationship to the accumulation of marine sediment bodies, reservoirs of hydrocarbons and other marine resources, which have formed a major focus for stratigraphic and geological work

(e.g. Wilgus *et al*. 1988; Seibold and Berger 1996). A large body of literature shows how diverse even the late Quaternary history of sea-level appears to be in different parts of the globe (see for example Smith and Dawson 1983). This diversity of sea-level history in fact shows that there are multiple controls on RSL, and different histories of sea-level against the land have often been produced regionally under the dominant influence of a particular control or group of controls. Before a good theoretical understanding of these controls had been developed, especially the significance of the geoid mentioned above, progress in the correct interpretation of the field evidence was hampered. A very substantial body of descriptive literature thus exists, which it would be impossible (and unhelpful) to review here. This literature reflects an enormous amount of fieldwork, site description, coring, sampling, and radiometric dating of diverse materials such as shells, corals and plant remains. Although it provides the basis for much of what we shall discuss, we will be unable to consider much of this evidence here. Rather, our goal in this chapter will be to provide an overview of the sea-level history of the late Quaternary, focusing mainly on the mechanisms responsible for sea-level fluctuation. An understanding of these mechanisms will then enable field evidence from any region to be viewed and interpreted in an informed way.

CAUSES OF SEA-LEVEL FLUCTUATION: A REVIEW

Many processes can cause sea-level to vary. Some are exceedingly slow, but others can act almost instantaneously. We will review some of the most important mechanisms here, and examine what is known of the timing and size of their effects on sea-level during the Quaternary. Other reviews may be found in Tooley (1993) and in Smith and Dawson (1983).

Tectonism and changes in gauge sea-level

The largest and most gradual fluctuations in sea-level are caused by changes in the volume or holding capacity of the ocean basins themselves. A principal factor causing such change is the nature of activity along the mid-ocean ridge system which spans the globe. Along these ridges, volcanic activity results in warmth and youthfulness of the oceanic crust, and hence the rocks stand isostatically high. Sea-floor spreading carries the new rock materials away from the ridges at rates measured in cm per year, and as this takes place, slow cooling results in subsidence of the sea floor. The broad profile of the oceanic ridges is thus defined by thermal isostatic processes.

In the modern oceans, the profile of the oceanic ridges, forming a broad arch, largely determines the depth of the ocean floor below the geoid (water surface). Near the margins of the continents, thick piles of sediments blanket the sea floor, but away from

these zones, the depth of the ocean bears a close relationship to the age of the rocks forming the sea floor (Kennett 1982). For example, Parsons and Sclater (1977) showed that the relation below was a good fit to empirical data for sea floor ages up to 70 Ma:

$$\text{ocean depth (m)} = 2500 + 350 \sqrt{t} \qquad (6.2)$$

where t is the sea-floor age in Ma.

Any acceleration in the rate of sea-floor spreading, reflecting increased delivery of heat to the ridge system, carries the new crustal rocks away from the ridge more quickly, and hence they remain isostatically high over greater distances. This alters the shape and dimensions of the ridge system so that it occupies a greater volume within the ocean basin, and sea-level must be displaced upwards. In a similar way, if the rate of sea-floor spreading slows, oceanic ridges become narrower and occupy smaller volumes in the ocean basins, and sea-level is lowered. Such fluctuations in oceanic ridge volume are capable of producing the largest shifts in sea-level, perhaps spanning ±500 m; the fluctuations are also the slowest, typically occurring over tens of millions of years. Quaternary sea-level fluctuations, then, must relate to other causes which we shall discuss shortly.

Major marine transgressions in the Cretaceous, however, were produced in the way just described by an episode of rapid sea-floor spreading, leading to *tectono-eustatic* transgression (a *gauge* sea-level rise as well as a relative one). Sediments of this age occur over large areas of Australia, whose low elevation permitted widespread marine transgression, and are now important aquifers in the artesian groundwater system that supports the modern pastoral industry. The magnitude of the tectono-eustatic rise involved in the Cretaceous is considered to have been up to 350 m (Mörner 1987; see also Schenk 1991). In the US, an extensive late Cretaceous shoreline has been mapped at the geological transition of marine and non-marine facies. After corrections for isostatic and flexural changes to the evolution of this feature, McDonough and Cross (1991) were able to estimate a sea-level of about 270 m above present level at 93 Ma BP.

Major plate tectonic events at the margins of the continents also exert a significant control on sea-level in the long term. Part of the Tertiary trend of falling sea-levels relates to enlargement of the Atlantic Ocean (where the continental margins are all of the cool, passive, non-volcanic type) at the expense of the Pacific, which is dominated by active margins (Dockal and Worsley 1991).

Quaternary ice growth and decay: glacio-eustatic control of sea-level

The growth and melting of land ice provides the second major cause of sea-level change, and the one dominant in the Quaternary. This is the mechanism classically thought of as producing eustatic changes in sea-level. Evaporation from the oceans occurs principally in the low latitudes where the water is warm, but the resulting fall in sea-level when the water is stored as land ice in high latitudes is distributed globally. As glacial conditions develop, sea-level falls very slowly (perhaps averaging 1 m per thousand years, but reaching 5 m per thousand years during short periods of rapid ice growth). The 'eustatic' lowering reached at the last glacial maximum amounted to 120–150 m below present level, as we calculated earlier. Glacial terminations, though involving up to several steps or interruptions (see Chapter 3), are more rapid, with sea-level rising to present levels over no more than 10 000–20 000 years (i.e. at an average rate of 5–10 m per thousand years).

Isostasy and sea-level changes

The *glacio-eustatic* fluctuations are associated with a redistribution of mass among the ocean basins and the continents. When loaded with ice, the continents subside (and the oceanic crust beneath the partly emptied ocean basins rises); deglaciation results in isostatic uplift of the unloaded land, and subsidence of the ocean floors. This introduces complications in unravelling the history of sea-level; the ice-loaded continents occupy a much smaller area than the ocean basins, and the ice only occupies part of the most northerly and southerly continents (or land at high elevations elsewhere). Therefore, the amount by which the continents subside or rise must be greater than the amount by which the oceanic crust moves (Bloom 1967). Furthermore, in subsiding or rising, the crust must displace material in the underlying mantle, which is enormously viscous. Thus the time-scale over which the continents subside or rise is different from that over which the oceanic crust does, since different amounts of vertical motion are involved, and because the mantle below continents has different thermal characteristics to that below oceanic crust. Hence the trace of sea-level left on the land will be a complex encoding of a host of processes acting at different rates. An interesting control is exerted by the tectonic setting of the mantle flow from ocean to continent. If this flow is obstructed by the descending slab of lithosphere at a subduction zone, quite a perturbed crustal response (and hence sea-level response) must ensue (Chappell 1983).

All isostatic subsidence and uplift of continents and ocean basins involves the compensating movement of displaced material in the mantle, just as the water in a bath must flow and redistribute itself when a block of wood is floated in the water, and causes displacement in accordance with Archimedes' Principle. However, land ice grows and melts, initiating isostatic subsidence and uplift too rapidly for the viscous flow of the mantle to keep pace. Thus the subsidence of an ice-loaded continent produces deformation in the mantle, and the disturbance is slowly transmitted by viscous flow. The volumetric transfer of material required as the Earth readjusted repeatedly between the glacial and non-glacial conditions in the late Quaternary is very large; Chappell (1974a) showed that it exceeds the aggregate rate of mass transfer at all tectonic subduction zones globally. The mantle deformation is complex, because the Earth is essentially spherical, and depressing the crust in one place causes upward displacement in a surrounding zone (called a *forebulge*). This disturbance spreads, like a ripple, around the globe to affect even distant areas that bore no ice. Changes in mantle configuration continue to occur in distant locations long after the ice or water load has stabilised (in a process termed *relaxation*), so that not even the timing of sea-level fluctuations can clearly be seen in the field evidence. Sites located on forebulges produced by ice loading (such as the Netherlands and other areas facing the North Sea, on the forebulge of the Fennoscandian ice sheet, or the eastern seaboard of the US, affected by the Laurentide ice sheet) experience a very large and prolonged relative sea-level rise during deglaciation, as the glacial forebulge subsides. Simultaneously, the deglaciated land areas experience rapidly falling relative sea-levels, as isostatic uplift occurs in response to the removal of the ice load. There were of course multiple forebulges during the Quaternary, including one produced by the subsidence of Antarctica under its enlarged ice load. The interaction of multiple collapsing forebulges across the surface of the globe contributed to the apparently different sea-level histories revealed at study sites worldwide. The magnitudes of the isostatic effects described here are considerable. Collapse of the North Sea forebulge after the last glacial period is estimated to have lowered the affected areas by about 170 m; drowned coastal features consequently occur seaward of the present coast. Isostatic uplift of the glaciated areas of Scandinavia has left shorelines standing up to almost 300 m above the present coast. Although the ice load was removed thousands of years ago, the uplift is continuing at rates of up to 9–10 mm per year (Balling 1980; Devoy 1987b). Likewise, uplifted shorelines from the last deglaciation stand at up to 140 m above present sea-level in western Greenland

(Ten Brink and Weidick 1974) because of isostatic rebound.

The isostatic uplift of the unloaded lithosphere in glaciated regions has in turn allowed the collapse of the forebulges. Consequently, in ocean basins where forebulge collapse has taken place, increased volume is available between the geoid surface and the ocean floor. Seawater fills this space to maintain the equipotential water surface. The source for the additional water is the oceans far removed from the ice-loading effects, largely in low latitudes where ice growth was minimal. Consequently, during the collapse of the forebulges, water drained from the distant oceans (termed far-field oceans) to those nearer the ice sheets (Tushingham and Peltier 1991). Mitrovica and Peltier (1991) termed this process 'equatorial ocean syphoning'. The syphoning generated a fall of relative sea-level in the far-field, such as the equatorial Pacific Ocean. While additional water was being supplied to the oceans by melting ice (that is, while deglaciation was still underway), the net effect on far-field sea-level was small. But once deglaciation was completed, the movement of water toward the former forebulge sites could cause net sea-level fall in the far-field. Many sites in the equatorial Pacific (e.g. Palau, Guam, New Caledonia) show evidence of this, in the form of mid-Holocene shoreline markers located 1–2 m above present sea-level. Similar records have been derived from other sites also, such as the Strait of Magellan off South America (Porter *et al.* 1984). These raised shorelines exist because the meltwater influx ceased by about 6 ka BP. The shoreline markers have since been left above sea-level by the equatorial ocean syphoning, and we can thus interpret raised evidence dated to the mid-Holocene not as a sign of higher sea-levels then, but of ongoing water redistribution within the oceans.

In contrast, the isostatic response to glacial unloading dominated sites around the great ice sheets. In such areas, deglaciation was associated with rapidly falling sea-level, despite the enormous volume of meltwater being returned to the oceans. For instance, in Spitsbergen, Forman (1990) has mapped and dated strandlines and other evidence of Holocene sea-levels. Emergence of the land evidently began some time prior to 13 ka BP, and was slow until about 10.5 ka BP, occurring at rates of 1.5–5 m/ka. Minor stillstands or regressions corresponded to periods of renewed glacial advance. In the period 10.5–9 ka BP, emergence was much faster, at 15–30 m/ka, indicative of rapid deglaciation, and occurred despite the rapid return of water to the oceans at this time which would have been generating sea-level rise in far-field oceans. This period of rapid deglaciation was also inferred from the ages of moraines in western Greenland by Ten Brink and Weidick (1974).

In the North Sea region, a slightly different postglacial scenario was modelled by Lambeck (1995a). According to the geophysical model of sea-level used by Lambeck, the crust under northern England and Scotland rose isostatically just as in Spitsbergen and Greenland, although more slowly. In consequence, the eustatic sea-level rise was able to keep up with the land, and the coastline remained roughly stationary until about 12 ka BP. In this model, the English Channel was inundated quite late in the deglaciation, and finally reverted to a marine waterway at about 7.5 ka BP.

Changing water loads and their effects

A process responsible for local and regional changes in relative sea-level that must be mentioned here is *hydro-isostasy*. Marine transgressions caused by eustatic rise lead to an additional load of water being placed on the continental shelves, which carry a diminished water load during glacial stages. Underlying the shelves is continental crust; being more rigid than mantle material (elastic rather than viscous), the continental crust *flexes* under the changing load, which is greatest on the outer shelf where the water depth is greatest, and least near the continental shore. This flexing leads to relatively small changes in relative sea-level along the continental margins, amounting to perhaps a few metres. Clearly, the amount of flexure will relate to the rigidity of the rocks underlying individual continental margins, and the load of water covering the continental shelf (which will depend primarily on the width of the shelf). Geographical variability in these parameters will once again ensure that 'eustatic' sea-level changes are recorded by field evidence at varying elevations worldwide. In general, however, postglacial transgression caused by eustatic sea-level rise is associated with the up-warping of continental margins, as the adjacent sea floor is warped down under the heavier water load. This means that raised shoreline features are to be anticipated, although, just as in the case of the Mid Holocene raised shorelines of the far-field oceans, such features need not record a higher *gauge* sea-level. Rather, they reflect the subsequent hydro-isostatic tilting and elevation of the physical evidence marking a former position of the shore.

The geoid: gravitational controls on sea-level

Land emergence (falling relative sea-level) during deglaciation is only partly due to isostatic rebound of the crust. A component of the relative sea-level fall that is observed near the areas loaded by glacial ice is caused by perturbation of the geoid (Mörner 1976). Let us consider this important effect a little further.

A record of land emergence such as might be pieced together from dated strandlines does not show the actual rate of isostatic uplift of the land, because of the synchronous eustatic rise caused by meltwater influx to the oceans. The total amount of uplift of a shoreline marker in such a location is given by its elevation above sea-level plus the amount by which sea-level itself has risen since the shoreline marker was set down. But near to the ice masses, the 'eustatic' sea-level change is greatly perturbed because of the loss of the ice mass itself, because when it is present the ice mass exerts a gravitational attraction that warps sea-level upward toward the edge of the ice. This gravitational force disappears with the ice during deglaciation.

Simple calculations of the size of this effect can readily be made (see J.A. Clark 1976; Clark *et al.* 1978). To focus on the geoid effect alone, we assume that the Earth itself is rigid and does not respond isostatically to the loss of the ice load. Under this assumption, the sea-level change, *s*, at some distance from an ice mass is given by

$$s = \frac{GM}{2ag \sin\left(\frac{\theta}{2}\right)} \qquad (6.3)$$

where θ is the angular separation of the shoreline and the ice mass in degrees, M is the ice mass, G is the gravitational constant, a is the radius of the Earth, and g is the normal mean acceleration due to gravity at the Earth's surface.

If, for example, we consider the glacial maximum Greenland ice sheet to have had a mass of 2×10^{18} kg, and we use a = 6370 km and G = 6.67×10^{-11} N m^2/kg^2, and approximate the gravitational effect of the ice to be that of a point-mass located at the centre of Greenland, the above equation shows that at the coast, 4.5° away from the centre of mass, the sea-level perturbation would have been

$$s = (6.67 \times 10^{-11} \times 2 \times 10^{18}) /$$
$$(2 \times 6.37 \times 10^{6} \times \sin(4.5/2))$$
$$= 27 \text{ m approx.}$$

The actual perturbation at any point along the coast would depend on the radial distribution of ice mass around that point, so that the effect would have varied around the coastline near any major ice mass. Note that this is not a eustatic change, nor does it include isostatic rebound; it represents only the change in the level of the ocean surface (the geoid) drawn up by the gravitational attraction of the ice mass.

J.A. Clark (1976) provided a calculation of the size of this geoid effect (assuming a rigid Earth) across the global oceans, following the last deglaciation. The geoid effect perturbed the notional 'eustatic' change (an average 85 m sea-level rise in the calculation done by Clark) such that sea-level would actually *fall* by 85 m in Hudson Bay, but *rise* more than 100 m in the South Pacific. This unexpectedly large rise (more than the equivalent sea-level rise related to the ice volume) represents the aggregate effect of the increased water volume fed to the oceans by melting ice, together with the water released from its elevated position in the gravitational field around the ice mass. Clearly, therefore, the effect of the changing gravitational attraction of the great ice sheets is a substantial one, with the greatest effect close to the ice itself, but worldwide in impact none the less. (Indeed, it is of the same order of magnitude as the equivalent sea-level change induced by meltwater influx, as noted by Clark *et al.* 1978.) Most importantly, the influence exerted on sea-level change is effectively instantaneous, so that the geoid perturbations arising from ice loss would have occurred synchronously with deglaciation. Further slow geoid perturbations would follow, however, as a result of the isostatic response of the crust, which constitutes an additional change in the mass distribution that controls the form of the geoid.

Summary of sea-level controls

In summary, the major factors influencing sea-level and its fluctuation may be placed in three groups:

1. Those producing notional 'eustatic' changes:
   glacial eustasy (loss or gain of water from ice growth or melting)
   accretion of new water (juvenile water from igneous activity; discussed below).

2. Tectonic and isostatic processes which cause relative sea-level change essentially worldwide without actually altering the oceanic water volume (so that the changes are not strictly *eustatic*):
   geoidal eustasy (changed configuration of the sea surface as a result of relocations of rock, ice or water masses, or astronomical effects)
   tectono-eustasy (oceanic ridge or tectonic uplift or subsidence effects, including the subsidence associated with the fragmentation of supercontinents discussed below)
   sedimento-eustasy (the filling of parts of the ocean basins by marine or terrestrial debris and sediments).

3. Local isostatic and tectonic effects which primarily affect a restricted area (and hence are in no sense eustatic) but whose effect may in principle be transmitted to all parts of the globe with reduced amplitude via deformation of the mantle:

glacio-isostasy (ice loading and unloading effects)
hydro-isostasy (water loading and unloading effects).

FORMALISING THE DESCRIPTION AND ANALYSIS OF SEA-LEVEL

The major components of Quaternary sea-level change may be expressed in the form of a sea-level equation. This is commonly written

$$\Delta\zeta(\varphi,\lambda:t) = \Delta\zeta^e(t) + \Delta\zeta^{ice}(\varphi,\lambda:t) + \Delta\zeta^w(\varphi,\lambda:t) \qquad (6.4)$$

which expresses the sea-level at a location on the Earth given by the geographical coordinates (φ,λ) and at a time t. The first term on the right-hand side is the equivalent sea-level, the 'eustatic' component, defined as noted earlier as

$$\Delta\zeta^e = \frac{\text{change in ocean water volume}}{\text{ocean surface area}}$$

The second and third terms on the right express the changing loads of ice and water that are carried by the crust, both varying in time as well as with geographic location. The 'eustatic' term, $\Delta\zeta^e$, is a function of the timing and rate at which meltwater is added to the oceans, and does not have a geographical location attached to it since the effect is transmitted globally.

The global sea-level responses to the combined changes embodied in this equation are used to subdivide the response into several realms where there is a distinct outcome.

Within areas occupied by the Quaternary ice sheets, or close to these areas, the change in weight arising from ice growth and decay, which is all borne by a relatively small area of land, triggers a crustal isostatic response that exceeds the response elicited in the sea floor. This is because the changed loading of seawater is borne by the sea floor globally. Thus, in this realm, termed the *near field*, $\Delta\zeta^{ice} >> \Delta\zeta^w$. On the other hand, at locations distant from the actual ice load, the change in load felt by the crust is dominated by the water load as sea-level changes eustatically, so that $\Delta\zeta^w > \Delta\zeta^{ice}$. This condition applies to the bulk of the world's oceans, and is termed the *far field* (see Lambeck 1993a).

The various terms in the sea-level equation can be evaluated if appropriate data can be obtained. Such data include the history of relative sea-level worked out from various dated shoreline markers (reefs, beach deposits, mangroves, etc.) and perhaps the history of ice volume. This can be estimated independently from the isotopic composition of marine forams (see discussion later in this chapter and in

Chapter 3). It is also necessary to know how the Earth deforms under the changing loads, and especially how fast it is able to respond. Reliable geophysical data are thus required. One of the uses of dated Quaternary sea-level markers has in fact been to use the sea-level equation to estimate parameters such as the thickness of the rigid lithosphere, and the depth distribution of viscosity within the underlying mantle. A key step in this process is to recognise that differences in the recorded sea-level history within a particular region cannot be eustatic in origin, but must reflect the loading terms in the sea-level equation. The properties of the lithosphere and mantle can then be found by an iterative procedure which seeks the combination of properties that yields the best correspondence of observed and predicted geographical patterns of relative sea-level response to the time-and-space variation in the ice and water loads (see for example Mitrovica and Peltier 1991; Lambeck 1993a, b). In the case of the British Isles, where the record of apparent relative sea-level is quite variable (see Fig. 6.2), Lambeck (1993b) has inferred that, to account for the observed variation, the lithosphere must be about 65 km thick, and underlain by a mantle whose viscosity is 4–5×10^{20} Pa s above 670 km depth, and $> 4 \times 10^{21}$ Pa s below. In the near-field, the approach used to interpret the sea-level equation can even be used to constrain estimates of the thickness of the Quaternary ice sheets, because in this realm the loading response arising from the weight of the ice is the dominant term in the equation. Estimates of ice load that are significantly too high or too low then predict that shoreline markers will be found respectively higher than or lower than they are actually observed to be. In this way, estimates of the thickness of the ice sheets can be tied down using geophysical inference. This has been done, for instance, by Lambeck (1995a), who has shown that sea-level records from eastern Scotland are inconsistent with the suggestion that a continuous ice cover may have linked Scotland to Fennoscandian ice across the North Sea. The sea-level model developed by Lambeck further constrains the last glacial ice thickness over Scotland to about 1500 m and that over Ireland to a maximum of about 500 m. The equivalent sea-level change contributed by the melting of these ice masses can then be estimated.

SOME FACTORS INFLUENCING LOCAL RSL HISTORY

In addition to global tectonic, isostatic, and geoidal influences on global sea-level, there are several local circumstances that can influence the RSL history at a site. Let us consider some of these.

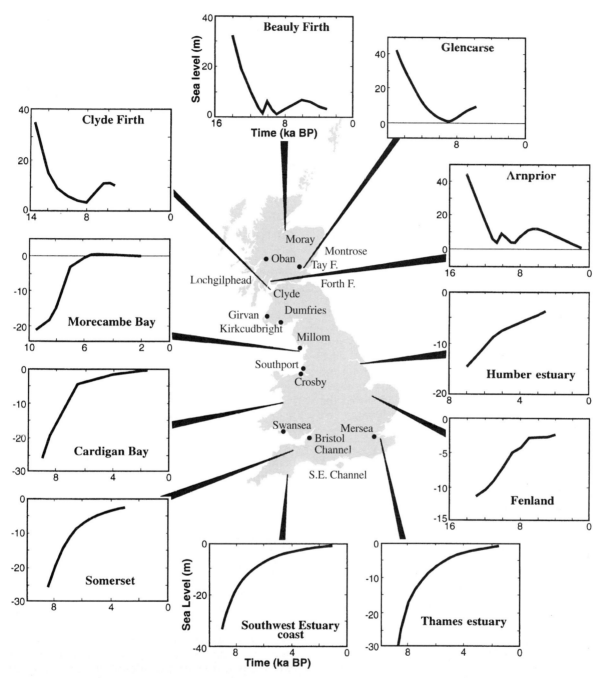

Fig. 6.2 The pattern of regionally varying Holocene trends in relative sea-level around the British Isles. Such diversity cannot reflect a eustatic influence, and rather demonstrates the variability in crustal response to changing ice and water loads. (After Lambeck 1995a)

The shape of the coastline and its effect on the sea-level record

An interesting illustration of the role of crustal movement on apparent relative sea-levels is provided by considering events that took place during the Holocene marine transgression, which we discuss in more detail later. This event resulted from the melt-

ing of the last great Quaternary ice sheets, causing the > 100 m equivalent sea-level rise calculated earlier.

The additional water load resulting from this return of water to the oceans had marked effects along the margins of the land, especially in areas not near the ice sheets. At such locations, the load borne by the continents was not altered, while the nearby oceans received the additional water load. This imbalance led, as we have seen, to relative sinking of the ocean basins, displacement of mantle material toward the continents, and thus to tilting of the continental margins upward. Furthermore, sinking of the oceans has continued even after the addition of meltwater ceased in the Mid Holocene. The ocean basin sinking causes a decline in relative sea-level.

The amount of subsidence after the cessation of melting (all else being assumed equal) is greatest near parts of the crust most heavily loaded – like promontories reaching out to sea. The subsidence is least in parts of the crust bearing the least water load, like long embayments running inland (see Fig. 6.3). Consequently, the Mid Holocene shoreline on a promontory continues to sink with the oceanic crust, lowering the evidence for emergence. In contrast, the equivalent shoreline features around the head of an embayment do not sink as much, because the crust is not loaded. This leaves higher relict shoreline features there as the sea-level falls, and this appears to suggest higher relative sea-levels around embayments than near promontories. This kind of geographic variation in sea-level records must have contributed to the great variety of apparent Holocene highstand elevations noted in the Taiwan–Japan region by Pirazzoli (1978), who concluded that the diversity of apparent sea-levels in this region argued against the occurrence of a true 'eustatic' high sea-level stand. The effect of the geometry of the coastline and the continental shelf offshore has also been demonstrated from the region of northern Australia, where the Gulf of Carpentaria provides a large and relatively shallow embayment with relatively little water loading (e.g. see Smart 1977; Chappell *et al.* 1982). Holocene shorelines are also elevated around the uppermost parts of the narrow Spencer Gulf in the coast of South Australia and are in good accord with models of crust and mantle behaviour (Nakada and Lambeck 1989).

Coastal sedimentation and its effect on the shoreline

Coastal areas are of course the sites where river-transported sediments are delivered to the oceans. The condition of the coastline reflects the interplay of marine erosion and sediment transport and fluvial sediment delivery. Under appropriate conditions of both coastal morphology and trend in relative sea-level, coastal progradation may result from continued fluvial and marine sediment delivery to the shore zone. Extensive sets of beach ridges, low-gradient chenier plains (Augustinus 1989), and related deposits, then alter the pattern of water loading, sediment loading and water depth along the coast. These materials also provide a store of datable remains from which a chronology of the coastal plains, and perhaps of relative sea-level against the coast, can be built up. A full discussion of these matters is beyond the scope of this chapter.

THE TIMING OF SEA-LEVEL FLUCTUATIONS

Over the enormous period of time during which the Earth has possessed oceans, their depth may have varied in a systematic way as the volume of water at the surface changed. Little firm evidence on the volume of ocean water far back in geological time exists, but it is often considered that this may have increased through time at a very slow rate. Initial condensation of water is likely to have proceeded in the early history of the Earth, and it seems reasonable to suppose that during the Phanerozoic (the last 600 Ma) at least, the total oceanic water volume has been more or less fixed. Certainly over Quaternary time-scales, the addition of further *juvenile* water must have been so slight that it may be omitted from consideration as a factor significantly influencing sea-level.

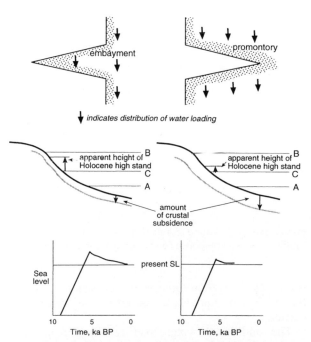

Fig. 6.3 Differing records of apparent Mid Holocene high sea-levels arising at narrow embayments and promontories in deep water from the differing amounts of crustal subsidence produced by water loading

TERTIARY AND EARLIER SEA-LEVELS

The dominant influence of tectono-eustasy can be seen in inferred sea-level histories spanning the Phanerozoic (Fig. 6.4), and based upon the sedimentary sequences around the continental margins. These show broad fluctuations of 300 m and more occurring over perhaps 50–100 Ma, with higher frequency (more rapid) fluctuations superimposed.

Essentially, two periods of high sea-level can be recognised in such reconstructions; one occurred in the early–middle Palaeozoic, at about 450 Ma BP, and the other at around 100–150 Ma BP, in the late Mesozoic. These high stands are paralleled by peaks in records of global igneous activity (Fisher 1984). The most probable explanation is that these *supercycles*, as they have been called, relate to the repeated fragmentation and aggregation of supercontinents. Pangaea, the most recent supercontinent, began to break up in the Jurassic (during the interval of low sea-level separating the two supercycle high stands). Supercontinents act to trap geothermal heat in the mantle; eventually, convectional overturning commences, and continental fragmentation begins. This sets off two processes which cause relative sea-level to rise. First, the moving continents, being smaller, allow mantle heat to escape more readily: material deep in the crust and in the upper mantle thus cools, becomes more dense, and subsides isostatically. Second, as new oceans are produced along the fractures separating the continental fragments, a growing proportion of the sea floor is young, thermally buoyant, and standing isostatically high, displacing sea-level upwards. Not surprisingly, apparent worldwide

transgression follows, persisting over a period of time which appears to be about 75 Ma. As the continental fragments converge once more, the sea floor becomes old and subsides, and the new supercontinent once more traps mantle heat and begins to ride high isostatically; thus, regression commences. This cycle evidently takes about 300 Ma in total, so that only two supercycles are revealed in the Phanerozoic record. It thus appears likely that repeated fragmentation and aggregation of the continents, occurring in a broadly cyclic manner, may exert a fundamental, long-term control on the global sea-level (Worsley *et al.* 1984).

During the Tertiary, sea-level underwent a progressive but irregular decline from perhaps +300 m to the present level. The decline was apparently punctuated by a series of major, very rapid regressions and transgressions, perhaps of as much as 100 m in only 1–2 Ma; these remain to be fully documented. The overall falling trend presumably relates to the mechanism just outlined: the rate of sea-floor spreading declines as continental fragmentation proceeds, and the mean age of the sea floor consequently increases, producing steadily larger capacity in the ocean basins, and a gauge fall in sea-level. Thus, we see that the Cainozoic trend of declining sea-level is just a late stage of the second of the Phanerozoic supercycles.

QUATERNARY SEA-LEVELS

The record of sea-level is only really acceptably known for part of the upper Quaternary, perhaps the last 400 ka, and even then only from a few locations

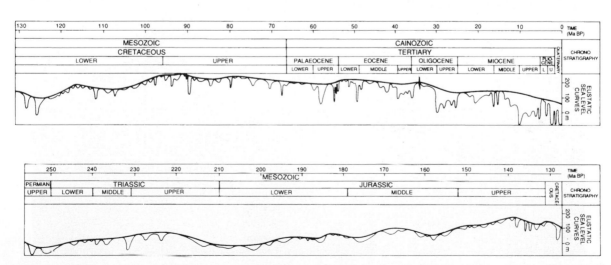

Fig. 6.4 Long-term trends in sea-level over the Phanerozoic (after Haq *et al.* 1987). Note the rise in relative sea-level in the Mesozoic, associated with subsidence of the dispersing fragments of Pangaea

where suitable records exist. The older evidence is dated by uranium disequilibrium techniques, with radiocarbon employed in the most recent 40 ka. As will be readily understood from the foregoing discussion of the controls on sea-level, almost everywhere, tectonics or isostasy of some form has affected the elevation of field evidence. Surprisingly, the best records come from some of these unstable areas, including uplifted sites in Barbados, Haiti, Papua New Guinea, Timor and elsewhere (e.g. Chappell and Veeh 1978; Bender *et al.* 1979; Dodge *et al.* 1983; Bard *et al.* 1996).

An excellent example is the record derived from fossil coral reefs found in a staircase-like array above the present coastline of the Huon Peninsula in Papua New Guinea (e.g. see Chappell 1974b; Bloom *et al.* 1974). This area lies in the collision zone between the advancing Pacific and Indo-Australian lithospheric plates, and is consequently subject to continual tectonic influences. The result has been fairly steady uplift of a fault-bounded block of land at a mean rate of 0.5–3 mm per year (varying systematically along the coast). Because this uplift relates to major ongoing plate tectonic processes, it has not fluctuated widely during the late Quaternary.

The present Huon coastline, located at about 6° S, supports fringing coral reef communities. These do not flourish, however, because coral communities are continually elevated and suffer exposure. Consider, however, the situation during a glacial termination; meltwater, returning to the ocean basins, produces a worldwide transgression. The sea-level rise, occurring at a typical rate of some millimetres per year, then effectively keeps pace with the rising Huon Peninsula, and the coral communities are able to grow larger in size. Eventual stabilisation of sea-level brings conditions back to something like the present. The flight of fossil coral reefs running up to 700 m and more (Fig. 6.5) thus represents the sequence of glacial terminations, the crest of each reef dating from the peak of a postglacial transgression. Falling sea-levels are excessively hostile to coral growth, and the Huon Peninsula does not provide a record of the elevation of the lowest stands of the oceans during glacial times.

To extract a sea-level history from these reefs involves subtracting from their present elevation above sea-level the part due to the steady uplift. The remainder represents the sea-level at the time of coral growth. Dating the coral remnants then allows a full chronology to be established (see Aharon and Chappell 1986).

The record created in this way is shown in Figs 6.6 and 6.7, together with an oxygen isotope palaeotemperature record derived from marine microfossils; the similarity between the two records reinforces the conclusions drawn from them.

What does the record tell us about the history of sea-level at this site?

Major reef complexes were produced by rapidly rising seas (rates of up to 8 mm per year) at 8.2 ka and 118–138 ka. These represent the Holocene (Flan-

Fig. 6.5 Oblique aerial photograph of the coral terraces of the Huon Peninsula in Papua New Guinea. The prominent terrace in the middle of the photo dates from the last interglacial high sea-level stand. (Photo: D.L. Dunkerley)

drian) and previous Sangamon postglacial transgressions. The last interglacial sea-level stood about 6 m higher than the present. After the last interglacial high-stand, sea-level trended downward, but with a

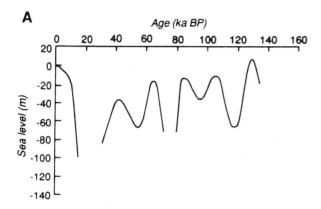

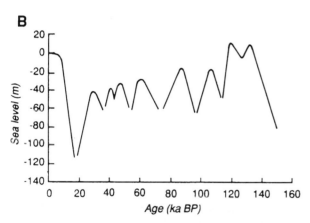

Fig. 6.6 Sea-level records for the late Quaternary derived from (A) coral terraces at Barbados and (B) Huon Peninsula, Papua New Guinea. Both records lack evidence of the lowest levels reached by sea-level during stadial and glacial phases, as explained in the text. (After Steinen *et al.* 1973; Aharon 1984)

series of reversals during short warming episodes (*interstadials*) when smaller reefs were built. These have ages and relative sea-levels of 107 ka (-12 m), 85 ka (-19 m), 60 ka (-28 m), 55 ka (-32 m), 45 ka (-38 m) and 40 ka (-42 m). The whole episode of falling sea-level in the period 80–20 ka BP represents the last major Quaternary glaciation (the *Wisconsin* of America and the *Würm* of Europe). The fact that even the interstadial high sea-levels recorded at the Huon Peninsula were 12–42 m below present indicates that there was significant (and growing) land ice through the *whole* period. This finding is consistent with records derived from Bermuda, which is a stable site. From caves that formed below present sea-level during low-stands, Harmon *et al.* (1983) sampled and dated cave deposits (speleothems). One sample from a cave location 15 m below present sea-level indicated continuous growth (i.e. air-filled cave conditions, and thus a sea-level lower than the cave) for the whole period from 110 ka BP to 10 ka BP. (A statistical compilation of numerous other coral dates derived for sea-levels in this period was presented by Smart and Richards 1992.)

It is important to observe why sea-level records such as those from Papua New Guinea or Bermuda are important. Being at sites distant from the changing Quaternary ice loads, that is, in the far field, the sea-level variations derived from the records (once the effects of steady uplift are removed in the case of tectonically unstable sites) largely reflect the 'eustatic' component of the sea-level record, disturbed only by the relatively minor far-field effect of the changing water-load. The size of this effect only amounts to a few per cent of the glacial–interglacial eustatic change.

The physical sea-level record found at stable sites is very different in character to records created on tectonically rising coasts. While active tectonic uplift separates and preserves the benches, reefs, or other

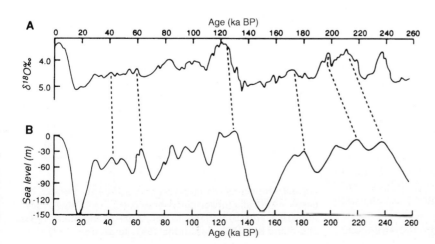

Fig. 6.7 Late Quaternary sea-level and associated oxygen isotope record derived from coral terraces, Huon Peninsula, Papua New Guinea (Aharon and Chappell 1986). (A) Oxygen isotope record; (B) sea-level record

features formed during the Quaternary, so that the record is spread out in a staircase fashion, in stable sites the record is vertically compressed. One feature that often does stand out at stable sites is the record of the last interglacial sea-level, which is very widely found standing significantly above present sea-level. For example, the last interglacial record sits at +4 to +6 m in Bermuda (Harmon *et al.* 1978), +5.6 m in the Bahamas (Neumann and Moore 1975), +7.6 m on the Hawaiian island of Oahu (Ku *et al.* 1974; Jones 1993), and at +9 to +10 m on the Atlantic coast of the US (Toscano and York 1992). From the stable Bermuda site, Harmon *et al.* (1983) were also able to estimate even the height of the penultimate interglacial, at around 200 ka BP, which they put at about 2 m higher than present. Chen *et al.* (1991) have dated closely-sampled materials of last interglacial age at sites in the Bahamas. They confirmed a sea-level like those just listed (6 m above present) and their data indicate that a high sea-level at this elevation persisted for about 10 ka (from 130 to 120 ka BP). Interestingly, this corresponds well to a period of high summer insolation predicted by the Milankovitch hypothesis (see Chapter 5).

The height of the last interglacial sea-level is of critical importance to unravelling the sea-level history from tectonically unstable sites, such as the Huon Peninsula coral terraces. The elevation of the last interglacial evidence (>100 m on the Huon coast) can be divided by its age to yield information on the uplift rate. Knowing this, the elevations of other fossil reefs can be corrected for uplift occurring since their emplacement, and the record of varying sea-level unravelled.

THE HOLOCENE TRANSGRESSION

The termination of the last glacial phase poured vast quantities of water into the ocean basins. Ice melting took place at rates that varied from site to site. In northeast Canada, ice sheets lasted until at least 7 ka BP, but were completely gone from northwest Europe 1.5–3.5 ka earlier (Devoy 1987b). The Barents Sea ice sheet is generally thought to have decayed early (see Chapter 3), being unstable in the face of sea-level rise (Elverhøi *et al.* 1993). (Lambeck (1995b) has used geophysical modelling to support the inferred behaviour of the Barents ice. He deduced an ice thickness of up to 2 km and essentially complete deglaciation by 15 ka BP.) Reduction in the volume of Antarctic ice is estimated to have contributed 25 m to the transgression (Clark and Lingle 1979), but the bulk of the water was released from the Laurentide ice sheet.

Detailed stratigraphy and dating of the Holocene

reef show that the maximum rate of sea-level rise in the Holocene occurred at the Huon coast at about 9–10 ka BP (Chappell and Polach 1991).

The same transgression affected many coastal areas worldwide, but there is a great diversity of apparent Holocene sea-level records from sites around the world (Fig. 6.8). Some of this results from tectonic movement of the sites, from hydro-isostatic warping, and from dating problems. Taking an envelope which includes the bulk of the data (corrected for known major isostatic effects) reveals a generally clear picture of the behaviour of sea-level. It is important to remember, however, that the range of effects described earlier, such as local geoid perturbations and isostatic warping, means that in reality different histories of sea-level rise against the land *did* occur at various locations; at some sites, sea-level might still have been at −30 m at 10 ka BP, while at others at the same time it had reached higher, say to −25 m. No single history of sea-level necessarily applies *in exact detail* to any other place. What follows is thus a generalised account.

Following the commencement of deglaciation at some time after 20 ka BP, sea-level rose rapidly, relative rates of rise of > 30 mm per year (30 m/ka) being indicated at some sites. Present sea-level was reached at around 6 ka BP in many areas, although for Bermuda, Harmon *et al.* (1983) deduced that the present level was only reached at 2 ka BP. Fast though the rate of rise was, it was small in comparison with the rate at which the *shoreline* must have migrated landwards over the gently sloping continental shelves, which were progressively inundated as the transgression proceeded. For example, taking the width of the shelf off the NW coast of Australia

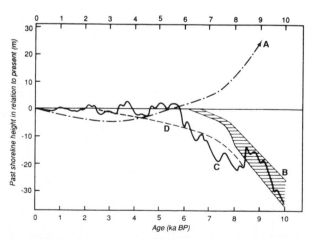

Fig. 6.8 Holocene sea-level records for (A) Baffin Island, (B) eastern Australia (uncertainty shown), (C) a synthesis of several areas, and (D) Holland. (After Andrews and Miller 1985; Thom and Roy 1985; Fairbridge 1961; Jelgersma 1961)

as 200 km, and the duration of the transgression as 10 ka, the rate of shoreline advance comes to 20 m per year (20 km/ka), or about 40 cm per week! This is about 1000 times more rapid than the rate of vertical sea-level rise. This shoreline advance is a strikingly rapid and sustained process which, through the continual migration of habitat, must have been particularly taxing for coastal flora and fauna (as well as human coastal populations).

Closely dated records provide a more detailed picture than the broad outline just presented. Remains of the reef-crest coral *Acropora palmata* are found in Barbados and other Caribbean islands at various elevations, having suffered tectonic uplift. On Haiti and Barbados, last interglacial corals are located at 40–60 m above sea-level. Given the radiometric ages of these corals of 120–130 ka, mean uplift rates in the intervening period can be found (approximately 0.4 mm/a). These rates can then be applied to correct the elevations of younger coral remains.

Using this method, Bard *et al.* (1990a) reconstructed a sea-level curve for Barbados covering the last 130 000 years. Their record for the last 20 ka is quite detailed, and shows that the rate of sea-level rise underwent significant fluctuations during this interval. The last 20 ka have also been dated in great detail from the Barbados record by Fairbanks (1989) (see Fig. 6.9).

According to these records, the sea-level rise in fact began slowly at about 20 ka BP, at a rate of about 5 m/ka. At about 14 ka BP, sea-level rose much more rapidly, reaching a rate of 37 m/ka. A rapid sea-level rise at 12 ka BP (24 m in less than 1000 years) provides evidence of very rapid ice melting, and has been termed 'Meltwater Pulse IA' by Fairbanks (1989). The rate then slowed again, to about 8 m/ka, and then jumped once more at 11–10 ka BP, to 25 m/ka. This rapid rise culminated in 'Meltwater Pulse IB', involving a rapid rise of about 28 m. The rate of rise then slowed to 9 m/ka. (The source of the meltwater responsible for the meltwater pulses is considered in Chapter 3.) From the structure of *Acropora palmata* reefs in the Caribbean, Blanchon and Shaw (1995) have inferred rates of sea-level rise even more rapid than those just mentioned, ranging up to > 45 m/ka. As shown in Chapter 3, stepped deglaciations, and hence changing rates of sea-level rise during marine transgression, appear to have been the norm during the late Quaternary.

Low Quaternary sea-levels are not well resolved in records from either unstable sites like the Huon Peninsula or stable sites like Bermuda. During lowstands, for example, the sea was below the elevation of the platform upon which higher sea-levels were able to build the coral cap of Bermuda. Consequently, no record was set down. Evidence located well below sea-level is also much harder to investigate.

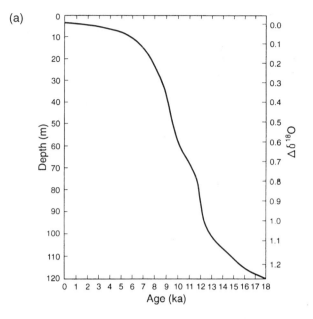

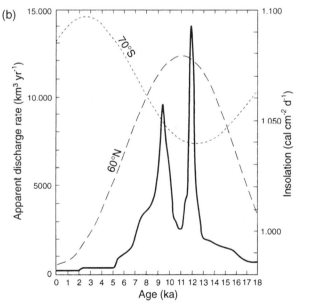

Fig. 6.9 (A) Pattern of Holocene sea-level rise and (B) corresponding rates of meltwater influx to the oceans, derived from Barbados coral records. (From Fairbanks 1989)

With greater difficulty, sea-level records that fill this gap can in principle be derived independently from the isotopic composition of foram tests, since these reflect the seawater isotopic composition. The major control on the latter is ice volume (see Chapter 3), so that the marine oxygen isotope record can, under special conditions, be used to infer both ice volume and sea-level records. The changes in seawater isotopic composition reached their most extreme during the lowest stands of sea-level, thus facilitating subsequent analysis and interpretation of the isotope record. This approach was employed by Bard *et al.* (1989), in a study of the planktonic foram *Globigerina bulloides* collected from cores in the northeast Atlantic Ocean. From the temporal variation in the isotope record, they were able to estimate a mean sea-level depression of about 70 m at 12.2 ka BP, and a rapid sea-level rise of about 40 m at 14 ka BP that matches the one subsequently found in the Barbados coral record noted earlier. The late Quaternary ice volume/sea-level record is illustrated more fully in Chapter 3 (see Fig. 3.5).

Analysis of the isotopic composition of the shells of the giant clam *Tridacna gigas* collected from the coral terraces of the Huon Peninsula has also provided isotope data (see Aharon and Chappell 1986). The clams only grow at depths of less than 10 m, and may be used to infer the temperature of the surface layers of the tropical oceans; microfossils in deep sea cores reflect more the temperature of the deeper waters. The results from the Huon Peninsula are shown in Fig. 6.7.

Clearly, the curves of isotopic composition and sea-level undergo synchronous changes, and the trend of falling sea-level through the period 80–20 ka BP is paralleled by a trend of decreasingly negative $\delta^{18}O$ values (i.e. progressive ^{18}O enrichment).

The Holocene behaviour of sea-level has also been examined using the kinds of geophysical models already referred to in the case of the British Isles and the interpretation of far-field changes in the equatorial oceans. The history of the refilling of the Persian Gulf, for instance, has been modelled by Lambeck (1996). The mean depth of the Gulf is about 35 m, so that it must have been above sea-level during the glacials (although perhaps containing some lakes). The geophysical modelling suggests that the sea reoccupied the Strait of Hormuz by about 14 ka BP and flooded into the central parts of the Gulf by about 12.5 ka BP, finally reaching the present shoreline by about 6 ka BP. Relative sea-level then rose further, inundating parts of Mesopotamia (the lower Tigris and Euphrates valleys, now mainly in Iraq). An understanding of this history of inundation is clearly pertinent to the reconstruction of the movements of Palaeolithic and Neolithic peoples in the area. Similarly, the linking of separate landmasses at times of low sea-level, which we consider briefly below, is a key issue in the migration of plants and animals as well as of people in many parts of the globe.

LAND BRIDGES: A PRODUCT OF QUATERNARY LOW SEA-LEVELS

During Quaternary low sea-levels, the geography of the planet was considerably altered. The shelves were exposed around the major continents, with new river courses being cut across them as the enlarged land areas drained to the more distant sea. Additionally, dry *land bridges* connected many areas now separated by straits (Fig. 6.10). The Australian mainland was connected by such bridges both to Papua New Guinea in the north and to Tasmania in the south. Britain and Ireland were connected to the European mainland, Alaska and Siberia were joined by the Bering land bridge, and many islands of southeast Asia were linked and formed an extension of the Asian mainland. These transitory land bridges were exploited by the peoples of the time to journey among these landmasses. Many sites recording the occupation of the land bridges must now lie submerged and inaccessible.

The major coral reef communities of the present day would also have been exposed during glacial phases. Thus, the crest of the Great Barrier Reef of northeastern Australia would have stood about 150 m above sea-level (Carter and Johnson 1986). Acid rainwater falling on the exposed limestone materials would have attacked them readily, forming a rugged landscape of jagged pinnacles with cave development occurring below the surface as water drained downward. Some of the caves which evolved at these times experienced roof collapse, a process common in present terrestrial cave environments; now drowned, these caves appear as *Blue Holes*, deep openings into the ancient reef below the contemporary reef crest, which can easily be seen from the air. Torres Strait, once part of the land bridge there, displays similar cave openings which are now drowned, as do other present-day reef environments, including the Caribbean.

RECENT AND HISTORIC CHANGES IN SEA-LEVEL

There is good reason to consider the contemporary trend of sea-level, in view of the size and important trading role of coastal cities, and the survival of the habitats of low-lying coastal wetlands. Nations with significant land area lying near sea-level have a particular interest in possible future trends (Wyrtki 1990; Milliman and Haq 1996; Warrick and Ahmad 1996) and in possible consequences for coastal erosion (Bird 1996). Global warming is seen as a potential cause of rising sea-levels, with various mechanisms potentially involved. This subject is

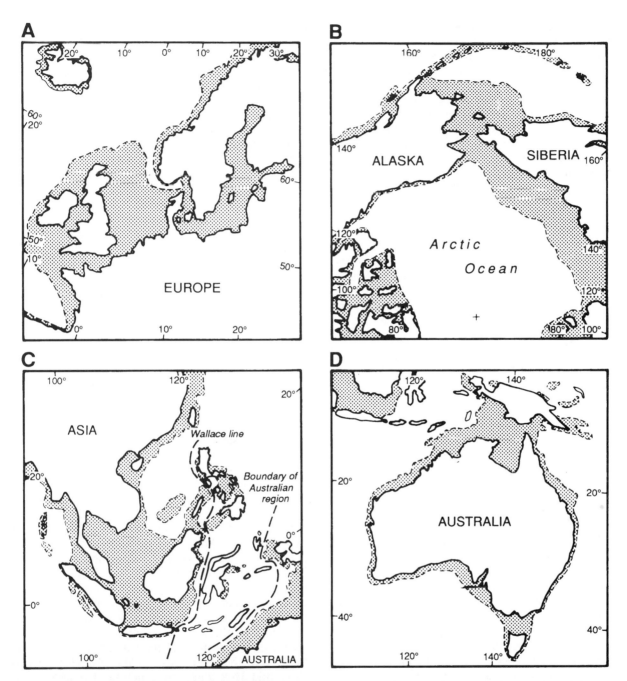

Fig. 6.10 Approximate extent of selected land bridges (shaded areas) at the last glacial maximum (~ 21–17 ka BP) drawn at the 183 m isobath. (A) Europe; (B) Beringia; (C) southeast Asia; (D) Australia and Papua New Guinea

reviewed regularly and forecasts made by the Intergovernmental Panel on Climate Change (e.g. see Wigley and Raper 1992; Wigley 1995; Warrick *et al.* 1996). A review of evidence and causal mechanisms is provided by Raper *et al.* (1996).

Considerable difficulty is attached to working out how equivalent sea-level is behaving at present, in relation to any possible melting of ice sheets or glac-

iers. The ongoing hydro-isostatic relaxation of the ocean floors and continental margins has the potential to swamp any signal related to anthropogenic sea-level rise that might be present in the global array of tide gauges (Lambeck 1990; Douglas 1991; Davis and Mitrovica 1996; Zerbini *et al.* 1996). To separate the effects of relaxation requires that this process be well embodied in the kinds of geophysical models referred to earlier in this chapter. The very rapid iso-static uplift rates in areas that carried Quaternary ice sheets makes it virtually impossible to isolate any small greenhouse-related effect in these areas (e.g. see Barnett 1984; Parker 1996). An example of an attempt to isolate the effect and reveal trends in sea-level is provided by Peltier and Tushingham (1989, 1991). These authors concluded that there remained a coherent sea-level signal of about +2.4 mm/a that might relate to the effects of global warming.

It appears from work such as this and from historical and contemporary records that sea-level is rising steadily but slowly; a rise of about 10–15 cm over the past century is indicated (Gornitz *et al.* 1982; Gornitz 1993), and the present rate of rise is taken to be 1–3 mm per year. These figures refer to the mean gauge rise, with *relative* sea-levels still of course falling in areas experiencing rapid tectonic uplift.

Various geophysical parameters have been employed to look for a signal of rising sea-level and diminishing ice volume. The length of the day, for example, would vary if the Earth's moment of inertia changed (Etkins and Epstein 1982). This could be brought about by a transfer of mass from polar areas to lower latitudes, in the form of meltwater. Historical observations show that indeed the day has lengthened (the rate is low – perhaps 3 ms per 100 years), but the amount can satisfactorily be accounted for by tidal friction alone (Barnett 1983) and does not establish that high-latitude ice is presently shrinking.

Global warming also has the potential to affect sea-levels because of the thermal expansion of sea-water (Wigley and Raper 1987, 1993). Records of sea surface temperatures collected from ships show a good correspondence with the sea-level record (Robock 1983), and this effect must therefore be quantified and understood before any meltwater contribution can be isolated. Many estimates suggest about 10–20 cm of sea-level rise from this cause alone by 2050.

The exact causes of the present sea-level rise are thus not known. Mantle readjustments following the last deglaciation are undoubtedly still taking place, but it is nevertheless possible that the major cause is global warming related to human activity.

Another consequence of warming is that mountain glaciers undergo retreat, feeding additional meltwater into the rivers and hence to the sea. Small glaciers are more sensitive to environmental change than large ice sheets. It has been estimated that this process may have raised sea-level by 2–7 cm this century, and that a further 30 cm rise is possible in the next century (Meier 1984). Larger volumes of meltwater may, in the much longer term, be contributed from the massive Greenland and Antarctic ice caps.

Antarctic ice has the potential, however, to pro-duce a significant sea-level rise in the relatively short term. The main East Antarctic ice sheet lies on bedrock above sea-level. The 10% of Antarctic ice forming the West Antarctic ice sheet, however, is grounded below sea-level. It is stabilised by adjacent ice shelves, such as the Ross Ice Shelf. It has been suggested that relatively little warming and sea-level rise would be required to destabilise this arrangement, and allow melting of the West Antarctic ice (e.g. Thomas and Bentley 1978), although others envisage that very substantial warming would be required to trigger collapse (Vaughan and Doake 1996). Mercer (1978) suggested that the ice shelves could collapse in about a century, with complete disintegration of the West Antarctic Ice sheet following over a further few centuries. The water so released could lift sea-level by about 5 m, and there would additionally be geoid perturbations. Interestingly, sea-level during the last interglacial was about 6 m higher than present (as indicated earlier). Elevated shorelines at about +6 m are found in Antarctica, and are about 100 ka in age. It thus seems possible that the West Antarctic ice sheet did not survive the warming of the last interglacial, and that the additional meltwater from its disintegration accounts for the higher last interglacial sea-level; this adds weight to the idea that it might not survive the present one either, given the extra anthropogenic warmth! Holocene disappearance of the enlarged glacial ice mass over Antarctica is considered to have involved the rapid collapse of a marine-based 'Ross ice sheet' covering the area of the present-day Ross ice shelf (Thomas and Bentley 1978), with collapse triggered by sea-level rise.

It is also possible that just the opposite effect might arise in Antarctica if the oceans warm. As noted in Chapter 3, warm oceans nourish ice sheets by promoting precipitation and an increase in the mass balance. Therefore, global warming could conceivably initiate a sea-level decline, if the change in the mass balance of the Antarctic ice overshadowed the effects of thermal expansion of sea water and of ice melting elsewhere. There is a suggestion that this might have been the case in the orbitally-induced early–mid Holocene warm period (Domack *et al.* 1991). Mapping of sediments around the Antarctic margin has indicated ice expansion then, and this may form an analogue to the Earth warmed by the greenhouse effect (although the two mechanisms may differ in other important respects). The same effect has been foreshadowed for some of the northern hemisphere ice (Miller and de Vernal 1992), aided by the declining summer insolation related to orbital changes which is now underway. The Antarctic ice shelves, however, may indeed be threatened by global warming (Vaughan and Doake 1996), and it appears that warming will cause net ablation of the large Greenland ice sheet (Oerlemans 1989, 1993b). Modelling studies of these differing mass balance changes have suggested that in fact the net effect on global sea-level may be slight (Ohmura *et al.* 1996).

The time required for melting to occur gives us the opportunity to plan ahead, and to adapt our coastal settlements accordingly (and to moderate our production of greenhouse gases!). Some deliberate control of sea-level is possible; it has been estimated that water storage in dams has prevented 1–2 cm of sea-level rise this century (Gornitz *et al.* 1982). Water storage on a much larger scale would be required to halt a continuing rise, and the deliberate flooding of low-lying areas such as the rift valley between Israel and Jordan has been contemplated, but would bring with it severe environmental costs. Whether or not we seek to intervene deliberately in this way, the evident contemporary sea-level rise, perhaps the precursor of the larger change which may follow, obliges us to continue our observation of the sea, and to refine our knowledge of its past as well as its contemporary behaviour.

CHAPTER

7

EVIDENCE FROM THE OCEANS

I should have been a pair of ragged claws scuttling across the floors of silent seas.

T.S. Eliot (1888–1965)
Morning at the Window

Oceans cover 71% of the surface of the planet and contain the largest component of the planet's biosphere. It is therefore essential to understand the interactions within the oceans in addition to the influence they have on the climate of our planet. Two-thirds (67%) of the land area of the globe is located in the northern hemisphere. Oceans occupy 61% of the northern hemisphere, and 81% of the southern hemisphere. The southern hemisphere is often termed 'maritime' in contrast to the northern hemisphere which is described as the 'continental hemisphere'. The implications of this asymmetry need to be examined in palaeoclimatic and palaeoenvironmental studies (Tchernia 1980; Emiliani 1981).

Oceans cover a very uneven topography ranging from unusually narrow, gently sloping continental shelves to deep and narrow, elongated trenches, some as deep as 10 km or more. Three-quarters of the oceans (77%) are deeper than 3 km, so that we need to consider the influence exerted by deep oceanic water masses on global climate (Labeyrie *et al.* 1987). A third of the oceanic water mass by volume (30%) has a temperature between 1 and 2°C. Consequently, it is important to investigate the impact of this cold water mass on overall global temperatures. We need to know how the oceans have changed during the Quaternary before trying to assess the effects they may have had and may have in the future on global change (Open University Course Team 1989a, b, c). In addition, sea-levels have fluctuated throughout Quaternary time (Chapter 6), sometimes influencing the chemical and physical attributes of the oceans. The purpose of the present chapter is to examine the changing nature of the ocean environment during the course of the Quaternary.

PROPERTIES OF SEAWATER AND OF THE DIFFERENT WATER MASSES

Seawater is salty because it contains dissolved salts (solutes) which are the weathering products of continental and oceanic rocks and derive in part from sea-floor gases. Sodium and chlorine are the principal ions found in seawater because they are among the most soluble constituents of weathering. Salinity is a measure of the total dissolved salts in seawater, and the units commonly used are expressed as grams of dissolved salts per litre of water or parts per thousand (ppt or ‰). It is important to recognise that salinity values differ between oceans, or parts of them, and between different water depths. Although salinity differences may be small, they play a significant role when water densities are also taken into account. Surface-water salinities range from 33.5‰ in the Antarctic Weddell Sea to 37.5‰ in the North Atlantic. The latter value is indicative of high evaporation whereas the Weddell Sea is diluted by ice meltwater. A value of around 35‰ is found throughout most oceans, and is thus accepted as average for oceanic water. Several dissolved gases in ocean water are important owing to their abundance, and because they affect biological activity. These gases are oxygen, carbon dioxide and nitrogen. Their significance is discussed later.

As mentioned earlier, a third of all ocean water has a temperature between 1 and 2°C, rendering this water rather dense, especially when its salinity is high. For denser water to form in an ocean, it has to originate at some stage near the surface either through cooling or through evaporation, or both. Hence, cold water masses must originate near the poles, and this explains why the North Atlantic Deep Water (NADW) originates in the Norwegian and

Greenland seas, and the Antarctic Bottom Water (AABW), which is colder (usually around 0.5°C) and has a characteristic salinity of 34.7‰, originates all around the Antarctic continent, especially in the Weddell Sea, Ross Sea and Prydz Bay areas (Bleil and Thiede 1990). This water mass is the most widespread in the world and is denser than the NADW which characteristically has a temperature of around 2°C and a salinity of 34.95‰. On approaching the Antarctic continent (Fig. 7.1), the NADW rises progressively closer to the surface to become the Antarctic Circumpolar Water (Corliss 1983). This water reaches the surface at the Antarctic Divergence, where surface waters diverge (Fig. 7.1). It is in this area of divergent water masses that biological activity is among the most prolific in the world (Hedgpeth 1969).

Above the deep cold-water masses is a series of 'intermediate' water masses such as the Antarctic Intermediate Water (AAIW) (Fig. 7.1). The AAIW, which in the Atlantic Ocean travels past the equator at a depth of about 1–2 km, is in fact the most widespread intermediate water mass in the world. Its temperature is usually between 2 and 4°C, and it has salinities lower than the AABW (34.4‰).

Characteristically, the 'upper water masses' are more localised because they are under the influence of surface currents, so that temperature and evaporation/precipitation ratios may differ markedly between seasons. Consequently, the lateral and vertical extent of these water masses will also vary quite substantially. The upper water masses consist of waters which have a steep vertical gradient in temperature, salinity and dissolved oxygen content. There is thus a mixed surface layer which sits above a zone called the thermocline within which the temperature decreases significantly with depth, and density increases. Below the thermocline are the 'central waters' which usually form the subtropical belts or gyres. The thickness of these central waters depends on their location. For example, where two water masses converge at the surface, as is the case at the subtropical convergence, warm water sinks so that the thermocline is lowered, and the upper layer increases in thickness.

Nevertheless, since water density is controlled by temperature as well as by salinity, some very cold water (under 1°C), which is less saline than the average ocean water, may sink down to the abyss simply because of its low temperature. Hence several water

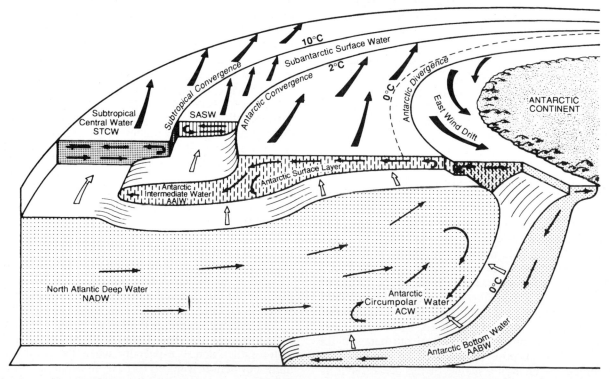

Fig. 7.1 Schematic diagram showing the various water masses in the Southern Ocean adjacent to Antarctica. Note the temperatures of the different water masses which influence their respective densities. Also observe the direction of movement of the different currents, bearing in mind that where two water masses diverge from one another, there is upwelling of water to the surface; converging water masses induce downwelling. Eventually, the AABW will continue flowing into the Atlantic Ocean. (Adapted from Hedgpeth 1969)

masses, each with different densities, are superimposed upon one another in the oceans. These different water masses are not only characterised by different densities but also have other properties of importance to physicochemical processes in the oceans and to biological activity.

Deep, cold water is characteristically rich in nutrients and dissolved CO_2 and, if such water is brought to the surface, as would occur in a zone of divergence, an 'upwelling' phenomenon occurs which is beneficial to fisheries. The best documented regions where this phenomenon occurs are along the southwestern coast of Africa, off Namibia, and along the west coast of South America, off Peru. Upwelling of cold water, rich in nutrients, causes high organic productivity at the surface and results in rich fishery catches. It is also the area where venting of CO_2 occurs. Such venting areas occur preponderantly near the equatorial belts where surface currents diverge. Nevertheless, climatic links with areas of upwelling in the oceans and adjacent land surfaces are significant, as cold upwelling offshore is often associated with aridity onshore. It is important to remember that water density is a key factor that will dynamically control the movement of water masses in the oceans below the surface. Thus, a better understanding of temperature and salinity changes in the oceans should help to define the functioning of the oceans during the Quaternary. It is also necessary to become aware of the Coriolis effect on water mass movements under the influence of winds. At the ocean's surface first, water movement is deflected towards the right in the northern hemisphere, and in the opposite direction in the southern hemisphere, and layers below the surface become progressively deflected with depth. Such a spiral pattern is called the Ekman spiral. It explains the change of oceanic current direction which eventually forces some surface currents to move in the opposite direction to those that occur several hundreds of metres below the surface down to the sea floor.

The extent of sea ice in polar regions also plays a significant role in controlling water density, since the water below the sea ice becomes more saline as a result of ice formation at the surface. In addition, the presence of sea ice, even if it is for the winter season only, helps to prevent a CO_2 sink from occurring, in comparison with the period during which there is no ice cover (Martinson 1990).

Earlier we introduced the concept that there are different water masses in the oceans, and that these are characterised by specific salinities and temperatures. Combination of the latter two factors produces a particular water density, which is the controlling parameter with respect to water mass movement. Examination of Fig. 7.1 shows directly that many of the water masses must be in continuous motion, and

this led several oceanographers to postulate the global pathway taken by oceanic currents, not only at the surface, but also at different depths in the oceans. Stommel (1957) was the first to visualise a general circulation both at the surface and at depth for at least the Atlantic Ocean. More recently, however, it was Broecker (1987) who popularised the idea of a global system of oceanic circulation, which he termed the 'global conveyor belt' (GCB). The idea of the latter global system became possible only after numerous water samples (literally thousands) had been taken at sea during the international programme called GEOSECS (Geochemical Oceans Sections) (see Ostlund et al. 1987). This allowed the identification of the age of various water masses as well as the sites in the oceans where water downwells fairly readily, as recognised by isotope markers, resulting from atomic bomb tests, showing progressive contamination of water masses. The GCB is basically controlled by density gradients, with the most important site being the North Atlantic where dense saline water is produced and downwells before undertaking a long journey into the deepest portions of the global ocean. Because of the importance of salinity and temperature controlling the behaviour of water masses in this circulation system, the GCB is also referred to as the 'thermohaline circulation' system. Broecker's models have recently been reviewed by Schmitz (1995) who proposed several circulation models, including a three-layered ocean. For simplicity, his two-layered model is reproduced here in Fig. 7.2 with slight modifications. It shows that the NADW, after its journey from the North Atlantic into the deep Pacific Ocean, will see at least part of that water mass upwell in the North Pacific prior to passing through the Indonesian Archipelago, while becoming much less dense (being very warm and less saline due to the substantial local precipitation in the area), and will eventually reach the Atlantic Ocean, either by circumnavigation around the African Cape, or through Drake Passage south of South America (the latter situation is preferred and receives more emphasis in Fig. 7.2 because the strength of the subtropical convergence south of Africa is such that it acts as a barrier to most currents going westward along the southern tip of Africa, and which otherwise would have the potential to circumnavigate this continent). It is this GCB system that has received much attention in the last few years because it can explain the links between all the oceans and the leads and lags between them. Nevertheless, it is necessary to know that to close the loop of the GCB, close to 2000 years are necessary under the present-day scenario. Through time, the GCB may have operated under a different regime, with evidence that its rate of movement may have been slower, or even completely shut off.

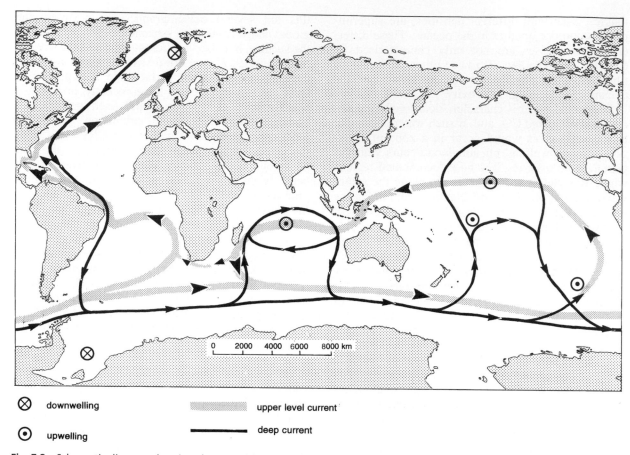

Fig. 7.2 Schematic diagram showing the approximate pathway of the 'global conveyor belt' in a two-layered global ocean. The bold line shows the path taken by the deep current that is first generated through downwelling (circle with inner cross) of cold, saline water in the North Atlantic. The grey line represents the path taken by the upper water currents which result from upwelling (circle with dot) of some of the cold, substantially old and deep water in the central Pacific Ocean. Note the direction of the currents, and especially the path taken for the return flow into the North Atlantic. Not all texts indicate a small component of the return flow along the Cape, but instead a major component of the surface water circumnavigates the southern tip of Africa. (Adapted from Schmitz 1995)

The processes of thermohaline circulation that operate in the oceans with respect to different water masses are basic to deciphering past changes. Surface temperatures, controlled by solar radiation and heat losses or gains through evaporation and associated atmospheric processes, affect ocean currents and other processes below the surface, and form part of the driving forces controlling the earth's climate. In addition, biological productivity in the oceans is determined by the physicochemical interactions which occur throughout the ocean water column, especially carbon dioxide and oxygen uptake or discharge. There is 50 times more CO_2 stored in the oceans than in the atmosphere and the biosphere, so that the storage of CO_2 dissolved in the oceans or used by organisms secreting $CaCO_3$ skeletons is vital to an understanding of CO_2 fluxes relevant to global climate (see Varney (1996) for a recent summary). It is

principally in high latitudes that CO_2 is taken up more effectively by the very cold, dense surface waters. Through the sinking of dense water, CO_2 will eventually end up in the deep part of the oceans where it is stored on a long-term basis (Anderson and Malahoff 1977). It is mainly in the tropics that CO_2 is vented through the surface to the atmosphere, especially where water masses diverge.

Seawater composition is kept nearly constant because of the conservative nature of its major elements, and because biological activity and other physicochemical processes do not affect it significantly. Other elements present in smaller amounts (minor or trace elements) display what is called non-conservative behaviour because they are frequently influenced by biological processes (such as when trapped in, or forming part of, the skeletons of organisms), or by physical processes such as where two water masses mix, or by dissolution of organically or inorganically produced precipitates. Changes in these 'unstable' elements will reflect changes in physicochemical processes within the water column or at the sediment–water interface. Dissolved gases also play an important part in oceanic processes. Dissolved oxygen, for example, is strongly affected by biological activity in the oceans, although temperature and salinity play an integral role in controlling the amount of dissolved oxygen in seawater. Dissolved oxygen levels increase with a decrease in temperature, as does the amount of dissolved CO_2. The latter, of course, is of great importance for the preservation of calcium carbonate skeletons of organisms, and its effect needs to be taken into account when interpreting the fossil record, simply because species indicative of particular environmental conditions may have been destroyed through dissolution. Nevertheless, as water masses move away from the zone of downwelling, oxygen levels progressively decrease due to biological activity leading to oxygen consumption. Hence waters a long way away from their source of origin, like those in the deep Pacific Ocean, are poorer in oxygen compared with those near the surface in the North Atlantic.

The composition of seawater is such that with a change in water temperature and pressure, and also in the partial pressure of dissolved CO_2, calcium carbonate in contact with that water may dissolve, more particularly in the deeper parts of the oceans. Calcium carbonate, being the principal component of many planktonic organisms, dissolves through the water column as water depth and pressure increase and temperature decreases. It is principally the carbonate ion concentration in seawater, determined by these three parameters, which controls the preservation of calcium carbonate shells or tests. Conse-

quently, it is necessary to determine the depth in the ocean where calcium carbonate returns to solution. This level is called the *calcite compensation depth* (CCD), and varies between and within oceans (Takahashi 1975). Figure 7.3 shows depth contours which delineate the levels in the oceans where calcite (the most stable form of calcium carbonate) becomes soluble. It is below these depths that under normal circumstances no calcium carbonate is found in sediment on the sea floor. The CCD levels are much deeper in the Atlantic Ocean because the total amount of dissolved CO_2 is greater there; a large supply of terrigenous material brought by rivers into the oceans will cause an increase in the amount of total organic carbon in seawater, causing the CCD level to rise near the edge of continents. Principally, below the CCD, clays and to some extent remains of siliceous organisms (see section below) predominate.

The CCD is important for ocean studies because it affects the preservation and/or dissolution of organisms that secrete a calcitic shell or test. The principal organisms of concern are the microscopic unicellular foraminifera (usually under 1 mm) and the much smaller organisms of algal affinity called coccoliths (of the order of a few micrometres in diameter).

So far, we have discussed the level in the ocean where calcium carbonate (principally calcite) is totally dissolved. We also need to take into account the level in the ocean where calcium carbonate *starts* to dissolve. Some organisms have a thinner and smaller skeleton than others, and so are more prone to dissolution. Recognition of this phenomenon is important when using fossil taxa to interpret palaeo-environments, as some taxa are 'missing' from the record because of dissolution processes which operated while the remains of the organisms descended through the water column or were lying on the sea

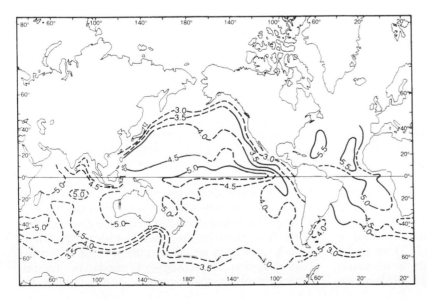

Fig. 7.3 Map showing contours (km) in the oceans defining the calcite compensation depth (CCD). This was obtained by examining sediment on the sea floor that contained calcite; below these depths, there should be little or no calcium carbonate preserved on the sea floor. Solid contours represent more than 20 samples per 10° square, and dotted contours represent fewer than 20 samples. (From Open University Course Team 1989b)

floor. The level in the ocean where calcium carbonate starts dissolving preferentially for thinner and fragile tests is more than 1.5 km above the CCD. However, it is the level below this, called the lysocline, where only the resistant forms of organisms are found. The lysocline level usually parallels the CCD level, and usually occurs at about 500 m above the CCD. For further information, consult Varney (1996).

There is another type of calcium carbonate, called aragonite, which is secreted by organisms and precipitated directly from the water. Corals are among the best known organisms with an aragonitic skeleton. Most corals are found in the tropics because it is in those regions where water temperature is sufficiently high to allow easy precipitation of aragonite. Another group of organisms called pteropods or sea butterflies, and which are related to gastropod molluscs, have a shell made of aragonite. They can frequently reach up to 1 cm in length. Aragonite is a less stable form of calcium carbonate and thus dissolves much more readily than calcite. Consequently, the aragonite compensation depth (ACD) in the oceans is much closer to the surface than the CCD, and can be used as a more sensitive recorder of CO_2 in the oceans through time, especially during periods of oceanic changes when water masses alter their proportion significantly, or when upwelling of cold, CO_2-richer waters occurs. Identification of the location and timing of calcite and aragonite dissolution through the study of deep-sea cores is indicative of processes occurring in the oceans, and can help to define their relationship with climatic change. Microplankton, especially foraminifera and coccoliths, are so abundant in some parts of the oceans that they constitute a large proportion of the biomass and so help to regulate the amount of carbon produced by the biosphere. It becomes vital, therefore, to put a lot of emphasis on a better understanding of the production and preservation of biogenic carbonates in the oceans, since they form a major sink as well as storage of carbon, obviously of importance to the global 'greenhouse effect'. It is necessary to determine the depth of dissolution and the CCD in the oceans through time in order to identify the role the oceans have played in the ocean–atmosphere budget.

Silica (SiO_2) forms the lattice of several important groups of organisms, and is another significant component of the oceanic chemical budget. Among the algae are organisms called diatoms which secrete a variety of siliceous, pill-box shaped tests called 'frustules'. Radiolarians comprise another group of unicellular organisms related to the foraminifera. Silica, frequently labelled as opaline silica because of its amorphous and porous texture, is secreted by these micro-organisms (less than 1 µm in diameter) which commonly inhabit zones of high productivity.

The remains of siliceous organisms are found in the oceans where carbonate concentration is low (as in the polar regions where the dissolved concentration of CO_2 is high), or where carbonate is absent, or below the CCD. The study of biogenic siliceous remains thus becomes important for the reconstruction of past conditions in these parts of the oceans. Silica also dissolves in the oceans, especially in the upper 50 m of the water column (the solubility of silica increases as the temperature increases), and within the sediment as well, especially at the sediment–water interface, and this will affect the fossil record and its interpretation.

Dissolution of biogenic silica is highest within the waters at the surface and progressively decreases down to approximately 500–1000 m depth, depending on the ocean, before reaching a steady state. Sedimentation down to the deep sea of small diatoms, and of the coccolith plates which are so minute, may take a long time: from several weeks to possibly several years. This gives plenty of time for dissolution to occur. However, a rapid sedimentation rate may prevent the slow process of dissolution from occurring at the sediment–water interface. One way of rapidly transporting microscopic remains or organisms is through the settling of faeces of various organisms which fed on these organisms and whose skeletons are kept somewhat intact. Copepod crustaceans are a common example of a large producer of faecal pellets in the oceans.

MICROFOSSILS AS TOOLS FOR PALAEOENVIRONMENTAL RECONSTRUCTION

Foraminifera (also called forams) are ubiquitous marine organisms which secrete a test consisting of a series of small chambers. Most forams have a calcitic test. There are two main types of foraminifera, characterised by their mode of life (Funnell and Riedel 1971). One is labelled planktonic because the organism is able to control its position in the water column, although most individuals occur near the surface and migrate up and down during their life cycle, sometimes as much as several hundred metres. The other group of foraminifera, referred to as benthic forams, lives on the sea floor, or sometimes within the upper few centimetres of the sediment at the bottom of the ocean. Numerous benthic forams have a test made of an agglutination of debris from the sea floor and thus differ from those in the other benthic group which have a calcareous test. A wealth of information has been obtained on ecology of planktonic forams, especially about their modern distribution in the oceans (Bé and Tolderlund 1971). Since there are only about 40 species of planktonic forams, it is possible to define their respective biogeographical boundaries. Figure 7.4 shows the distribution of these organisms in the present-day oceans.

Studies of the fossil remains of the organisms on the sea floor greatly elucidate the characteristics of ocean surface waters for the entire Quaternary, during which most of these species existed. Several planktonic species of forams are characterised by a different direction of coiling for the chambers made by the individual throughout its ontogeny. Coiling is either sinistral (to the left) or dextral (to the right) and the direction of coiling in some taxa seems to be broadly related to a particular temperature regime. Different coiling types for distinct species are now used to relate to portions of the oceans with specific surface temperature regimes. The best known example is the foram *Neogloboquadrina pachyderma* which characteristically has a sinistral coiling direction in cold water. A change to dextral coiling in this species seems to occur in waters with a mean annual temperature greater than 9°C. Several other species listed in Fig. 7.4 have different coiling directions in regions with different thermal regimes. Study of chamber arrangement in planktonic forams thus allows us to detect a shift in surface water masses during the Quaternary. Nevertheless, one should be

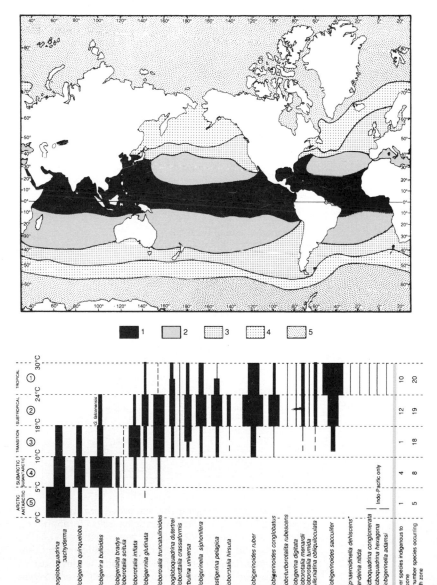

Fig. 7.4 Map showing zoogeographical zones in the oceans defined by planktonic foram assemblages. Taxa belonging to the five zones are shown below. The thickness of the bars in the lower figure represents the relative abundance of each taxon in each zone. (Adapted from Bé and Tolderlund 1971)

aware that the coiling direction is not always entirely consistent with temperature signals.

Several authors have tried to relate foram size, shape, surficial texture and other architectural factors to particular temperature regimes and/or geographical zonations. All of these features may be related to the capability of individual forams to secrete a test (so that for some species it is easier for crystallisation to occur at a certain 'optimum' temperature), or to cope with variations in water density (controlled by salinity and temperature). Temperature should also control the rate of calcification and the chemical composition of the foraminiferal test.

Benthic foram morphology reflects its mode of life (Corliss and Fois 1990). Narrow, elongated and cone-shaped forams are often burrowers (infauna), whereas flat and broad ones live on the surface (epifauna). Observations of benthic foram morphology and diversity are starting to provide information on organic productivity at the surface above the site where the forams are living, since the amount of organic matter supply, through its 'showering', can affect foram diversity and rate of growth. It now appears possible to relate the infaunal/epifaunal ratio to the interaction between phenomena that occur at the surface and at the bottom of the ocean.

OXYGEN ISOTOPES

Foraminifera have also been successfully used in palaeoceanography by studying their isotopic composition with respect to the stable isotopes of oxygen and carbon (Savin and Yeh 1981; Vincent and Berger 1981; Bradley 1985). The principle behind the use of stable isotopes in foraminifera is that the ratio between the two isotopes of oxygen (^{16}O and ^{18}O) taken up by the organism during test formation is controlled by temperature and by the isotopic composition of the ambient water. With knowledge of the latter two variables, the isotopic composition of foraminifera can be used to reconstruct palaeotemperatures. The formula commonly used to determine the isotopic difference ($\delta^{18}O$) between these isotopes of oxygen is:

$$\delta^{18}O = \frac{(^{18}O - ^{16}O)_{sample} - (^{18}O - ^{16}O)_{standard}}{(^{18}O - ^{16}O)_{standard}} \times 1000 \quad (7.1)$$

The units are in parts per thousand (‰). The standard commonly used for forams is PDB, a Cretaceous belemnite from the Pee Dee Formation in North Carolina. Standard Mean Ocean Water (SMOW) is the standard for present-day water and is given a nil value (also in ‰).

Fractionation occurs between the two isotopes of oxygen of interest here. The lighter isotope ^{16}O prefer-

entially escapes as vapour during water evaporation, so that rainwater is isotopically lighter than the ocean water from which it originated, and the $\delta^{18}O$ of carbonate shells decreases as water temperature increases (see Figs 7.5 and 7.6).

During glacial times, when sea-level had dropped by approximately 120 m (Chappell 1987; see also Chapter 6), much water was locked up in ice caps and mountain glaciers (Chapter 3). This water, in the form of ice, was isotopically enriched in ^{16}O, so that ocean water became proportionately enriched in ^{18}O (Shackleton 1987). About 0.11‰ change in the $\delta^{18}O$ in planktonic carbonate fossils, such as forams, represents a 10 m change in sea-level, although estimates vary somewhat between authors. The difference between the isotopic composition of seawater during a glacial period and today is approximately 1.2‰, with the ocean water being isotopically heavier during glacial times. Measurements of the isotopic composition of foraminifera from Quaternary cores are represented as curves of the isotopic composition of the water and its inferred temperature. Many workers have opted to study cores taken in tropical regions where it has been postulated that very little temperature change occurred between glacial and interglacial

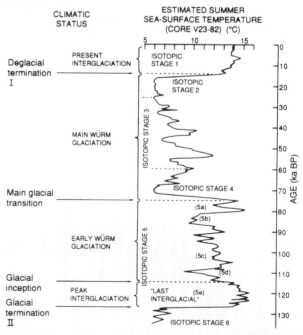

Fig. 7.5 Diagram showing the estimated sea-surface temperature (SST) for the North Atlantic based on foraminiferal assemblages of core V23–82 in the northwest of the Atlantic. Dating is obtained through correlation with other cores (based primarily on tephra layers and sea-level curves from Barbados), and the various, commonly used oxygen isotope stages are also indicated on this figure. (Modified from Bradley 1985)

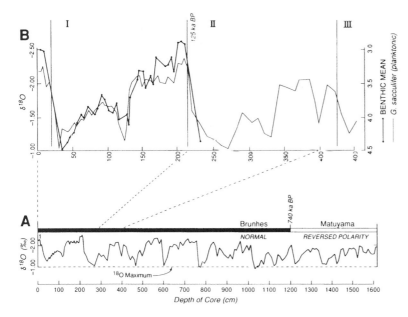

Fig. 7.6 Oxygen isotope stratigraphy of core V28–238 from the western equatorial Pacific. (A) $\delta^{18}O$ values for the planktonic foraminifer *Globigerinoides sacculifer* in parallel with the palaeomagnetic stratigraphy (the Brunhes–Matuyama polarity boundary is placed around 740 ka BP). Note the cyclicity of the record, with the most negative values returning to the same points. (B) is an enlarged portion of (A), covering approximately the last 250 000 years of the record to show variations in the $\delta^{18}O$ content of *G. sacculifer* and the mean values of benthic forams. Note the transition between isotope stages 6 and 5e placed around 127 ka BP and the sharp nature of this transition. Note also the difference in $\delta^{18}O$ values between the benthic and planktonic forams. Roman numerals represent the timing of the last three glacial maxima. (Slightly modified from Vincent and Berger 1981)

episodes. The record of isotopic change in the tropics has been interpreted as relating almost entirely to the effect of global ice volume and hence sea-level change, amounting to approximately 120 m, equivalent to an isotopic shift of 1.2‰. However, elsewhere in the oceans an isotopic change in the forams found in cores results from both temperature and ice volume changes (Mix and Ruddiman 1985). Nevertheless, the amplitude of changes recognised in cores from nearly anywhere in the oceans between glacial and interglacial periods is sufficient to allow the patterns of changes through time to be used for correlation between different cores at different locations.

If one examines a much longer record spanning the entire Quaternary, it becomes obvious that fluctuations in the isotopic values found in planktonic foraminifera are somewhat cyclical. Figure 7.6A displays such a pattern which is repeated about every 125 000 years. In the core V28-238 taken in the west equatorial Pacific (Shackleton and Opdyke 1973), compaction of the sediment in the core is such that the upper portion of the core appears thicker than an equivalent time period lower down the core. The difference in isotopic composition between the benthic and planktonic forams analysed from the same core is used to identify the differences between both the surface temperature as well as the smaller differences in $\delta^{18}O$ between bottom and surface water (Birchfield 1987). At present, there is still disagreement about these differences because the isotopic composition of deep ocean water during glacial times is not precisely known. We still do not know how much the bottom water, which has a slightly different isotopic composition, mixes with surface water and thus alters the isotopic value at the surface (Shackleton 1987).

Several other problems also occur because it is necessary to calibrate the isotopic data against other factors which influence the isotopic composition of foraminifera. For example, one has to determine the fractionation between the two isotopes of oxygen for different taxa, and in addition, to determine the size fraction of forams because different sizes grown in the same water apparently have different isotopic signatures (Vincent and Berger 1981).

Some foram species are in closer *isotopic equilibrium* with seawater than others which diverge from the expected value (see below), and which consequently register an enrichment in one of the two isotopes. In addition, some planktonic foram taxa have a mode of life which permits them to live at different water depths (where temperature may differ markedly, and water isotopic composition less so), so that different chambers of the one specimen may have different isotopic values. There is some evidence that it is the rate of calcification which controls the oxygen isotopic fractionation in forams. If correct, this would explain the different isotopic values for the different foram sizes.

The life-style of the forams can also influence the isotopic composition of their tests. To allow gametogenesis to occur, some forams add an extra calcitic layer which is often in isotopic equilibrium with the ambient water, itself at a very different temperature to that of the surface. (Gametogenesis is the period when gametes are released in the deeper parts of the water column where the forams live, prior to dying and sinking to the sea floor.) We need to know more about the mode of life of these organisms before interpreting their isotopic signatures.

Isotope stratigraphy, as it is called, is used to provide a basic age for cores taken from anywhere in the world's oceans (Jansen 1989). For example, the last major change in $\delta^{18}O$ in a core, with values becoming more enriched in ^{16}O, is interpreted as a result of the melting of the ice caps after about 20 ka BP. This phase is labelled as the transition between isotope stages 1 and 2 (see Fig. 7.5). A previous similar pattern in cores occurs at the transition between isotopic stages 5 and 6. Such cyclicity has now been recognised in all deep-sea cores and it is accepted that the cycles relate to astronomical cycles which govern the position of the Earth with respect to the sun, the tilt of the Earth and the wobble of the Earth along its axis (see Chapter 5). The importance and amplitude of some of these cycles have been examined for several marine cores, and it has been recognised that three main cycles lasting about 100, 41 and 23/19 ka consistently occur (McIntyre 1989). The cycles correlate fairly well with astronomical forcing, but it is important to realise that the effect of the forcing may be felt differently at different latitudes. For example, the 41 ka cycle has a more pronounced effect at high latitudes, compared with the 19 ka cycle which is more important in middle to low latitudes. The 41 ka cycle has been the main driving force behind the changes in glaciated areas, and so has the 100 ka cycle for the North Atlantic. However, the amplitude of some of these cycles has been such that their influence on climatic variation during the entire Quaternary has not always been the same. For example, the 100 ka cycle is considered to have only played a dominant role during about the last 650 ka of the Pleistocene, whereas the 41 ka cycle had a significant amplitude during the entire Matuyama chron, spanning 2.5–0.7 Ma. These cycles have been deciphered mainly from isotopic changes in foraminifera and from variations in the total $CaCO_3$ content in marine cores.

Carbon isotopes (^{13}C, ^{12}C) are usually analysed in conjunction with oxygen isotopes, and have been used to obtain information on the origin of organic matter used by planktonic organisms, since there is a marked difference between the carbon isotopic ratio of ocean water compared with organic matter of continental origin. It has not been possible to use carbon isotopes in foraminifera to correlate global events, as has been done with the oxygen isotopic record. On the other hand, isotopic shifts in carbon isotopes have been used to infer substantial changes in planktonic blooms at the ocean surface, as well as changes in primary productivity (causing large amounts of $CaCO_3$ to precipitate), and changes in the supply of usually fresh water of continental origin to the ocean. For example, an isotopic shift was detected in the latter part of the Pleistocene in the Gulf of Mexico, and was interpreted as a sudden influx of glacial meltwater carried into the Gulf by the Mississippi River (Broecker *et al.* 1989).

In addition, methanogenesis, which occurs within the ocean and more significantly within the sediment, may also alter the carbon isotopic ratio. Nowadays, it has also been possible to detect past changes in bottom-water ventilation by analysing the $\delta^{13}C$ of specific species of benthic foraminifera, especially if taken from several cores along various depth transects in the oceans.

Nitrogen isotopes (^{15}N, ^{14}N) have also been applied in combination with carbon isotopes to determine the origin of the organic matter in deep-sea cores. Upwelling of cold, dense and nutrient-rich water near the ocean surface can be detected from the carbon and nitrogen isotopic signature of organisms which live near the surface, such as forams and nannoplankton. François *et al.* (1993), for example, have demonstrated the applicability of $\delta^{15}N$ to evaluate change in nitrate utilisation in surface waters affected by palaeoflux variations in the western Indian Ocean.

TRANSFER FUNCTIONS

Another technique used to reconstruct conditions in the oceans through time has been to establish the relationships that exist between assemblages of species and the ecological conditions which control them, especially sea-surface temperature (SST). Thus, faunal association and composition have been used to compute a palaeoclimatic index (see Imbrie and Kipp 1971). In a sense, it is like plotting a ratio of selected warm-versus-cold species to obtain an indication of the temperature in which the faunal assemblage lived. Multivariate statistical analysis has been successfully used to quantify the past conditions in the oceans through correlation of modern-day species with established oceanic conditions. Consequently, several research groups have come up with 'transfer functions' for different oceans after having related SST to the presence of faunal assemblages recovered from the tops of cores collected in different oceans. Naturally, an important uncertainty pertains with regard to interpreting the information, since the remains of organisms found on the sea floor are not necessarily modern. Phenomena such as reworking by bioturbation or bottom-current activity, or slow sedimentation or preferential dissolution of some taxa, may affect the accuracy of the correlation.

Four major faunal assemblages (from polar, subpolar, subtropical and tropical regions) have commonly been chosen to establish the equations or transfer functions used to calculate sea-surface temperature, and to distinguish between summer and winter values. In fact, information obtained from these transfer functions has been used in parallel with oxygen isotope curves, and a good correlation between the two was established, confirming the usefulness of transfer functions despite the uncertainties they contain. Fossil groups used for the transfer functions are principally foraminifera, but also include radiolarians, diatoms and coccoliths. The use of transfer functions climaxed during the CLIMAP (Climate: Long-Range Investigation Mapping and Prediction) project (McIntyre 1981; CLIMAP Project Members 1981). One important objective of the CLIMAP project was to establish the conditions on the globe during the last glacial maximum, nominated as 18 000 years ago. For this project, sea-surface temperatures were reconstructed and are presented for August in Fig. 7.7. This figure represents the state of knowledge in 1981 when the CLIMAP maps were

first published. Some of the temperature reconstructions presented by CLIMAP are being re-examined and are now being checked by other means.

The CLIMAP transfer function method proved extremely popular but had failed to link faunas from the past for which no analogues could be found. Thus, the *modern analogue technique* (MAT) was developed to define the degree of dissimilarity between modern faunal assemblages and fossil ones within specific temperature ranges. Prell (1985) provided a comparison between Imbrie and Kipp's (1971) approach and the MAT. Since then, several other techniques have been developed, and rely on better quality core material than the ones used for the CLIMAP investigations. The work of Pflaumann *et al.*(1996) should be consulted so as to understand better the intricacies of using MATs.

This approach has also been applied to other periods of the Quaternary. For instance, the ocean during

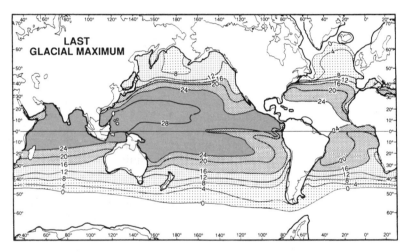

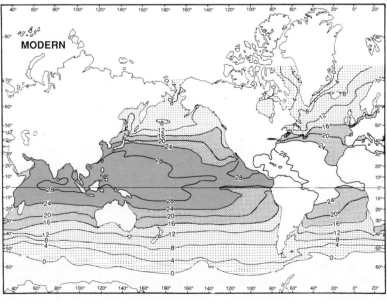

Fig. 7.7 Reconstruction of surface ocean temperature in August for the glacial maximum at 18 ka BP, and comparison with the present day. This reconstruction by the CLIMAP group (see text) was based on transfer functions from several planktonic organisms and on foraminifera oxygen isotopic compositions. (Modified from McIntyre 1981)

the last interglacial has been compared with the present day, because sea-level is considered to have been approximately the same. A map representing the differences in SST for February and August was also compiled in order to determine past conditions and to validate the use of transfer functions using faunal/floral assemblages and stable isotopes (Ruddiman 1984).

Information about past conditions in the oceans, especially for surface water, can be used in combination with continental reconstructions to model past climates. One aim is to model oceanic processes through time in order to detect how ocean currents interacted with the atmosphere (Broecker 1987a; Broecker and Denton 1990). This type of work is claiming more interest nowadays, especially because of the concern for the future 'greenhouse effect'. Modellers need to understand better the role played by the oceans in the global CO_2 system; reference to past events is much needed in order to comprehend the range and rates of variability of the ocean–atmosphere system.

PALAEOCHEMISTRY

Several other chemical analyses of marine organisms have recently been used to supplement the isotopic record of the oceans. The trace element cadmium, and the more recently studied barium, have been recognised as important indicators of nutrient levels in the oceans. Cadmium, being directly linked with the amount of phosphorus in the oceans, has been used through the study of the Cd/Ca ratio in foraminifera to detect the nutrient content in the oceans, especially for different water masses when cores are taken at different water depths. This field of research can help us understand better the past and present processes in the oceans, including the location and timing of ocean ventilation which controls CO_2 release into the atmosphere, and is also relevant to the calibration of the radiocarbon record (see Appendix). This record will eventually be used for correlation with that of the air bubbles found in ice cores (see Chapter 3). Cadmium is thus a useful tracer in palaeoceanography, especially when used in combination with $\delta^{13}C$ in foraminifera (Boyle 1990, 1992).

Barium has been used in parallel with cadmium analyses in forams to reconstruct the nutrient levels in the deep water (Lea *et al.* 1989). For example, it has been postulated that circulation in the deep oceans differed during parts of the Pleistocene, and that nutrient levels were higher during glacial times in the North Atlantic than today. Barium has also been analysed in corals to determine a change in nutrient levels in the surface ambient water. A change

in nutrient level could be explained by the shoaling of the thermocline which enabled deeper water, richer in nutrients and also in barium, to emerge at the surface where corals grow (Lea *et al.* 1989). This preliminary work was on modern corals from the Galapagos Islands, which come under the occasional influence of upwelling induced by SST anomalies caused by El Niño–Southern Oscillation (ENSO) events (see Chapter 8). Cadmium, on the other hand, was analysed in the same project to demonstrate the effect of upwelling, but the results were not as obvious, probably because cadmium is more readily scavenged by organisms. Techniques using trace elements have tremendous potential for improving our understanding of oceanic processes such as the behaviour of different water masses and currents through time.

The content of germanium in biogenic silica (diatoms and radiolarians) has been used to determine the amount of silica present in ocean water. In a sense, germanium behaves like a heavy isotope of silica, and although only a small proportion of germanium takes the place of silicon atoms (one in a million), it is possible to reconstruct the Ge/Si ratio of seawater through time. This ratio has been used to show that productivity had apparently dropped during glacial periods in some parts of the oceans, on the basis that silica uptake, and consequently accumulation of biogenic silica on the sea floor, was lower then. Very recent investigations by Sanyal *et al.* (1996) were aimed at determining the level of concentration of boron in seawater through intricate analyses of the isotopes of this element, so as to determine possible changes of pH in the oceans that may clarify past changes in alkalinity levels. These may link directly with the pCO_2 of the oceans. This work is in its infancy but promises to provide exciting insight into the role of the ocean in the global climatic system through time.

AEOLIAN DUST AND POLLEN

A substantial programme of research aimed at correlating oceanic and continental events has been accomplished in the Atlantic Ocean. This analysed the record of airborne dust originally deflated from Africa and deposited out at sea (Hooghiemstra *et al.* 1987). This work sought to establish a record of arid events on the continent and to correlate this record with oceanic conditions. The principle has been to distinguish between the supply of quartz grains coloured red by iron oxides and considered typical of a desert origin, and to compare it with white quartz usually originating primarily from fluvial material deposited near the continental edge. It has also been possible to establish the prominence of trade winds for different latitudes through time from the study of those terrigenous remains originating from the deserts (Chapter 9). Similarly, the study of pollen originating from the continents and recovered from marine cores has enabled palaeoecologists to reconstruct the vegetation record spanning long periods of the Quaternary record. These are frequently not found on land because the relevant deposits have been eroded by glacial activity. The study of marine planktonic organisms (through faunal/floral associations and their chemical composition) and of marine sediment composition, in combination with the study of aeolian dust and pollen, provides a means of establishing past environmental conditions of global importance (Prell 1984). This kind of multidisciplinary study is essential to provide the necessary correlation between the global climatic fluctuations and changes in the biosphere on land and at sea. It has been important to identify the time lag, if any, between events that occurred on land from those that occurred in the oceans. For quite some time it has been possible to define past rates of climatic change on land, principally through vegetational changes via pollen analyses. It has only recently been shown, through the identification and study of deep-sea cores registering rapid sedimentation rates, that the oceans did change quite dramatically over short periods of time, with respect to sea-surface temperature for example, but also returned to the previous conditions within decades or centuries. Such changes can be caused by the rapid meltdown of large ice sheets, such as the Laurentide ice sheet, releasing vast numbers of icebergs at sea or providing a huge quantity of glacial meltwater to the ocean (for more details see below).

ICE-RAFTED MATERIAL

Ice-rafted debris transported by icebergs and sea ice can clarify the extent of ice transport and the direction of oceanic currents (Keany 1976). It is now well known that particular clay minerals, such as chlorite, can originate principally from the weathering of rocks in the polar regions. The presence of such a clay may once again be indicative of the extent of ice-rafted material. Various isotopes, such as those of Nd and Sr (Revel *et al.* 1996), can help to identify the source of material, especially ice-rafted debris found in deep-sea cores in the north and central Atlantic Ocean. Studies of clay particles and mineralogies are becoming more intricate as different water masses, which move at different rates and in various directions in the oceans (see Fig. 7.1), are now known to transport clays from different sources before they settle on the sea floor (Petschick *et al.* 1996). The

same phenomenon applies to pollen before settling on the sea floor, thus posing some challenge to the interpretation of pollen spectra. Similarly, studies were also made of the distribution of diatoms that grow on sea ice in order to determine, from the distribution of the diatom frustule remains in cores, the extent of sea ice through time.

Heinrich (1988) was the first to demonstrate the link between the presence of lithic fragments in several deep-sea cores in the Atlantic Ocean with the calving of icebergs in the North Atlantic, depositing ice-rafted debris (IRD) on the sea floor. These iceberg drifts, later called Heinrich events (see Fig. 7.8), correspond to temperature oscillations recognised in Greenland ice cores and which are spaced on average every 7–10 ka (see Bond *et al.* 1993). More recent work by Bond and Lotti (1995) identified warm–cold oscillations in North Atlantic cores which are spaced at intervals of 2–3 ka, and which are also recognised in Greenland ice cores; these are referred to as Dansgaard–Oeschger temperature cycles (D–O events). Note that such cyclicity has also been recognised on land through a record of strong aeolian activity reported by De Deckker *et al.* (1991) from a lacustrine core taken in the now-flooded Gulf of Carpentaria, North Australia.

THE ATLANTIC AND ITS SIGNIFICANCE

To describe in detail events which occurred in individual oceans and at particular periods of time is beyond the scope of this chapter. Because of its proximity to the largest concentration of research institutes, the Atlantic Ocean has been more extensively studied than any other ocean. However, caution is needed in extrapolating from the Atlantic to the other oceans, since the Atlantic is a somewhat unusual ocean in several respects. It is fairly narrow and is bounded by large landmasses which set a particular pattern of ocean circulation. Atlantic water is unusually saline in comparison with the other larger oceans. Deep water in the Atlantic is under the influence of cold water originating in the Norwegian and Greenland Seas to the north and in the Antarctic Seas to the south. The Pacific and Indian Oceans do not have such deep cold-water formation sites in their northern region. Nevertheless, this saline water which originates in the Atlantic may have a profound effect once it enters other oceans via the deeper waters. The Atlantic may have a significant effect on global climate because saline water can help drive more water vapour into the atmosphere. The Atlantic is a driving force for numerous other processes such

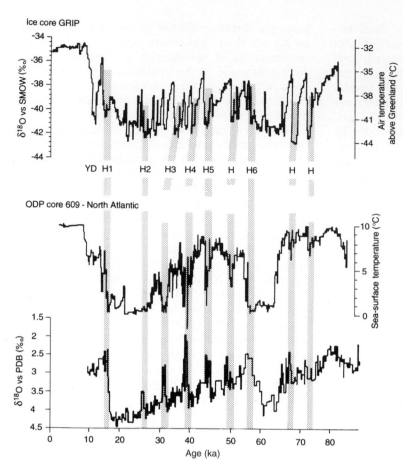

Ice core GRIP

ODP core 609 - North Atlantic

YD H1 H2 H3 H4 H5 H H6 H H

Fig. 7.8 Diagram to show a comparison for the last 85 ka between the $\delta^{18}O$ record of the GRIP ice core in Greenland, transformed into air temperature values, and sea-surface temperature values from the Ocean Drilling Program core 609 from the North Atlantic, south of Greenland, obtained from calculation of the percentage of the left-coiling form of the foraminifer *Neogloboquadrina pachyderma*, plus the $\delta^{18}O$ record of the same taxon. The presence of Heinrich events (labelled H) is recognised by the substantial increase in lithic fragments in the core which were brought in over the coring site as ice-rafted debris by icebergs. YD stands for Younger Dryas, the short period of cooling which principally affected northern Europe. This figure nicely illustrates the close link between the continental/atmospheric record and that of the oceans. (Modified from Duplessy 1996 and Bond *et al.* 1993)

as the control of deep cold water. There is still a great need to determine the links that may have existed between the different oceans with respect to different water masses (for nutrient levels, salinity and total dissolved CO_2 levels) before we can establish the precise impact oceans have had upon climate on land and upon the biosphere in general. The areas which will receive more attention in the future are the Southern Ocean, which characteristically circumnavigates Antarctica and plays a significant role in global circulation (see Fig. 7.2) and the uptake/release of CO_2, and the 'Warm Pool', which sits at the junction between the western Pacific and the eastern Indian Ocean (the area contoured with a sea-surface temperature >28°C in Fig. 7.7 for the modern oceans). This is the site where the oceans are typically the warmest in the world and where the largest transfer of heat and moisture to the atmosphere occurs. The size of the Warm Pool (often referred to as the Western Pacific Warm Pool) is directly affected by the El Niño–Southern Oscillation and also controls the strength and presence of some boundary currents in both oceans. Therefore, more concentration of research in these vital areas will help us to understand patterns of climatic variability, especially those

linking the oceanic, atmospheric and continental systems.

CORALS AS HIGH-RESOLUTION RECORDERS OF ENVIRONMENTAL CHANGE

In the context of oceanic temperature change in the tropics, much debate still pertains with respect to SST changes through time, especially during the last glacial maximum. In effect, there is ample evidence for high altitude regions in tropical regions, and in particular in Papua New Guinea (see Hope 1984 for pollen and geomorphological evidence) and in Peru (see Thompson *et al.* 1995 for isotopic evidence from glacial ice) among others, that temperatures there were substantially lower during the last glacial maximum in comparison with today's values. This evidence is in conflict with the CLIMAP results, and more recent ones which indicate that SSTs departed only very slightly from modern-day values. Several modellers believe that this discrepancy requires explanation and that SSTs reconstructed from planktonic foraminifers need to be 'revisited'. This approach has required another technique to reconstruct past SSTs. This is the reason for which the compositions of aragonitic skeletons of corals have received much attention during the last few years. Some particular coral species which live near the surface of the oceans have since been thoroughly investigated, and in particular their skeletons because they display a good record of microlaminations, some of which are formed on a diurnal basis. These colouration bands are now also recognised to be affected by flooding events caused by rivers discharging various fulvic and humic acids into the oceans (see Fig. 7.9). These characteristic bands are recognised by U-V fluorescence (Isdale 1984). X-ray radiography is now commonly used to define annual banding in sections of coral heads, and these are sampled on a microscopic basis so as to detect short-term changes in the composition of the corals. A combination of stable isotope analyses, combined with trace-elemental

records, such as Sr/Ca, Ba/Ca, Cd/Ca and Mn/Ca, have enabled us to establish changes in the oceans where the corals grew. The $\delta^{18}O$ record in laminations of corals can reveal water temperature and salinity changes (including the supply of river water), and Sr/Ca the temperature changes (an increase in this ratio relates to a temperature drop only if one assumes that the Sr/Ca of seawater remained constant) as well as the supply of river water. For an example, refer to Fig. 7.9. Changes in nutrient supply, especially in a region of upwelling, can be detected through the analysis of Ba, Cd and Mn in the coral skeletons. Other elements, such as B and U, are also being investigated. For a very comprehensive review of the use of corals as recorders of oceanic and atmospheric interactions, refer to Dunbar and Cole (1993). It is certain that coral palaeothermometry will generate more discussion on the behaviour of the oceans in the tropical regions through time.

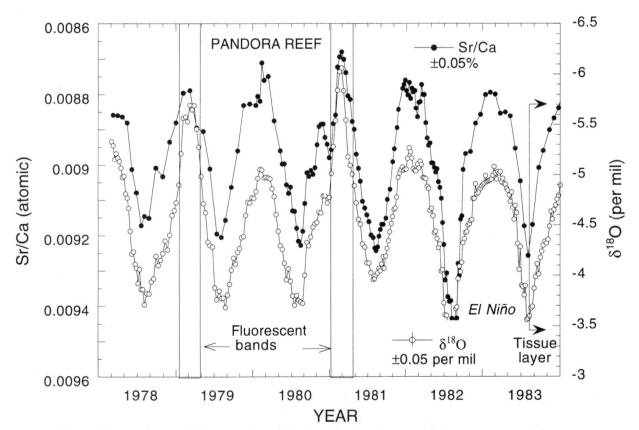

Fig. 7.9 Diagram to show a high-resolution comparison over a five-year period (1978–83, plus with some outer tissue material) of the Sr/Ca ratios and the $\delta^{18}O$ values for the coral *Porites lutea* from Pandora Reef on the Great Barrier Reef in Australia. Highest Sr/Ca and most positive $\delta^{18}O$ values register the coolest temperatures (in August in the southern hemisphere). The vertical lines show the presence of U-V fluorescence bands produced by organic-rich waters during extensive periods of floods from the adjacent River Burdekin which occur in the austral summer. In 1982–83, there was an El Niño event which is detected by a substantial drop in ocean temperature (~3°C) as seen in the Sr/Ca change in winter. Note that it is possible to equate the Sr/Ca of the coral into a temperature signal, and therefore separate this latter signal from the $\delta^{18}O$ record of the coral. (From McCulloch *et al.* 1994)

CHAPTER
8

RIVERS, LAKES AND GROUNDWATER

Suppose, now, that the Nile should change its course and flow into this gulf – the Red Sea – what is to prevent it from being silted up by the stream within, say, twenty thousand years? Personally, I think even ten thousand would be enough.

Herodotus (c. 485–425 BC)
The Histories, Book Two
(*Trans. Aubrey de Sélincourt, 1954*)

Rivers, lakes and groundwater are treated together in this chapter for a number of reasons. Clearly, all represent forms of surface and near-surface water, and must have experienced changed environmental controls as continental water balances shifted during the Quaternary. Rivers, lakes and groundwater are, however, linked in more direct physical ways. Where the water balance is normally positive (i.e. where a surplus of precipitation over losses to evaporation is available) rivers and lakes are often fed by slow seepage from the groundwater stored in higher parts of the landscape. Such seepage sustains river flow during rainless periods and similarly helps maintain lake levels. In arid regions, the reverse may be the case: river flow is lost through the porous stream-bed into the underlying alluvium, where it recharges the groundwater store. Elsewhere, lake behaviour is closely related to the state of the regional groundwater store. For instance, near-surface, saline groundwater may crop out at the surface in shallow salt lakes which episodically dry out in the summer or during a series of drier years. Such saline lakes affect the surrounding vegetation through, for example, the deflation of salts from the dry lake bed on to the surrounding landscape. A final reason which may be mentioned for linking the subjects of this chapter is that, especially in low-gradient alluvial plains such as the Riverine Plain of southeastern Australia or the Gezira Plain of the central Sudan, infilled river channels left by previous episodes of river incision act in a number of ways to guide the movement of groundwater in the sedimentary basin. In tributaries of the Murray River in Australia, present-day and infilled, buried palaeochannels form a linked three-dimensional network that conducts water at the surface and in the subsurface (Schumm *et al.* 1996).

Rivers, lakes and groundwater then are merely the major manifestations of an integrated system of pathways through which surface and subsurface water move through the landscape. An important aspect of this network of water flow pathways is that it also constitutes a major avenue through which dissolved rock materials, plant nutrients and other solutes are transported. Groundwater draining from the continents can be very significant in carrying dissolved nutrients to the oceans, because of its long residence time within the aquifers (Moore 1996). This aspect, however, will not be pursued in detail here.

RIVERS OF THE PRESENT DAY

Riverine environments globally constitute a vast, complex, and incompletely understood set of landscapes (Schumm and Winkley 1994). The range of simple physical characteristics spans the permanently flowing *perennial* streams, such as those of the tropical rainforests and other humid areas, through *intermittent* streams whose seasonally varying flow relates to regular monsoonal rain or spring snowmelt, to the shortlived *ephemeral* streams of the desert areas whose brief floods after local rain rapidly give way to dry, sandy or rocky channels. In any of these environments, floods of widely varying size may occur, resulting in potentially major changes in channel size, in bank scour and collapse, and in alteration to floodplains, bars and islands. Across this range of environments, the stream channel may adopt one of a number of recognisable basic forms, including the *braided* form which displays multiple, small intertwining channels and which is characteristic of streams supplied with large quantities of sandy and coarser debris. Streams carrying finer silty materials often display the smoothly sinuous *meandering* form which is very characteristic of lowland floodplain sites. Figure 8.1 illustrates some of the river channel types mentioned here. Streams in high latitudes are greatly affected by ice, which forces banks to deform during the winter; glacial meltwater streams draining major mountain belts are fed enormous loads of sediment as moraine is dumped at the glacier snout, and proceed in turn to set down great depths of alluvium, often rapidly filling lowland valleys (Fig. 8.2). Steep mountain streams carry very coarse sediment particles, while those on the lowland alluvial plains may be able to move only silts and clays. Sediment particles must thus be set down, sorted according to size, as streams flow from upland areas on to lowland plains. Rivers of macrotidal coasts may experience tidal influence up to 100 km inland, and may undergo daily tidal depth fluctuations in their lower reaches of 5 m and more. Rivers of tectonically active areas in many cases erode their courses downwards as the terrain is elevated, leaving suites of terraces and the remains of old stream beds high above the modern channel; elsewhere, tectonic stability and low gradients result in streams whose rate of change is almost immeasurably slow. In some areas, streamflow removes water to the oceans (*exorheic drainage*), while in others, such as the enormous Lake Eyre basin of Australia, flow is directed toward an inland lake (*endorheic drainage*). Large areas exist that possess no surface streams at all (they are *arheic*), while in others the *drainage density* may be very high. The course followed by streams may be determined largely by the underlying rocks and their structural arrangement, or it may be essentially random in nature. The course of

many streams has nowadays been set deliberately by engineering works designed to straighten or alter their course or depth for purposes such as navigation and flood control.

All of these influences contribute to the vast diversity of riverine landscapes. However, there are certain environmental factors which exert a similar influence on all rivers and streams. These include the nature of the drainage basin that delivers water and sediment into the river, the external climate experienced by the drainage basin, and the tectonic stability of both the landscape which is drained by the stream and the lake or ocean level (the *baselevel* of the drainage system) to which it drains. Major Quaternary sea-level fluctuations, for example, affected the baselevel of *all* exorheic streams worldwide, although the effect on an individual river would be moderated by other local factors such as the resistance of the rocks into which the stream was incising its valley, and whether the catchment area was icebound or escaped Quaternary glaciation. Also, reaches lying upstream of major knickpoints or escarpments would be relatively unaffected. However, it can safely be said that at least all coastal river systems of the world were repeatedly thrown out of equilibrium by the environmental dynamism of the Quaternary.

CHARACTERISTICS OF SOME CONTEMPORARY RIVERS

The principal characteristics of some modern river systems are listed in Table 8.1. This tabulation reveals the wide range in sizes, sediment loads and discharges which exists. Analysis of data on global

Table 8.1 Principal physical characteristics of some modern river systems

River	Location	Annual discharge (km³/a)	Dissolved load (10⁶t)	Suspended load (10⁶t)	Mainstream length (km)	Basin area (10⁶ km²)
Amazon	S. America	6300	223	900	6300	7.18
Congo	Africa	1250	36	43	4700	3.82
Orinoco	S. America	1100	39	210	2500	0.99
Yangtze	China	900	226	478	5000	1.94
Ganges–Brahmaputra	Asia	971	136	1670	2900	1.48
Nile	Africa	91	17	57	6700	3.349
Murray–Darling	Australia	12	2	6	3770	1.072
Fly	Papua New Guinea	190	–	115	1120	0.076

river-loads, and drainage basin denudation rates derived by calculation from them, have provided many interesting results. It has been found that drainage basin relief exerts a primary control on denudation rates, with the highest aggregate rate (made up of mechanical and chemical forms of removal) for a large river system being in the Brahmaputra River basin, which has 6.7 km of relief and a mean denudation rate of 688 mm/ka (Summerfield and Hulton 1994).

Australian rivers are relatively minor in most respects by world standards, reflecting the overall dryness and flatness of the continent. Global average figures for river flow are always distorted by the characteristics of the Amazon, which is so large in flow volume that this single stream carries about 20% of all the water flowing off the land surfaces of the globe. The characteristics of many Asian rivers (such as the Huang Ho or Yellow River) reflect their peculiar environmental history combined with human use and modification of the catchment area.

FACTORS INFLUENCING RIVER ENVIRONMENTS DURING THE QUATERNARY

Before turning to some case studies of Quaternary river adjustment, we will examine some of the principal factors that must have affected riverine environments during the Quaternary. The factors considered are

- baselevel change and tectonic effects
- catchment water balance and erosional processes
- catchment fluvial (*river channel*) processes
- glacial meltwater outburst floods.

It is also necessary to mention the influence of people on river form, which has been especially great in the last few centuries. Many of the deposits recording the Quaternary features of rivers lie buried beneath Holocene sediments. More easily seen are the many changes wrought by people, seeking to drain marshy floodplains, or control the flooding of productive lands. In many areas, stream disturbance has followed from alteration of land cover and landscape hydrology by activities such as the introduction of agriculture or pastoralism. Stream incision and bank destabilisation have often resulted (e.g. Hereford 1993) (Fig. 8.3). Damming of rivers has also led to widespread changes in the sediment loads carried by rivers, and hence on patterns of upstream and downstream erosion and sedimentation. It has also led to major ecological consequences, with biota often directly affected, and also indirectly affected through altered fluxes of nutrients, changed levels of

oxygenation, and altered seasonality of temperature changes. Although it has wrought many significant changes in channel form, the history of such human manipulation of rivers is not considered here. Details may be found in works such as Petts *et al.* (1989), Bravard and Petts (1996), Collier *et al.* (1996) and Gornitz *et al.* (1997). The Mississippi River, whose huge catchment has been affected by many large dams, provides a well-studied instance. In this case, the sediment load reaching the delta has shown a historical decline in annual tonnage together with a declining sand fraction and an increase in silt (Keown *et al.* 1986). In many instances, steady low flows capable of sustaining river habitats, known as environmental flows, are deliberately released from large dams; periodically, it may also be necessary to release large artificial 'floods', as has been done experimentally on the Colorado River in the US (Collier *et al.* 1997).

Baselevel changes

For exorheic streams and for groundwater flow, a major series of disturbances during the Quaternary was caused by repeated shifts in baselevel as sea-level rose and fell with the glacial–interglacial cycles. The magnitude of the fluctuation exceeded 120 m (see Chapter 6). During glacial periods, the course of rivers would have been lengthened by the distance across the continental shelves. By falling through the additional 120 m or so, significant additional erosional work would be possible for the river flows, so that canyons were cut in many places through the sediments lying on the exposed shelves, such as the narrow gorge of the Mississippi River (called the Mississippi Trench) running across the shelf in the Gulf of Mexico (Bloom 1983). The disturbance would have been felt progressively through much of the inland drainage basin also, with all points along the course of the stream lying higher above baselevel when sea-level was lower. Incision of channels was the common (but not invariable) consequence. Such incision leaves remnants of the original valley floor standing above the newly incised streams to form *river terraces*. Many streams display multiple river terraces which trace an earlier path of the stream. Terraces may also, of course, be produced by river incision that results from tectonic uplift; this *rejuvenates* the lower reaches of the stream which formerly were nearly at baselevel, and leads to an upstream progression of erosion. Terraces may be produced in other ways as well, as we shall shortly see.

Effects almost opposite in tendency are produced as sea-level rises and the continental shelves are once more inundated during deglaciation. River canyons on the shelves are flooded, and the seas occupy the

Fig. 8.1 Natural and anthropogenic variations in river channel form: (a) The meandering channel of the Murray River, Australia (b) The braided channel of the Markham River, Papua New Guinea (c) An ephemeral desert stream channel, Fowlers Creek, NSW, Australia (d) River terraces in a tectonically active region, North Island, New Zealand

Fig. 8.2 The braided outwash train of the Tasman Glacier, South Island, New Zealand. Here, abundant bed-load materials are supplied from glacially scoured debris which is released at the melting glacier snout

lower reaches of the stream valleys. Regions formerly well above baselevel are then located almost at sea-level, and the ability of the stream to carry sediment is reduced. River mouth regions would thus slowly fill with deposited sediment at times of high sea-level (like the present day). These materials would be cut into during the next time of low sea-level, so that events of this kind must have been repeated every 100 ka or so during the upper Quaternary. The *stratigraphy* of the lower reaches of rivers must therefore be expected to show a confusing array of sediment bodies of varying ages. Eroded materials will be laid down in a similarly complex sequence on the continental shelves and below the canyons on the shelf slope and rise. A schematic diagram showing the formation of terraces produced by low sea-level stands is included in Fig. 8.4.

There are some well-dated reconstructions of the effects of the late glacial–Holocene sea-level rise on the lower reaches of rivers. Kilden (1991) provided a series of Holocene longitudinal profiles of the River Scheldt in Belgium, which drains to the North Sea. Here the sea-level rise greatly reduced the stream

gradients, especially in the lowermost 100 km of the river course, and this in turn led to the accumulation of thick peat deposits flanking the river. The river may also have undergone metamorphosis, adopting an anastomosing or multichannelled form. Kilden (1991) described the sandy, braided form that was displayed by the river in the late Weichselian glacial, and which was supplanted, probably prior to the Holocene, by the development of a very large meandering channel, whose dominant discharge may have been 3–5 times larger than the present river discharge. This channel in turn was aggraded, as sea-level rose and climate changed, and the modern channel is set within the deposits left by the former large, meandering river system. Models of the timing and nature of river transformations accompanying interglacial–glacial and glacial–interglacial changes, and of terrace formation, have been put forward by Vandenberghe (1995) and Merritts *et al.* (1994).

In the new course across exposed continental shelves during glacial low sea-level stands, opportunities for additional stream junctions, and thus for the formation of even larger integrated drainage basins,

Fig. 8.3 Channels disturbed during the historic period by human activity, including the introduction of grazing herds, within the drainage basin: (a) Large arroyo on Rio Puerco, New Mexico, US (b) Arroyo cut in Quaternary sediments, western NSW, Australia

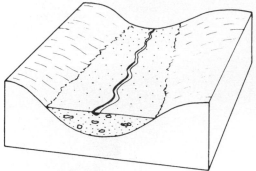

A. 120 ka BP: SEA AT +8 m

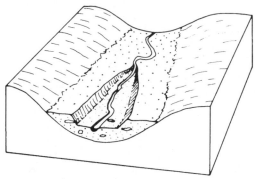

B. 20 ka BP: SEA AT -150 m

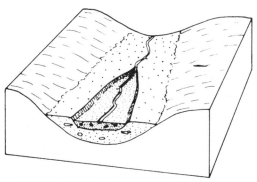

C. TODAY: RIVERS NOW GRADED TO 0 m

Fig. 8.4 Schematic representation of the valley aggradation and subsequent incision and terrace formation which result from baselevel change

must have arisen. The products of such integrations may in certain instances have created what Mulder and Syvitski (1996) termed 'giant rivers' (Fig. 8.5). They illustrated the kinds of transformations of the drainage system with the case of the Rhine, which, as an enlarged system, would have drained southwest through the English Channel. The aggregate area drained by this system would have increased three-fold, to 840 000 km^2. Effects of these reorganisations include an increase in river sediment delivery to the oceans, but from a smaller number of major river systems discharging a lower average concentration of sediment to the oceans (Mulder and Syvitski 1996).

This would represent a significant influence on patterns of sedimentation in the ocean basins.

Catchment water balance and erosional processes

During the Quaternary, as temperature and precipitation fluctuated, the water balance of stream catchments worldwide underwent repeated change. Undoubtedly, complex patterns of change were involved, with reduced evaporation combining with rainfalls regionally shifted by varying amounts to produce unevenly distributed water surpluses. Seasonal distribution of rainfall must have altered in some areas; lower temperatures also affected the vegetation cover, and hence the protection offered to the soil against erosion. Sediment washing into streams must thus have been altered both in quantity and kind, with perhaps more sand and coarser materials being stripped from relatively unprotected slopes during glacial phases.

Stream channels, which were approximately adjusted to carry the sediment loads fed to them during interglacials, underwent changes in form and function as a consequence of the onset of cold conditions. These changes of river form and process are grouped together under the title *river metamorphosis* (see Schumm 1977). Streams fed with quantities of coarse debris probably became wider and shallower, with a somewhat less sinuous course being slowly developed. Streams that perhaps had a meandering form (and which had been carrying fine silts or clay materials) would become more like contemporary braided channels during glacial times, and undergo the reverse form change as interglacial conditions evolved. Where the river flow was strongly seasonal, the exposed bed sediments in some cases were blown away by the wind (*deflated*) to form sand dunes in areas adjacent to the channel. Once again, changes of these types happened repeatedly during the Quaternary so that we must expect the present landscape to be composed of a rather fragmented assemblage of deposits of varying age and character.

Those streams that drained the melting margins of the great Quaternary ice sheets must all have experienced significant and repeated changes in discharge and sediment load through the glacial–interglacial cycles. In particular, great volumes of sediment must have been contributed to them in the early stages of deglaciation. The Mississippi river system of North America experienced such *proglacial* conditions as the Laurentide ice retreated (see Chapter 3 for the chronology of this retreat). This river system was one of the main outlets for meltwater and glacially-derived sediments related to Laurentide glaciation. River sediments preserved in various areas show that during this phase, the central Mississippi became a

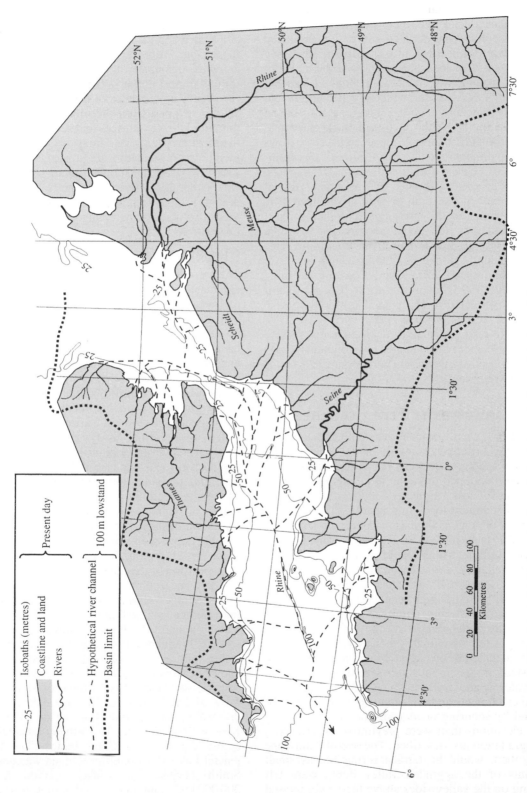

Fig. 8.5 The configuration of the 'giant river' inferred to have drained through the English Channel during Quaternary low sea-level stands. This was formed by the merging of the Rhine, the Thames and other rivers to yield a much larger combined catchment area. (After Mulder and Syvitski 1996)

large braided channel because of the loads of sands and gravels fed to it (Baker 1983). A radiocarbon chronology developed by Royal *et al.* (1991) shows that in Missouri, the sandy braided Mississippi of the Wisconsinan glacial maximum persisted, and acted as a source for wind-deflated sands that produced dunes nearby until about 11.6 ka BP. The declining flow and sediment load finally allowed the development of a meandering channel by about 11.5 ka BP. In areas such as Australia, which escaped significant Quaternary glaciation, global and regional climate swings essentially synchronous with those elsewhere nevertheless had a marked impact upon river flows and river sediment loads. Using a series of thermoluminescence (TL) and uranium–thorium (U/Th) dates, Nanson *et al.* (1992) showed that periods of fluvial activity more intense than present are indicated by sediment bodies in the landscape. The data suggest that the intensity of fluvial activity in Australia, which is associated with interglacial conditions, lags 5–10 ka behind global temperature and sea-level maxima. In the Lake Eyre basin of central Australia, Croke *et al.* (1996) found that these patterns of fluvial activity related to the glacial cycles were also reflected in patterns of lacustrine activity.

Dramatic changes in river environments are also known from other non-glaciated environments which are now arid (e.g. ancient Egyptian drainage systems discussed by Issawi 1983).

Fluvial processes and river channels

The changed conditions over the catchment area produced by Quaternary environmental change had more involved consequences for river channels than those just mentioned. Imagine, for example, a catchment whose less protected slopes shed much increased sediment loads into nearby streams in the early part of a glacial phase, as former soils and weathered materials were rapidly eroded. This increased sediment load would overtax the stream's carrying capacity, and some material would be set down (the downstream channels would be forced to *aggrade*, or build up their beds). Valleys lower down in the drainage system might be *alluviated* in this way to substantial depths as the excess materials were deposited. Eventually, however, the available weathered materials on the catchment slopes would be exhausted. In this circumstance, relatively sediment-free water would begin to be shed from the bare upland catchments and collect an appropriate load of material by scouring or *incising* the valley floors at lower elevations than were previously experiencing the aggradation just described. The overall landscape effect, then, would be terrace formation, as small remnants of the aggraded valley floors were left standing on the valley sides above the newly incised

streams. This terrace formation would happen some time *after* the climatic swing into glacial cold conditions, and would in fact occur at a time when no *external* or *extrinsic* environmental change was taking place. Landscape changes of this sort, occurring at a time when there is no evident external trigger for the change, represent examples of what is termed *complex response* in the riverine environment (Schumm and Parker 1973). This complex response (involving unknown time-lags as weathered materials are stripped from the uplands) makes the interpretation of river terraces and sediment bodies one of the most problematical areas of Quaternary environmental reconstruction (e.g. see Bettis and Autin 1997). It is all the more difficult because baselevel was generally varying at the same time as the drainage basin climate, vegetation and hydrology were also undergoing change.

Glacial lake outburst floods

The termination of Quaternary glacial periods was associated with the release of enormous quantities of meltwater which had to find a way back to the oceans. Huge discharges were associated with some of these flows, and as noted in Chapter 3, some may have been sufficiently large to dilute surface seawater and so interfere with the major thermohaline circulation of the world oceans. In this way, the meltwater flows became a cause of climate variability in their own right, and were not merely a straightforward consequence of climatic warming.

The flows did not always arise as a steady release of water, however. Ordinarily, the flows would have had a strong seasonal signal. But periodically, as the ice fronts retreated, temporary meltwater lakes were created between the ice front and a moraine ridge or a mass of stagnant ice (Kehew and Teller 1994). The formation of these lakes was facilitated by the topographic depression created by the weight of the ice sheet, with topographic uplift occurring isostatically far more slowly than ice retreat.

The catastrophic failure of a meltwater lake released huge volumes of water. This outburst flooding often cut major floodways across the landscape, leaving a mantle of boulders spread across the landsurface. Frequently there was a cascade of effects as water spilling from one lake caused the failure of another downstream (Teller and Kehew 1994), partly because of the massive volume of sediment influx. West of Lake Agassiz, 13 large lakes formed and drained in the interval 12–10.7 ka BP (Teller and Kehew 1994), leaving 18 spillway channels running across the northern Great Plains. The history of glacial Lake McConnell in Canada was described by Smith (1994). This lake covered more than $200\,000\,km^2$, and discharged major flows at about

9.9 ka BP, triggered by a massive influx of water from Lake Agassiz. This flow, estimated to have had a discharge of $1.2–7.38 \times 10^6 \, m^3/s$, released $21\,000 \, km^3$ of water over a period of months. The water influx to the oceans raised global sea-level by about 6 cm (Fisher and Smith 1994). The $6000 \, km^2$ Lake Souris in North Dakota was similarly breached as a result of an influx of $74 \, km^3$ of water and $25 \, km^3$ of sediment from Lake Regina in Saskatchewan (Lord 1991). The floods so generated reached discharges of $6 \times 10^4 – 8 \times 10^5 \, m^3/s$, and velocities of up to nearly 12 m/s. Glacial Lake Missoula, in Montana, which contained at maximum about $2500 \, km^3$ of water, was breached repeatedly and set down massive rhythmite deposits (Waitt 1985).

EXAMPLES OF QUATERNARY RIVERINE ENVIRONMENTAL CHANGE

In order to illustrate some of the processes explained in general terms above, and to see how they are manifested in contrasting environments, we will examine the Quaternary record of a sample of drainage basins spanning the globe. River systems which provide suitable examples for our purpose are:

- the Nile in northeast Africa
- the Murray–Darling system of southeastern Australia
- the Amazon system of South America.

The locations of these river systems and their catchment areas is shown in Fig. 8.6.

The Nile river system

The Nile river system is presently an *exotic* or *exogenous* one for much of its course; that is, it flows through dry areas which would not normally be expected to display such streams. The Murray–Darling system in Australia, which we shall consider shortly, is also of this kind.

The Nile system consists of three major branches: the White Nile, which drains large lakes in the southern headwater areas of Uganda; the Blue Nile, which drains the high northern and central plateau areas of Ethiopia; and the Atbara, which also drains northerly areas of Ethiopia. The Blue Nile and the Atbara sup-

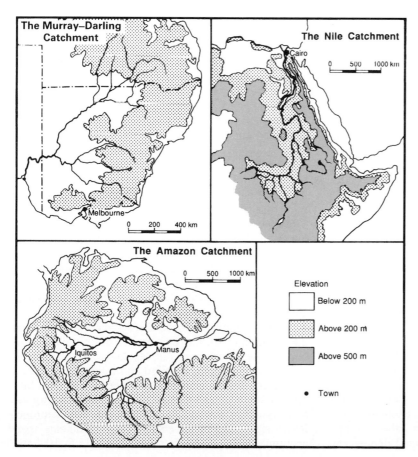

Fig. 8.6 Location and catchment relief of the Nile, Amazon, and Murray–Darling river systems

ply the bulk of the sediment carried by the lower Nile, but the White Nile is critically important in sustaining the Nile flow during the dry winter of Ethiopia. From the headwaters of the White Nile just south of the equator, the system flows nearly 7000 km north to its delta in the eastern Mediterranean, crossing 35° of latitude (Adamson *et al.* 1980). Clearly, therefore, the Quaternary fluctuations in the behaviour of this river must have related to the altered environments of areas whose climates ranged widely (presently, equatorial to hyperarid); *river metamorphosis* of the middle and lower Nile in Egypt can be expected to relate to the altered environments of the distant Ethiopian uplands, in particular, as these supply much of the water and sediment load of the present-day river, and thus probably also its Quaternary predecessors.

In fact, the middle and lower Nile display just such a complex record of river sediments as we might expect. Sediments covering parts of the modern landscape, or partly filling valleys in the system, have been much studied for the information that they contain relating to the development of human settlement along the Nile, and dated using isotope techniques and thermoluminescence (Schild and Wendorf 1989; Wendorf and Schild 1989).

Two major periods of alluviation along the Nile are recognised from the Late Quaternary, but more are likely to have occurred and remain unidentified because of burial or removal of the sedimentary evidence by later erosion. The episodes of alluviation are taken to reflect altered conditions in the Ethiopian uplands during Quaternary cold phases. During these times, the highest areas were transformed from alpine grasslands to bare soil and rock with ground ice and periglacial processes; at lower elevations, tree and shrub cover was replaced by grassland. The treeline was probably lowered by 1000 m during the coldest times. In association with the colder conditions, rainfall was also lower than that of the present, but still delivered in a fairly intense wet season. Smaller volumes of water would have been fed into the rivers, but with a *larger* sediment load (and a *coarser* one also) because of the reduced groundcover. A *lowering* of the stream competence or carrying capacity in this way, with a simultaneous *increase* in the load of sediment fed from the catchment area, not surprisingly resulted in alluviation along the Nile.

In the best-dated of the periods of alluviation, spanning the period 20–12.5 ka BP (and thus coinciding partly with the coldest part of the last glacial), the sediments along the Nile reveal that the river underwent metamorphosis to adopt essentially the braided form, with a sandy floodplain occupied by a much smaller water flow than that of the modern river (perhaps only 10–20%). Continued deposition

of sediments scoured from the upland catchment built up the floodplain progressively during the period of alluviation. The distant Mediterranean would have been about 100 m lower, but would have had no influence on river incision or deposition upstream of the most northerly cataract. Outside of the Nile valley, the environment was hyperarid during the cold conditions, and the valley became an especially important habitat for people and animals. Even within the valley, the dry conditions allowed extensive dune development, and it must still have been a relatively harsh environment.

As the climate warmed in the period 15–5 ka BP, the vegetation in the upland catchments recovered, and despite the increased rainfall associated with the warming, loss of sediments was reduced, and the eroded materials were restricted to finer grain sizes. Deposits of these materials left along the Nile by floods reflect a metamorphosis to a more typical suspended-load stream, with higher sinuosity.

This picture of the Late Quaternary behaviour of the Nile is supported by other essentially independent methods of environmental reconstruction. For example, during the glacial cold conditions, the headwater lakes which feed the White Nile were much shallower, and did not overflow as they presently do. The White Nile itself may have been blocked by dune sands. In this situation, the lower Nile would not only have had a much lower discharge; the sediments it carried would no longer reflect the geology of the White Nile source areas. In interglacial conditions, with the White Nile flowing, the composition of the materials carried by the lower Nile would shift accordingly.

Analysis of materials drilled from the Nile delta (Foucault and Stanley 1989) shows a composition of heavy minerals (used as fingerprints of the geology of the Blue and White Nile catchments) which fluctuated through the last 40 ka BP. In the period 20–10 ka BP the mineral assemblage was consistent with the major mineral source area being the Ethiopian uplands; following this, materials consistent with an increased White Nile flow appeared. The White Nile also appears to have been active in the period 40–20 ka BP. These dates are in good accord with those based on fossil materials within the Nile valley itself, and with the model of Nile behaviour already outlined.

A final confirmation of the scenario comes from the sediments of the eastern Mediterranean. Drill cores of sediment here display layers of black mud rich in organic matter, called *sapropel*. These layers reflect conditions in which the ocean becomes layered or *stratified*, so that vertical mixing is restricted and organic materials do not oxidise. Two sapropel layers are found in the eastern Mediterranean; their accumulation is placed in the intervals 11.8–10.4 and

9–8 ka BP (Rossignol–Strick *et al.* 1982). The formation of these sapropels may be related to the occurrence of major flooding along the Nile as the climate warmed after the last glacial phase. Such floods would deliver large volumes of fresh water which, being less dense than seawater, floats as an upper layer and so stratifies the ocean. The floods in the periods when these sapropels were formed would have been very large indeed; the rainfall at the time was higher than that of the present, and there may have been great releases of water as the lakes in the White Nile headwaters began to overflow once more. Indeed, the high rainfalls may have been a phenomenon of the equatorial areas especially, so that it was only the extraordinary Nile, crossing such a wide latitudinal belt, which carried the influence of these rains as Nile floods all the way to the distant Mediterranean. It has been calculated that the Nile in the late glacial phase when these floods occurred may have carried a discharge 250% larger than that of today, and could have delivered in 15 years sufficient fresh water to cover the whole Mediterranean to a depth of 25 m.

A brief period of aridity over equatorial Africa interrupted the period of flooding and allowed the development of two separate sapropel layers.

In summary, the Late Quaternary history of the Nile river system includes: a dry phase at 20–12.5 ka BP associated with cold conditions; rapid erosion of the Ethiopian uplands and valley alluviation; a terminal Pleistocene to Mid Holocene moister interval at 12.5–5 ka BP associated with stabilisation of upland slopes; and river metamorphosis to the large single channel associated with the modern Nile. Since about 5 ka BP, conditions have become increasingly arid, and today no water flows into the Nile at all during its 1000 km path through Egypt. There is, of course, a more extensive history of valley incision and alluviation related to tectonic events and to the 3000 m drop in the level of the Mediterranean during the Messinian salinity crisis, but the influence of these more remote events on the Nile is at present less well documented.

The Amazon river system

The Amazon river system of South America drains an enormous catchment area spanning Peru, Guyana and Brazil, which amounts to more than $7 \times 10^6 \text{km}^2$. The river system runs approximately west–east just south of the equator, spanning 33° of longitude but only 12° of latitude, draining the Peruvian mountains and flowing into the Atlantic Ocean. Because all of this area lies within the tropics, and receives high rainfalls averaging 2300 mm per year, the Amazon carries an astonishing volume of water: the mean discharge is 175 000 m³ per second (Sioli 1984). This

amounts to nearly 20% of the entire flow of fresh water carried by the combined rivers of the world! The Amazon is also exceeded in total length only by the Nile system, and through much of its middle and lower sections runs on a very low gradient. Despite this, rapid erosion in its rugged Andean headwaters combined with the enormous water flow result in the Amazon carrying the third largest tonnage of suspended sediments globally (exceeded only by the Huang Ho river and the Ganges–Brahmaputra). It is estimated that the Amazon carries 9×10^8 tons of suspended sediment per year (Sioli 1984).

The history and Quaternary behaviour of this enormous river system are not well known (Irion 1984; Bigarella and Ferreira 1985; Colinvaux 1989). Fractures or depressions in the underlying Precambrian crystalline basement (perhaps related to stress patterns associated with the break-up of Africa and South America) are the fundamental control on the orientation of the Amazon valley. This depression contains sediments varying widely in age, including large deposits of Tertiary materials.

Some sediments from the Amazon are carried out across the continental shelf at the Atlantic seaboard and spill down the continental slope to form a large sediment accumulation known as the Amazon cone. Sampling of the materials composing the cone has revealed that it has only been accumulating there for about 10–15 million years, since about the end of the Miocene (Damuth and Kumar 1975). This is interpreted to mean that before this time, at least part of the present Amazon basin drained westwards, into the Pacific. Subsequent uplift of the Andean mountains must have diverted the system into its present valley, guided by the underlying structural features.

The present-day Amazon is associated with a wide variety of environments, which include the low-lying forest areas that are flooded annually, known as the *várzea*. Higher areas that lie above flood level are known as *terra firme*. In the *várzea* areas, the Amazon and its tributaries meander in complex loops through young sediments, and are flanked in many areas by flights of terraces ranging up to 90 m or so above river level. Even higher level surfaces, up to 180 m above river level, are formed in eroded Tertiary materials (Klammer 1984).

The exact origin of the *várzea* materials remains to be discovered. Much of this material must be Holocene in age, laid down during and since the last marine transgression. During periods of glacial low sea-level, the Amazon must have incised its course, cutting a channel out across the continental shelf. The rate at which this took place may, however, have been exceptionally slow in the case of the Amazon because of its enormous sediment load. Postglacial transgression would have inundated the deep valley system, causing aggradation to begin. The enormous volume of the system again probably means that infilling would be very slow to complete, and there is evidence that inland, perhaps 1000 km from the river mouth, infilling of the pre-Holocene valley is still actively proceeding. Tributaries in some subcatchments of the Amazon that lack high ground carry only very small suspended sediment loads (these are termed the *clear water* rivers, in contrast to the turbid, sediment-laden *white water* rivers like the main Amazon itself). Even slower valley aggradation could be anticipated in these systems, and many show very clear flights of river terraces.

The terraces of the Amazon system indeed pose unanswered questions. It may be that they are related to baselevel fluctuations during the Quaternary glacial–interglacial alternations. This would be particularly probable because of the very low elevation of so much of the Amazon catchment, which lies less than 100 m above sea-level over vast areas. At the mouth of the Rio Negro, 1500 km from the sea, low water level is only 15 m above sea-level; even at Iquitos, low-water level is only about 100 m above sea-level. The gradient of the valley over great distances could thus have been *doubled* when sea-level stood 100 m lower than today, because the length of the river would only change by the width of the continental shelf, resulting in considerable impetus to incision. However, the formation of terrace systems in such an enormous river basin requires that there be substantial time for migration of the effects from the coast inland, and it may be that the repeated baselevel swings occurred too rapidly for this process to be completed.

It is also possible to imagine that the Amazon terraces relate to climate change affecting the catchment in just the way already described for the Nile, with drier conditions causing overloading of rivers with sediment, and forcing aggradation downstream. A subsequent return to wetter conditions could then promote renewed incision and terrace development. There is some palynological evidence suggesting that drier conditions have periodically affected parts of the Amazon basin. Damuth and Fairbridge (1970), who studied the sedimentary record of the Amazon cone and adjacent areas, concluded that the mineralogy of the sediments is consistent with the existence of arid to semiarid climates over much of equatorial South America during the last glacial. However, the true relative importance of baselevel fluctuation and catchment climate change remain to be resolved in the case of the Amazon, as does the part played by tectonic activity, especially uplift. Once again, however, significant changes in riverine form, flows and sediment loads can be seen to have occurred during the Quaternary, and perhaps many times.

The Murray–Murrumbidgee–Darling river system of southeastern Australia

The rivers that flow westwards from the uplands of southeastern Australia (such as the Darling, Murrumbidgee and Murray) descend on to a vast alluvial plain which is partly composed of sediments laid down by marine sedimentation during a series of major transgressions into the Murray basin during the Tertiary. Stream gradients are extremely low, partly because the plains are flat and additionally because these rivers have sinuous, meandering courses which require the water to travel much further than the direct distance to the sea.

During the Quaternary, the courses of these rivers changed repeatedly as bank erosion occurred or as flooding resulted in new channels being cut on the floodplains. The plains now preserve evidence of these former channels as infilled *palaeochannels* which are not easy to see at ground level but which are visible in aerial photographs. Some of the palaeochannels relate to glacial times, when rainfall and erosional processes in the highlands were different to those of today, and others to interglacial times, which might have been not unlike the present.

In fact, three different major kinds of channel are visible on these plains, which are collectively known as the *Riverine Plain* (Bowler 1978). The present-day channels represent one kind. They are generally very sinuous, and carry dominantly fine suspended sediment loads (Fig. 8.7). In cross-section, the channels are broadly dished, and in the case of the Murray, for example, about 6 m deep in the middle section of the river course.

Palaeochannels of two distinct forms can also be identified (Schumm 1977). These include a set of channel remnants which resemble the present river (i.e. having a highly sinuous meandering course) but which are much larger; these often form large oxbow lakes on the floodplain of the present river. These old channels were about twice as wide as the modern one, as well as deeper, and their meanders were correspondingly larger in size. They must have carried a considerably greater discharge, and also been able to shift more and coarser sediment because of the higher flow velocities which must have been generated in them.

A second variety of palaeochannel of very different form can also be identified. These display a much less sinuous channel, and hence flowed on a steeper gradient (moving more directly down the slope of the alluvial plains). On the basis of the channel remnants, they must have been nearly three times as wide as the modern rivers, but much shallower. From this and the preserved channel sediments it can be concluded that these were essentially *bedload* channels, moving vast quantities of sand.

Fig. 8.7   The channel of the Darling River near Pooncarie, NSW: a typical large, meandering, suspended load channel on the Riverine Plain of SE Australia. (Photograph courtesy of Ellyn Cook)

The palaeochannels similar to, but larger than, the modern rivers must relate to times when the catchment shed more water than at present. Their form is essentially that of the present channels, but scaled up. The absence of coarse sediments in these channels suggests that the highlands, as the source area of the bulk of both the water and the sediment load, must have been well protected by vegetation as a result of humid conditions.

In contrast, the wide, low-sinuosity palaeochannels are likely to be the product of a drier environment. Under such conditions, the highlands would have been less well protected by plant cover, and storms would have been able to generate both large floods (from the barer surface) and strip the larger quantities of coarser sand which now fills these channels. The straighter course must have evolved in response to the greater sediment load, and provides another clear example of *river metamorphosis* resulting from environmental change. The change from a wide, shallow, sand-transporting stream to a narrower, more sinuous, suspended-load stream as the climate moderated provides an important example of a *constraint* on metamorphosis. Because of the great length of the river system, incision or aggradation as a consequence of a change in incoming sediment load becomes a less probable outcome for the stream than does gradient change. In these systems gradient was apparently adjusted by a change in planform sinuosity. Thus, halving the sinuosity doubles the channel gradient, and vice versa. River metamorphosis must thus have been associated with dramatic bank erosion but relative stability of bed elevation. In the case of shorter rivers, the same catchment change might well be accommodated more completely by bed scour or aggradation, and less by planform change. The particular kind of metamorphosis experienced will relate to aspects of the catchment geomorphology, as well as to the nature of the external environmental change itself. Once again we see, therefore, that the morphological response of rivers to environmental change can be complex indeed. The link between the form and sediments of palaeochannels and the regional environment at the time they were active is an indirect one, modified by parameters which we may as yet not fully understand.

Metamorphosis is revealed in the alluvial history of other river systems in Australia, many of which remain to be dated and explored in detail, and similar large alluvial plains elsewhere. For example, the multiple anastomosing channels of the Cooper Creek system in southwestern Queensland presently carry dominantly fine suspended sediments, and deposit very fine alluvium as floods wane. Underlying the present channels, however, is a system of sand-rich, sinuous meandering palaeochannels (Rust and Nanson 1986). The age of these is not known with any certainty, but it has been suggested that repeated change from sandy to muddy channel systems through the Quaternary may reflect dominantly the effects of climatic change. The actual fluvial processes involved in the meandering-to-anastomosing channel metamorphosis remain to be unravelled.

The Gulf coastal plain in North America shows sets of low-sinuosity, gravel-transporting channels as well as high-sinuosity channels which carried sands and silts. The metamorphosis reflected in these different channel forms is taken to reflect the change in sediment and water loads resulting from the change from relatively arid to more humid conditions of the late Quaternary (Baker 1983). A history of changing fluvial environments has also been described for the Son river in north-central India (Williams and Royce 1982; Williams and Clarke 1984). Here again the sedimentary record, with a chronology based upon ^{14}C dating of shells, charcoal and carbonate, shows a relatively wide, seasonal, and low-sinuosity channel in the last glacial transformed to a narrower, less seasonal and more sinuous channel, carrying and depositing a finer sediment load. In response to the wetter climate of the Holocene, the Son river has incised its Pleistocene floodplain to a depth of about 30 m. Once again, the transformation of the river can be interpreted in terms of the change from a sparsely vegetated watershed during the cold and dry glacial conditions, to a better vegetated one which consequently sheds reduced amounts of sediment during the milder and wetter Holocene. An understanding of the mechanisms behind such transformations can usefully be applied to the development of forecasts of the effects of future environmental changes (see Chapter 13; Knox 1984).

EVALUATING RIVER HYDROLOGY FROM PALAEOCHANNEL EVIDENCE

A most attractive possibility is raised by the preservation of Quaternary channels, terraces and riverine sediments. This is to attempt to reconstruct, on the basis of these kinds of evidence, the former flow conditions of the rivers and hence to make inferences about the environments of their drainage basins in a quantitative way. Certainly, it is possible to infer from preserved river sediments something of the speed of the flowing water responsible for carrying them. If suitable cross-sections of the old channels can be measured, it is then possible in principle to estimate the volumetric water flow rate that formerly existed. This requires estimates of the former channel gradient, and of factors like the channel roughness which may have affected the velocity of flow along the channel. In a similar way, by using relationships established between channel geometry (say, the form of meander bends) and modern discharge, and extrapolating these to palaeochannels, it is possible to say something of the former flow conditions. A range of methods for this kind of study was reviewed by G.P. Williams (1984) and Rotnicki (1991). However, the relationships involved in such palaeohydrological inference are complex and only incompletely understood. The complex response of river systems outlined earlier makes the interpretation of river terraces equally involved. As our knowledge of river behaviour develops, it will be possible to make further use of the evidence of palaeochannels. At present, however, the difficulties in making firm quantitative inferences from them confound most attempts. Rotnicki (1991) estimated that the standard error of estimates of past discharge made from data gleaned from former channels might be up to 20%. However, even if the former flow rate can be estimated, obstacles remain before this can provide information on palaeoclimatic parameters such as rainfall or evaporation. Lakes, however, offer more readily available sources of data on the environmental history of their surroundings, and we turn now to consider some of these.

LAKE MORPHOLOGY AND ORIGIN

Lakes and other non-flowing water bodies owe their origin to a great variety of geological circumstances. Some of the largest lakes in the world occur in depressions left in the landscape as a result of glacial erosion. Commonly, these lakes are deep (with depths of more than 100 m) and are found in glacial valleys in mountainous regions. Numerous lakes in the Alps belong to this category. Other large lakes are relict depressions left in the landscape by a retreating ice sheet, such as the Laurentide ice sheet in North America (see Chapter 3 for more details). The Great Lakes are the best example. The sedimentary record in those lakes formed by glacial ice begins from the time of the most recent retreat of ice from the lake basin. Hence, most of these lakes have a record of sedimentation that spans less than 18 000 years. More discussion on the record of these lakes will be given later in this chapter.

There are several other types of large lakes which owe their origin to phenomena other than those associated with glaciation. Principally, those lakes occur in large and sometimes blocked depressions such as Lake Eyre, which occurs in the lowest portion of the Australian continent; Qinghai Lake, which occurs in a vast depression on the plateau in northern China behind the Himalayas; the Aral Sea, which occupies an extensive sedimentary basin once connected to a large seaway that became closed as a result of tectonic activity; or in large tectonic depressions like the very long and deep (over 1 000 m) Lake Baikal in SE Siberia, Russia and Tanganyika in East Africa. Because these latter two depressions in which the lakes occur were not formed as a result of glacial activity, the sedimentary records in the lakes extend much further back in time than those of the glacial lakes. In fact, the sedimentary sequences below Baikal and Tanganyika extend to several kilometres, as recently recognised from seismic profiles run over both lakes (Hutchinson *et al.* 1992; Scholtz and Rosendahl 1988).

Among other types of lakes, it is necessary to distinguish those that receive most of their water from surface drainage and directly from rainfall, from those that are principally fed by groundwater. The former type encompasses crater lakes, which are perhaps the best type of lake for the reconstruction of climatic records. These lakes can either be formed as a result of the impact of extraterrestrial material (e.g. meteorite impact) or as the result of volcanic activity, either by a volcanic eruption or the result of an explosion below ground caused by basalt coming into contact with an aquifer, or by a pocket of gas (e.g. CO_2 originating from the mantle) expanding on its way to the surface. In each case, a crater is formed at the surface. Under normal circumstances, crater lakes act like gigantic rain gauges because most of the water that enters the lakes results from precipitation directly over the lakes. There is usually no river and little groundwater input into the lakes which are concentrated there.

The second type of lake is usually referred to as a 'groundwater window' because it is most often found in arid and semi-arid regions where evaporation is greater than precipitation, and this causes the groundwater to rise and then to seep into the lake (see Chapter 9). In fact, the level of these lakes can also be a good indicator of changes that occur on a regional basis, reflecting, for example, the amount of water entering the lake's catchment. A great proportion of the groundwater-window lakes yield saline water simply because of the excess of evaporation in the area, and the solutes (soluble mineral components) that are picked up and transported by the water within the sediment interstices before emerging above the floor of the lake.

Other small lakes, sometimes with a long record of deposition, are karstic lakes which occur in regions where dissolution of the local lithology (e.g. limestone, gypsum, laterite) operates substantially. Although these lakes are usually characterised by a small catchment area compared with the others mentioned above, the lake records are often disturbed by further dissolution of the lake margins, thus causing slumping of the sedimentary sequences along the margins. Similarly, some lakes that are found in shallow depressions behind dunes along the coastline, or behind aeolian deposits deflated from lake floors, characteristically have a small catchment and also a short-lived record due to the frequent migration of coastlines and associated dunes. This is in contrast to the other types of lakes with a large catchment, which have the potential to clarify hydrological, and thus climatic, regimes for extensive regions, and which provide information of greater relevance on a continental scale. Nevertheless, only major events are recorded accurately in large lake basins, whereas minor events/changes are usually more faithfully recorded in the small lakes which readily respond to (and thus register) small hydrological changes.

There are also less significant lacustrine systems such as small pans, oxbow lakes (occurring in cut-off meanders of rivers), and even recently established artificial lakes, such as dams and farm dams, which usually yield a short record of sedimentation, but which are of relevance to changes occurring nearby and commonly resulting from anthropogenic interaction with the environment (especially for dams). Nevertheless, some of these small aquatic systems and their deposits can provide relevant information about short-term climatic change and provide a high-resolution record of change. High resolution implies that it is possible to decipher a record of change from the sedimentary record of a waterbody that represents short-spaced events, such as on a seasonal level or even less, compared with the other records recognised in lakes or the oceans which cannot, under normal circumstances, provide records spanning episodes of less than centuries.

Lakes can therefore be considered as valuable sites from which a record of environmental change can be obtained. It is within the sediments found at the bottom of lakes or associated features (see below) that it is possible to retrieve information on past changes that directly affected the lakes or that occurred within the lake's catchment.

ASSOCIATED FEATURES

In order to reconstruct the history of lakes it is necessary to obtain indications of lake-level changes or of water quality. The former is of importance because a change in hydrology in a particular region, often best represented by an alteration of the ratio of precipitation to evaporation, will cause the lake volume to change, sometimes dramatically (Forester 1987). For example, a lake may have retained water for centuries (and hence be called a *permanent* lake) and change status to become an *ephemeral* lake, commonly dry but filling up occasionally, as a result of a change in the climatic/hydrological regime. However, before discussing geomorphological features indicative of lake-level change, it is necessary to point out that lake water budgets also cause water chemistry and aquatic biota to change accordingly. These features are discussed later in this chapter because they occupy an important place in the reconstruction of palaeoenvironments.

The best way to describe the various features characteristic of different lake phases is to examine a full, deep lake with permanent water which progressively changes to a shallow, ephemeral lake as a result of a change of hydrological regime (De Deckker 1988). Emphasis is placed on features that are distinctive of the different lake phases, but which can also frequently be recognised in ancient lake deposits so as to allow reconstruction of lake histories (Fig. 8.8).

The large, deep lakes are characterised by a substantial, deep water column which has an important effect on the occurrence of biota in the lake and the preservation of their skeletal remains on the lake floor, as well as the composition and preservation of inorganic sediment. Several of the large lakes are so deep that the amount of dissolved oxygen is so low that common invertebrates and vertebrate organisms cannot live below a certain water depth. The consequence of this 'de-oxygenation' of the water is that a layering or stratification of the water column occurs. Those lakes become stratified as a result of chemical

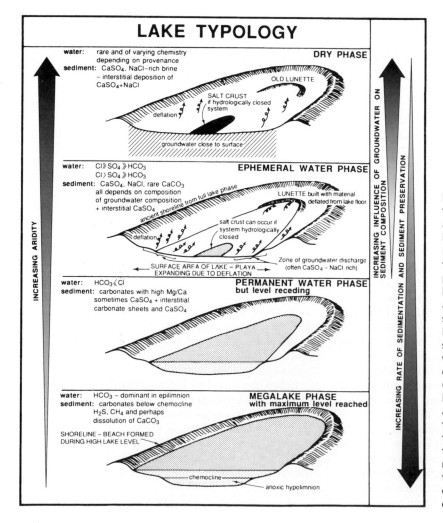

Fig. 8.8 Geomorphological, sedimentological and chemical features associated with different types of water budget affecting a lake. The diagram is arranged to distinguish the characteristics of a large lake during a wet period and evolving progressively to become a dry playa partly covered by a salt crust. Note that in this particular system where a lake ends up with NaCl–dominant water, a change in sediment mineralogy is also registered with a dominance of gypsum ($CaSO_4$) crystals, and finally with a halite (NaCl) crust. Note also that the deep-lake phase (called a *megalake*) is characterised by a stratified water column that does not mix at least for substantial periods of time, and that leads to the preservation of laminated sediments because of the absence (due to the lack of oxygen) of burrowing organisms that would disturb the layering. (After De Deckker 1988)

changes (e.g. changes in oxygen, pH and composition levels) or physical changes (e.g. changes in density and temperature with depth: (note that these two features interact); thus, the bottoms of these lakes are devoid of organisms which otherwise would disturb the sedimentary record by burrowing and other forms of bioturbation. Consequently, lakes which are stratified retain an undisturbed sedimentary record, which may sometimes be laminated. The latter consists of a thinly layered sequence caused by colour or composition change in the sediment that usually results from alternation with biological remains (e.g. siliceous skeletons of algae) falling to the bottom of the lake. An algal bloom at the lake's surface can in turn engender a change in the chemical composition of the surface water, which may force the precipitation of some inorganic minerals (the commonest being tiny crystals of calcium carbonate such as calcite and aragonite, typically a few μm long) that subsequently settle on the lake floor above the remains of siliceous algae, forming a layered sedimentary sequence. Frequently such a laminated sequence is characterised by an alternation of pale and dark layers, and it has the potential to provide a high-resolution record if the laminations can be assigned to particular phenomena such as seasonal/annual changes. Such layered sequences, if rhythmically deposited, form extremely important records which can be compared against other records such as historical ones (for calibration) and tree-ring data (see Chapter 10). In some lakes, not only the nature of the laminations but their thickness can relate to environmental changes affecting the lakes themselves.

The shorelines of large lakes (and not necessarily deep ones) are characterised by beach deposits (Fig. 8.8) consisting of coarse material that frequently shows signs of reworking. During a phase of lake-level recession (regression), material previously reworked and deposited at a shoreline will remain as part of the landscape, and thus leave evidence of a former lake level. On the other hand, during a lake-level rise (transgression) across previous levels, former shorelines/beach deposits may be eroded away, and thus a loss of previous records occurs. Commonly, remains of aquatic organisms mixed with those of terrestrial material (e.g. plants) are found in shoreline deposits.

As the water level recedes, the dissolved salts in the water become concentrated, and a chemical evolution of the water ensues (see Fig. 8.8). Depending on the nature of the rocks surrounding the lake and in the catchment, the soluble weathered components of the rocks end up in the lake and their variety will control the chemical composition of the lake water. Of the two principal types that are commonly found in lakes, one at the earliest stages of solute concentration (in other words, as salinity progressively increases) has bicarbonate as the dominant ion, whereas in the other type of water, the bivalent cations calcium and/or magnesium are dominant. The progressive evaporation of the lake water will force some precipitation of minerals such as the ubiquitous calcium carbonate which is also one of the least soluble. Calcite, a common form of $CaCO_3$, will frequently precipitate first at the expense of the minor components, forcing a further dominance of others. In other words, once calcite has precipitated (this is called the *calcite branch point*), the chemistry of the water will follow one of two pathways: one evolves from a water enriched in bicarbonate to the detriment of the bivalent cations (Mg and Ca), whereas the other is bicarbonate depleted and enriched in either, or both, of the above-mentioned cations. The bicarbonate-rich waters will see a further enrichment, and finally precipitation of carbonate minerals, with an increase in salinity compared with the other type of chemical pathway which will eventually register the precipitation of gypsum ($CaSO_4.2H_2O$, labelled only as $CaSO_4$ for simplicity in Fig. 8.8) and halite (NaCl, which is the same composition as table salt), with a progressive salinity increase. (This second type of chemical pathway has a dominance of some ions in similar proportions to that of oceanic water, especially if it were to be evaporated.) It is important to distinguish between the two types of chemical pathways, water chemistries and salinities as they both have a significant effect on the presence and absence of aquatic organisms. By reversing roles, the remains of aquatic organisms recovered from the sediments of a lake can yield information on the water quality of the lake, and a relationship between water quality and salinity can be used to reconstruct climate. For example, the evaporation/precipitation ratio can be estimated depending on salinity levels if there is good control of salt budgets, and also the amount and source of weathering (thus leading to discovery of a change in lake catchment) can be postulated from a change in water chemistry.

Once a lake level has receded sufficiently to uncover a large portion of the lake floor, deflation of surficial material (namely, sediments as well as remains of organisms) can occur during dry and windy periods (see Chapter 9). Hence, this deflated material may be transported for long distances, and may even be finally deposited in the oceans some thousands of kilometres away. A large portion of this deflated material eventually accumulates to form dune deposits. The best example of this phenomenon is represented by the crescent-shaped dunes formed downwind of lakes. With a further lake-level drop, salinity continues to increase and the diversity of aquatic life (plants and animals) decreases, and the final step will generally see a salt crust forming on the lake floor. This salt crust usually prevents further deflation of lake floor material. It is during low-level lake phases or when a lake is dry that several of the sedimentary features that originated during previous high-level lake phases are completely or partially destroyed (through deflation, bioturbation or non-biological, mechanical processes such as the formation of mud cracks), thus blurring the record of lake history.

Another type of lake, and associated water chemistry, is frequently found in wet and often cold environments; these are characteristically mountain lakes and those in tundras or forested areas such as peat bogs. These aquatic waterbodies are characteristically rich in organic content, be it in the sediment or in the water. Carbonate minerals (inorganically precipitated or as skeletons of organisms) are usually nonexistent in those lakes which commonly have a low pH. Thus, it is the remains of siliceous organisms (chrysophytes, diatoms, sponges) and perhaps crustaceans such as cladocerans recovered from the sedimentary pile of these aquatic waterbodies which can help decipher patterns of environmental change. As mentioned earlier, pigments left in the sediment and pertaining principally to algae or bacteria can help to distinguish a change in lake status and trophic level. The effects of human activity in a catchment, such as those caused by land clearance, fire activity, addition of fertilisers, acid rain and organic pollution, can be determined from a change in sediments and associated organic remains.

Many of the lakes that occur in regions with substantial seasonal contrasts will undergo characteristic changes that directly relate to atmospheric conditions: lakes may become thermally layered (stratified) as a result of significant temperature changes between seasons; lakes may freeze at the surface; water may mix as a result of a change in density gradients through time, for instance, lakes 'overturn' as a result of mixing. Many of these phenomena that affect lakes will often have a profound impact on the nature of the sediment deposited at the bottom of the lakes and/or on the biological activity in the lakes. For further details, see De Deckker (1988).

LAKE STATUS

Lake status, as previously defined, provides the best possible indicator of the climatic history of a region. Nevertheless, it is also necessary to recognise that large lakes with a large catchment area have the potential to provide information on the evaporation to precipitation (E/P) ratio over a very large area. In fact, a slight change of this E/P ratio may not necessarily be recognised easily by features such as shoreline or water salinity changes. On the other hand, it is

principally the dramatic changes in lake levels that are to be recognised, and to be correlated only with a substantial change in the hydrological regime affecting either the entire region, or at least part of it.

It is the small lakes, with a small catchment area, that provide a more accurate account of slight changes of the E/P budget over the lake's catchment (Fig. 8.9). The disadvantage of these small lakes, on the other hand, is that they do not usually exist for a long period of geological time, and often if there is a dramatic change in the E/P ratio, such as that due to an excess of water, the lake may overflow and information relevant to the climatic change may be lost if solutes are flushed out of the catchment. Similarly, no shoreline features that would accurately record the extent of the lake could form. An excess of evaporation would cause the lake to dry, and loss of sediment caused by deflation would again suppress some of the information of value for a reconstruction of the lake history.

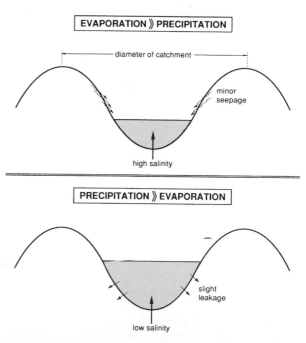

Fig. 8.9 Schematic diagrams showing characteristic features associated with crater lakes under two specific hydrological regimes. Top: When evaporation is greater than precipitation, the water level remains low and solute concentration is high, assuming that most of the water in the lake originates mostly from within the lake's small catchment, as defined by the periphery of the crater rim. Bottom: This represents the situation where the evaporation to precipitation ratio is reversed, the water level is then high and the solute concentration low. Consequently, crater lakes can be considered as gigantic rain gauges, and thus either lake-level or salinity changes can be used to monitor evaporation–precipitation changes

Climate is one of the prime factors that help to change the nature of a lake, and this is reflected in the nature of the sediment and the biological remains found within it. Therefore, a combination of the study of the latter two will help us to understand environmental changes affecting any lake.

AQUATIC ORGANISMS

Aquatic organisms are numerous and diverse. They inhabit almost any type of aquatic environment, from a shallow ephemeral pool to a hypersaline lake (Fig. 8.10). Organisms can also survive drought by burrowing in the mud, or survive a long desiccation phase by producing drought-resistant eggs. Other organisms such as fish only survive and reproduce in permanent lakes; others, like some snails or freshwater mussels, can survive short periods of drought by closing/sealing their conch or tightening up their valves and remaining in a state of torpor. The recognition of species which have very specific ecological requirements enables the palaeoecologist to use the fossil remains of organisms for the reconstruction of ecological changes in lakes, and thus infer physico-chemical changes that directly translate to hydrological and climatic changes (Frey 1969; Meriläinen *et al.* 1983; Löffler 1987; Gray 1988). It is also important to recognise that numerous species of aquatic organisms are to be found in very specific water chemistries, and that it is therefore possible from the identification of fossil remains to reconstruct changes through time in chemical pathways in a lacustrine system. The latter can give information on the particular types of water chemistry in which the organisms lived.

The most commonly available organisms that enable us to reconstruct water chemistry are diatoms (algae with a siliceous skeleton), and invertebrates such as ostracods (microscopic crustaceans related to shrimps), gastropods (snails) and bivalve molluscs, which all have a calcareous shell made of $CaCO_3$. All

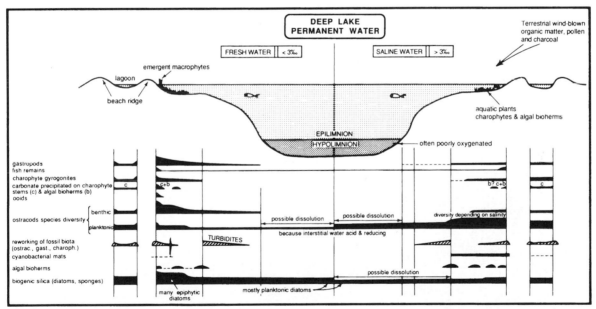

Fig. 8.10 Schematic diagram showing some of the organisms most commonly used to reconstruct the history of lakes. The profile shows a deep lake in the centre and shallow ones on either side. Two types of lakes are recognised, one fresh and the other saline. Note that most organisms can be found in both fresh and saline ecosystems, but diversity would naturally decrease with a salinity increase. Some organisms such as gastropod molluscs are restricted to shallow portions of a lake, as they are frequently herbivores and feed on aquatic plants (emergent macrophytes, some of which produce pollen and seeds) and algae fixed to the lake floor (e.g. charophytes which secrete calcareous egg cases called gyrogonites) that require light to grow. Ostracods are calcareous, bivalved microcrustaceans which are ubiquitous aquatic organisms with well-defined ecological requirements. Algal mats, made principally of blue-green algae (cyanobacteria) forming small mats, but occasionally large mounds (bioherms which are occasionally lithified), grow in the lakes at shallow depths. Siliceous remains of algae (principally diatoms and chrysophytes) and sponges can also be used to reconstruct the history of lakes since they are excellent ecological indicators. Diatoms especially have very particular salinity, chemical and nutrient requirements, and thus provide information on water quality. Care should be taken when assessing a lake's record from fossil remains as some may be reworked at shorelines or along steep flanks by turbidity flows, or may be partly dissolved by physico-chemical processes after death of the organisms. (After De Deckker 1988)

of these organisms are commonly found in fresh and saline lakes, and other physico-chemical parameters control or influence their presence and abundance in lakes. Some of the important parameters are: water temperature (which principally controls hatching and reproduction), pH, salinity, dissolved oxygen, and trophic/nutrient levels.

Figure 8.10 shows the types of organisms, and their remains, that are to be found in lakes. This diagram places emphasis on the occurrence of many of the different types of organisms living in lakes, be they saline or fresh. Discussion here only refers to the organisms that are commonly represented in lakes and which have remains that preserve readily. Not mentioned in this diagram are several other organisms that have been used by palaeoecologists to reconstruct lake histories, such as aquatic insects (remains of larvae and adults), vertebrate remains (e.g. fish, frogs), remains of unicellular organisms such as the cestate thecamoebians (which resemble the well-known marine foraminifers), some foraminifers brought into lakes by birds (and which can survive as well as reproduce in the lakes provided the salinity and water chemistry are similar to that of seawater), and remains of crustaceans among which are crabs and other decapods and isopods. More commonly used are the cladocerans (commonly known as water fleas), rotifer eggs and algal

remains (some of which are even distinguishable by the different pigments recovered in the sediments).

Of importance too are pollen, spores and seeds of aquatic plants which may be used to help to reconstruct conditions in a lake and define its status. Naturally, the pollen blown into lakes from elsewhere can also indicate the vegetation in the vicinity of the lakes (see Chapter 10 for more details), and thus complement the reconstruction of the history of a region through a definition of the hydrological budget of a lake and of the effect of the water regime on the vegetation in the area.

QUATERNARY LACUSTRINE RECORDS

Features characteristic of large, deep lakes have been recognised on all continents. In North America, the high lake-level phases are referred to as *pluvial lake* phases, whereas in Australia they are called *megalake* phases. The latter term is preferentially used in Australia and should be worldwide because the fact that the lakes retained water may not necessarily be the result of more rainfall (as the word 'pluvial' would infer) in the lake catchment. Instead, a high lake-level stand may be more the result of less evap-

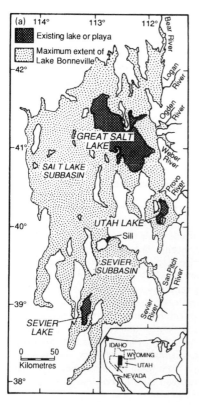

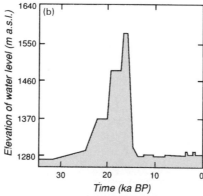

Fig. 8.11 (a) Map showing the maximum extent of the palaeolake Bonneville which existed 18 ka ago, and the present configuration of the Great Salt Lake, Sevier and Utah lakes in the western US. These three lakes represent the shrunken remnants of Lake Bonneville under the present climatic and hydrological conditions. (After Benson and Thompson 1987) (b) Schematic reconstruction of lake-level fluctuations for the Great Salt Lake for the last 32 000 years derived from sedimentological, mineralogical and microfossil studies carried out by Spencer *et al.* (1984). Note the step-like rise in lake level prior to the glacial maximum registered at 18 ka, followed by the extremely rapid lake-level drop. (After Spencer *et al.* 1984)

oration or more cloud cover, or even a change in runoff conditions (e.g. due to a change in vegetation cover). The best-documented example of large lakes having formed during the Quaternary in the US is Lake Lahontan in the now arid southwest. The Great Salt Lake (GSL) (see Fig. 8.11a; Benson and Thompson 1987) in the state of Utah is now very saline (with a NaCl-dominant chemical pathway) and is a relic of the once much larger Lake Bonneville, which during its high-lake phase reached an altitude close to 1600 m compared with the present-day GSL level of 1270 m (see Fig. 8.11b; Spencer *et al.* 1984). In fact, a drop in lake level of some 300 m occurred during approximately 2000 years, i.e. at a rate of 15 cm per year assuming a constant rate of lake-level drop. The maximum high lake-level stand occurred at 16 ka BP at a time when glaciers started melting in North America (and when most of Australia was extremely dry).

There are now sufficient data available to be able to reconstruct the history of lake-level changes for large North and South American, African and Australian lakes, covering the last 18 000 years since the last glacial maximum had a significant effect on the landscape in the northern hemisphere. Examination of maps of the world detailing lake status (low,

intermediate and high lake-level) indicates that not all lake levels were of the same status, even on a continental-wide basis at the same time. Figure 8.12 shows as an example the lake-level status for the African continent for four selected time-frames which demonstrate the differences that exist on a continental basis between major lacustrine systems, and the need to be cautious before attempting to infer broad climatic generalisations from lake-level changes (Street-Perrott *et al.* 1989). Different regions register a drop in lake level whereas others may have high lake levels. This phenomenon relates to changes in precipitation regimes that may be affected by large-scale intercontinental atmospheric trends (Open University Course Team 1989b). For example, today when the South-East Trade Winds are strong near the coast of South America (Fig. 8.13), thus causing upwelling (see Chapter 7 for further details) along the Peruvian coast, rainfall is recorded in northern Australia, even affecting the level of lakes in central Australia, whereas the western part of the US is usually undergoing drought conditions. This phenomenon of intercontinental connection is now quite well documented and warns against comparing lake-level status (e.g. highs) from different sites on a global scale, or between two continents, in order to

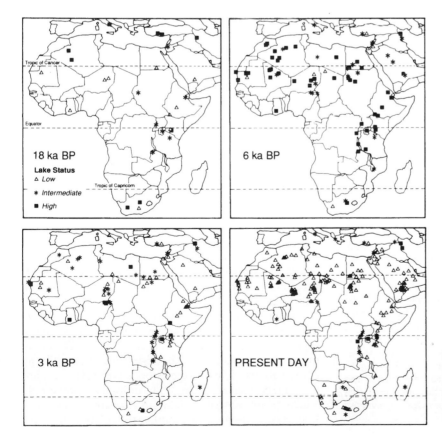

Fig. 8.12 Composite diagram showing the status of lake levels registered on the African continent for four specific time-frames: 18 ka, 6 ka, 3 ka and present. (After Street-Perrott *et al.* 1989)

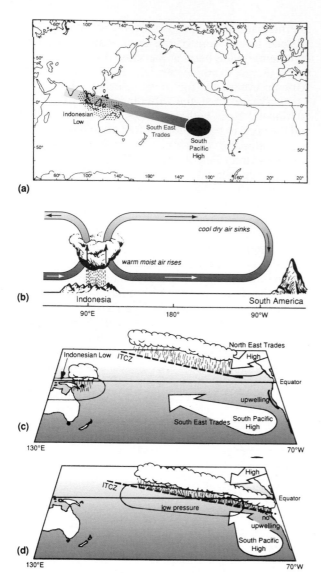

Fig. 8.13 (a) Map showing the position of a typical atmospheric high-pressure cell (High) in the South Pacific above Tahiti which corresponds to a Low above Indonesia and northern Australia, and which would both correspond to an excess of rainfall in the latter area as indicated in diagram (b). (c) Schematic diagram to show the effect of the High in the South Pacific under normal circumstances and the corresponding effect on ocean temperature. In such a situation, upwelling of cold water and nutrients occurs along the Peruvian coast, and a large portion of the eastern equatorial Pacific, including the southwestern portion of the US, remains dry (see cross-hatched area). Rainfall is substantial over northern Australia under that scenario. Ocean surface temperatures greater than 28°C are circled by a stippled line. (d) During El Niño years, there is no upwelling, the South-East Trades are weaker, ocean temperature is lower and the Intertropical Convergence Zone (ITCZ) is closer to the equator in the eastern Pacific. As a consequence of the above conditions, there is a rainfall deficit over northern Australia and most lakes are dry. (After Open University Course Team 1989b)

demonstrate a globally wetter climate. In fact, it is the prevailing winds over the equatorial Pacific that form the link between the different rainfall patterns on either side of the Pacific. This is defined as the Southern Oscillation Index, which relates to a difference in atmospheric pressure measured in Tahiti and Darwin in northern Australia, and which defines an overall set of atmospheric conditions across the Pacific. A high pressure cell in the South Pacific is paralleled by a low-pressure cell over northern Australia which strongly affects summer rainfall there. In the opposite case, drought conditions prevail over a large portion of the Australian continent. Such a transcontinental relationship in climate which affects lake level needs to be recognised to explain the changes that are registered in lake records across the globe. More details of the El Niño and the Southern Oscillation Index and global teleconnections affecting climate are presented in Chapter 12.

LAKE HISTORIES

The record of lake-level fluctuations in large lakes can be obtained through several methods of investigation. The first is the study of ancient shorelines, which can help document past lake-level highs. Such geomorphological features are sometimes so obvious and extensive across the landscape that they are distinguishable on aerial and even sometimes satellite photographs, especially in extensive lacustrine systems. It must be remembered, though, that only the regressive phases leave shorelines on the landscape. A transgressive lake phase will tend to rework ancient shorelines except for the one that is formed during the maximum extent of the lake. In a sense, the way shoreline sequences can be interpreted with respect to lake-level fluctuations is the same as for glacial moraines left on the landscape after a glacial retreat. Concretions formed in association with algae, such as benthic diatoms or cyanobacteria, often grow in abundance underwater on the edge of some lakes. Such organically precipitated concretions are frequently found along ancient lake shorelines and therefore can be used to determine the extent of past lake levels (see Benson (1994) for an excellent documentation of such features at Great Salt Lake, Utah).

Another method for reconstructing lake-level histories is to examine the sedimentary and palaeontological record of the large lakes. Different mineralogies, sediment chemistry and faunal/floral compositions of fossiliferous beds will relate to particular sedimentological and biological processes, which relate directly to lake-level histories. In addition, recent research to decipher the relationship that exists between water chemistry (e.g. the particular chemical pathways discussed previously in this chapter) and several taxa of calcareous-shelled ostracods, and also the siliceous diatom frustules, has allowed the reconstruction of the chemical evolution of the lake waters through time, and has thus related chemical changes to climatic change (Forester 1987).

Naturally, one should also be aware of the potential reworking of sediment and fossil remains along lake margins where wave activity is predominant, and also of possible slumping along lake flanks where turbidity flows are frequent (Fig. 8.10). In addition, dissolution and diagenetic effects can alter the sedimentary and palaeontological record.

Several sophisticated techniques have recently been used to decipher the history of lakes. Geochemical analyses of microfossil remains of the sort now commonly applied to the marine record (see Chapter 7 for further details) have successfully been used on lacustrine material. Because of the almost complete absence of foraminifers in lakes, investigations have had to focus on different organisms, and more frequently on the calcareous-shelled ostracods and gastropod and bivalve molluscs. Stable isotopes of oxygen have been analysed from the calcareous remains of organisms mentioned above with the purpose of defining past physico-chemical parameters that existed in lakes, such as temperature and water composition, with respect to isotopes of oxygen (^{16}O and ^{18}O) and carbon (^{12}C and ^{13}C) (for further details on the isotopes refer to Chapter 7).

An outstanding example study documenting the composition of stable isotopes of organisms, mainly ostracods and bivalve molluscs, was carried out by von Grafenstein *et al.* (1992, 1997) from specimens extracted from cores from two Bavarian lakes. This enabled the past oxygen isotope variation of atmospheric precipitation that entered the lakes during the Holocene to be reconstructed. Examination of Fig. 8.14 shows that, for Lake Ammersee, a sharp positive shift registered in the $\delta^{18}O$ of the ostracod valves on several occasions (in the diagram, this shift is registered by an increase in the amount of ^{18}O to the detriment of ^{16}O). For example, the shift which occurs at the Younger Dryas/Preboreal transition, just close to 10.2 ka BP, is interpreted as a dramatic climatic improvement that was then maintained for the rest of the Holocene, except for a short period around 8.2 ka which lasted approximately 200 years (von Grafenstein *et al.* 1997). Note the similarity in Fig. 8.14 between the Greenland GRIP record and the Lake Ammersee record, suggesting a close correlation of atmospheric events at quite different latitudes on land along the eastern Atlantic Ocean. Ostracod $\delta^{18}O$ records can shed light on a variety of phenomena affecting lakes which are of climatic significance, including a change of inflow into a lake probably caused by isotopically light water originating from the melting of glacial ice (see Lister 1988) or a change in salinity (Chivas *et al.* 1993). The $\delta^{13}C$ of biogenic carbonates, on the other hand, can be used to interpret the palaeoproductivity of a lake, but the interpretation of the data is usually quite complex. For a good introduction on the combined use of oxygen and carbon isotopes in gastropods, refer to Abell and Williams (1989).

Recent investigations of trace elements, namely magnesium and strontium, which replace calcium atoms in the calcite lattice of ostracod shells from non-marine environments, have permitted a better definition of physico-chemical conditions in lakes, such as temperature, ionic composition and salinity. Chivas *et al.* (1986), for example, were able to relate a change in the atomic ratio Sr/Ca in the valves of single ostracods from a Holocene sequence from a crater lake, the volcanic maar Lake Keilambete in southeastern Australia, to salinity changes in the lake (Bowler 1981; Chivas *et al.* 1986). An increase in salinity in a lake with a well-defined/constricted catchment, such as that of a crater lake (Fig. 8.9), for example, as a result of an increase in evaporation, should only register an increase in solute concentration and not a chemical change such as modification in the Sr/Ca ratio of the water. However, in the case of Lake Keilambete, the Sr/Ca of the ostracods (and thus of the water too) did change and could be directly related to a change in salinity (Fig. 8.15), since the lake water must have remained supersaturated with respect to the bicarbonate ions despite the fact that some carbonate precipitates frequently formed within the lake (A.R. Chivas, pers. comm.). The best approach nowadays is to combine the study of trace elements and stable isotopes on the same ostracod valves, or any other kind of biogenic carbonates, to reconstruct past conditions in lakes (see Chivas *et al.* 1993). Those studies provide the best possible definition of environmental changes, especially if combined with palaeoecological analysis of faunal assemblages which can further elucidate past events in the lakes. Another 'tool' relates to the use of isotopes of strontium (^{86}Sr and ^{87}Sr) to determine a change in the origin of the water entering a lake. The ratio of these two isotopes, expressed as $\delta^{87}Sr$, is determined by the provenance of the water, i.e. lithologies with which the water has been in contact, or the origin of the Sr cations, such as from sea spray-salt. Therefore, the determination of the $\delta^{87}Sr$ from carbonate minerals (often measured on the skeleton of calcareous organisms) can determine the origin of the water in which the organisms lived. A direct application of this technique is to be able to define a marine connection to a lake previously isolated from the ocean.

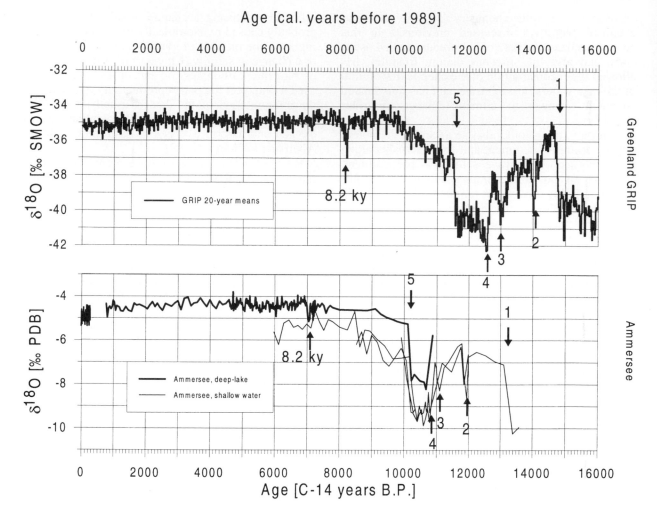

Fig. 8.14 The upper diagram represents the $\delta^{18}O$ record of the GRIP ice core from Summit, central Greenland, for comparison against the diagram below which documents the $\delta^{18}O$ record of ostracods from Lake Ammersee in Bavaria, Germany. In the lower diagram, the bold line represents the $\delta^{18}O$ record of juvenile ostracods belonging to the genus *Candona,* and the thin line represents the record of the ostracod *Cytherissa lacustris.* The Ammersee record obviously matches the Greenland ice core, especially when similar events are identified at both sites. These are: 1, Oldest Dryas/Bølling transition; 2, older Dryas; 3, Gerzensee Oscillation; 4, Allerød/Younger Dryas transition; 5, Preboreal/ Younger Dryas transition. Note that the $\delta^{18}O$ scale differs for the two diagrams, but this results from the different chronological methods used. The age differences, increasing to more than a thousand years around the Younger Dryas, are the expression of variations of the ^{14}C content of the past atmosphere. The radiocarbon-based ages for the Ammersee core have not been corrected because corrections from tree-ring chronologies (which are the most accurate) only go back to the end of Younger Dryas. For further details on isotope terminology, refer to Chapter 4. (From von Grafenstein *et al.* 1998)

Techniques now used to reconstruct lake histories are becoming more sophisticated, but it is necessary to be aware that the selection of lacustrine sites is very important in order to obtain the relevant information. Two lakes adjacent to one another frequently can react differently to the same climatic conditions, for example, if their morphologies differ (e.g. a conical waterbody will evaporate at a different rate compared with a lake that is large and extremely shallow) or the quality of their water differs (e.g. water turbid-ity affects the trapping of solar heating, and the biota can also affect the water transparency).

The most outstanding example of a multidisciplinary study carried out on any lake is the one done on Elk Lake in Minnesota, US. Bradbury and Dean (1993) have edited a volume which demonstrates the importance of choosing a lake well before applying numerous techniques to decipher, in their case, a high-resolution Holocene record of climatic changes. The location of this lake with respect to some partic-

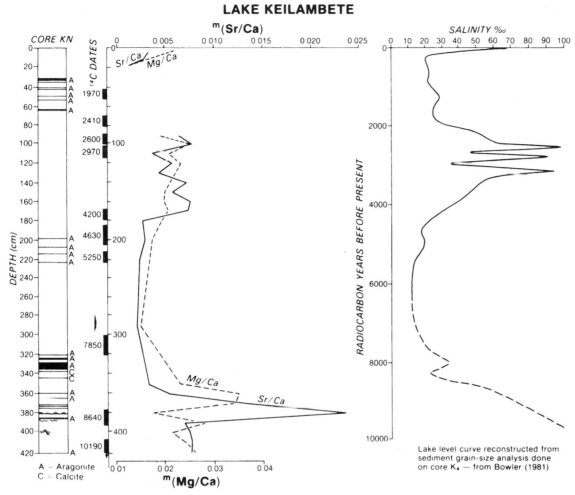

Fig. 8.15 Plots of the mean value for individual layers of the molar ratios of Sr/Ca and Mg/Ca measured on ostracod shells from a 4.2 m long core (representing the entire Holocene) from the maar lake Keilambete (data from Chivas *et al.* 1986) compared with the smoothed palaeosalinity curve of Bowler (1981) based on sediment grain-size (which is related to water depth in the lake) from some 40 layers from an adjacent core. A schematic stratigraphical log on the left margin of the diagram shows the presence of calcitic and aragonitic layers that testify to the saturation nature of the lake waters through time with respect to carbonates. Sr/Ca values in the ostracod shells are related to salinity changes in the lake (see Chivas *et al.* 1986)

ular climatic boundaries, and the physiography and physico-chemical nature of the lake, have engendered the formation and preservation of fine laminations for the entire Holocene. Thus, Elk Lake is an exemplary recorder of environmental change. It is the presence of those annual laminae on the lake floor that enabled the detection, for very specific and short periods of time, of significant climatic shifts in the vicinity of the lake. It is this kind of multidisciplinary approach that will help to make substantial advancements in palaeoclimatic research; this can

only be achieved, however, through the judicious choice of a lake worthy of palaeoenvironmental study of the highest quality.

In this chapter we have seen that different lake types provide diverse sorts of palaeoclimatic information; and different organisms, their chemical composition, and associated sediments yield information on various types of conditions and phenomena. A multidisciplinary approach to the study of lacustrine deposits will therefore provide the most complete array of information.

CHAPTER
9

EVIDENCE FROM THE DESERTS

Ubi solitudinem faciunt, pacem appellant.
(*They create a desert and call it peace.*)
Tacitus (c. AD 55–117)
Agricola, 30

Deserts are remarkable repositories of palaeoclimatic information. Many desert landforms reflect the operation of processes which are no longer active today. Every desert in the world has its legacy of dry or saline lakes, many of them part of once integrated but now defunct and segmented drainage systems. The fossil fauna of great deserts like the Sahara is an eloquent witness to a time when the presently parched wilderness was able to sustain abundant plant and animal life, including such large tropical herbivores as elephants and giraffes, as well as a widespread aquatic fauna of turtles, hippos, crocodiles and Nile perch (Williams 1994a; Vernet 1995). In sheltered valleys in the Aïr Mountains of Niger, in the southern Sahara, there are remnant populations of savanna primates: anubis baboons (*Papio anubis*) and patas monkeys (*Cercopithecus patas*), which reached the mountains during wetter times when the West African savanna woodland was considerably more extensive than it is today (Monod 1963; Newby 1984). The dwarf crocodiles which apparently still inhabited some of the permanent waterholes in the Tibesti Mountains of the south-central Sahara in the 1950s (A.T. Grove, pers. comm.), but which may well now be extinct (see Lambert 1984), are (or were) part of this palaeoclimatic heritage, prompting us to ask a number of questions. When and why were our present-day deserts once able to sustain such an abundance of plant and animal life, and why can they not do so today?

Many of the animals that once roamed the Sahara were observed by wandering bands of Late Stone Age hunters, and they recorded what they had seen in a multitude of remarkable naturalistic rock engravings and rock paintings (Muzzolini 1995). These Upper Palaeolithic paintings and petroglyphs depict most of the larger savanna mammals and are scattered throughout the Sahara wherever suitable rock

walls were available. With the advent of Neolithic animal domestication in the early to Mid Holocene, the subjects of the paintings changed, and great herds of domesticated cattle are depicted on smooth rock faces in mountains as far apart as Jebel Uweinat ring-complex in the far southeast of Libya, the famous Tassili sandstone plateau in southern Algeria (Lhote 1959), and the Aïr Mountains in Niger (Roset 1984). The great herds of brindled cattle depicted with their herdsmen in the Neolithic rock-art galleries of the Sahara, together with the occasional archaeological find of the bones and horn cores of prehistoric domesticated cattle (Smith 1984; Gautier 1988), raise a further interesting question. To what extent did Neolithic overgrazing by large herds of hard-hoofed cattle, sheep and goats accelerate soil erosion by wind and water, and initiate humanly-induced processes of desertification, especially in the drier second half of the Holocene (M.A.J. Williams 1982, 1984a; Stiles 1988)? In this chapter we try to answer some of these questions, but we return to the subject of desertification in Chapter 13.

PRESENT-DAY DISTRIBUTION OF ARID AND SEMI-ARID REGIONS

Before attempting to identify the relative impact of former climatic and human influences upon desert environments, it is appropriate to begin with a brief analysis of the causes of aridity and the reasons for the present-day distribution of arid and semi-arid areas depicted in Fig. 9.1.

Deserts are regions where the rainfall is too low and erratic and the evaporation too high to enable many plants and animals to survive. Those species that do survive in arid areas are physiologically and

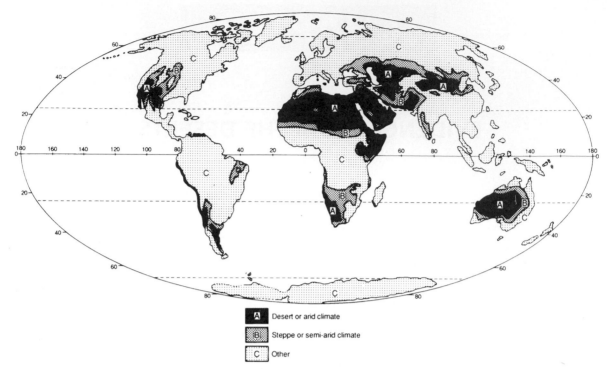

Fig. 9.1 Present-day distribution of tropical and temperate deserts and semi-deserts. (After Tricart and Cailleux 1972; *The Times Atlas* 1980)

behaviourally well adapted to use scarce water efficiently. Low rainfall is not in itself an adequate criterion of aridity. In many cold areas of the world where the rates of evapotranspiration are very low, a relatively dense vegetation cover may exist even when local precipitation is quite low. Another, more utilitarian definition of a desert is a region where sustainable agriculture is not possible without irrigation. If we accept this working definition, then roughly 36% of the land area of the globe is either arid or semi-arid and provides a home for about 13% of the world's present human population.

Figure 9.1 shows that the major deserts of the world today either lie astride (or very close to) the tropics of Cancer and Capricorn, or else are situated in the interior of mid-latitude continental regions, often in the rainshadow of high mountain ranges such as the Rocky Mountains of North America and the Altai, Tien Shan and Kunlun Mountains in western China. What factors are responsible for this very particular distribution pattern?

CAUSES OF ARIDITY

Global atmospheric circulation is primarily controlled by latitude. Insolation is at a maximum over

the equator, so that the surfaces of both land and sea become warm and they in turn heat the air above them by convection. The warm air rises, expands and cools adiabatically. Since warm air can store more water vapour than an equivalent volume of cold air, the rising air soon becomes saturated with respect to water vapour and any excess water vapour condenses to form clouds. Convectional uplift induces further cooling, leading to precipitation of water droplets which eventually coalesce into larger drops and fall as rain. As the air aloft becomes colder and denser it begins to subside. The tropical latitudes are zones of atmospheric subsidence. By now the air has shed much of its excess water vapour so that the air which descends over the two tropics is habitually dry.

The hot tropical deserts are in latitudes where the atmospheric pressure is high for much of the year and sustained by dry subsiding air. As the air subsides it is compressed and becomes warmer so that its capacity to absorb additional water vapour is increased. The result is that the relative humidity of desert air is usually very low, although the absolute amount of water vapour retained in the air may be quite large, becoming evident only when the night temperatures fall sufficiently for the evanescent desert dew to precipitate on chilled rock surfaces.

Since the tropical anticyclonic deserts are a direct result of global atmospheric circulation, their location is determined by latitude and has very little to do with the regional distribution of land and sea. The two polar deserts (not shown on Fig. 9.1) are also under the influence of semi-permanent anticyclones and of cold, dry subsiding air. As a result of this latitudinal distribution of high pressure cells, the oceans in both polar and strictly tropical latitudes also receive very little precipitation, and are in effect the marine counterparts of the continental deserts.

The mid-latitude deserts, including those of western China and Uzbekistan, are dry because of their geographical location in the centre of large continental land masses far removed from the influence of moist maritime air. This factor of 'continentality' or distance inland applies to all the larger deserts, including the Sahara and Australia. Indeed, rainfall tends to decrease rapidly away from the coast in all parts of the world except those close to the equator. In the case of the hot tropical deserts, the effects of continentality therefore serve to reinforce those of latitude.

Three additional factors may enhance the aridity resulting from latitude and continentality, or may themselves be direct causes of reduced precipitation. These factors are the presence of cold sea surface water offshore, the 'rainshadow' effect, and low relief inland. They may operate individually or in concert.

The presence close offshore of cold upwelling water or a cold ocean current is an effective cause of coastal aridity in tropical latitudes. The Atacama desert in Chile and the Namib desert in southern Africa are flanked offshore by the cold Peru and Benguela currents, respectively. The western borders of all the tropical or 'trade wind' deserts are washed by cool ocean currents associated with the ocean gyres which flow clockwise in the northern hemisphere and anticlockwise in the southern hemisphere. When moist and relatively cool maritime air masses move onshore they encounter a land surface which is warmer than the adjacent ocean surface. The result is a reduction in the relative humidity of the former maritime air, so that it develops an increased capacity to absorb rather than to shed moisture. The major source of moisture in these narrow coastal deserts is the coastal fog which blows inland in winter, and this effect is greater if there is high relief close to the shore. The mist oasis of Erkowit in the Red Sea Hills of the eastern Sudan is a case in point.

The 'rainshadow' effect is a globally universal phenomenon and is not peculiar to deserts. Wherever there are hills or mountains close to the coast, the incoming moist maritime air will be forced upwards. As the moist air rises it is cooled adiabatically, attains vapour saturation, and sheds its precipitated water vapour as rain or snow. The air then flows over the coastal ranges and downhill, becoming warmer and drier for the reasons outlined earlier. The region inland of the coastal ranges is said to lie in their 'rainshadow' and will always remain drier than their coastal flanks. Patagonia lies in the rainshadow of the Andes. Other examples of rainshadow deserts are the arid or semi-arid areas immediately to leeward of the Rockies, the Himalayas, the Ethiopian uplands and the Eastern Highlands of Australia. Extreme examples are the Dead Sea Rift and the Afar Depression, both of which are flanked by high mountains while they themself lie close to, and in places well below, sea-level.

The 'rainshadow' effect is increased when the region inland of the humid coastal ranges is low-lying and devoid of any significant relief. Some of the driest deserts are those in which the landscape consists almost entirely of extensive plains or low plateaux, such as the stone-mantled 'gibber' plains of central Australia and the gravel-strewn *serir* of southern Libya. The converse is also true and the high mountains of the central and southern Sahara, such as Tibesti (3415 m), the Hoggar (2918 m) and Jebel Marra (3042 m) receive sufficient rainfall for them to have served as refugia for plants, animals and humans throughout the Quaternary (Rognon 1967; Messerli *et al.* 1980; Williams *et al.* 1980).

There is one additional factor that deserves mention, for it also provides a link between present and past. Late Cainozoic uplift of the Tibetan Plateau (Ruddiman and Kutzbach 1991) resulted in the development of the easterly jet stream which now flows from Tibet across the Arabian peninsula towards Somalia, accentuating the aridity in those regions and helping to create a desert right on the equator (see Fig. 9.1; Rognon and Williams 1977).

The distribution of our present-day deserts therefore reflects the lingering influence of past tectonic events as well as the combined effects of at least six other major factors. These factors are the prevalence of dry subsiding anticyclonic air masses over deserts (itself controlled by latitude and global atmospheric circulation), a vast land area, coastal ranges, low relief inland, cool ocean water close offshore, and a subtropical jet stream aloft. Within the time-frame of the Quaternary, these factors did not change very drastically, although the degree to which they influenced aridity did vary in response to changes in oceanic and atmospheric circulation associated with orbital perturbations and global temperature fluctuations.

THE DESERT ENVIRONMENT

Before embarking on the delicate task of trying to reconstruct the Quaternary environmental changes to which all deserts were subjected, it is useful to describe some of the characteristics of existing desert environments. For clarity and simplicity we group desert landforms somewhat arbitrarily into erosional and depositional. We begin by considering the desert landscape as a whole.

Perhaps the single most characteristic attribute of deserts throughout the world is their lack of a perennial and integrated system of drainage. Desert streams are ephemeral. They flow episodically, for variable distances, depending upon the intensity and duration of sporadic rainstorms in their upper catchments (Walther 1924; Berkey and Morris 1927; Jutson 1934; Cooke and Warren 1973; Mabbutt 1977; Rognon 1989; Thomas 1989b, 1997; Cooke *et al.* 1993). Of course, allochthonous rivers like the Nile may flow through parts of a desert, but in their desert sojourn they do not gain any additional water from perennial tributaries. On the contrary, rivers which traverse deserts constantly lose water by seepage to the often deep-seated regional groundwater table. Most desert rivers are *endoreic* (Chapter 8) and flow into closed depressions like the Tarim basin in China or the Lake Eyre basin in Australia. The fossil river valleys of the Sahara, the Gobi and Western Australia have long interested geologists. Today they are broad linear depressions filled with Cainozoic alluvium, which is often cemented with iron, silica or calcium carbonate to form low erosional remnants or sometimes extensive sheets of resistant ferricrete, silcrete or calcrete (Lamplugh 1902). In Mauritania, Namibia and Western Australia these valley-fill calcretes may also contain variable amounts of secondary uranium minerals precipitated out of slowly moving groundwater, originating from the Precambrian host rocks which form the valley interfluves.

The fauna and flora of streams in semi-arid regions are well adapted to the spatial and temporal variability of rainfall and runoff in semi-arid stream catchments (Dahm and Molles 1992; Grimm 1993; Stanley *et al.* 1997), so that a sound appreciation of present-day aquatic ecology is a prerequisite for any attempt to interpret past environments using aquatic plant and animal fossils. Many dryland streams are highly sensitive to minor climatic fluctuations. In a number of seasonally-wet and semi-arid regions, including southern and eastern Africa, northeast Brazil, New Mexico, eastern and northern Australia, central India, and northeastern China, the interannual flow regime is strongly influenced by the incidence of El Niño–Southern Oscillation (ENSO) events (Williams and Balling 1996). It also appears that some and possibly many of these dryland rivers show a runoff response which may considerably amplify the initial precipitation input from ENSO events, so that the hydrological response of dryland rivers is highly sensitive to global sea-surface temperature anomalies (Molles and Dahm 1990), and doubtless was equally so in the past (Nott *et al.* 1996).

A further attribute of all desert landscapes is their clarity and starkness, for their erosional landforms invariably show a high degree of adjustment to rock type and geological structure. In more humid regions, evidence of similar structural control is usually well camouflaged by deep soils and a more or less continuous cover of vegetation. Desert soils are usually skeletal lithosols on rocky hillslopes, or almost unweathered deposits of gravelly alluvium or wind-blown quartz sand (Dunkerley and Brown 1997). Nevertheless, in sheltered mountain valleys in the heart of even our greatest deserts, and especially along their semi-arid margins, relict deep weathering mantles and well-developed palaeosols are often sufficiently well preserved to allow a sequence of formerly wetter climatic regimes to be reconstructed, sometimes in considerable detail (Williams *et al.* 1987).

Many desert landforms are exceedingly old. The vast desert plains of the central and western Sahara have been exposed to subaerial denudation for well over 500 million years (Williams 1984a), as have the Precambrian shield deserts of the Yilgarn Block and the Pilbara in Western Australia. It is misleading to consider such well-known desert monoliths as Ayers Rock (Uluru) in central Australia, or the granite inselbergs of the Sahara, as diagnostic of aridity, for they owe their present morphology to prolonged and repeated phases of weathering and erosion under a succession of former climates, few of which were particularly arid.

Linked to this longevity of most erosional desert landforms is another somewhat paradoxical attribute of many desert landscapes: the close juxtaposition of very young depositional features with very ancient erosional landforms. It is these young landforms and sediments, whether aeolian, fluviatile or lacustrine, which best retain the imprint of past environmental changes, most notably the rapid climatic fluctuations of the late Tertiary and Quaternary, to which we now turn.

LATE CAINOZOIC COOLING AND DESICCATION

The onset of late Cainozoic aridity and the slow emergence of the deserts portrayed in Fig. 9.1 were associated with the global tectonic events discussed in Chapter 2. As a result of the lithospheric plate movements shown in Fig. 2.2, a number of changes in global atmospheric circulation resulted from the changing horizontal and vertical distribution of land and sea, and certain continents or regions also moved into dry tropical latitudes, most notably North Africa and Australia.

The origin of the Sahara as a desert was associated with several independent tectonic events. Slow northwards movement of the African plate during the late Mesozoic and Cainozoic saw the migration of much of North Africa from wet equatorial into dry tropical latitudes (Williams 1994a; Partridge *et al.* 1995a, b). A slight clockwise rotation of Africa during the Miocene and Pliocene brought Africa into contact with Europe, and was accompanied by crustal deformation and rapid uplift in the Atlas region, and by volcanism and updoming in Jebel Marra, Tibesti, the Hoggar and Aïr Mountains (Williams 1984a, 1994a).

Two additional factors were responsible for accentuating the late Cainozoic desiccation of North Africa. One was the gradual expansion of continental ice in high latitudes associated with the post-Eocene cooling of the Southern Ocean and the North Atlantic (Schnitker 1980; Kennett 1995). This cooling was initiated by the break-up of Laurasia and by the separation of Australia from Antarctica (see Chapter 2). It culminated in the establishment of a large ice cap on Antarctica by 15 Ma ago (Chinn 1996), and in a sudden increase in the volume of northern hemisphere ice caps towards 2.5 Ma ago (Shackleton and Opdyke 1977; Shackleton *et al.* 1984). One effect of the progressive build-up of high latitude ice sheets was to steepen the temperature and pressure gradients between the equator and the poles, resulting in increased trade wind velocities. Faster trade winds were better able to mobilise the alluvial sands of an increasingly dry Sahara and to fashion them into desert dunes. The first appearance of wind-blown quartz sands in the Chad basin, for example, is towards the end of the Tertiary, when they occur interstratified among Plio-Pleistocene fluvio-lacustrine sediments (Servant 1973). The associated lacustrine diatom flora indicates temperatures cooler than those now characteristic of this region (Servant and Servant-Vildary 1980), reinforcing the notion that the late Pliocene was both cooler and drier along the tropical borders of the Sahara (Leroy and Dupont 1997). Diatom and pollen evidence from Pliocene Lake Gadeb in the southeastern uplands of Ethiopia is also consistent with this conclusion (Gasse 1980; Bonnefille 1983), suggesting that intertropical cooling and desiccation may have been closely bound up with the expansion of northern hemisphere ice caps towards 2.5 Ma ago (Suc *et al.* 1997).

A further factor contributing to the late Cainozoic desiccation of the Sahara, and briefly alluded to earlier, was the Neogene uplift of the Tibetan plateau (Liu *et al.* 1996; Derbyshire 1996) and the ensuing creation of the easterly jet stream which brought dry subsiding air to the incipient deserts of Pakistan, Arabia, Somalia, Ethiopia and the Sahara. In this context, it is interesting to note that carbon and oxygen isotopic analyses of Neogene palaeosols and fossil herbivore teeth collected from the Potwar Plateau of Pakistan reveal a dramatic change in flora and fauna between 7.3 and 7.0 Ma ago (Quade *et al.* 1989). Prior to 7.3 Ma, there was a dominance of C3 plants (i.e. those which follow the C3 photosynthetic pathway; see Chapter 11) indicative of forest and woodland. After 7.0 Ma, C4 plants (i.e. those which follow the C4 photosynthetic pathway; see Chapter 11) were the most abundant, indicating a rapid expansion of tropical grassland at the expense of forest. Quade and his co-workers have interpreted this change as being consistent with a major strengthening of the Indian summer monsoon during the very late Miocene, if not indeed with the actual inception of the monsoon.

Later work has indicated that three other factors may have been equally critical in contributing to intertropical cooling and desiccation during the past 30 million years. One involved the progressive shrinkage of the Parathethys, a warm, shallow epicontinental sea stretching across Eurasia which shrank gradually from its large areal extent during the Oligocene to an ever-more limited area during the Miocene (Ramstein *et al.* 1997). The outcome of this was an ever-more seasonal rainfall regime at the expense of the previously well-distributed precipitation throughout the year. The second factor was the decrease in atmospheric CO_2 associated with increased erosion, weathering and concomitant consumption of CO_2 caused by the late Cainozoic uplift of the Himalayas, the Rockies, the Andes, the Ethiopian uplands and perhaps also the Transantarctic Mountains (Yemane *et al.* 1987; Quade *et al.* 1995; Cerling *et al.* 1997). The global increase in plants using C4 photosynthesis and the dramatic but time-transgressive reduction in C3 photosynthesising plants between about 8 and 6 Ma ago is certainly consistent with a decrease in the concentration of atmospheric CO_2 (Cerling *et al.* 1997). The threshold for C3 photosynthesis is higher at warmer latitudes, and so it is not surprising that the initial change from C3 to C4 plants was in the lowland tropics first (Cerling *et al.* 1997). A third possible factor was the climatic cooling triggered by the eruption of the voluminous Ethiopian flood basalts over a period of no more than a million years around 30 Ma ago (Hofman *et al.* 1997).

Changes in the Cainozoic flora and fauna of the Sahara show a similar trend to that inferred for the Himalayan foothills of Pakistan (Williams 1994a; Leroy and Dupont 1997; Cerling *et al.* 1997). During the Palaeocene and Eocene much of the southern Sahara was covered in equatorial rainforest, and there was widespread deep weathering at this time. During the Oligocene and Miocene much of what is now the Sahara was covered in woodland and savanna woodland, but by Pliocene times many elements of the present Saharan flora were already present. Maley (1980) reviewed the evidence from pollen preserved in scattered localities in northern Africa, concluding that replacement of tropical woodland by plants adapted to aridity was already under way during the late Miocene and early Pliocene, a conclusion consistent with the pollen evidence preserved in deep-sea cores off the northwest coast of Africa (Dupont and Leroy 1995; Leroy and Dupont 1997).

Late Cainozoic cooling and desiccation was not confined to the vast tropical arid zone which extends from the western Sahara across Arabia as far as northwestern India (Fig. 9.1). Nor was it peculiar to the northern hemisphere. The loess of northwestern China first began to accumulate around 2.4–2.5 Ma ago, although the onset of deposition was time-transgressive (Heller and Liu 1982; Burbank and Li 1985; Kukla and An 1989; Ding *et al.* 1993; Rutter and Ding 1993). Summer aridity first became apparent around the Mediterranean basin at about this time (Suc 1984; Suc *et al.* 1997). In the tropical Andes there was a major change in the flora towards 2.5 Ma (Hooghiemstra 1989, 1995). The onset of aridity in South America, South Africa and Australia is harder to pinpoint with precision, but the evidence from geomorphology and geochemistry and from the fossil flora and fauna is all indicative of progressive Neogene desiccation on land and of a post-Eocene cooling of the oceans to the south and west of all three southern land masses (Zinderen Bakker 1978; Barker and Greenslade 1982; Dingle *et al.* 1983; Williams 1984c, 1994; Zarate and Fasano 1989; Kukla 1989; Williams *et al.* 1991; Archer *et al.* 1995).

QUATERNARY GLACIAL ARIDITY

From about late Pliocene times onwards, the great tropical inland lakes of the Sahara, Ethiopia and Arabia began to dry out. The formerly abundant tropical flora and fauna of the well-watered Saharan uplands became progressively impoverished as entire taxa became extinct, and a once-integrated and efficient network of major rivers became increasingly obliterated by wind-blown sands. Allowing for local differences in timing linked to regional climatic and tectonic factors, a similar sequence of events was also true of the deserts of China, India, Australia and southern Africa. With the advent of the Quaternary, a global pattern of climatic oscillations now became established, apparently linked to and modulated by orbital perturbations, although other mechanisms may also have played a role (Berger 1981c,

1989a; Peltier 1982; Mörner 1989; Huggett 1991; Shackleton 1995; deMenocal and Bloemendal 1995).

In Chapter 5 we saw that the astronomical cycles vary in amplitude and frequency, and are responsible for variations in the amount of solar energy received at different times in the past, and at different latitudes. The 41 ka cycle is linked to changes in the obliquity of the Earth's axis, which is now 23°27′ but oscillates between 22°30′ and 25°. The 100 ka cycle is controlled by variations in the eccentricity of the Earth's orbit around the sun. The present eccentricity value is such that the Earth now receives 6% more energy when closest to the sun than when furthest from it; at maximum eccentricity the value is closer to 27%. The precession of the equinoxes varies with the changing distance between the Earth and the sun. The 21 ka precession cycle has two peaks: one of 19 ka, the other of 23 ka. The relative influence of each of these cycles has varied during the course of time, so that the duration of glacial–interglacial cycles has also varied, with concomitant repercussions for all of the Earth's surface, including the deserts.

Prior to 2.4 Ma (and the rapid expansion of North American ice caps), the dominant cycles evident from magnetic susceptibility measurements of deep-sea cores from the Arabian Sea and the eastern tropical Atlantic were the 23 ka and the 19 ka precession cycles, but after 2.4 Ma the 41 ka obliquity cycle became dominant (Bloemendal and deMenocal 1989; deMenocal and Bloemandal 1995). Oxygen isotope records from deep-sea cores spanning the entire Quaternary reveal that the early Pleistocene from about 1.8 Ma to 0.9 Ma was subject to frequent low-amplitude fluctuations in the oxygen isotope differences between glacial and interglacial maxima (Williams *et al.* 1981). Since changes in the oxygen isotopic composition of benthic foraminifera broadly reflect changes in global ice volume (Shackleton 1977, 1987), the changing magnitude and frequency of Quaternary glacial cycles should also be reflected in the severity and frequency of former cycles of Quaternary aridity and of desert expansion and retreat. Let us now confront this hypothesis with evidence from the deserts.

Recent evidence from North Atlantic cores shows that the 41 ka obliquity cycle was dominant during the Matuyama magnetic chron from 2.60 Ma to 0.78 Ma (Ruddiman and Raymo 1988). (Note that these authors used the older time-scale of 2.47 Ma and 0.73 Ma for the end of the Matuyama and Brunhes chrons respectively. The recalibrated dates are 2.60 Ma and 0.78 Ma (Shackleton 1995). During the last 0.78 Ma of the Brunhes magnetic chron the 100 ka orbital eccentricity cycle was dominant. The present-day Atlantic exerts a powerful influence upon snow accumulation in North America and Europe as well as upon rainfall and drought in northern Africa, and a similar influence is discernible in the geologically recent past.

Earlier work had convincingly demonstrated that during the last 600 ka, at least, maximum concentrations of Saharan desert dust in equatorial Atlantic deep-sea cores coincided with times of glacial maxima and low sea-surface temperatures (Parkin and Shackleton 1973; Parmenter and Folger 1974; Bowles 1975; Sarnthein et al. 1981). A similar pattern of glacial aridity was also evident in the Gulf of Aden and the Red Sea. The isotopic composition of planktonic foraminifera from deep-sea cores in this region showed that during at least the last 250 ka, glacial maxima were times of extreme aridity, with increased sea-surface salinity reflecting even higher rates of evaporation than prevail there today (Deuser et al. 1976).

The work in the Atlantic has since been confirmed and extended back to the Pliocene by Tiedemann and co-workers (1989, 1994) and by deMenocal and Bloemendal (1995), although Rognon and Coudé-Gaussen (1996) have recently suggested that some of the fine quartz particles observed in deep-sea cores off the northwest coast of Africa, which other workers had interpreted as wind-blown dust from the Sahara, may have been transported from the north by cold currents fed by glacial meltwater during and after the last glacial maximum some 18 ka ago. Sirocko and colleagues (1993) have recently demonstrated that the relative inputs from the three main sources of desert dust which are evident in marine sediments from the Arabian Sea have also varied over the past 24 000 years, highlighting the complexity of desert responses to changes in monsoon intensity and incidence.

Despite such arguments of detail about the former strength and location of the trade winds, westerlies and monsoonal air masses, there seems strong consensus that glacial maxima were drier than today and interglacial maxima as wet or wetter. The northern hemisphere evidence of enhanced aeolian dust flux during glacial maxima is strongly supported by the evidence from Australia (McTainsh and Lynch 1996; Hesse 1994) and Patagonia (Yung et al. 1996) Nor should we forget that aeolian dust may itself have an impact upon local and regional climates. Maley (1982) noted that in tropical north Africa from 15 ka to about 7 ka the rivers were mainly depositing clays, and after 7 ka they mostly carried sands. He attributed this abrupt hydrological change to a change in the size of raindrops, arguing that fine rains were associated with considerable atmospheric dust which provided nuclei for many small raindrops to form, and large, highly erosive drops were related to a reduction in atmospheric dust load. Yung et al. (1996) pointed out that the very high inputs of aeolian dust to central Antarctica and Greenland during

the last glacial maximum, which are clearly evident in the Vostok and GRIP ice-core records, are consistent with shorter dust washout times and a weaker global hydrological cycle. They note too that a high concentration of atmospheric dust may in itself have contributed to a lowering of sea-surface temperatures in the tropical western Pacific, especially in the warm shallow seas immediately to the north of Australia. Given the growing recognition of the interactions between present-day desertification processes, dust generation, and the impact of dust particles in scattering incoming solar radiation (Williams and Balling 1996), it seems highly plausible that wind-blown dust would be both a cause and an effect of Quaternary climatic fluctuations. We turn now to the crucial question of what was the impact of these alternating wetter and drier Quaternary climatic phases upon the deserts.

GLACIAL AND INTERGLACIAL DESERT ENVIRONMENTS

To equate glaciations with aridity and interglacials with an increase in desert rainfall is to oversimplify. The reality is both more complex and more interesting. We have long known that during the last glacial maximum towards 18 ka, aridity was more widespread in the intertropical zone (Fairbridge 1970; Williams 1975), trade winds were stronger (Parkin 1974), dunes were active well beyond their present limits (Grove and Warren 1968; Sarnthein 1978; Talbot 1980) (compare Figs 9.2 and 9.3), and considerable volumes of desert dust and loess were deposited on land (and far out to sea: Parkin and Shackleton 1973), in regions where such dust mantles are now vegetated and relatively stable (Figs 9.4 and 9.5; Liu 1987; Liu et al. 1989).

We noted earlier that the maximum concentrations of desert dust in equatorial Atlantic deep-sea cores coincided with glacial maxima. Such dust is easily recognised by its high degree of sorting, shown in the very characteristic cumulative frequency curves of dust particle size illustrated in Fig. 9.6. Although dust mobilisation presupposes aridity and appropriately strong seasonal winds to entrain the dust plumes aloft and offshore (Morales 1979; McTainsh 1980, 1985; Pye 1987), the glacial loess deposits of Eurasia and North America (Figs 9.4 and 9.5) reflect the former presence of unvegetated Pleistocene glacifluvial outwash sediments upwind of the loess mantles, rather than simply aridity and strong winds. Although the parna sheets of Australia (Butler 1956) may resemble the loess deposits of north-central India (Williams and Clarke 1984) in terms of their physical characteristics, a fine, well-sorted grain-size

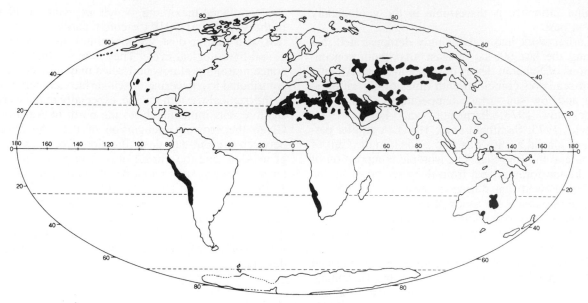

Fig. 9.2 Present-day global distribution of active sand dunes. (After Goudie 1983)

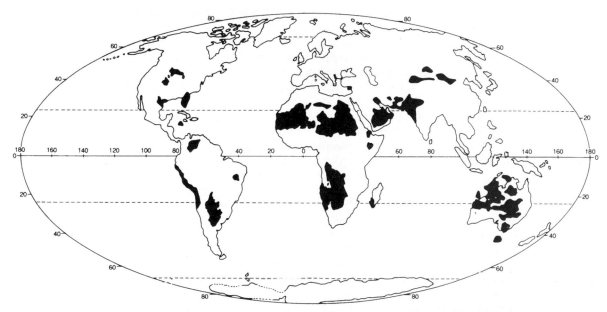

Fig. 9.3 Global distribution of active sand dunes during the last glacial maximum (18 ka). (After Sarnthein 1978)

distribution is not in itself indicative of either a glacial outwash or a desert dust origin.

Until very recently, it was difficult to date phases of dune mobility directly, so that the radiocarbon ages of organic sediments above and below dune sands were used to bracket episodes of dune activity. The increasing reliability of thermoluminescence dating (see Appendix) has meant that individual sand laminae within dunes may now be directly dated, so that there has been a resurgence of Quaternary dune studies in the deserts of Australia, India, Africa and

North America within the last few years (Nanson *et al.* 1995; Magee *et al.* 1995). Dunes, alone, however, are not diagnostic of aridity, since dune mobilisation depends upon such factors as sand supply, vegetation cover, moisture content and surface roughness, as well as wind velocity and turbulence (Wasson and Hyde 1983; Wasson and Nanninga 1986; Haynes 1989; Pye and Tsoar 1990; Lancaster 1995; Wiggs *et al.* 1995; Livingstone and Warren 1996). The source-bordering dunes referred to in the previous chapter are most frequently found in semi-arid regions. Pre-

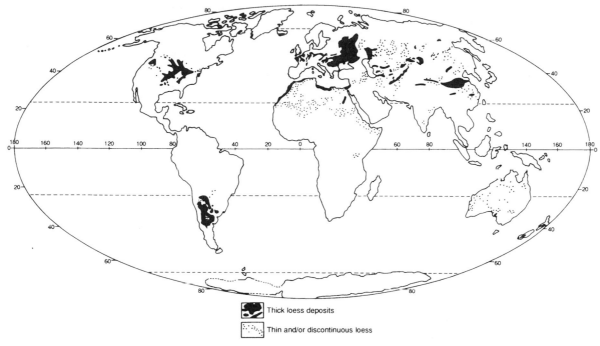

Fig. 9.4 Present-day global distribution of loess. (After Snead 1980; Pye 1984, 1987; Thomas 1989)

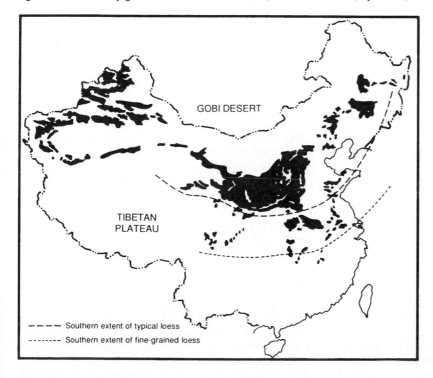

Fig. 9.5 Present-day distribution of loess in China. (After Liu 1985; Pye 1984)

requisites for their formation are a regular (usually seasonal) replenishment of river channel sands or sandy beaches by longshore drift in deep lakes, as well as a strong seasonal unidirectional wind and a lack of riparian or lake-margin vegetation (Williams 1985a). The first prerequisite – regular renewal of the

sand supply from seasonally-active rivers – effectively precludes a fully arid climate. It is also worth emphasising that only a small proportion of all deserts are covered in sand dunes, only a fifth in the case of the Sahara, and two-fifths of the much less arid Australian desert, and many of the Australian

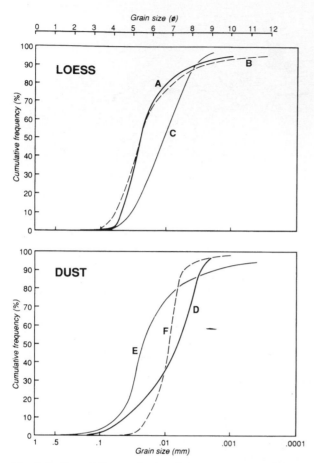

Fig. 9.6 Cumulative grain-size frequency curves of loess and desert dust. (A) Sanborn loess (after Swineford and Frye 1945); (B) Siberian loess (after Péwé 1981); (C) Malan loess (after Liu *et al.* 1981); (D) Mongolian dust recorded in Beijing (after Liu *et al.* 1981); (E) Arizona dust (after Péwé 1981); (F) Saharan dust recorded in London. (After Wheeler 1985)

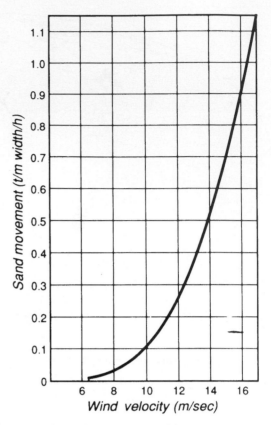

Fig. 9.7 Relationship between wind speed and volume of desert sand transported. (After Bagnold 1941)

dunes are only active along their ridge crests (Ash and Wasson 1983). The pioneering and now classic empirical and experimental work of Bagnold (1941) demonstrated that the volume of transported desert sand increased exponentially with wind velocity above a certain threshold value (Fig. 9.7). Where sand supply and wind speed are not limiting factors, dune mobilisation will increase as vegetation cover decreases. Since plant cover in dry areas is governed primarily by rainfall and evapotranspiration, there is a close relationship between the amount of rainfall and the average outer limit of active dunes in such deserts as the Thar desert of Rajasthan (Goudie *et al.* 1973) or the Sahara, where the southern limit of presently mobile dunes coincides remarkably closely with the 150 mm isohyet (Fig. 9.8).

When all the *caveats* mentioned above are taken into account, it becomes very clear that in many of

the world's hot deserts, including the Gobi and Uzbekistan deserts, the dominant climate during the last glacial maximum (18 ± 3 ka) was drier, windier and colder than today, although the summers may still have been very hot (Bowler 1978; Sarnthein 1978; Wasson *et al.* 1983; Bowler and Wasson 1984; Williams 1985a; Thomas 1989b; Dong 1991; Sanlaville 1995; Petit-Maire *et al.* 1995; Stokes *et al.* 1997). Previously fixed and vegetated dunes along what are now the semi-arid margins of these deserts became mobile, so that the effective range of the Sahara extended 400 to 600 km further south (Grove and Warren 1968), and that of the Rajasthan desert some 350 km further to the southeast (Goudie *et al.* 1973). Many of the desert lakes, which immediately prior to about 18–20 ka had occupied deflational hollows or tectonic depressions and were full and fresh, now dried out or became hypersaline (Gasse *et al.* 1980; Harrison *et al.* 1984; Street-Perrott *et al.* 1985; Fontes and Gasse 1991), as previously perennial desert margin rivers became seasonal, while seasonal rivers became, at best, highly intermittent or ephemeral streams (Rognon 1989).

With glacially lowered sea-levels, the desiccating influence of greater continentality was also enhan-

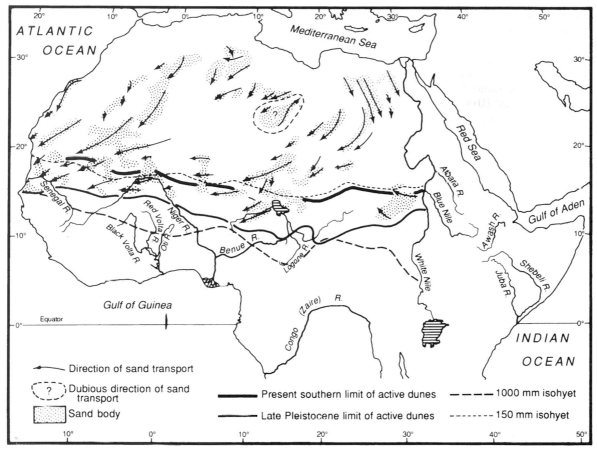

Fig. 9.8 Directions of sand transport in the Sahara and its southern margins based on alignments of present-day and later Pleistocene desert dunes. (After Grove 1980; Mainguet and Cossus 1980; Mainguet *et al.* 1980)

ced. Stronger trade winds associated with steeper pressure gradients between the equator and poles caused increased upwelling of cold water close offshore, further accentuating the aridity of coastal deserts.

The contrast between the terminal Pleistocene aridity of the Sahara and its verdant early Holocene status is hard to imagine today, but it had an enormous impact on the Late Stone Age and early Neolithic peoples who witnessed these changes (Chapter 11). As postglacial temperatures and sea-levels rose around the world, evaporation from the intertropical oceans also increased, and the previously weakened summer monsoons of northern Australia, India and West Africa once more became reliable sources of seasonal rainfall (Pastouret *et al.* 1978; Williams 1984b, 1985a). Throughout the previously dry tropics, groundwater levels rose, aquifers were replenished, lakes refilled, hitherto mobile dunes became vegetated and stable, and savanna woodland and grassland re-occupied what are today the semi-arid regions of the world. A remarkable and well-integrated drain-

age network became established in many parts of the Sahara, Arabia and Rajasthan, all of which were studded with innumerable freshwater lakes and ponds reaching a peak towards 9 ka (Faure 1966; Singh *et al.* 1974; McClure 1976; Gasse and Fontes 1992; Kropelin 1993; Hoelzmann 1993; Szabo *et al.* 1995; Ayliffe *et al.* 1996; Pachur and Altmann 1997). This was the time of 'le Sahara des Tchads' (Balout 1955) when 'aqualithic' Upper Palaeolithic hunter–fisher–gatherer communities (Sutton 1977) used barbed bone harpoons to obtain Nile perch and other large fish from the Saharan lakes; hippos, crocodiles and turtles also featured in their diet (J.D. Clark 1980).

The climatic conditions of the warm and wet early Holocene were very similar to those of the last interglacial at around 125 ka (oxygen isotope stage 5e of the deep-sea core record), as was the global distribution of land and sea. The desert environments no doubt oscillated between these two extremes, with the interglacials (125 ka and 9 ka) being mostly slightly warmer and very much wetter than today (Szabo *et al.* 1995; Crombie *et al.* 1997), and the glacial maxima (140 ka and 18 ka) colder and mostly drier. However, not all arid phases coincide with glacial maxima, any more than do all humid phases with interglacial times. For instance, Lake Chad in the southern Sahara (Servant 1973) and Lake Abhe in the Afar desert of Ethiopia (Gasse 1975) were both very high for at least 10 000 years before 18 ka, when they fell rapidly. They were then intermittently dry (Lake Chad) or dry (Lake Abhe) until 12 ka, rising rapidly thereafter to reach peak levels at 9 ka. Since about 4.5 ka both lakes have remained low apart from occasional brief transgressions. Very schematically, we could consider the >30 ka to 18 ka phase of high lake levels as a humid glacial phase; the 18 ka to 12 ka regression as an arid glacial phase; the early Holocene transgression as a humid interglacial phase; and the late Holocene interval of low lake levels as a dry interglacial phase. This four-fold subdivision does caricature reality, and also ignores local hydrological and geomorphic controls over rainfall, runoff, evaporation, seepage losses and groundwater inflow (Fontes *et al.* 1985; Abell and Williams 1989).

QUATERNARY SOILS IN ARID AND SEMI-ARID AREAS

Early studies of desert regions (Walther 1924; Berkey and Morris 1927; Jutson 1934) tended to focus upon specific desert landforms such as dunes, alluvial fans, deflation hollows, wind-abraded fluvio-lacustrine deposits, and erosional surfaces cut across unconsolidated sediments or bedrock (for modern reviews see Mabbutt 1977; Rognon 1989; Cooke *et al.* 1993; Thomas 1989b, 1997). (Since then, Quaternary research has increasingly concentrated upon using alluvial and lacustrine deposits and their associated plant and animal fossils to reconstruct the alluvial history of desert lakes and rivers such as the Afar lakes of Ethiopia (Gasse 1975) or the depositional history of the Nile (Williams 1966; Butzer and Hansen 1968)). These pioneering and essentially descriptive geomorphological studies revealed the presence of relict deep-weathering mantles and more or less truncated fossil soil profiles sometimes characterised by indurated secondary accumulations of calcium carbonate, hydrated iron oxides or silica (*duricrusts*) in regions now too dry for any significant chemical weathering. Initial attempts to interpret these fossil weathering profiles, duricrust remnants and buried palaeosols in terms of the standard A, B and C soil horizons recognised in present-day Russian, European and North American soils soon ran into difficulties (Yaalon 1971; Hunt 1972).

One reason for this difficulty concerns the role of the soil fauna, especially that of ants and termites, in

generating texture-contrast soil profiles in seasonally-wet tropical and subtropical regions (Williams 1968; Lee and Wood 1971; Paton *et al.* 1995). The resulting sandy top-soil overlying a stone layer which directly overlies weathered bedrock fits uneasily into the orthodox A, B, C soil horizon nomenclature. Failure to recognise the very considerable role of the soil fauna in top-soil replenishment has led to some curious interpretations of the sandy top-soil and the underlying stone layer as reflecting a change from a dry to a wet climate, the stone layers being considered diagnostic of aridity for reasons that have more to do with preconceived palaeoclimatic notions than with any independent evidence of the hypothesised aridity (Fairbridge and Finkl 1984).

Part of the problem lay in the apparently incomplete state of many of the soil profiles (Catt 1995), but a larger difficulty arose from an inability to specify with confidence the relative role of each of the major factors of soil formation, notably parent material, topography, climate, time and biota, despite vigorous efforts to try to quantify the role of each of these factors in contributing to the processes of soil formation (see, for example, Birkeland 1974, 1984; Paton 1978; Wright 1986; Johnson and Watson-Stegner 1987; Retallack 1990; Mack and James 1994; Paton *et al.* 1995). If we consider parent material and topography as passive factors, and climate and the biota as active factors, we might expect that over time the relative influence of the active factors will tend to outweigh those of the passive factors, provided the soil profiles are located on well-drained gentle slopes characterised by low rates of both erosion and deposition. In practice, very few soils can meet these criteria, although some obviously will. As a consequence, soil scientists are more likely to be consumers than producers of Quaternary palaeoenvironmental data. The role of time is critical, so that a reliable chronology of the onset and close of pedogenesis is essential to understanding how the various soil-forming factors have interacted through time to produce the soil profiles preserved in Quaternary formations. Unfortunately, very few good studies of long soil chronosequences yet exist for dryland regions, the exception being the loess palaeosols of central China which we now discuss.

THE LOESS–PALAEOSOL SEQUENCE IN CHINA

We noted earlier that fine-grained wind-blown dust or loess had accumulated at intervals throughout the Quaternary on the downwind margins of glacifluvial outwash deposits in North America and Eurasia. We also saw that in terms of their sorting (Fig. 9.6) and mineralogy, the desert dust mantles of Africa, Australia, South America and Asia were indistinguishable from the classical central European loess mantles, so that the term loess can apply generically to any fine-grained aeolian deposit irrespective of its original provenance. Nowhere in the world have the loess deposits been studied in such detail as in the Loess Plateau of central China, (Fig. 9.5) where the remarkably thick and extensive loess deposits cover an estimated area of 275 000 km^2 and attain thicknesses commonly in excess of 100 m and more than 300 m near the city of Lanzhou (Liu 1985, 1987, 1991). Detailed studies by Professor Liu Tungsheng and his colleagues over the past three decades have made the Chinese loess sequence, with its alternation of unweathered loess and loess palaeosols, the most important and informative long Quaternary terrestrial sequence on Earth (Heller and Liu 1982; Liu 1987; Kukla and An 1989). There are promising signs that the Tadjikistan loess–palaeosol sequences in Central Asia may prove to be equally informative (Bronger *et al.* 1995).

Detailed sampling of a number of key stratigraphic sections located on a series of west–east and north–south transects has revealed a sequence of 37 loess–palaeosol couplets (interpreted as cold to warm climatic cycles) spanning the past 2.5 Ma (Ding *et al.* 1993; Rutter and Ding 1993). A palaeosol was defined as weathered loess showing at least as much pedological organisation as the ubiquitous Holocene soil at the top of the loess sequence, and a very complete section at Baoji in the southern Loess Plateau was chosen as the type pedostratigraphic section for Chinese loess (Rutter *et al.* 1991). Interpretation of the loess–palaeosol couplets is based on high-resolution sampling and detailed analyses of grain size, magnetic susceptibility, organic carbon, sediment micromorphology and mineralogy, calcium carbonate content and mollusc species. The clay-rich loess palaeosols are interpreted as indicating a weaker winter monsoon and a stronger summer monsoon. Conversely, the coarser-grained unweathered loess with generally much weaker magnetic susceptibility values is regarded as evidence of a stronger and more extensive winter monsoon and a weaker summer monsoon (Maher and Thompson 1992, 1994; Liu Xuming *et al.* 1993; Zhang *et al.* 1994; Derbyshire *et al.* 1995). Chronological control is based on the palaeomagnetic time-scale, cross-correlation with the marine oxygen isotope record (Fig. 9.9), and a combination of radiocarbon and thermoluminescence dates for the more recent part of the sequence (Zheng *et al.* 1995). Current work is concerned with comparing the Chinese loess record with evidence from deep ice cores as well as with the marine record, to which we now turn (Porter and An 1995; An and Porter 1997; Chen *et al.* 1997).

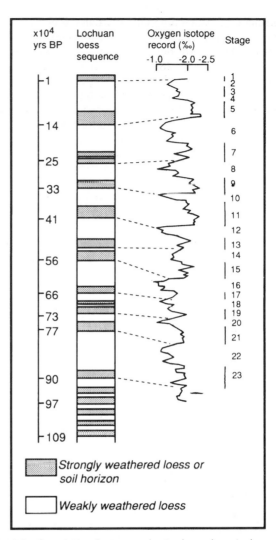

Fig. 9.9 Correlation between the Lochuan loess/palaeosol sequence of China and the North Pacific oxygen isotope record of core V-28-238 during the last million years (core record stops at 970 ka). Soils formed or the loess became strongly weathered when the relative concentration of ^{18}O was high, indicating warmer and wetter climatic conditions. (After Liu and Yuan 1987)

CORRELATING THE TERRESTRIAL AND MARINE RECORDS

Earlier in this chapter we noted the influx of Quaternary desert dust into the oceans. Most Quaternary land records are discontinuous in space and time and many contain very big stratigraphic gaps, in contrast to the long, continuous records from deep-sea cores. An outstanding exception to the general rule that most Quaternary desert records are fragmentary is the unique Chinese loess deposits, which we have just described. Figure 9.9 represents one attempt to correlate the palaeomagnetically dated loess deposits and associated palaeosols at Lochuan in north-central China with the North Pacific oxygen isotope record from core V-28–238 for roughly the last one million years. Cold phases (even isotope stage numbers) discernible in the North Pacific generally coincide with times when loess accumulation was rapid and soil formation was at a minimum. Warm phases (odd numbers) mostly coincide with weathering of the loess mantles and with times of soil development. Ignoring the smaller fluctuations (0.5‰ or less) evident in the oxygen isotope record, it seems reasonable to accept the conclusion that interglacials or interstadials evident in the North Pacific deep-sea core record were synchronous with intervals of warmer and wetter climate in North China (Liu and Yuan 1987). Conversely, glacials and stadials coincided with colder and drier intervals during which accumulation of unweathered loess was rapid and extensive. Of particular interest is the recent demonstration that the multi-proxy climate records from Chinese loess for the past 75 ka coincide in detail with the North Atlantic Heinrich events and Bond cycles (see chapters 4 and 7) as well as with interstadials recorded in the Greenland and Antarctic ice cores (Porter and An 1995; An and Porter 1997; Chen *et al.* 1997) The Chinese loess record is entirely consistent with the glacial aridity model proposed earlier in this chapter, but, as with every broad generalisation, it is the exceptions and the local variations which must ultimately be considered if we are to understand the full complexity of the Quaternary history of our deserts.

CHAPTER
10

EVIDENCE FROM TERRESTRIAL FLORA AND FAUNA

They attacked the cedars . . . and while Gilgamesh felled the first of the trees of the forest Enkidu cleared their roots as far as the banks of the Euphrates.

Anon (c. 2700 BC) *The Epic of Gilgamesh*
(Trans. N.K. Sanders, 1960)

The fossilised remains of terrestrial flora and fauna are to be found in almost all global environments but, because of their abundance and the large number of sites available for preservation, the major focus of this line of investigation has been on humid temperate and tropical regions, which are poorly represented in the other lines of evidence for the reconstruction of Quaternary environments considered in this book. In order to survive under the relatively harsh conditions experienced on land, most terrestrial plants and animals possess some hard parts that preserve within sedimentary deposits. However, the usefulness of different organisms as palaeoenvironmental indicators varies as a result of a number of factors, including the abundance and ease of identification of preserved parts, the degree of evolutionary change during the Quaternary, and the extent of their habitat ranges.

Since the pioneering work of Von Post in Scandinavia, early this century, the analysis of pollen and spores derived from higher land plants has become the most widely used technique for Quaternary ecological investigation. This is because these particles are well represented in most accumulating sediments, their wide dispersal provides a broad picture of surrounding vegetation, and their intricate and distinctive patterning allows identification to parent plants at a reasonable taxonomic level. Broad regional vegetation–climate relationships were established from early pollen studies in northwest Europe, providing an important relative dating tool for archaeological and other events. The value of pollen analysis within this area declined in the 1950s with the development of numerical dating techniques, particularly radiocarbon dating (see Appendix), but the ready application of this dating technique to pollen sequences opened up far more opportunities for it in palaeoenvironmental and palaeoecological investigation.

Other biotic materials generally serve to refine interpretations from pollen analysis or provide evidence from the few areas or sites where pollen is not preserved. Of significance are other plant remains, such as leaves, seeds and phytoliths, which can frequently be identified to a lower taxonomic level than pollen and provide more detail on the composition of particular communities, and animal remains, particularly beetles and vertebrates, which can respond more quickly than most plants to environmental change. Charcoal is valuable in providing direct evidence of fire, while detailed environmental information can be provided by the nature and variation in thickness of tree rings preserved in the trunks of both living and dead trees.

This chapter is designed to reflect the relative importance of the different materials used for palaeoecological reconstruction and the techniques used in their analysis. The focus is primarily on the methodology and applications of pollen analysis. Other major floral and faunal remains are briefly examined in their role as providing additional and generally complementary information to that from pollen. The chapter concludes with an attempted overview of global vegetation during the Quaternary.

POLLEN

Pollen analysis or palynology are general terms embracing the study of a variety of plant microfossils of which pollen grains and spores (particularly fern spores) are the most important. Pollen grains are produced in the anthers of male flowers of angiosperms (flowering plants) and gymnosperms (conifers), and their function is the transfer of genetic material to female flowers to effect fertilisation and seed production, while spores are the dispersal units of ferns and lower plants that give rise directly to the next generation. Both these types of microfossil are small, about 5–100 microns, and their external coating is composed of a resistant substance known as sporopollenin. Many other types of microfossil can be analysed in association with pollen, but apart from charcoal and phytoliths, these provide evidence mainly of local site conditions rather than dry land environments.

There are a number of processes involved in the production of final pollen assemblages from the original vegetation that need to be taken into account in the reconstruction of this vegetation. These processes are identified in Fig. 10.1 and outlined below.

Production

Plants are adapted to producing different amounts of pollen and, in general terms, those producing the greatest quantities will be best represented in pollen samples. However, for any given species, production will vary according to a number of factors including climatic conditions and its position within the vegetation. Most plants will flower most prolifically under optimal growing conditions, but others may only flower or spore when stressed.

Dispersal

Only a proportion of pollen fulfils its intended function, and much of it is available for transport to a suitable depositional site for accumulation in the fossil state. Dispersal is effected mainly by wind (anemophily), by nectar-feeding insects and birds (zoogamy) and water (hydrogamy). As animal pollination is generally efficient, the majority of pollen grains reaching a deposition site are those dispersed by wind, and also by water if there are streams running into it or there is easy inwash from surrounding slopes. There are many other variables involved in pollen transport. These variables include attributes of the pollen itself such as size, shape and density, the position of plants in the vegetation, and climatic conditions. Representation of pollen will also be influenced by the proximity of the plants to a depositional basin and to the size of that basin.

A generalised model has been constructed that illustrates the ways in which different airborne pollen components can reach a depositional basin. This model is shown in Fig. 10.2 for small lakes within two different types of forest. The transported pollen

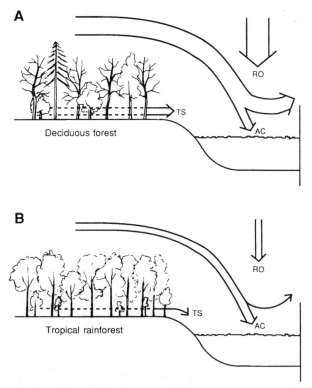

Fig. 10.2 A diagrammatic representation of the relative importance of different modes of pollen transfer to a lake basin from (A) mixed deciduous forest and (B) tropical rainforest. (Based on results from Tauber 1967; Kershaw and Hyland 1975)

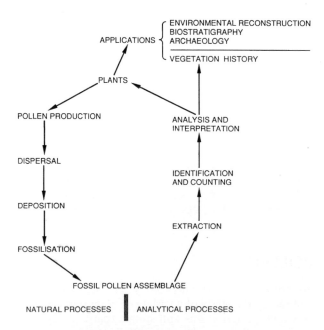

Fig. 10.1 Processes involved in the production of fossil pollen assemblages from parent plants and subsequent analysis and interpretation

is caught in specially designed traps floating on the surfaces of the lakes. There are three main modes of transport: by wind above the canopy, through the trunk space, and by rain after mixing in the atmosphere (Tauber 1967).

In deciduous forest, values for rainout (RO) and above canopy (AC) components are relatively high because most species have wind-dispersed pollen, there are strong winds to transport the pollen, and regular showers to wash the pollen from the atmosphere. The trunk-space (TS) component is also large because, although wind speeds and therefore carrying-capacity are lower than outside the forest, the absence of foliage in early spring when many trees flower does allow significant air flow. In addition, in autumn and winter when many trees have lost their leaves, there is a substantial reflotation of pollen that was trapped by the foliage earlier in the year.

In tropical rainforest, all components record much lower values because the majority of species are animal pollinated and release little pollen into the atmosphere, and wind speeds are lower. The trunk-space component is further reduced by the maintenance of a dense vegetation cover throughout the year, while the rainout component is insignificant because of very regular showers which inhibit significant atmospheric mixing, or because of the existence of long dry seasons.

The three components tend to represent vegetation from different distances away from the site of deposition. The trunk-space component reflects vegetation growing closest to the site, the canopy component is derived mainly from canopy species within the general area, while the rainout component may include pollen from communities growing some distance away. The larger the lake and the further away from the lake edge that samples are taken for pollen analysis, the more important will be the pollen from the rainout component relative particularly to the trunk-space pollen.

One important limitation to this model is that it fails to take account of waterborne pollen. Where there is significant stream inflow into a lake and particularly into marine environments, most pollen can be from this source and will emphasise stream valley vegetation. In small lakes, a substantial component can be derived from inwash of pollen originally deposited on lakeside soil surfaces.

Awareness of the importance of the differential representation of vegetation at varying distances from a pollen site has led to the definition of distance components. These are local pollen, derived from communities growing at or close to the pollen site; extra-local pollen, most easily defined as pollen from plants growing round a depositional site and influenced by site hydrology; regional pollen, representing the vegetation characteristic of the area; and long-distance pollen, from vegetation types further afield (Jansson 1966).

The relative importance of these components at pollen sample sites in four similar-sized basins and catchments with varying site and vegetation characteristics is illustrated in Fig. 10.3. In the two lake examples (A and B), pollen from regional vegetation is dominant. Extra-local pollen from lakeside communities has reasonable representation because of

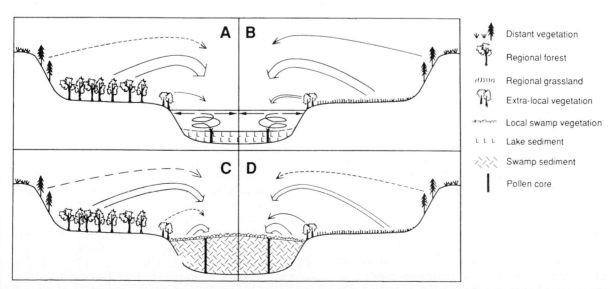

Fig. 10.3 An indication of the relative representation of pollen from different components of two different vegetation landscapes deposited in lake and swamp basins (see text for details)

easy inwash of pollen into the lake. In the absence of plants growing in the basins, there is no local pollen component. There is a small component from distant forests, and proportions for this and for extra-local pollen are larger in case B because the low-growing regional grassland vegetation has lower production and dispersal capability than regional forest. There are similar relative proportions of extra-local, regional and long-distance components in the bog examples surrounded by grassland and forest (C and D), but all are smaller than in the corresponding lake situation because of the abundance of locally derived pollen from plants growing on the bog surface.

Deposition

Figure 10.3 also serves to demonstrate different depositional processes within lakes and bogs. In bogs, deposited pollen tends to be caught effectively on the accumulating surface, whereas in lakes there can be movement of pollen on the surface and within the lake due to wind and water circulation patterns. These movements affect pollen types differentially, producing different mixes in different parts of a lake. There may also be concentration in the centre of steep-sided lakes due to sediment focusing, while mixing of deposited pollen may be caused at times of lake overturn and by bottom-dwelling organisms.

Post-depositional changes

Once pollen has been deposited and incorporated into accumulating sediments, the very resistant outside wall or exine can be preserved almost indefinitely. However, if the depositional surface is subject to periodic drying, destruction by oxidation can occur, and if incorporation into the sediment body is slow, pollen grains can be exposed to microbial attack. Similarly, changes to pollen assemblages can occur if sediments are subsequently subjected to drying out or disturbance. Caution has to be exercised in reconstructing vegetation from samples that show signs of deterioration since pollen grains can be destroyed differentially, and some are more readily identifiable in a corroded state than others.

Some post-depositional effects are shown in Fig. 10.4, a section through a deposit in New Zealand. The section was originally prepared to reveal a selection of the large number of skeletons of large, extinct flightless birds known as moas, preserved in the swamp. Although not shown here, these skeletons interrupt the pollen sequence and it is obvious that a great deal of sediment mixing would have occurred as the birds became trapped in the surface sediment. Similarly, the tree remains in the lower part of the sequence would have influenced sedimentation patterns. They demonstrate that the swamp was forested,

and this forest invasion may mean that the surface became sufficiently dry to allow some oxidation of the pollen. The topmost sediments have also been disturbed, most likely by tree-root penetration, and record distortion which could be caused if younger sediments have filled the space originally occupied by these roots.

Site selection and sampling

There is probably no ideal site for most pollen studies, since places of sediment accumulation are generally unrepresentative of the pollen catchment area. However, sites vary greatly in their suitability for various studies. For vegetation reconstruction, the choice is generally between lakes and mires (swamps and bogs). Deep lakes are frequently preferred because of constancy in the depositional environment, and therefore sediment sequences are likely to be continuous. Swamps or bog sites, on the other hand, are more likely to have had a more varied history because of their sensitivity to changing hydrological conditions, and because of successional changes resulting from the accumulation of sediments to maximum attainable heights for particular communities. Pyramid Valley Swamp (Fig. 10.4) provides a good example of some of the problems that can be encountered with swamp sites. The stratigraphy (see next section for terminology) clearly shows changes in the nature of the site from an initial swamp (indicated by the accumulation of herb detritus) to a lake, characterised initially by organic lake mud and then carbonates, and finally back to herb detritus. The presence of wood also indicates invasion of the swamp phases by trees that may be inseparable from those dominating the regional vegetation. As previously mentioned, the trees would also have disturbed previously accumulated sediment.

However, swamps do provide better pollen-receptive surfaces than lakes, and where there is continuous accumulation they have the potential to provide a much more detailed vegetation record. In environments where the most suitable kinds of sites are rare, it is necessary to examine a range of sites with different characteristics to allow reconstruction of a regional vegetation picture.

For some studies, neither deep lakes nor bogs may be appropriate. For example, in archaeological studies, caves that show signs of habitation may be preferred. Although pollen is not likely to be as well preserved as under constantly waterlogged conditions, and there may be difficulties in relating the pollen to regional vegetation, insights may be gained into the plant materials used by cave dwellers, and into the impact of these people on the local vegetation.

2620 BP

2930 BP

3720 BP

4280 BP

Fig. 10.4 Section through Pyramid Swamp, New Zealand, showing variation in accumulated swamp sediments

Once a site has been selected, it is important to undertake a stratigraphic survey in order to find the most appropriate spot for sampling. This will generally be close to the site centre, which is likely to contain the longest and most continuous sequence. An exposed sediment face provides the ideal basis for selection, since local disturbances such as root penetration, animal skeletons and wood remains can be avoided, as indicated in Pyramid Valley Swamp. However, in basins that are waterlogged it is necessary to rely on the examination of discrete cores, and a variety of samplers have been designed to collect material from the range of lake and swamp sites encountered.

Description of sediment stratigraphy

Both sediment core correlation and interpretation of palaeoecological records are facilitated by a standardised method of sediment description. Unfortunately there are a variety of systems in operation, ranging from those that emphasise the genesis of deposits as a basis for classification to those that are adaptations of descriptive systems devised for predominantly inorganic sediments. Ideally, a system should not assume that the investigator has a strong background in any particular natural science, should be simple enough to encourage its use in any geographical and depositional environment, and should be sufficiently flexible to incorporate the level of detail in description required by the investigator. The approach devised by Troels-Smith (1955) possesses many desirable characteristics, and his system or modifications of it (Aaby and Berglund 1986) have been the most widely applied in the description of lake and particularly swamp sediments over recent years. The scheme outlined here is one such modification of the Troels-Smith system which has been progressively simplified in an attempt to overcome student resistance to its use. Its major features are that it has dispensed with Latinised names for sediment characteristics and with its Eurocentric flavour.

Troels-Smith incorporated three major properties into his system: physical properties including the appearance and mechanical qualities of the deposit; humification, which is the degree of decomposition of the organic material present; and components which include the nature and proportion of the elements comprising the deposit. A listing and brief description of the modified characteristics is shown in Table 10.1. For most features a five-class scale (0–4) is used for characterisation, with 0 denoting the absence of, and 4 the maximum value of the feature in question. In the case of sediments composed of a number of components, the total value of the individual components must not exceed 4. However, very slight changes in the value of any character can be indicated by the use of plus and minus signs. In the case of undefined components or 'accessory elements' such as shells or artefacts, their abundance can also be accommodated by the five-class scale, but outside the sum of defined components. Additional information on specific elements such as insect remains, ostracods, diatoms, and so on, can be provided from microscopic study using a five-point scale of abundance (i.e. 0 = absent, 1 = rare, 2 = occasional, 3 = frequent, 4 = abundant).

An illustration of the application of the system is provided by the hypothetical example shown in Table 10.2, and Fig. 10.5, where a basin has been progressively infilled with inorganic, lake and swamp sediments. This kind of succession is characteristic of site development since the last glacial period except that the degree of variation has been exaggerated here to allow inclusion of all defined sediment components. One major characteristic, humicity, has been excluded because the defined measure can be accommodated by the humus component.

The proforma (Table 10.2) is designed for field description, although sediments can also be used for description in the laboratory. Space is allocated for measures of accessory elements, in this case shells, as well as for comments on general or specific features of each sedimentary unit.

The data are graphed in Fig. 10.5. The stratigraphic column shows the major components of each unit by use of symbols derived from those of Troels-Smith, and available in the computer graphics program TILIA2 (see data presentation section), together with a measure of the sharpness of the boundary between units. In contrast to the Troels-Smith scheme which illustrates the proportion of each component by changing the density of the symbol, proportions are indicated here by graphing individually the values of various components, a method introduced by Shaomeng *et al.* (1986). This system allows for minor components to be shown and for other characteristics to be displayed in a similar manner. Additional data may be added from more detailed, microscopic examination of the sediments. It is often convenient to determine the nature and abundance of accessory elements of interest from low-powered microscopic examination of sieved material resulting from preparation of samples for detailed palaeoecological analysis. In the example shown, abundance measures are considered to have been estimated from the greater than 100 micron fraction discarded in the preparation of pollen samples at 10 cm intervals along the core. The macro-charcoal values provide a more detailed estimate of local fires than the bulk sediment charcoal measures, while the detailed ostracod values have replaced the shell determination made in the field. If desired, grain-size analysis of the inorganic sediment components could be undertaken to refine or augment the inorganic component measures of silt, clay, sand and gravel. Conversely, only the stratigraphic column, or selected individual characters, need be illustrated if the data are to be graphed alongside the results of detailed microscopic examination of pollen, diatoms, and so on.

Extraction

Samples for pollen analysis extracted from the sediment sequences are subjected to various physical and chemical treatments in order to reduce the sediment matrix and make the pollen clearly visible for microscope examination. A range of treatments is possible because pollen is more resistant to destruction than most other components of the sediment. Commonly the following sequence of treatments is undertaken: addition of potassium hydroxide to break down the structure of the organic matter and remove humic acids; sieving to remove coarser plant remains and larger inorganic fragments; hydrogen fluoride treatment to dissolve fine siliceous matter; and acetolysis (a mixture of acetic hydroxide and sulphuric acid) to digest cellulose and other polysaccharides and to darken the grains for easy recognition. In addition, treatment with Shultz solution (a mixture of nitric acid and potassium chlorate) may be necessary to remove woody material or lignin, while the use of a heavy liquid such as zinc bromide to separate

Table 10.1   Characteristics of a proposed system of sediment description based on that of Troels-Smith (1955)

Physical features

Degree of darkness (*Nigror*)	Varies from 0 in the lightest occurring shades (e.g. clear quartz sand and lake marl), through 1 (e.g. calcareous clay), 2 (e.g. fresh swamp peat), 3 (e.g. partly humified peat).
Degree of stratification (*Stratificatio*)	Visual or structural horizontal banding or layering. Varies from 0 where the deposit is completely homogenous or breaks in all directions, to 4 which consists of clear thin layers or bands.
Degree of elasticity (*Elasticitas*)	The sediment's ability to regain its shape after being squeezed or bent. Varies from 0 in plastic clay, sand, disintegrated peat, etc., to 4 in fresh peat.
Degree of dryness (*Siccitas*)	Deposits fall between 0 (clear water) and 4 (air dry material). 1 indicates very wet runny sediment such as surface lake muds, 2 represents saturated sediments, the normal condition below the water table, while 3 indicates moist, unsaturated sediments.
Colour	Best determined by reference to Munsell soil colour charts. Changes in colour with exposure to air should be noted.
Structure	The dominant structural feature (e.g. fibrous, homogeneous).
Sharpness of boundary (*Limes superior*)	The boundary can be diffuse (> 1 cm: lim. 0), very gradual (< 1 cm to > 2 mm : 1), gradual (< 2 mm to > 1 mm: 2), sharp (< 1 mm to > 0.5 mm: 3) or very sharp (< 0.5 mm: 4).
Humicity (*Humicitas*)	The degree of humification or disintegration of organic substances. It is measured by determination of the nature and amount of material passing through the fingers on squeezing; 0 (fresh peat yielding clear water), 1 (slightly decomposed peat yielding dark coloured, turbid water), 2 (decomposed peat yielding half its mass), 3 (very decomposed peat yielding three-quarters of its mass) and 4 (totally decomposed peat yielding almost all its mass)

Components

Mosses (*Turfa bryophytica*)	*Sphagnum* is the most common peat-former.
Woody plants (*Turfa lignosa*)	Roots of trees and shrubs together with attached stumps and branches, frequently in growth position.
Herbs (*Turfa hebacea*)	Roots of herbaceous plants together with attached stems and leaves, frequently in growth position.
Woody detritus (*Detritus lignosus*)	Fragments of woody plants > 2 mm.
Herb detritus (*Detritus herbosus*)	Fragments of herbaceous plants > 2 mm.
Fine detritus (*Detritus granosus*)	Fragments of woody or herbaceous plants < 2 mm.
Charcoal	Carbonised fragments of predominantly woody plants.
Organic lake mud (*Limus detrituosus*)	Homogeneous organic lake sediment composed of remains of microplankton and humified remains of macrophytes.
Humus (*Substantia humosa*)	Completely disintegrated organic substances and precipitated humic acids.
Organosilicates (*Limus siliceous*)	Siliceous skeletons or skeleton fragments of diatoms, sponges, etc.
Carbonates (*Limus calcareus*)	Calcium carbonate or marl. Similar in colour and texture to *L. siliceous* but soluble in hydrochloric acid.
Iron oxides (*Limus ferrugineus*)	Iron oxides of various types and colours
Clay (*Argilla steatodes*)	Mineral particles < 0.002 mm.
Silt (*Argilla granosa*)	Mineral particles 0.002–0.06 mm.
Sand (*Grana minora*)	Mineral particles 0.06–2 mm.
Gravel (*Grana majora*)	Mineral particles > 2 mm.

remaining organic from inorganic material may be employed if the organic component containing the pollen is relatively small. Between stages, samples are centrifuged to concentrate residual material and washed free of strong acids and alkalis in miscible liquids. In addition to pollen, charcoal is preserved in this preparation process.

Prepared samples are mounted in a suitable medium such as glycerine jelly or silicone oil on microscope slides ready for examination.

Table 10.2 An example of a field description sheet for sediment description

Site: Quaternary Bog   Transect Number 3   Core Number: 5   Distance from site edge: 120 m   Date: 19/5/1997

Level (cm) Upper	Lower	Darkness (0–4)	Stratification (0–4)	Elasticity (0–4)	Dryness (0–4)	Colour	Structure	Upper boundary	Mosses	Woody plants	Herbs	Woody detritus	Herb detritus	Fine detritus	Charcoal	Organic lake mud	Humus	Organosilicates	Carbonates	Iron oxides	Clay	Silt	Mud	Sand	Gravel	Comments
0	18	2	0	3	2	Reddish-brown	Fibrous	3	2	1	1	+	+													Woody roots from epacrids
18	34	3	0	2	2	Dark brown	Sub-fibrous	4	+		2		2				+									
34	40	4	0	0	3	Black	Granular	4	+				+		3		1									A well-defined charcoal layer
40	50	3	0	1	3	Dark red-brown	Heterogeneous	4			+	1	1				2									
50	72	3	0	1	2	Dark brown	Sub-fibrous	2			+		2				2									Very moist in places
72	90	2	0	1	2	Dull brown	Sub-fibrous	0			+		4													Darkens on exposure to air
90	120	2	1	2	2	Yellowish-brown	Heterogeneous	4					2			2		1								Minor contamination, 98–100 cm
120	144	2	1	2	2	Yellowish-brown	Homogeneous	1					+	+		3										
144	160	2	1	2	2	Yellowish-brown	Homogeneous	1					+	+		4										
160	200	1	1	2	2	Olive yellow	Banded	3						+		2			2				+			Concentration of $CaCO_3$ in bands
200	220	1	0	0	3	Yellowish-grey	Heterogeneous	4					+				+				3			1		Difficult to penetrate
220	238	1	0	0	3	Dull orange	Coarse	0														1		2	1	Rusty mottles
238	252	1	0	0	3	Yellowish-brown	Very coarse	2											+				+	1	3	

Physical features   *Components (total = 4)*

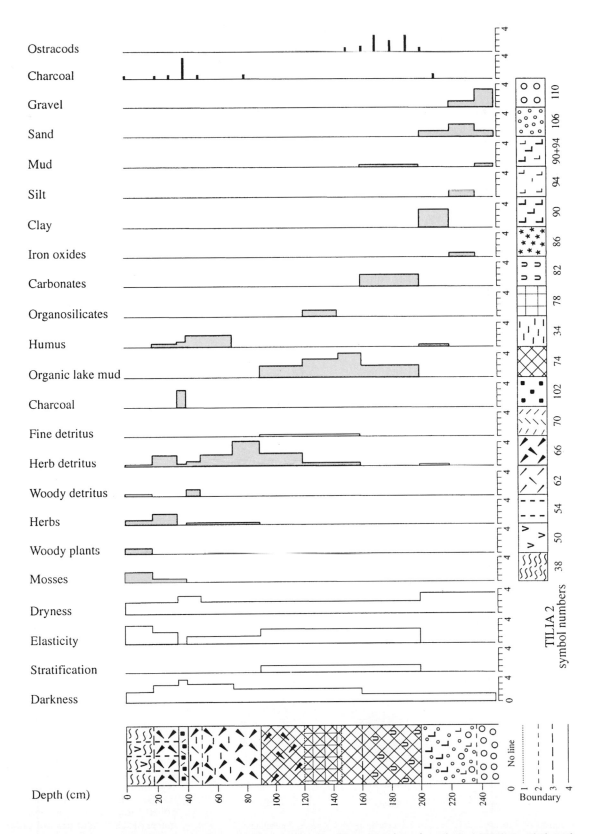

Fig. 10.5 Hypothetical example of the proposed system of sediment description based on that of Troels-Smith (1955)

Identification and counting

Pollen grains and spores are identified from morphological features of the exine. Most pollen types have a symmetry that relates to their tetrad position during formation in the anther (see Fig. 10.6A). Some of the major morphological features of pollen grains are illustrated in Fig. 10.6 B–D.

Many other forms are illustrated in Fig. 10.23. Those that are very different to the described forms include the gymnosperms *Pinus* and *Podocarpus* that have wings or bladders to assist in wind dispersal, and *Chenopodiaceae*, whose pores are evenly distributed over the whole grain surface. Identification to genus or family level and species level for selected taxa is routinely achieved.

Pollen counting is undertaken along viewed transects of the prepared slides under a microscope with high magnification until the desired number of grains has been recorded. Two major considerations in count size are the achievement of relatively constant percentage levels of major recorded taxa, and inclusion of a substantial number of types present in the sample population. An illustration of how percentages for variously represented taxa in a sample level differ with increased count size is shown in Fig. 10.7A. Below a total count of about 600 grains variability is great, and if the count is too low, percentages will be unacceptably inaccurate. For poorly represented types such as *Lygodium*, significant variability is maintained in counts above 600 grains, and for most palynological studies necessary count levels would be unacceptably high to provide reliable estimates of percentages for these taxa. Statistical probability limits for calculated percentages of taxa with different count levels have been determined, but unfortunately are seldom applied.

The manner in which the number of taxa recorded increases with count size is illustrated in Fig. 10.7B . There is a great deal of difference in pollen taxon diversity between samples from the different community types and it is probable that, in samples from

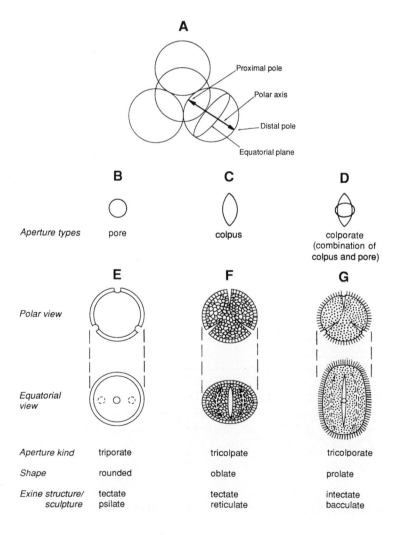

Fig. 10.6 Aspects of pollen morphology. (A) The symmetry of pollen grains related to tetrad arrangement in the anther. (B)–(D) Three types of aperture commonly found in pollen grains. (E)–(G) Schematic representations of pollen grain types showing selected morphological features in polar and equatorial views

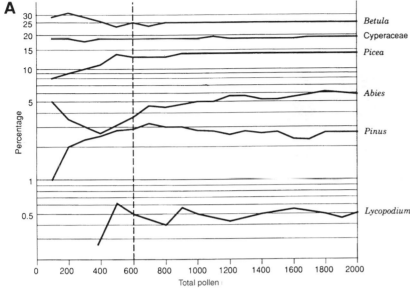

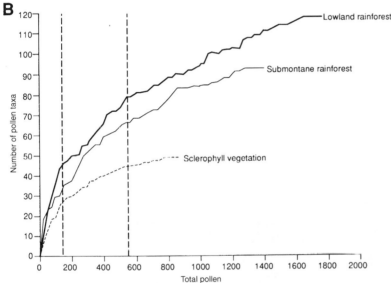

Fig. 10.7 (A) Changing percentages of selected taxa with increasing count size (from Birks and Birks 1981). (B) Number of taxa counted as a function of sample size from selected pollen samples from northeast Queensland, Australia (from Kershaw, unpublished data)

lowland rainforest, many thousands of grains would have to be counted to ensure the inclusion of most taxa. It is suggested from these results that counts of about 550 grains are required to ensure the inclusion of a reasonable proportion of types represented, with an absolute minimum of 150 grains.

Data presentation

Pollen counts from a site sequence are normally portrayed in the form of a pollen diagram where percentage values for recorded taxa are graphed in relation to stratigraphic depth. The diagram shown in Fig. 10.8a illustrates many features common to more traditional types of pollen representation. Here,

emphasis is on changes in the abundance of the dominant tree taxa as a basis for determination of gross vegetation changes. In line with this emphasis, the total pollen count for tree taxa from each sample forms the pollen sum on which all percentages are based. The counts for the understorey shrubs, *Corylus* and *Salix*, are omitted from the sum but are percentaged relative to the sum. This has resulted in percentages in excess of 200% in some samples for the very abundant taxon *Corylus*.

The diagram has been divided into zones based on changes in the representation of major taxa. Zonation simplifies description of the diagram and aids subsequent interpretation. The selection of changes in major taxa are those that have been found to occur

HOCKHAM MERE POLLEN DIAGRAMS

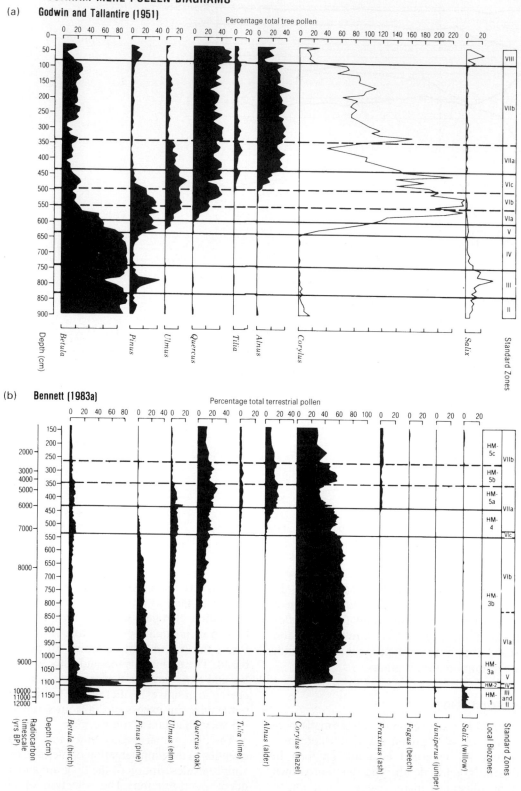

Fig. 10.8   Selected features of pollen diagrams from Hockham Mere, East Anglia, UK

over a wide area. These changes include the beginning of a sustained rise in *Corylus* at the base of zone V, a rise in *Ulmus* in zone VIa, the achievement of greater levels of *Quercus* than *Ulmus* in zone VIb, major increases in *Alnus* and *Tilia* at the base of zone VIc, and a decline in *Ulmus* at the zone VI/VII boundary. Adopting a standard zonation provides a means of correlating between diagrams and consequently a relative time-scale for dating associated events like changes in the nature of sediments, or materials such as archaeological artefacts, in the absence of any absolute dating control.

A more recently produced pollen diagram from the site (Fig. 10.8b) shows a number of differences. Some of these can be explained by selection of a core from another part of the site, demonstrating the significance of this variable, but the majority of differences are related to the application of more recently developed ideas on, and techniques of, pollen diagram construction. In the first place, the pollen sum is based on the total count of all taxa derived from dryland vegetation, rather than just the trees. This is an acknowledgement of the fact that the landscape has not always been forested and it also demonstrates increased interest in the ecology of all components of the system. Similarly, the use of local biozones, based on internal homogeneity in pollen composition of sections of the diagram independent of information from other sites, is a reflection of the importance of local site variation. Once individual diagrams have been interpreted independently, then comparisons with others can be made. The need for pollen as a stratigraphic tool has also been reduced with the development of radiocarbon dating. Finally, improved identification and much larger and more frequent counts have resulted in the recognition of a greater number of pollen types.

A high level of flexibility, combined with greater objectivity in construction and analysis of pollen and other microfossil diagrams, has been made possible by the recent development of accessible computer programs for graphical portrayal incorporating statistical methods. The most widely adopted is the program TILIA (Grimm 1988). Summaries of diagrams constructed from TILIA for a large number of sites over a large part of the globe can be perused at NOAA worldwide web site http://www.noaa.gov./palaeo/softlib.html.

Interpretation

The first step in interpretation of most pollen diagrams is the reconstruction of the original vegetation. Once this has been accomplished, then explanations for the vegetation patterns and changes can be sought. Vegetation reconstruction is a difficult process considering the large number of processes involved in the production of fossil pollen assemblages from the original plant cover. One way of accounting for many of the variables is to use modern pollen samples collected from known communities as a basis for interpretation. This not only provides a direct link between pollen and vegetation but, if modern and fossil samples are prepared and counted in a similar fashion, variation due to analytical processes is kept to a minimum.

An altitudinal sequence of modern pollen samples from the New Guinea Highlands (Fig. 10.9) provides a good example of some of the values and also limitations of this approach. It has been used to assist interpretation of a number of pollen diagrams from the area. The vegetation types are clearly distinguished by significant representation of one or a combination of pollen types. The lower altitude grasslands are dominated by grass pollen and can be separated from alpine grasslands on higher grass percentages and an absence of alpine indicators. Oak forest has the only high values of the 'oak' taxon *Lithocarpus/Castanopsis*, mixed forest is characterised by the only appreciable percentages of one component taxon, *Quintinia*, while the beech, *Nothofagus*, dominates beech forest. It is clear that most pollen is deposited close to its source but some wider dispersal does occur, particularly to higher altitudes. There is significant representation of *Nothofagus* in all samples of higher altitude than beech forest, while this taxon, and the equally well dispersed *Casuarina*, are the most important components of, and are characteristic of, the highest altitude bare area which has no local pollen sources.

One factor complicating interpretation of climatically induced altitudinal movements recorded in pollen diagrams is the changes that have been brought about by people within the last few thousand years. It is considered that the lower grasslands are a result of deforestation and establishment of agricultural practices, and that little evidence is preserved of the original forest cover. The effects of people are a common feature in pollen diagrams from many parts of the world and can prevent the application of modern pollen studies to the elucidation of natural causes of change, particularly climate. On the other hand, vegetation is a good indicator of the past impact of people, and in areas where vegetation alteration has not been so drastic, some assessment can be made of both natural and anthropogenically induced changes. In the New Guinea diagram (Fig. 10.9), the representation of the forest disturbance taxa, *Macaranga*, *Trema* and *Casuarina*, provides information on the degree of impact on the forests at various altitudes without destroying the basic altitudinal sequence.

Major refinements to interpretation have also been achieved by the use of absolute, rather than

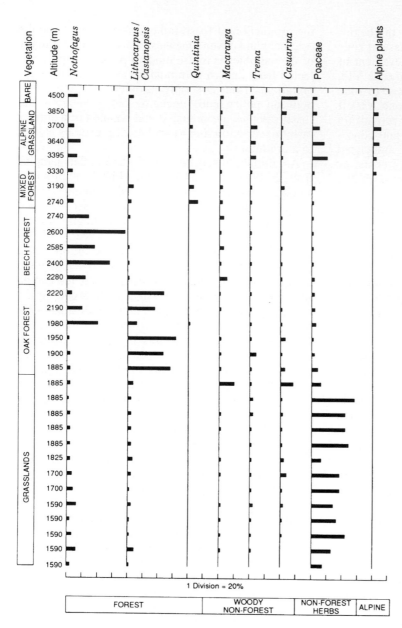

Fig. 10.9 Representation of major pollen taxa in relation to vegetation along an altitudinal surface sample transect in Papua-New Guinea. (Modified from Flenley 1978)

percentage, pollen data. With information in percentage form, a change in the proportion of any one taxon will influence the values of other taxa regardless of whether there have been concomitant changes in the distribution or abundance of parent plants. Where it is possible to calculate the number of grains deposited per unit volume of sediment, generally from sediments that have accumulated at a constant rate or where annual variations in sediment deposition can be detected, then measures of the real abundance of taxa can be derived. The usefulness of so-called pollen influx data is well illustrated by a comparison of the percentage and absolute pollen diagrams from Roger's Lake (Fig. 10.10).

The percentage diagram has much in common with those derived from Hockham Mere. Although from different continents, the basic composition of deciduous forests in these northern hemisphere mid-latitudes is similar, as is the pattern of forest development after the last glaciation. The Lateglacial period, prior to about 10 ka ago, was dominated by herbs with significant representation by a small number of tree taxa. The postglacial period shows a gradual increase in forest taxa, due largely to increasing temperature levels, that eventually formed the present forests. The pollen influx diagram demonstrates that some of the features of the percentage diagrams are apparent rather than real. There is no major change

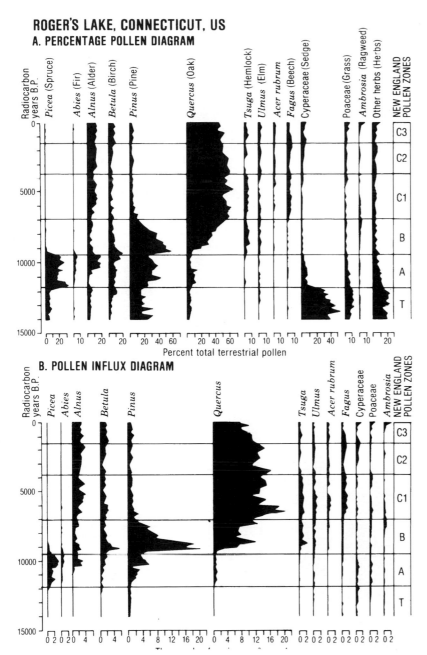

ROGER'S LAKE, CONNECTICUT, US
A. PERCENTAGE POLLEN DIAGRAM

B. POLLEN INFLUX DIAGRAM

Fig. 10.10 A comparison of percentage and influx pollen diagrams from Roger's Lake. (Modified from Davis 1967)

from herb to tree-dominated vegetation from the Late-glacial to the postglacial but rather the trees increase in abundance while the herbs maintain their values. The cause of misinterpretation in the percentage diagram is due to very low influx values in the basal sediments. These demonstrate that the herbaceous vegetation was very open and that trees were probably absent from the area, their pollen being derived from distant sources. The pollen influx data also provide a much more realistic picture of subsequent forest changes, with spruce woodland clearly dominant

about 12 ka ago, followed by mixed forest within which pine was very prominent before establishment of the present deciduous forest at about 7 ka BP.

FINE-RESOLUTION PALYNOLOGY

Palynological reconstruction and interpretation has traditionally been at a relatively coarse temporal level with little attempt made to resolve variation on time-scales of less than hundreds or even thousands of years. However, recent developments in palynological methodology, in association with increased precision in dating techniques and the application of sophisticated statistical methods, are allowing more detailed investigations with resolutions at decadal or even yearly intervals from suitable sites (Green and Dolman 1988; Turner and Peglar 1988). The major thrusts of fine-resolution palynology, as it has been termed, have been on the documentation of human impact, often in combination with historical records, and on the examination of ecological processes that cannot be resolved adequately from standard pollen

analytical methods or from the monitoring of extant vegetation.

An example that clearly illustrates the value of the method, even without statistical analysis, is provided by a study of the more recent seasonally laminated sediments from Crawford Lake, Ontario. The lake is situated within mixed deciduous forests which have experienced disturbance from the agricultural activities of both Indian and European peoples. Contiguous samples taken at 5 or 10 year intervals were prepared from 1000 to 1970 AD. Selected features of the percentage pollen diagram are shown in Fig. 10.11. The

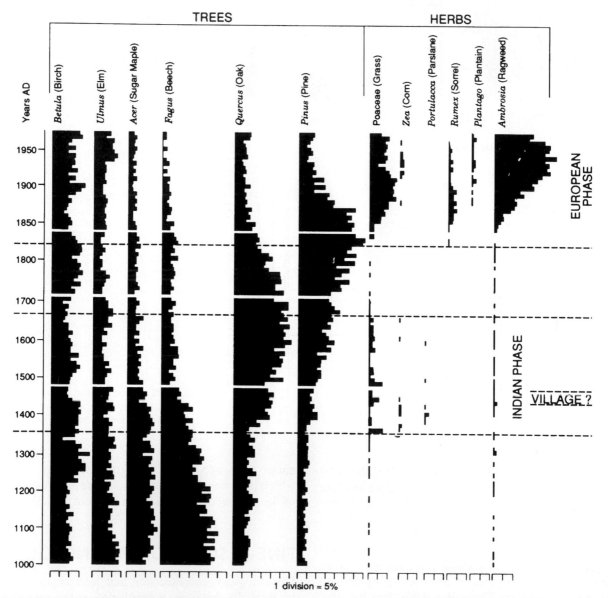

Fig. 10.11 Selected features of the fine-resolution pollen diagram from Crawford Lake, Ontario, Canada, illustrating agricultural indicators and changes in forest composition resulting from Indian and European occupation phases. (Modified from McAndrews and Boyko-Diakonow 1989)

Indian phase is clearly identified between about 1360 and 1660 AD by high grass levels, the presence of a few other weed species and by representation of pollen from cultivated corn. Variations in representation and abundance probably reflect the shifting nature of agriculture, with new clearings being made about every 20 years as the nutrient status of soils in old fields became depleted. It is likely that agriculture around Crawford Lake itself was practised from about 1440 to 1460 AD where there is high representation of weed pollen combined with unusually thick laminae, the result of soil erosion from these nearby corn fields. Excavation and dating of an Indian village close to the lake confirms the local presence of agriculture at this time.

The European phase is marked by much more extensive and permanent agriculture. Earliest evidence, about 1820 AD, is consistent with the documented time for initial land acquisition in the area, while the increase in weeds around 1840 AD corresponds with the beginning of occupation around the lake at this time. The peak in sorrel pollen at the same point as the increase in grass probably reflects the abundance of this weed in pioneer pastures, while subsequent high levels of native ragweed in association with the presence of corn and introduced plantain can be related to the introduction of mechanised corn cultivation.

Tree pollen values also provide evidence of the various agricultural activities. Initial Indian clearing is indicated by sustained decreases in the longer-living 'climax' forest trees, sugar maple and beech. Regeneration of forest after the abandonment of fields was marked by increases in successional trees such as oak and pine, with the latter subsequently being replaced by oak because of its longer life-span, before both succumb to selective logging and clearing by European people. Beech suffered a further decline during this European phase but some tree species maintained their values or even increased their representation with European farming.

MICROSCOPIC CHARCOAL AND ELEMENTAL CARBON

Small, angular carbonised fragments, generally presumed to be the products of biomass burning, have long been recognised in samples prepared for pollen analysis. However, it is only recently that charcoal counting has been included routinely in pollen analytical studies. This has come about partly as a result of an appreciation of the importance of fire as an environmental variable in many of the world's vegetation types, and partly due to a concerted effort to understand the processes involved in relating sedi-

mentary charcoal to individual fires and fire types (e.g. R.L. Clark 1983; Patterson *et al.* 1987). Many of the problems of charcoal analysis are similar to those of reconstructing past vegetation from pollen, but some are unique to charcoal. These unique characteristics include:

a. Difficulties in identifying charcoal to parent plants. Although large charcoal fragments can retain plant structures that allow identification in the same ways as other macroremains (see the next section), it is extremely difficult to identify the vegetation source of microscopic charcoal in many studies.
b. Identification of fire regimes. Charcoal, unlike pollen, is produced during relatively infrequent events and wide sampling intervals are likely to miss some fires and give a false impression of fire frequency. In addition, records are frequently blurred by substantial quantities of charcoal that wash into a basin from the surrounding soil surface after the fire events.
c. Post-depositional changes. Charcoal can break down into smaller fragments within the sedimentary matrix or during sample preparation, substantially altering the fire signal. This problem can be overcome by analysis of thin sediment sections (J.S. Clark 1988), although the method is somewhat labour-intensive for more routine studies.

Despite these problems, charcoal has proved to be very important in helping to understand gross vegetation changes and in some cases has provided good insight into the dynamics of past and present vegetation communities, when employed in association with fine resolution pollen analysis.

Palaeoecological studies from the conifer-hardwood forests of North America demonstrate the application of charcoal analysis at different scales of resolution and indicate the likely role of fire in facilitating the climatically – and anthropogenically – induced changes to these forests noted earlier from Roger's Lake (Fig. 10.10) and Crawford Lake (Fig. 10.11) respectively.

The period from 10 ka to 6 ka BP in the pollen influx diagram from Everitt Lake (Fig. 10.12) illustrates the major changes in the vegetation that accompanied climatic amelioration after the last glacial period. The large charcoal peaks suggest fairly regular intense fires which probably resulted from stress imposed on the vegetation by the changing climate. These fires would have opened up the vegetation and allowed invasion by tree species migrating into the area, which were better adapted to the 'new' climatic conditions. After about 4500 years BP there is little evidence of large fires and it was considered by Green (1981) that the vegetation had reached a degree of

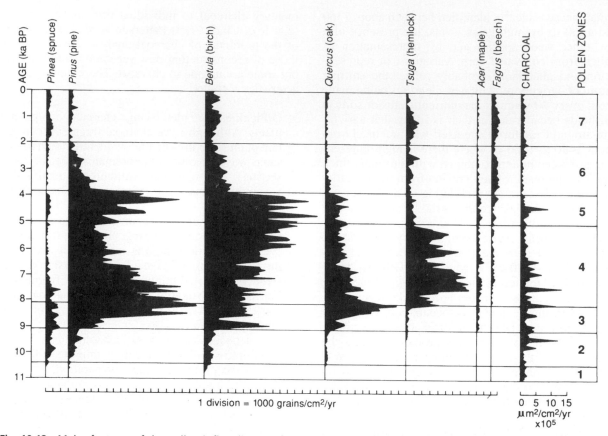

Fig. 10.12   Major features of the pollen influx diagram from Everitt Lake, Nova Scotia. (Modified from Green 1981)

equilibrium under relatively stable climatic conditions. The continued occurrence of small, less intense fires to the present day suggests that the vegetation is adapted to and maintained by this kind of fire regime. Through the use of time series analyses, Green (1981) was able to determine the detailed response of tree species to particular fire regimes. For example, from 4000–2000 years ago, small charcoal peaks suggest that fires recurred on average at 350 year intervals and that there was a progressive recovery of the vegetation after each fire, with relatively early successional taxa such as *Picea* and *Pinus* exhibiting pollen peaks about 50 years, later ones such as *Quercus* about 150–200 years, and latest successional plants like *Tsuga*, *Acer* and *Fagus* about 300–350 years after a fire. This knowledge of prehistoric fire regimes and vegetational successional responses is extremely important to future management of forests where fire control is now largely in the hands of people.

However, climatically induced succession has been complicated by variable but generally increasing human intervention within the late Holocene as indicated in the pollen record from Crawford's Lake. The role of fire as a major factor in vegetation disturbance over the last 2000 years is clearly illustrated for this site by a detailed study of contained charcoal (Fig. 10.13) (Clark and Royall 1995). The pre- and post-Iroquois periods are dominated by low charcoal accumulation rates, whereas the Iroquois and European periods have much higher charcoal values. Interestingly, the two phases of human occupation can be distinguished on charcoal patterns which indicate different patterns of burning. The Iroquois period is characterised by occasional very high values while the European period suggests more frequent burning but never of the same intensity as those reached during the Iroquois period.

Increasing interest is being shown in the study of elemental carbon (or black carbon) as opposed to charcoal particles (Clark *et al.* 1997). Elemental carbon represents total carbonised material present in sediments separated from other organic matter by chemical means. It provides a clearer indication of the residues of biomass burning than charcoal, in that it eliminates the confusion over whether a particle is really a product of burning or has some other origin, and the problem about what to do with partially burned particles. In addition, the method incorporates minute particles not clearly visible or identifi-

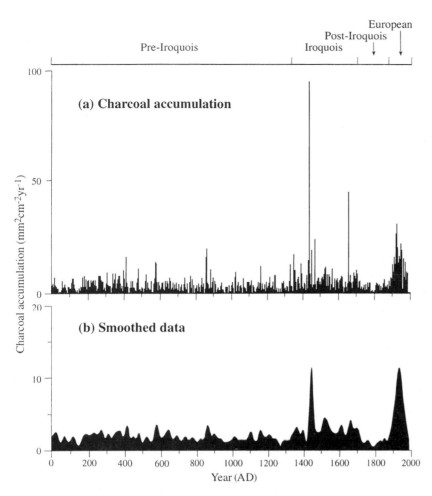

Fig. 10.13 Charcoal accumulation over the last 2000 years at Crawford Lake. (A) Annual accumulation rates. (B) Accumulation rates smoothed with a 30-year window to emphasise low-frequency variability. (From Clark and Royall 1995)

able under a light microscope. Chemical treatment also facilitates carbon isotope determination for use in identifying the origin of the carbon. The method has proved successful in separation of isotopic signatures from grasslands composed largely of plants using the C4 photsynthetic pathway and forests with C3 plants. Future use of elemental carbon is likely to be more common in marine sediment studies where, due to distance from the sources of biomass burning, charcoal size as an indicator of fire proximity is of limited value, and small particles will predominate because of their greater potential for long-distance transport.

PLANT MACROFOSSILS

The larger remains of plants, most commonly in the form of leaves, fruits and seeds, can be found in most sites suitable for pollen analysis. In fact, studies on macroremains preceded those on pollen. These studies tended to be concerned with past distributions of individual taxa rather than vegetation types, and lacked quantification and stratigraphic control (Watts 1978). Those studies associated with early pollen studies focused on the remains of local aquatic plants and contributed substantially to the documentation and understanding of bog and swamp development. More recently, there has been some focus on the examination of dispersed leaf cuticle rather than whole organs, as cuticle can be abundant in a variety of swamp and lake sediments and retains detailed and diagnostic stomatal features.

Increasingly, plant macrofossils are being examined with pollen in an attempt to refine records of terrestrial vegetation history. The preferred sites are either open lakes where dryland plant parts can be washed in or brought in by streams without impedance from swamp vegetation, or accumulations of animal refuse (middens) most commonly preserved in caves and rock shelters. One major value of macrofossils is that their presence can demonstrate conclusively that parent plants were growing in the vicinity of a site, an important consideration in reconstructing the vegetation of a small area or in the determination of migration rates where it is necessary to know exactly when a plant arrived at a particular site.

These features are often very difficult to determine from pollen that is widely dispersed. In certain instances, macrofossils can reveal the presence of a species whose pollen is rarely recorded. An additional value of macrofossils is that many parts can be identified to a more refined taxonomic level than with pollen. Of increasing interest is the value of macroscopic charcoal which, in addition to retaining detail of anatomical features, preserves extremely well. Concentrations of macroscopic charcoal can be very high in archaeological sites and provide good evidence of the use of materials for food, firewood, and so on, and in soils, where changes in forest composition associated with fire can be documented over large areas (see section on development of present vegetation patterns).

One of the most interesting and significant applications of macrofossil studies is in the elucidation of late Quaternary vegetation and environmental changes in arid areas. In these regions, where suitable continuous sedimentary sequences suitable for pollen analysis are often lacking, vegetation histories have been pieced together from the pollen and plant macrofossil analyses and radiocarbon dating of material accumulated by midden-building animals such as pack rats (*Neotoma* spp.) in the southwestern United States (Betancourt *et al.* 1990), hyraxes (*Procavia* spp.) in southern Africa (Scott and Bousman 1990), and stick-nest rats (*Leporillus* spp.) in central Australia (McCarthy *et al.* 1996). The middens, which are composed of plant material cemented to rock crevices by urine, can provide a remarkably accurate picture of a defined area of vegetation around a site because the animals forage within a limited area from their dens, and collect plant material from a whole range of species. As middens are constructed over a limited period of time, each provides a time-slice that, when radiocarbon dated, can be slotted into a sequence to provide a fairly continuous temporal record, sometimes extending beyond 40 ka BP in the case of the US, up to 20 ka in Africa, and through much of the Holocene in Australia.

To date, the poor preservation of plant macrofossils in South African and Australian middens has led to a focus on pollen analysis and, despite the localised nature of collected plant material, pollen spectra appear to provide regional vegetation and climate patterns. Southwestern US studies have focused on the analysis of plant macrofossils and detailed regional patterns have emerged through intensive spatial studies. Some relationship between pollen and macrofossil results is demonstrated by comparative results from the same site (Fig. 10.14). Similar overall patterns are revealed, but macrofossils provide additional information on the representation of species contributing to the juniper component.

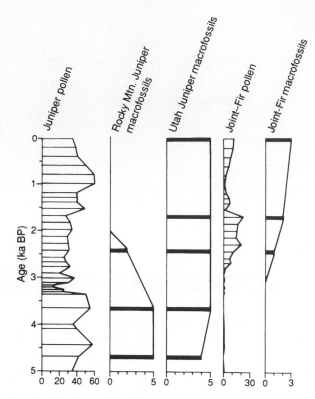

Fig. 10.14 Comparison between selected taxa from pollen spectra and macrofossil samples in packrat middens from Gatecliff Shelter, Nevada. (From Thompson and Krautz 1983, in Thompson 1988)

PLANT PHYTOLITHS

Phytoliths, composed of hydrous silica, are found in the leaves, stems and roots of many higher plants, possibly to help with plant strength and support, and are preserved in a whole range of sediments and soils. Their existence and potential for palaeoenvironmental study has been known since the last century, but it is only within the last few years that this potential is being realised. One feature that has limited exploitation has been the difficulty in establishing a classification of morphological types which has some taxonomic application. Phytolith form is controlled by the shape of plant cells and whether the phytoliths developed between or within cells. Diagnostic characteristics are limited and a variety of forms can be derived from the same plant.

Some major palaeoenvironmental applications are as a complement to pollen studies, particularly in tropical rainforests where the pollen of many plants is not deposited or preserved, or as a replacement for pollen analysis in areas such as arid environments or oxidising sediments where no or little pollen is preserved. Because of difficulties in identification of assemblages, the greatest value of phytoliths lies in more specific, targeted research such as the elucidation of the history of crop species. In this area, some success has been achieved, particularly with rice and other cereals, as the Poaceae are well-endowed with phytoliths. A simple illustration of the complementary value of pollen and phytoliths is provided by the identification of maize in a sedimentary record from the Amazon rainforest, which indicates the beginning of cultivation at about 5300 BP, some 2000 years before any previous report from the Amazon Basin (Fig. 10.15A). The pollen was identified by its large size and characteristic pattern compared with other Poaceae (Fig. 10.15B), while the phytoliths were identified on their 'cross-shape' (Fig. 10.15C), a morphological feature found only in maize (Piperno 1987). Support for maize cultivation is provided by the increase in representation of other herbaceous taxa around this time. The patchy representation of maize may indicate shifting cultivation, with attention focused around the lake site when lake levels were relatively low. The apparent abandonment of cultivation in the area, indicated after about 800 BP, could be explained by a change to a wetter climate (Bush *et al.* 1989).

TREE RINGS

Prior to the development of fine-resolution sediment-based studies with acceptable dating control, the major method available for constructing terrestrial palaeoenvironments over the last few hundred or thousand years in some detail was through dendrochronology. In a narrow sense dendrochronology can be equated with the counting of annual rings, formed from a combination of early and late growing-season cells in trees, to provide an estimate of tree age. Important applications of the technique to Quaternary studies include the dating of timber used in prehistoric structures by matching sequences of rings with living trees growing within the region today, and calibrating the radiocarbon time-scale (see Appendix). In a broader sense, dendrochronology includes dendroecology, which focuses on the dynamics of tree populations, and dendroclimatology, which is concerned with the reconstruction of past climates. These aspects are complementary in that a knowledge of past climatic conditions is necessary for a full understanding of population dynamics, while population characteristics such as competition can affect the response of trees to climate. Climatological information is revealed largely by annual variations in the thickness of rings within trees that are responsive to some seasonal component of climate, but can also be derived from an examination of wood density and the composition of carbon and

oxygen isotopes within identified rings (see Bradley 1985).

Tree ring samples from living trees are generally in the form of cores taken from the outside to the centre of the trees with an increment corer. These cores, smoothed with sandpaper to reveal the cellular structure, are examined under a microscope and the thickness of each ring is measured.

The development of a tree ring chronology for a particular area is based on the cross-matching of the records from a number of trees within a species population. Such a chronology is necessary to ensure that the chronology is complete, as individual trees may give incorrect ages due to missing rings, usually the response of the trees to extreme climatic conditions, or false rings when, due to variable conditions for growth, more than one ring may form in a year. Other problems involve unequal radial growth of trees that results in wedging or lobate growth. Some of these differences in growth patterns are illustrated in Fig. 10.16.

Most parts of the world that allow tree growth and have a seasonal component to the climate are suitable for tree ring studies (see Fig. 10.17). Within these areas the clearest and longest records have been obtained from conifers with the notable exception of *Quercus* in Europe and perhaps also *Nothofagus* in the temperate southern hemisphere rainforests. Most of these taxa have substantial pollen records and there is real potential to refine palaeoclimatic estimates from combined fine-resolution pollen and tree ring studies. The exceptionally long records are from long-lived trees – over 4000 years in the case of the bristlecone pine – extended by cross-matching with the preserved remains of dead trees. Preservation is facilitated in the bristlecone pine by the dry, cool desert environment in addition to very durable wood, and in the European trees by the anaerobic conditions existing in swamp and alluvial sediments. The importance of bristlecone pine is primarily in calibrating the radiocarbon time-scale, while the treeline species *Larix decidua* and *Picea abies* are affording insights into annual weather conditions in the European Alps. The *Quercus* sequences provide evidence of changing landscapes and river flow patterns as well as dating prehistoric dwellings and calibrating the radiocarbon record.

The reconstruction of past climates is potentially the most valuable and the ultimate goal in the majority of tree ring studies. Within those areas most suited to dendrochronology, it is the individuals close to the limits of the populations that are most responsive to some climatic variable by exhibiting greatest annual variability in tree ring thickness. The critical variable in higher latitudes and altitudes is generally summer temperature trends, while mean annual precipitation is more important in lower latitude, semi-arid regions.

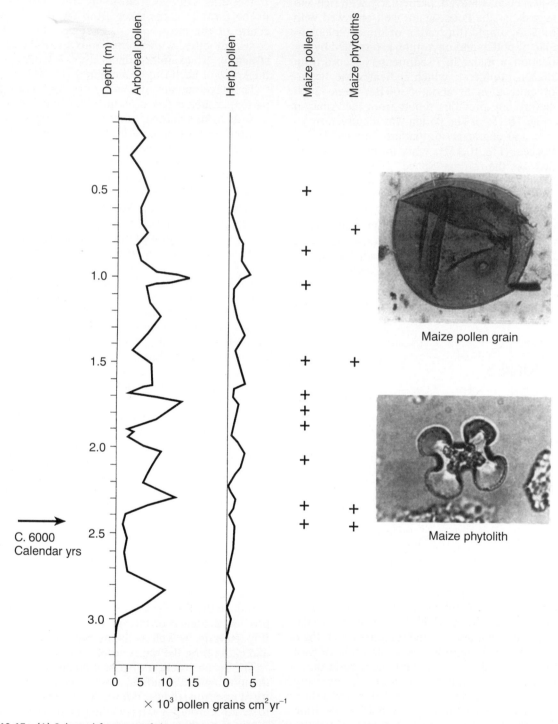

Fig. 10.15 (A) Selected features of the pollen and phytolith diagrams from Lake Ayauch, Equador. (B) and (C) Fossil pollen grain and phytolith of maize respectively. (From Bush *et al*. 1989)

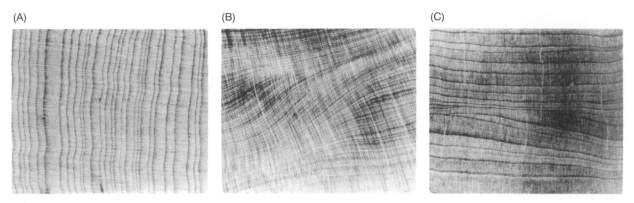

Fig. 10.16 Tree ring patterns of selected new Zealand conifers. (A) *Phyllocladus trichomanoides* showing extremely clear rings with high year-to-year variability: ideal for ring-counting, cross-dating and climatic reconstruction. (B) *Agathis australis* exhibiting clear rings but uneven radial or lobate growth. In extreme cases this effectively prevents cross-dating. (C) *Podocarpus totara* showing good ring definition but the wedging of individual rings, which provides problems in cross-dating. (From Dunwiddie 1979)

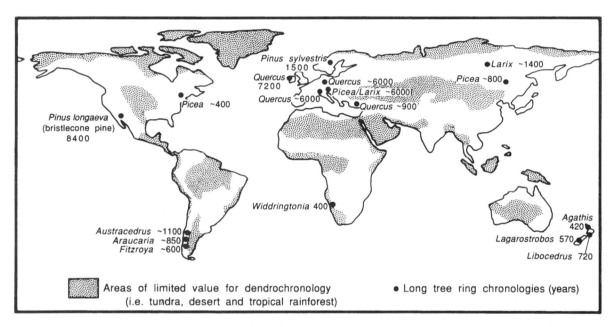

Fig. 10.17 Location of long tree ring chronologies in those areas of the globe most suitable for dendrochronology. (Modified from Schweingruber 1988)

Careful local site selection is also necessary to minimise non-climatic effects such as water-table fluctuations, competition, fire and disease. Ring widths also vary with tree age and this effect is routinely corrected by fitting a growth curve to each ring sequence and dividing each ring width by the corresponding value of the curve (Fritts 1976). Once a consistent chronology has been established from a number of individual sequences, this is compared against meteorological records and the relationships between tree ring variability and a particular climatic variable established. In ideal situations, the tree ring record can be calibrated from meteorological data and then used to provide quantitative climatic infor-

mation for periods beyond recorded history, or for areas lacking meteorological data.

In many cases it is difficult to extract a climatic signal from records, but this can lead to some useful insights into the operation of other environmental variables. One of the more important of these is forest fires which, as previously mentioned, are often difficult to isolate from charcoal preserved in sedimentary sequences. The tree ring record for ponderosa pine in northern Arizona provides a good example of the effect on trees of the adoption of a fire exclusion policy in the recent past (see Fig. 10.18). Before about 100 years ago, fires occurred approximately every 2–3 years but the vegetation has not

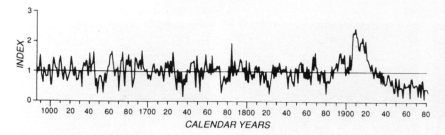

Fig. 10.18 Tree ring chronology for ponderosa pine in northern Arizona, US. (After Kennedy-Sutherland 1983, in Schweingruber 1988)

been burnt since. The trees responded to a period of high precipitation early this century by forming wide rings, but since that time growth rates have slowed substantially. This reduction in growth rates is explained by a number of factors resulting from fire exclusion, including competition for water from the many young individuals whose survival has been facilitated by the lack of fires, and a shortage of nutrients which are locked up in the very slowly decomposing litter instead of being rapidly recycled by frequent fires. This example serves to reinforce the historical record and strongly suggests that the fire regime operating prior to 100 years ago had been in operation for at least the last few hundred years. However, it does not reveal direct evidence for fire. In some trees, though, the presence of fire scars provides a more certain fire record. This is the case within the Australian treeline species *Eucalyptus pauciflora* (snowgum) where fire scars have been dated by ring counting (see Fig. 10.19). Here, the frequency of fires is shown to have increased from the later part of the 19th century with the advent of grazing by Europeans, and then decreased in the later part of this century with the establishment of a fire exclusion policy. The effect of European burning has changed the landscape in many areas from an open woodland of old trees to a denser scrub composed of young individuals.

VERTEBRATES

The preservation of bones and teeth in such sedimentary environments as semi-arid dunes, non-acidic lake margins and swamps, and particularly caves and rock shelters, has allowed the construction of fossil vertebrate assemblages from a wide variety of geographical areas. Although contributing significantly to a knowledge of Quaternary environments in general, the value of vertebrates for determination of detailed climatic and habitat change has been limited. This is because of a number of factors including relatively rapid rates of evolution, the nature or taphonomy of the assemblages, and problems in

identification of critical taxa and their ecologies (Behrensmeyer and Kidwell 1985; Baird 1991).

In comparison with plants and invertebrates, evolutionary rates including extinctions in many vertebrates appear to have been relatively rapid during the Quaternary. This has been of advantage in the construction of stratigraphies for terrestrial environments, but has effectively limited the use of modern analogues for interpreting fossil assemblages and for environmental comparisons from fossil assemblages of different ages. The clearest example is provided by the large mammals that grew in size through the Quaternary until suffering massive extinction towards the end of the Pleistocene period. The actual cause of these megafaunal extinctions is hotly debated and discussion of it is most appropriately left until later (see Chapter 11), when all relevant lines of evidence pertinent to this issue have been presented.

Despite the problems of evolutionary change in larger animals, one method of palaeoclimatic estimation has been devised that is based on morphological changes within large animals, particularly carnivores, which survived the extinction phase. This is based on the fact that body size for particular species tends to increase with decreasing temperature in order to help conserve energy. The explanation is provided by Bergmann's Rule, that as an animal increases in size, its volume (responsible for heat production) grows more rapidly than its surface area (responsible for heat dissipation) (Klein 1986). The application of measurements on present-day populations to glacial assemblages based on teeth, the best-preserved parts of the skeleton, has supported temperature-lowering estimates derived from other lines of evidence.

Taphonomy, which in a broad sense includes the processes involved in the production of a fossil assemblage, has been of major concern to interpretation of assemblages of small vertebrates. The majority of sites studied are caves or rock shelters where the composition of assemblages depends very much on how the remains got there – whether the animals lived in the shelters or were brought in by animal predators or people. Resolution of the origin of the faunas is not difficult with a knowledge of the ecology of the fossil vertebrates and of the eating habits

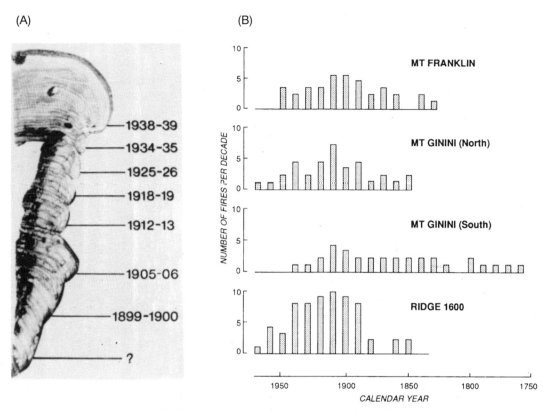

Fig. 10.19 (A) Section through a trunk of *Eucalyptus pauciflora* (snowgum) from the mountains of southeastern Australia showing annual growth rings and fires scars that were formed in particular years (from Clark 1981). (B) Fire frequencies determined from these fire scars, shown as number of fires per decade for a number of snowgum populations (from Adamson and Fox 1982). All information derived from the study of John Banks, Australian National University

of the predators, which can be determined from the state of the bones. The important point is that there are many different taphonomic pathways and only similar ones can be compared usefully.

A good illustration of the complexity and value of small mammal studies is provided by the detailed analysis of Dolina Cave in north-central Spain. The site is part of the Atapurca-Ibeas cave complex noted for its long sequence of human occupation. Dolina Cave itself recently revealed evidence of very early (Early Pleistocene) human occupation and the cave sediments straddle the Brunhes/Matuyama boundary. The small mammal remains have been investigated to help provide the environmental background to human occupation and lifestyle.

A summary of the sedimentary sequence and overview of results derived from it are shown in Fig. 10.20. There are remains of a large number of small mammals, most of which do not occupy caves and have been brought into the site by bird predators. The category of predator is determined by the degree and nature of digestion indicated by the bone remains of the prey, and varies from little modification (1) to extreme corrosion and breakage (5), while additional

features, including the composition of the bone assemblages which relates to the hunting strategy (range and prey preference) of the predator, have allowed the actual predator species to be determined. The nature of the predator and degree of predation have been determined, to some degree, by changing cave conditions. Initially there was indirect access to the cave and predator activity is evident through levels 3 to 6. As sediments accumulated this access was reduced and presumably no predator access was possible through the period represented by level 8. Subsequent roof collapse allowed direct entry into the cave in levels 10 and 11. The almost total lack of remains in some levels can be explained in various ways such as rapid infilling at the base and in level 6, the blockage of the cave outlet in level 9, and sub-aerial weathering towards the top of level 11. The interesting conclusion to the study is that the regional vegetation and climate remained fairly constant during the whole recorded period and that practically all variation in the data can be accounted for by taphonomic processes. The reconstructed dominance of temperate deciduous woodland is supported by studies of pollen, large mammals, birds and reptiles.

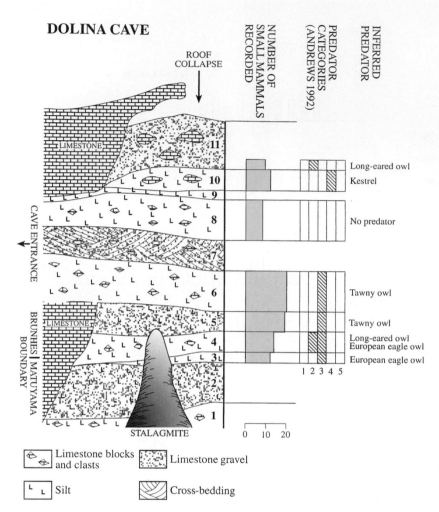

Fig. 10.20 Section through archaeological and palaeontological cave site sediments, Dolina, northern Spain, and summary information on small mammal species and their inferred predators recorded in fossiliferous layers. (Adapted from Fernandez-Jalvo 1996)

INVERTEBRATES

The majority of invertebrates used in palaeoenvironmental reconstruction live in aquatic environments (see Chapter 8 for details.) A notable exception comprises Coleoptera or beetles. This group contains many terrestrial members that possess a variety of features ideal for providing evidence of past climates (Coope 1970). They are taxonomically well-known, they exist in a whole variety of environments, and are preserved in abundance and diversity in a range of sedimentary deposits. In addition they appear to have been morphologically stable through much of the Quaternary. Their real importance, though, lies in their great mobility. Unlike plants which are generally long-lived or whose movements can be inhibited by unsuitable vegetation structure or soil conditions, and many animals that are 'tethered' to particular vegetation types, beetles can move rapidly and over long distances in response to changes in climate.

Their sensitivity to climate change in comparison with plants is well illustrated in a study from the middle part of the last glacial period in the British Isles (see Fig. 10.21). Here, there was a short phase, around 43 ka ago, dominated by thermophilous beetles which are now confined to southern Euro-Asia, wedged between a cool, moist arctic climatic phase characterised by beetles now largely found in arctic and subarctic regions, and a cool continental phase containing east-central Asian fauna. It is estimated that temperatures must have achieved levels as high as, if not higher than, those of today during the short period of amelioration, yet there is no indication of this in pollen data from this period. The preferred explanation for the lack of response in pollen assemblages is that the phase was too short for trees to be able to migrate into the area from their glacial retreats, and that there is no easy means of separation, on palynological grounds, of true tundra and higher temperature treeless vegetation.

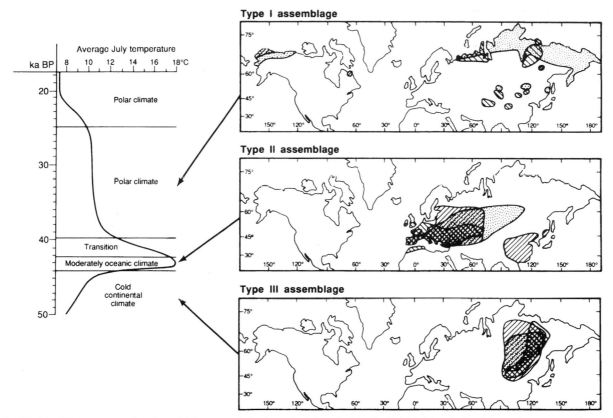

Fig. 10.21 Climate curve for the middle part of the last glacial period in the British Isles derived from the present-day distributions of fossil beetle assemblages. The maps show the modern distributions of selected species characteristic of three major periods recognised. (Adapted from Coope 1975)

A GLOBAL SYNTHESIS

So far in this chapter, information on Quaternary environments has been largely restricted to examples illustrating the different sources of evidence. Here an attempt is made to construct a broad regional picture particularly from studies in vegetation history.

Present-day distributions

Present-day distributions of plants and animals are largely a product of Quaternary environmental changes. They also constitute the starting point for the reconstruction and interpretation of fossil assemblages. Some indication of the distribution of plants at least can be gained from an examination of major vegetation types of the world shown in Fig. 10.22. Despite significant human modification of the landscape, the major vegetation types largely reflect present climatic variation. From the equator to the poles, where moisture is not limiting, there is a general gradient from tropical rainforest where temperatures are sufficiently high for year-round growth, through broad-leaf forest, where growth can be limited during the winter, and coniferous forest, where growth is confined to only a few months of the year, to tundra, where the growing season is too short to allow the survival of trees. This gradient is most complete in the land-dominated northern hemisphere. A similar gradient is evident in tropical montane areas, although this becomes progressively truncated with increased latitude. Major moisture gradients are evident in the tropics and subtropics where evergreen rainforest gives way to semi-deciduous forest, shrubland and grass-dominated savanna with lower and more seasonal rainfall, and in the Northern Hemisphere mid-latitudes where broadleaf forest is replaced by herb-dominated prairie and steppe in continental interiors. Where moisture is most limiting, desert communities occur. The remaining vegetation type identified at this scale is Mediterranean shrubland that occurs in warm temperate latitudes which have a marked winter rainfall maximum.

Superimposed on this structural pattern of vegetation variation is the taxonomic composition of the flora, which is a product of evolutionary history as

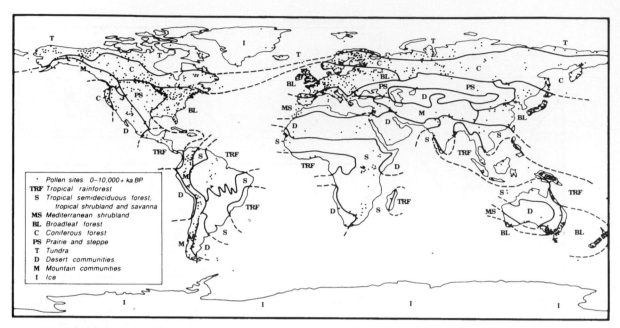

Fig. 10.22 The distribution of known palynological sites covering at least the Holocene in relation to major global vegetation types

well as present-day environmental conditions. Most evident are floristic differences between northern and southern hemisphere mid- and high-latitude forests which most clearly reflect origins on the separate ancient landmasses of Laurasia and Gondwana respectively. Subsequent evolutionary development and floristic mixing as a consequence of continental movements and its climatic implications as well as the Quaternary climatic fluctuations have led to further global differentiation of floristic patterns, particularly with woody plants. Dispersal of herbs has been generally effective so that differences at the taxonomic level identifiable from fossil data are less pronounced from one region to another.

Modern pollen spectra

The recognition of past vegetation is largely based on pollen signatures derived from modern pollen samples taken within identified communities. Despite the diversity of plants within many communities, most pollen samples are composed predominantly of only a few taxa. Fortunately for vegetation and climatic reconstruction, these are generally the easily recognised and well-known canopy dominants that most clearly reflect macroclimatic variation. The major exceptions are the floristically very diverse tropical rainforests, where the majority of trees are animal pollinated and little pollen disperses on to accumulating swamp and lake surfaces, and predominantly herbaceous vegetation where pollen spectra can contain a substantial component from surrounding forest

vegetation. However, in these areas, the total pollen spectra generally allow identification of existing vegetation. A selection of the most commonly occurring pollen types and the vegetation types in which they occur is shown in Fig. 10.23. These would probably allow the construction of broad vegetation histories from most parts of the world.

The fossil database

Although not claiming to be complete by any means, the known distribution of pollen records extending back from the present to at least the early Holocene (see Fig. 10.22) gives some impression of the extent of this kind of research within different environments. It is clear that there is relatively little information from many parts of the world. This is due to a number of factors including a lack of suitable sites for continuous sediment accumulation or pollen preservation in arid or seasonally dry environments, perceived difficulties in the application of pollen analysis to lowland tropical rainforests, and a lack of local interest or expertise. Sites are concentrated in traditional areas of research, particularly North America and Europe, where glacial erosion and deposition has resulted in the formation of numerous very suitable sites for investigation. The number and distribution of sites extending back to the last glacial period is very different (see Fig. 10.24). Many fewer sites provide a record of this period because conditions were generally much drier in low and middle latitudes, while the high latitudes of the northern hemi-

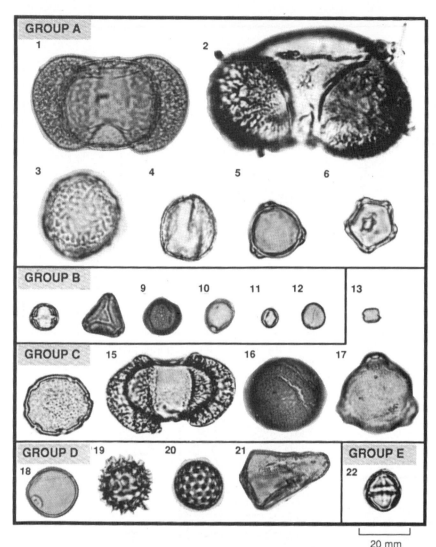

Fig. 10.23 A selection of the major pollen types recorded in Quaternary records. The identified groups are characteristic of environments in different parts of the globe. Group A, northern hemisphere mid-latitude forests: 1, *Pinus*; 2, *Picea*; 3, *Ulmus*; 4, *Quercus*; 5, *Betula*; 6, *Alnus*. Group B, southern hemisphere mid-latitude forests: 7, *Macaranga*; 8, *Myrtaceae*; 9, *Olea*; 10, *Celtis*; 11, *Elaeocarpus*; 12, *Moraceae*. Group C, tropical rainforests: 13, *Cunoniaceae*; 14, *Nothofagus*; 15, *Podocarpus*; 16, *Araucariaceae*; 17, *Casuarinaceae*. Group D, dry or cool herbaceous vegetation: 18, *Poaceae*; 19, *Asteraceae*; 20, *Chenopodiac*; 21, *Cyperaceae*. Group E, mangroves: 22, *Rhizophoraceae*

sphere were covered in ice. The highest frequencies of sites are found in the middle latitudes of both hemispheres and in the montane tropics. Only a handful of records from terrestrial environments cover the range of conditions experienced through at least one glacial cycle (Fig. 10.24). Of these, the longest records are from large subsiding lake basins and several of these, including those from the high plains of Bogota in South America, the Jordan–Dead Sea Rift Valley, Lake Biwa in Japan and Lake George in Australia provide reasonably continuous records through the whole of the Quaternary. Other substantial records, particularly from Europe, are not illustrated because they do not extend to the present day. Because of the limited number of sediment-accumulating environments on land, there is increasing interest in the production of marine palynological records which generally provide broad regional pictures of adjacent land vegetation.

The Late Tertiary/Quaternary transition

There is good evidence from many parts of the world of increasing replacement of forest by herbaceous vegetation through the latter part of the Tertiary period in response to a general trend towards drier and cooler conditions. This trend also led to the evolution of diverse assemblages of animals able to take advantage of expanding grazing land. This culminated in the megafauna of the Quaternary. The cooler conditions at higher latitudes, resulting from a steepening of the latitudinal temperature gradient (see Chapter 2) together with increased climatic variability accompanying the oscillatory movements of the developing ice sheets, caused a contraction of warmer elements towards the present tropical zone. Changes appear to have taken place at different rates and at different times depending on the attainment of

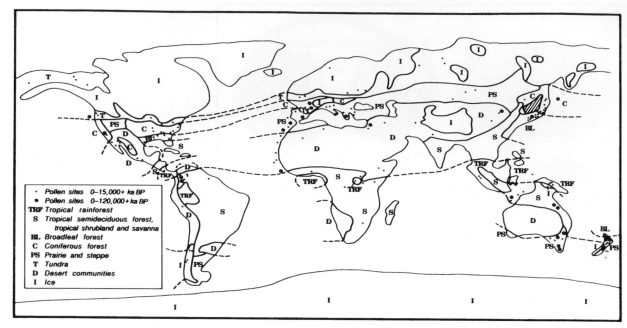

Fig. 10.24 The distribution of known palynological sites extending to the height of the last glacial period and those covering at least the last glacial–interglacial cycle in relation to inferred vegetation at the last glacial maximum. The vegetation distributions are based largely on the reconstruction of R.M. Adams *et al.* (1990)

critical threshold levels, although the data for this period are rather patchy and the dating often uncertain.

The most reliable information for this period is derived from the more continuous pollen records that can be tied into the palaeomagnetic time-scale. A number of these, particularly from Europe, indicate substantial changes around the proposed Pliocene/ Pleistocene boundary about 2.4 to 2.6 ma ago. Here a number of trees common in the Tertiary disappear with the establishment for the first time of tundra and steppe-like vegetation, indicating the first pronounced cold period. This is well illustrated by the Meinweg section from the Netherlands (see Fig. 10.25). Around the same time, dry periods are recorded in pollen records from the Mediterranean region (Suc and Zagwijn 1983) and in northwest Africa (Leroy and Dupont 1994), while Lake George in Australia registers the last appearance of rainforest within the region, a feature which appears to correspond with a change from a summer to winter rainfall regime (McEwen-Mason 1991). In the Colombian Andes, the record demonstrates a change to lower frequency and higher amplitude climatic fluctuations. Some sustained changes in the nature of the vegetation communities also occur at this time and subsequently, but these may have been more associated with the introduction of northern hemisphere elements into the flora with the closure of the Isthmus of Panama in the Pliocene than with actual climatic change (Hooghiemstra 1989).

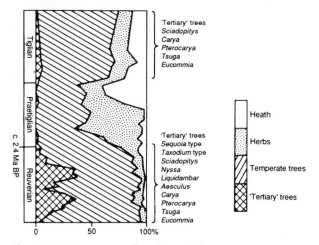

Fig. 10.25 Summary pollen diagram from Meinweg, showing the effect of the first major cool period around 2.4 million years ago on the impoverishment of the tree flora of Europe. (Modified from Zagwijn 1957)

Within Europe, the progressive disappearance of warmth-loving taxa continued with each cold period until the late Pleistocene. Many of these taxa can still be found at equivalent latitudes in North America and Asia, suggesting that temperature oscillations alone were not responsible for their demise. Instead, the major reason is considered to be the existence of east–west mountain barriers which prevented their southward retreat and consequent survival in suitable environments during the glacials.

Glacial/Interglacial cycles

There is evidence for cyclical fluctuations through the whole of the Quaternary period, but apparent differences between records and generally poor dating control have prohibited detailed correlation prior to the mid Pleistocene. Since this time the consistent low-frequency, high-amplitude nature of the cycles has allowed more realistic comparisons to be made between terrestrial sequences and general correlations with the established marine oxygen isotope stratigraphy. By this stage also the vegetation appears to have become adjusted to these cyclical oscillations, providing consistent climatic estimates from the pollen data. Summary diagrams for four terrestrial sequences are compared with the deep-sea core record in Fig. 10.26.

It is interesting to note that there are major similarities between these records despite the fact that the pollen components are very different and they have been interpreted with respect to different climatic variables. Lake Biwa and Funza are considered to reflect predominantly changes in temperature, Hula Basin moisture, and Lake George a combination of moisture and temperature. From these and other records it is evident that forests generally expanded during interglacial periods as a result of higher tem-peratures at higher latitudes and altitudes and an associated increase in effective precipitation at lower latitudes. The Hula Basin record is an exception and there are several parts of the world where driest conditions occurred during interglacials.

The Lake George core illustrates the changing role of fire within the recorded period. Through much of the sequence, charcoal peaks correspond with interglacials when it is considered that the vegetation, composed predominantly of sclerophyll forest or woodland, was more able to carry fire than during the cool dry glacials which supported only an open herbaceous vegetation. This burning pattern is used to infer the status of periods corresponding with deep-sea core stages 11 and 12 where pollen is not preserved. However, the pattern changes from the period considered to represent stage 5 (the last interglacial) to one of more intense and continuous burning within both glacial and interglacial periods. The major effect on the vegetation was to make it more fire-prone with the replacement of relatively fire-sensitive *Casuarina*, which had previously dominated the forests, by fire-promoting *Eucalyptus* and myrtaceous shrubs.

A more detailed picture of changing patterns of vegetation and burning within the last two glacial/

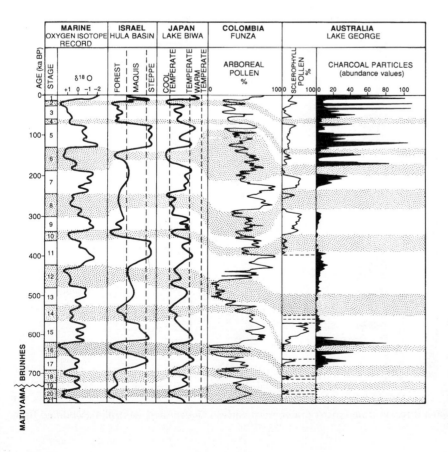

Fig. 10.26 Comparison of four Quaternary records – Hula basin, Israel and Lake Biwa, Japan (Fuji and Horowitz 1989), Funza, Colombian Andes (Hooghiemstra 1988), and Lake George, Australia (Singh and Geissler 1985) – in relation to the marine oxygen isotope record (Imbrie *et al.* 1984) through the last 800 000 years

interglacial cycles is provided by another record from Australia: Lynch's Crater (see Fig. 10.27). This site is a volcanic crater swamp situated within the largest expanse of rainforest in northeastern Queensland. Within the record, complex rainforest, defined by high levels of pollen from angiosperms, particularly *Cunoniaceae* and *Elaeocarpus*, dominates during the wetter interglacials. This alternates with the drier vegetation types Araucarian rainforest and sclerophyll woodland, identified by high percentages of the gymnosperms *Araucaria* and *Podocarpus*, and *Casuarina* and *Eucalyptus*, respectively, which become prominent during glacial periods.

Transitions between glacials and interglacials appear to be abrupt, and in the earlier part of the record they are accompanied by minor peaks in charcoal. It could be the case that the vegetation put

under stress from changing climate is more prone to fire and that burning accelerates the change to a vegetation type more in balance with the new climatic conditions. This scenario is similar to that proposed for early Holocene vegetation changes around Everitt Lake, as indicated in an earlier section (see also Fig. 10.12).

The general pattern of vegetation variation is altered between about 38 000 and 26 000 years ago, towards the end of the last glacial period, with the total replacement of Araucarian forest by sclerophyll vegetation. Although this corresponds with the beginning of the last glacial maximum, it is considered that climate alone would not have been the cause, since no similar change occurred towards the end of the penultimate glacial period. In addition, the gradual nature of the transition stands in marked contrast

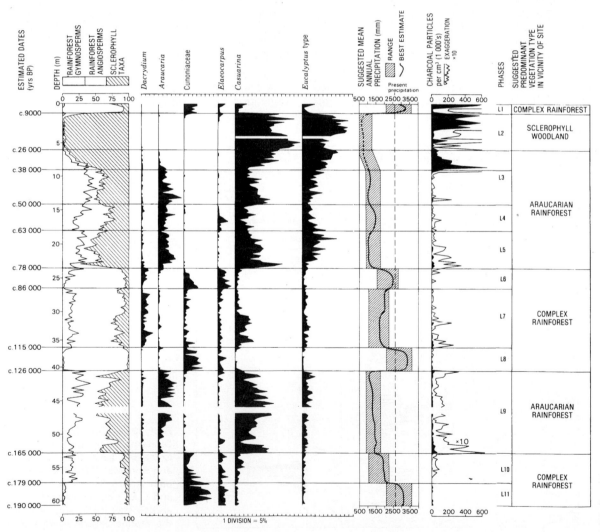

Fig. 10.27 Major features of the pollen diagram from Lynch's Crater, northeastern Queensland, Australia. (Modified from Kershaw 1986)

to other, more obviously climate-induced changes in the record. The sharp increase in charcoal abundance at this time points to fire being the critical factor, and this is supported by the change from fire-insensitive rainforest to fire-resistant sclerophyll woodland. This appears to have been a broad regional change as moist Araucarian rainforest now has a very restricted distribution, despite the Holocene expansion of complex rainforest. One important component, the conifer *Dacrydium*, is no longer present at all on the Australian mainland.

At both Lake George and Lynch's Crater, increased burning during the late Pleistocene has been attributed to the activities of Aboriginal people. This is despite the antiquity of the Lake George evidence for burning, well beyond the earliest dated evidence for the arrival of people. However, there is some controversy over the dating of the Lake George record, and the earliest date for Aboriginal presence is still being pushed further back in time. Despite these uncertainties, some explanation for major sustained changes in the vegetation of Australia, which have not been recognised elsewhere in the world, is required.

The development of the present vegetation pattern

The most intensively studied period of the Quaternary is that from the height of the last glacial period, about 18 000 years ago, to present. This is the period for which most sequences are available, dating is most exact, and which is of greatest relevance for understanding present-day patterns and processes. It is also the most complex in that the landscape has been substantially influenced by human activities. However, on a broad regional scale, climate is still considered to have been the major determinant on vegetation, even in the much-modified landscape of Europe (Huntley 1990).

At the height of the last glacial period when, on a global scale, both temperatures and precipitation reached their lowest levels, terrestrial ice-free landscapes were dominated by essentially treeless tundra and prairie-steppe at high and mid-latitudes, and open savanna woodlands and grasslands at low latitudes. Forest vegetation had a very restricted distribution, and although small areas of the major forest types are shown on Fig. 10.24, their extent in many places is inferred. Where evidence for forest types does exist, they bore only superficial resemblance to those existing today.

One major area of interest in Quaternary investigations is where and how the components of present-day forest types survived the last and previous glacial periods. With a dearth of fossil evidence, most discussion has centred on likely retreat areas or refugia,

constructed from present-day landscape features, particularly the modern distributions of plants and animals. Debate has been most intense over the situation in the Amazon region of South America where the identification of possible refugia has been applied not only to tropical rainforest survival but also to the explanation of high species diversity and endemism within tropical rainforest systems. In the refuge theory as originally envisaged by Haffer (1969), rainforest survived in areas now possessing high diversity and endemism within the great expanse of Amazonian rainforest. Inferred refugia determined largely from the overlapping centres of endemism for three rainforest taxonomic groups, butterflies, plants and birds, are shown in Fig. 10.28. Many of these refugia are in highland areas where it is thought that sufficiently moist conditions were maintained during the glacial periods to allow rainforest survival while savanna vegetation occupied the lowlands. Speciation would then be enhanced by geographical isolation during these glacial periods, and also by subsequent hybridisation during interglacials along contacts where forests from expanding refugia met (see Fig. 10.29).

However, fossil evidence suggests that this is probably an unlikely model for explaining patterns of diversity within rainforests (Bush and Colinvaux 1990). In the first place, palynological data indicate that mountain areas in Amazonia and elsewhere in the tropics suffered a depression of temperatures by some 4–6°C causing montane forests to descend to the altitudes considered to contain the lowland rainforest refugia. Consequently, although moisture levels are likely to have been suitable, at least some components of the rainforest would have been restricted to the lowlands as a result of low temperatures.

More recent palynological results from the major wet tropical regions of South America (Colinvaux *et al.* 1996a, b), Africa (Maley 1996), and southeast Asia (Sander van der Kaars, pers. comm.) have now established that rainfall reduction was either not sufficient or not regionally consistent enough to eliminate rainforest from many lowland areas. Some core 'refugial' areas of rainforest can be identified, but the nature of rainforest representation beyond these core areas is still a matter of speculation. In the wet tropics of Australia, refugial areas, postulated from concentrations of high plant diversity and primitive angiosperms, together with the pattern of rainfall lowering during the last glacial period determined from pollen records such as that from Lynch's Crater (Fig. 10.27), have been examined from the analysis of charcoal preserved within soil profiles (Fig. 10.30) All sites revealed charcoal dated to the last glacial period and was formed, almost exclusively, of *Eucalyptus*, indicating that fire had invaded even the wettest areas, resulting in the replacement

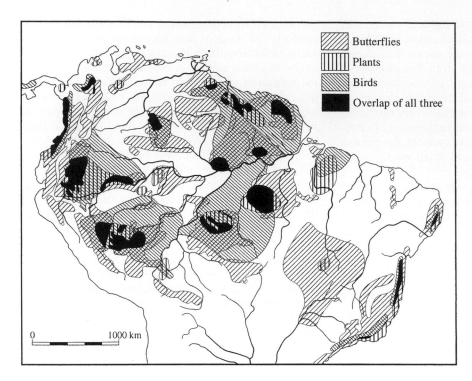

Fig. 10.28 Overlapping centres of endemism of butterflies, plants and birds in South American rainforests. (From Whitmore and Prance 1987)

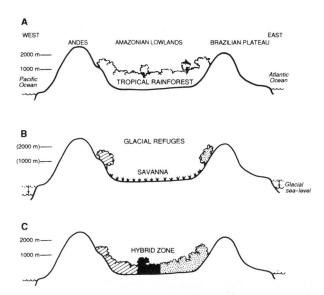

Fig. 10.29 Schematic diagram showing the major features of the refuge theory and modes of speciation.

(A) The preglacial situation: tropical rainforest covers the whole of the Amazonian lowlands to an altitude of approximately 1100 m above sea-level

(B) Glacial periods: the rainforest fragments and is restricted to isolated highland refugia where it experiences allopatric speciation

(C) Interglacial periods: rainforest expands from refugia with the possibility of hybridisation of related species occurring where refugial communities coalesce

of rainforest by open eucalypt woodland. It is proposed that rainforest must have survived within a network of corridors along streams and within other locally moist and/or fire-protected areas (Fig. 10.30). Such a pattern of survival for extra-tropical as well as tropical forests during the last glacial period is being increasingly recognised in many parts of the world.

It is difficult to determine the degree to which disturbance and association fragmentation resulted in genetic isolation of rainforest patches and the potential for evolution. However, it needs to be remembered that complex rainforests were much more extensive during the more stable Tertiary period, and regular disturbances or fluctuating climates may not have to be inferred to explain diversity in many rainforest components. Molecular studies on rainforest biota in north Queensland indicate that speciation rates have been slow during the Cenozoic period and that isolation for many millions of years is required for the production of new species (Moritz *et al.* 1997). These results are consistent with the conclusions of Bennett (1997, p.191) from a global analysis of vegetation that 'speciation is not a typical response to climatic change in the Quaternary since species persist for periods much longer than the periodicity of environmental change'.

Bush and Colinveaux (1990) also attack the basic concept of refugia on the grounds that it implies the existence of discrete communities that expand and contract as a whole, an idea which is no longer ten-

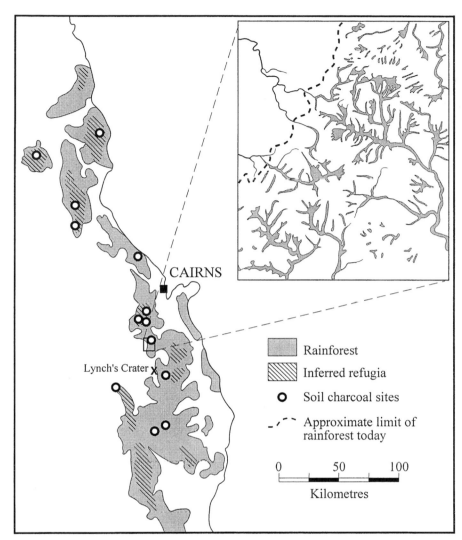

CAIRNS

Lynch's Crater

Rainforest

Inferred refugia

Soil charcoal sites

Approximate limit of rainforest today

0   50   100

Kilometres

Fig. 10.30 Distribution of soil charcoal sites of Hopkins *et al.* (1993) in relation to present extent and inferred major glacial refugia (Webb and Tracey 1981) of rainforest in the wet tropics region of northeastern Australia. Inset shows the likely nature of rainforest survival during the last glacial period based on results of soil charcoal analysis (Mike Hopkins and Andrew Graham pers. comm.)

able in the light of palynological and other evidence on the changing nature of plant communities. The most substantial evidence on long-term patterns of community change comes from analyses undertaken on pollen records since the last glacial, particularly those on the extensive data-sets in North America and Europe. Here the development of forests can be traced as the ice sheets waned and the climate ameliorated.

Davis (1976) plotted the rates of migration of individual trees that make up the present composition of mixed deciduous forests of eastern North America from the dates of arrival at each of the pollen sites. From the selected examples shown in Fig. 10.31, it is clear that each taxon has behaved individually, not as a member of an integrated community. Each appears to have had a different range during the glacial period and to have expanded at different rates since that time. In line with this, the composition of the vegetation has been changing continuously and there is little reason to believe that any kind of stability has been achieved with the present vegetation pattern.

The apparent lack of integrity of communities and differential migration rates have major implications for the reconstruction of climate from past vegetation. To what extent is the vegetation reflecting, or in balance with, prevailing climatic conditions at any point in time? It has already been suggested from the study of fossil insect assemblages that there may be a substantial lag in response of trees to climatic change. This could be caused by a number of factors including the limited dispersal capacity of seeds, competition from existing vegetation, and the unsuitability of environmental attributes apart from climate. Several studies have identified this latter factor as important. In an attempt to explain the curious late Holocene expansion of hemlock westwards within the last 5000 years (see Fig. 10.31), Davis *et al.*

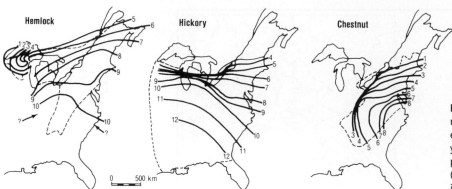

Fig. 10.31 Migration of selected mixed deciduous forest trees in eastern North America at 1000-year intervals from areas occupied during the last glacial period (from Davis 1976). Dashed lines indicate the present distributions

(1986) examined in detail a large number of sites within the Great Lakes region. It was concluded that this expansion was unlikely to have been the result of the climate becoming suitable; rather, it was most probably caused by problems experienced by hemlock in crossing Lake Michigan or xeric prairie around its southern margin that delayed expansion for up to 1000 years. In another example, a lag in colonisation of Britain by a species of birch after the last glacial period is attributed to the time required for the development of suitable soils on the glacial sediments (Pennington 1986). In a different environment, there appears to have been a lag of some 3000 years between the attainment of climatic conditions suitable for rainforest colonisation and its arrival at a site on the Atherton Tableland of northeast Queensland, after the sclerophyll woodland phase noted in the Lynch's Crater pollen diagram (Fig. 10.27). This delay was initially attributed to the slow intrinsic migration rate of rainforest from retreats occupied during the sclerophyll phase, but fine-resolution pollen and charcoal studies have indicated that fire probably inhibited the expansion process (Walker and Chen 1987).

In the same study, the mechanism of invasion by newly arrived tree taxa into established forest was investigated. From pollen influx data, it was determined that most populations increased exponentially, indicating little inhibition by other taxa to their expansion. Similar results were obtained from an examination of the expansion of deciduous forest taxa at Hockham Mere (Bennett 1983b). These studies suggest that competition from existing forest components may not seriously inhibit the spread of new species once climatic conditions become suitable for their establishment.

Although the problem of migration rates has not been fully resolved, there is growing confidence that, on time-scales of thousands of years, climate is the major determinant of plant distributions and that the complex and varying associations of plants through time is a reflection of continuously changing climate

(Webb 1986). This confidence has promoted attempts to refine palaeoclimatic estimates from pollen data. However, changing boundary conditions such as sea-level variation and the size of ice sheets, together with the growing realisation of the influence of Milankovitch forcing on climate, which demonstrates that each point in time will have a unique combination of climatic conditions at least within the late Quaternary (see Chapter 5), place uncertainty on the accuracy of palaeoclimatic estimates based on present vegetation/climate relationships.

Initial quantitative palaeoclimatic estimates used a transfer function approach similar to that developed for quantification of past sea-surface temperatures from assemblages of marine biota (see Chapter 7). The useful preliminary step is the construction of isopoll maps (contoured maps based on equal percentages) for individual pollen taxa at the present day, and the examination of the relationships between pollen abundances and climatic parameters. Those taxa showing climate-related variation are selected for the calculation of transfer functions which express the pollen–climate relationships. These transfer functions are then applied to the determination of climatic conditions from fossil pollen spectra, and climate curves are constructed for individual pollen diagrams. Selected results of this exercise on a summary pollen diagram from Elk Lake in Minnesota (Fig. 10.32A) using a subset of the American modern pollen database (Fig. 10.33) are shown in Fig. 10.34. The climatic variables illustrated – mean January temperature, mean July temperature and mean annual precipitation – represent the large-scale climatic controls of plant distribution.

In recognition of some of the uncertainties in palaeoclimatic estimation from pollen assemblages, there has been a move towards an analogue approach where the degree of similarity between modern and fossil assemblages can be more easily assessed (Overpeck *et al.* 1985; Guiot 1990). With this approach, each fossil spectrum is compared statistically against each modern spectrum in the

ELK LAKE, MINNESOTA

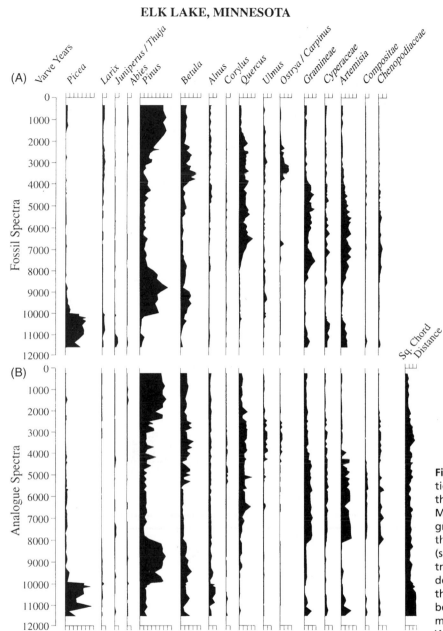

Fig. 10.32 Percentage representation of selected pollen taxa from (A) the pollen diagram from Elk Lake, Minnesota, and (B) the analogue diagram from the site constructed from the most similar modern spectrum (see Fig. 10.33) to each fossil spectrum. The right column shows the degree of similarity, as expressed by the squared chord distance value, between each fossil spectrum and the most similar modern spectrum. (From Whitlock *et al.* 1993)

data-set and allocated the climatic values of the closest match or calculated climatic values from a combination of matches.

Results for selected fossil spectra from the Elk Lake diagram (Fig. 10.32A), dated from the varve chronology for the lake, are shown in Fig. 10.33A–E. The site is from pine-hardwood forest in the northeast US. The oldest spectrum, 11 638 yr BP, which represents the early phase of deglaciation of the region, shows little similarity to any modern assemblage as indicated by the high squared-chord distance measure values. The closest matches are from

the interior of southern Canada where open *Picea* woodland exists under cold and relatively dry conditions. By 9584 yr BP, there is a pine-dominated assemblage with many more partial analogues. The majority of best matches are from upland areas to the southeast of the site and indicate that conditions had become much warmer and wetter. At 7232 yr BP, there were high grass and other herb values, and the best matches are from oak savanna and prairie to the west of the site existing under much drier conditions. Wetter conditions are suggested by the reduction in herbs at 3880 yr BP, with closest matches from mixed

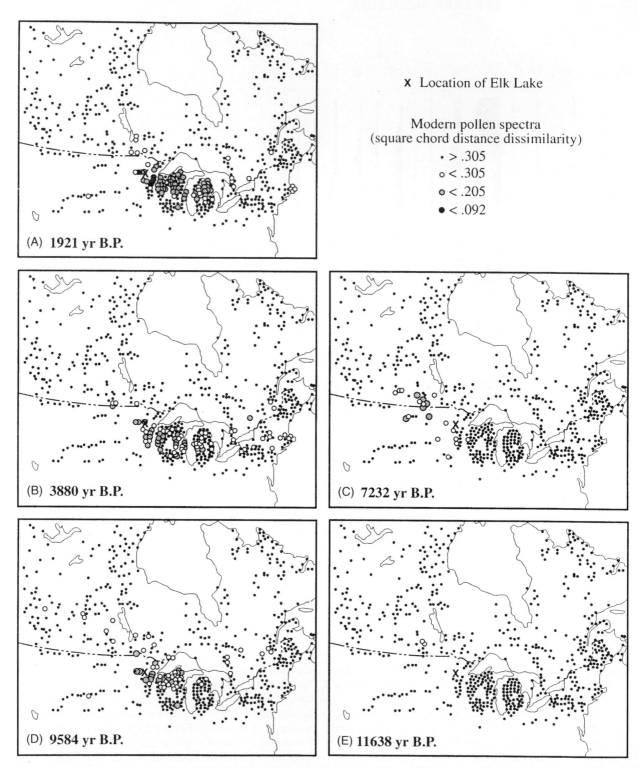

x Location of Elk Lake

Modern pollen spectra
(square chord distance dissimilarity)
- · > .305
- ○ < .305
- ◔ < .205
- ● < .092

(A) **1921 yr B.P.**

(B) **3880 yr B.P.**

(C) **7232 yr B.P.**

(D) **9584 yr B.P.**

(E) **11638 yr B.P.**

Fig. 10.33 Analogue maps showing the degree of similarity between modern pollen spectra in northeastern America with selected, dated pollen spectra from the Elk Lake pollen diagram. Increased size and shading of modern spectrum symbols indicate increased similarity with the spectrum from the diagram. The most similar modern spectra or best analogues, for all spectra, are used to refine interpretation of the pollen diagram in vegetation and climate terms. (Adapted from Whitlock *et al.* 1993)

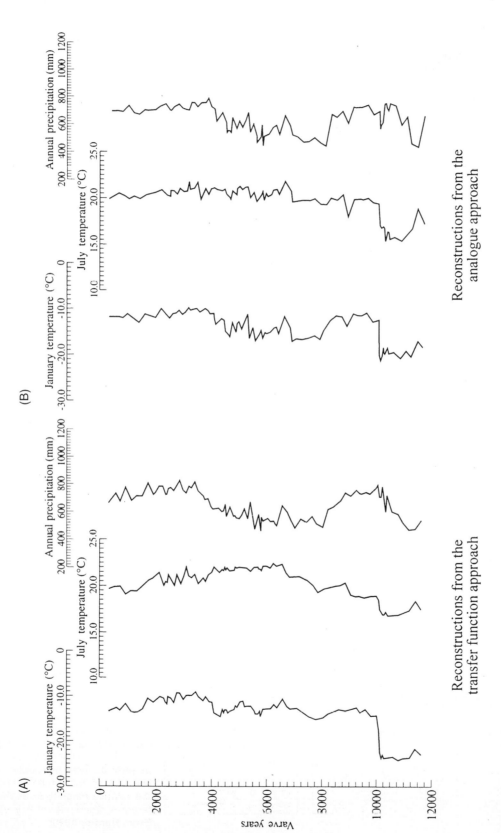

Fig. 10.34 Comparison of quantitative reconstructions for major climatic variables from the Elk Lake pollen diagram using the transfer function and analogue approaches. (From Bartlein and Whitlock 1993)

conifer-hardwood forests, while essentially modern conditions with the re-emergence of pine were established by 1921 yr BP.

Calculated climate values for all spectra are shown in Fig. 10.34B. These compare well with those from the transfer function approach (Fig. 10.34A). However, the synthetic diagram composed of the closest analogue spectra substituted for fossil spectra, and the squared-chord distance dissimilarity value of the closest analogue for each fossil spectrum (Fig. 10.32A), provide a clear visual impression of the nature and degree of difference between analogue and fossil spectra, and the caution that must be exercised in accepting the quantitative climate values. The general improvement in analogue matches

through time is seen in other such studies in the region and also in Europe, Africa and Australia, reinforcing the idea that modern communites have developed progressively since the last glacial period.

The application of isopoll mapping and numerical techniques to the large modern and fossil data-sets of North America and Europe is revealing substantial regional variation in late Quaternary climates which was previously either not suspected or difficult to substantiate from studies of restricted areas. A recent example from Europe, although expressing climatic conditions only in qualitative terms, illustrates the value of being able to extract information on seasonal climatic variation that was seldom possible with more traditional methods (Fig. 10.35).

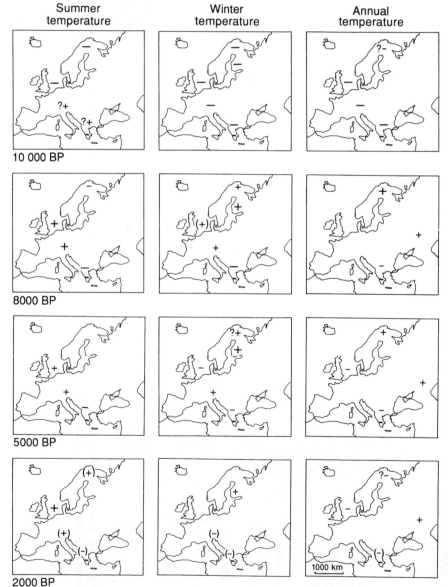

Fig. 10.35 Regional climatic changes during the Holocene of Europe derived from changing vegetation patterns deduced from pollen data. The size of the symbols indicates the relative magnitude of the differences from present-day values. (From Huntley 1990)

BIOME MODELS

The construction of extensive modern and fossil pollen data-sets from many parts of the world is placing palynological studies in a very useful position to improve predictive general circulation models. The importance of incorporating these data into the models is that vegetation not only responds to climate change, but through features such as its albedo, and transpiration processes, it substantially modifies temperature and precipitation. It is also an important carbon store influencing carbon dioxide levels in the atmosphere.

The use of taxonomic information is inhibitive to the reconstruction of global vegetation in that, as previously mentioned, there is significant variation in the distribution of taxa. For this reason there has been recent interest in the conversion of pollen data into structural or physiognomic vegetation types (*biomes*) which reflect climatic variation on a global scale and allow calculation of the amount of carbon stored in the vegetation biomass (e.g. Prentice *et al.* 1992). In the Prentice *et al.* (1992) system, conversion is achieved by the allocation of ecologically equivalent taxa to relevant plant functional types (e.g. boreal evergreen conifer, warm temperate sclerophyll shrub, steppe forb, grass, etc.) which have calculated ranges for key climatic parameters. Combinations of the dominant plant functional types, as determined by the major pollen types, then determine the biome type (e.g. tropical rainforest, temperate deciduous forest, tundra) of an assemblage.

The validity of the method has been established for Europe by comparing the results from modern pollen spectra with known patterns in the present-day vegetation, and by comparing a reconstruction for 6000 years ago with that from conventional pollen reconstructions (Prentice *et al.* 1996). The variation in carbon storage between the present and 6 ka has also been estimated (Peng *et al.* 1994). The use of plant functional characteristics alone has allowed realistic modelling of the present-day vegetation and demonstrated the potential of reconstructing past vegetation from very few pollen data-points. It is difficult to illustrate results from the method without resorting to colour. However, one ultimate aim is a refinement of the generalised vegetation distributions (or biomes) indicated for the present and 20 ka on Figs 10.22 and 10.24 respectively.

FURTHER READING

Bennett, K.D. 1997. *Evolution and Ecology: the Pace of Life.* Cambridge University Press, Cambridge.

Betancourt, J.L., Van Devender, T.R. and Martin, P.S. 1990. *Packrat Middens: the Last 40,000 Years of Biotic Change.* University of Arizona Press, Tuscon.

Birks, H.J.B. and Birks, H.H. 1981. *Quaternary Palaeoecology.* Edward Arnold, London.

Bradley, R.S. 1985. *Quaternary Palaeoclimatology.* Allen and Unwin, MA.

Clark, J.S., Cachier, H., Goldammer, J.G. and Stocks, B. (eds) 1997. *Sediment Records of Biomass Burning and Global Change.* Springer, Berlin.

Elias, S.A. 1994. *Quaternary Insects and their Environments.* Smithsonian Institution, Washington, DC.

Faegri, K., Kaland, P.E. and Krzywinski, K. 1989. *Textbook of Pollen Analysis.* (4th edn). Wiley, Chichester.

Fritts, H.C. 1976. *Tree Rings and Climate.* Academic Press, London.

Huntley, B. and Webb, T. III 1988. *Vegetation History.* Kluwer, Dordrecht.

Moore, P.D., Webb, J.A. and Collinson, M.E. 1991. *Pollen Analysis* (2nd edn). Blackwell, Oxford.

Piperno, D.R. 1987. *Phytolith Analysis: an Archaeological and Geological Perspective.* Academic Press, San Diego, CA.

Schweingruber, F.H. 1988. *Tree Rings: Basics and Applications of Dendrochronology.* Dordrecht, Reidel.

CHAPTER

11

HUMAN ORIGINS, INNOVATIONS AND MIGRATIONS

Knowledge may have its purposes, but guessing is always more fun than knowing.
W.H. Auden (1907–1974)
Archaeology

In earlier chapters we have considered some of the possible causes and consequences of the waxing and waning of the great Quaternary ice sheets. We have noted the impact of the associated fluctuations in sea-level upon the periodic emergence and submergence of shallow continental shelves and land bridges. We have also examined the impact of global climatic fluctuations linked to the growth and decay of the ice sheets upon the expansion and contraction of forests and deserts, upon river and lake behaviour, and upon the response of ocean waters to fluctuations in water temperature and salinity. In this chapter we discuss some of the interactions between evolving prehistoric human societies and their ever-changing local and regional habitats, starting with the remarkable fossil evidence for early human origins in Africa. We then trace the progressive migrations of small bands of hunter–gatherers out of Africa and into Eurasia, Australia and the Americas, and reconstruct some of the technical and cultural innovations which ultimately gave rise to plant and animal domestication and so to the great urban civilisations of the present-day. The following chapter deals with some of the regional impacts of the more recent and better-dated Quaternary climatic fluctuations of the last 20 ka, and the final chapter considers the global repercussions of accelerating human impact upon our biosphere, atmosphere and hydrosphere.

MIOCENE HOMINOIDS OF AFRICA AND EURASIA

Throughout most of the Miocene epoch, for nearly 20 million years, a number of quadrupedal, vegetarian hominoids roamed the forests and woodlands of Africa and Eurasia (Laporte and Zihlman 1983; Hill and Ward 1988; Alpagut *et al.* 1996; Moyà-Solà and

Köhler 1996). (The primate superfamily *Hominoidea* includes all modern and ancestral *Pongidae* and *Hominidae;* present-day pongids comprise the gorillas, chimpanzees and orang-utangs). The hominoids were cosmopolitan in their distribution (Fig. 11.1), suggesting that there were very few major geographical barriers inhibiting their ability to range freely across this vast region at that time. Among the better known of these Miocene hominoids was the large, forest-dwelling genus *Gigantopithecus*, and the smaller genus *Ramapithecus*, which appears to have foraged successfully across the ecotone between forest and grassland, at least during the later Miocene. The Miocene hominoids disappeared from the fossil record about 7 Ma ago, and there is a tantalising gap of several million years (Klein 1989) before the first appearance of the Pliocene hominids in Africa between 5 and 4 Ma ago (White *et al.* 1994; Woldegabriel *et al.* 1994; Leakey *et al.* 1995).

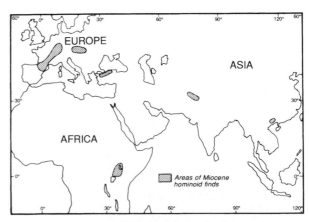

Fig. 11.1 Distribution of Miocene hominoid fossil discoveries in Africa, Asia and Europe. (After Scarre *et al.* 1988)

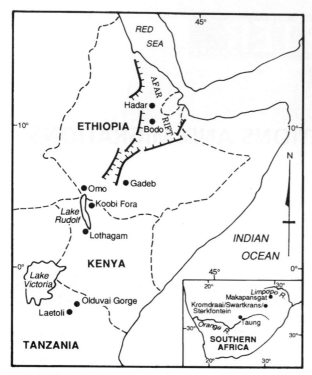

Fig. 11.2 Location of major hominid sites in Africa. Except for Gadeb (an Early Stone Age site in the Ethiopian uplands with possible early use of fire by *Homo erectus*) all the sites shown have yielded Australopithecine fossils ranging in age from 4 Ma to 1 Ma

PLIOCENE HOMINIDS OF AFRICA

Only in Africa have Pliocene hominids been found so far (White *et al.* 1981; Harris 1983; Johanson 1989). Claims for Pliocene hominids in Java remain unsubstantiated (Kramer 1994). The best-dated and most abundant Pliocene hominid fossils have come from the rift deposits of Ethiopia (the Afar Depression), Kenya (especially Koobi Fora near Lake Turkana) and Tanzania (most notably, Olduvai Gorge and Laetoli), but important early discoveries have also come from certain limestone caves in southern Africa (Fig. 11.2) as well as from the Bahr el Ghazal region of central Chad (White and Harris 1977; Walker *et al.* 1986; Johanson *et al.* 1987; Harris *et al.* 1988; Feibel *et al.* 1989; Tobias 1994; White *et al.* 1994; Leakey *et al.* 1995; Brunet *et al.* 1995; Tattersall 1995).

In connection with these hominid discoveries from Africa, it is interesting to call to mind some of Charles Darwin's comments about early human origins. In *The Descent of Man*, Darwin (1871, p. 520) wrote as follows:

In each great region of the world the living mammals are closely related to the extinct species of the same region. It is therefore probable that Africa was formerly inhabited by extinct apes closely allied to the gorilla and chimpanzee: and as these two species are now man's nearest allies, it is somewhat more probable that our early progenitors lived on the African continent than elsewhere.

Referring to the absence of fossil evidence in support of this suggestion, Darwin went on to say that 'those regions which are the mostly likely to afford remains connecting man with some extinct, ape-like creature, have not as yet been searched by geologists' (p. 521). He concluded by nailing his scientific colours firmly to the mast:

It has often and confidently been asserted that man's origin can never be known: but ignorance more frequently begets confidence than does knowledge: it is those who know little, and not those who know much, who so positively assert that this or that problem will never be solved by science.

A century later, Darwin's prediction has been amply vindicated, first with Raymond Dart's discovery of *Australopithecus africanus* at Taung in South Africa in 1924, then with Robert Broom's discovery of *Australopithecus robustus* in 1938, from the South African cave site of Kromdraai, then with the spectacular finds by Mary and Louis Leakey from Olduvai Gorge, which span nearly two million years of human evolution, and more recently with even older fossil discoveries from the Afar Rift of Ethiopia, from Laetoli in Tanzania, and from the region around Lake Turkana in northern Kenya (White 1984; Gowlett 1984a; Weaver 1985; Klein 1989; Leakey and Lewin 1992; Tobias 1994; Fagan 1995; Morell 1995; Suwa *et al.* 1997).

Recent advances in molecular biology are helping to clarify the vexed question of when the pongid and hominid lineages began to diverge (Tuttle 1988; Thomas 1993). Similarities in immunological response and in plasma protein structures point to divergence between the pongid and hominid families somewhere between 6 and 3 Ma ago (Sarich and Wilson 1967), an inference entirely consistent with the evidence from fossil hominids discovered in East Africa during the past 20 years.

In November 1976, Tom Gray, Don Johanson and their Ethiopian colleagues discovered the fossil remains of a small-brained bipedal hominid, *Australopithecus afarensis,* at Hadar in Ethiopia (Johanson and White 1979; Johanson and Edey 1981; Tattersall 1995). Hadar is a seasonal right-bank tributary of the Awash River in the southern Afar Rift of Ethiopia (Fig. 11.2). The Hadar fossils are dated between 2.9 and 3.2 Ma (Kalb 1993; Kimbel *et al.* 1994; Walter 1994).

The oldest securely dated *Australopithecus afarensis* fossils, recovered in late 1981 from the Middle

Awash valley of the southern Afar Rift in Ethiopia, come from 11 metres beneath a primary airfall tuff dated by $^{40}Ar/^{39}Ar$ and zircon fission-track analyses as 3.8–4.0 Ma old (Clark *et al.* 1984; Hall *et al.* 1984). Although small-brained, with a cranial capacity roughly similar to that of a modern chimpanzee, these early Pliocene *A. afarensis* hominids were indisputably bipedal, a conclusion based on detailed anatomical observations (White and Suwa 1987) and strikingly confirmed by the remarkable 3.5 Ma old fossil footprints of *A. afarensis* in carbonatite ash at the prehistoric site of Laetoli in Tanzania (Leakey and Hay 1979; Leakey and Harris 1987; Deloison 1992).

A number of thoughtful and often highly ingenious suggestions have been put forward to account for the emergence of the bipedal posture so characteristic of many of the African Pliocene hominids. Explanations include liberating the hands for carrying; better vision in savanna grassland; more efficient heat regulation (Wheeler 1993, 1994); a more efficient feeding posture (Hunt 1994, 1996); sexual selection (Sheets-Johnstone 1989); adaptations to changes in the body's centre of gravity (Jaanusson 1991); and bipedal displays to reduce violence and enhance the sharing of scarce resources (Jablonski and Chaplin 1993). However imaginative, none of these unicausal explanations are particularly convincing in themselves. There were probably a number of causes, not all of which necessarily operated at the same time.

It is a great deal easier to reconstruct the diet of early hominids using strontium-calcium ratios and strontium isotope ratios (e.g. Sillen *et al.* 1995) than it is to attempt the complex and difficult task of unravelling the influence of environmental and social factors upon hominid evolution and behaviour (Foley and Lee 1989; Vrba *et al.* 1989; Coppens 1989; Foley 1994; Oliver *et al.* 1994; Bromage and Schrenk 1995). Part of the problem lies in the fact that although an impressive number of hominid fossils have now been recovered (over 500 specimens have been excavated from Sterkfontein alone), the total number of relatively complete skeletons is small when compared with the great length of time during which the hominids were evolving. A great many more hominid fossils will need to be found before we can tackle with any confidence such questions as to numbers of genera and species and their temporal and evolutionary relationships (Hennenberg 1989; Wood 1989, 1991, 1994; Turner and Wood 1993; Hennenberg and Thackeray 1995). For these reasons we have adopted a very simple, conservative and schematic representation of some of the major stages in hominid evolution over the past 4 million years (Fig. 11.3), although this involves leaving out some

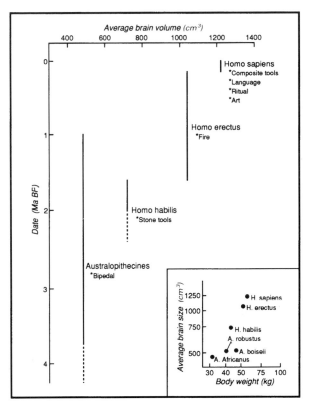

Fig. 11.3 Changes in hominid physical and cultural development from early Pliocene to late Pleistocene (after Fagan 1989). The inset shows that the *Homo* line has a much higher ratio of cranial capacity to body weight than any of the Australopithecines. (After Pilbeam and Gould 1974)

of the recent discoveries from Kenya and Ethiopia. For a more detailed discussion of possible hominid phylogenetic relationships, see Wood (1994).

The recent discoveries of even older Australopithecine fossils from Aramis in Ethiopia and from Kanapoi and Allia Bay in Kenya have extended the hominid record back to about 4.0–4.5 Ma, although there are some fragmentary but poorly dated hominid remains which may prove to be slightly older. The Aramis fossils from the Middle Awash valley of Ethiopia are a new species of Australopithecus designated *Australopithecus ramidus* (White *et al.* 1994), and are surface finds from immediately above a volcanic ash layer dated at 4.39±0.03 Ma (Woldegabriel *et al.* 1994). The Kenyan fossils are bracketed between tuffs dated at 4.1 and 3.9 Ma and also belong to a new species of Australopithecus: *A. anamensis* (Leakey *et al.* 1995). Both *A. ramidus* and *A. anamensis* lie close to the inferred point of divergence between earliest hominid and ancestral pongid, and both have anatomical traits similar to some of those found both in modern chimpanzees and in *A. afarensis*. *A. anamensis* from Kenya was indisputably bipedal (Leakey *et al.* 1995), but as yet there are no diagnostic bones to confirm whether or not *A. ramidus* was walking upright (White *et al.* 1994). The Aramis hominids seem to have occupied a closed wooded habitat (Woldegabriel *et al.* 1994), in contrast to both the *A. afarensis* hominids from Laetoli and Hadar, and the *A. anamensis* hominids from Kanapoi and Allia Bay, who occupied a range of habitats including grassland, scattered trees, woodland and gallery forest, very much like the present-day environment of the Middle Awash valley in the Afar Rift of Ethiopia (Bonnefille *et al.* 1987; Leakey *et al.* 1995).

Very considerable effort has been devoted to reconstructing the pattern of global tectonic and climatic events which may have been associated, directly or indirectly, with the emergence of the Australopithecine hominids in Africa (for comprehensive reviews see Axelrod and Raven 1978; Brain 1981a; Behrensmeyer 1982; Laporte and Zihlman 1983; Van Zinderen Bakker and Mercer 1986; Adamson and Williams 1987; Hill 1987; Singh 1988; Prentice and Denton 1988; Kingston *et al.* 1994; White 1995; Partridge *et al.* 1995a, b; deMenocal 1995; Feibel 1997). If one is seeking a single major cause, it is tempting to blame the Messinian salinity crisis of 6–5 Ma (Ryan 1973; Hsü *et al.* 1977; Hsü 1983) for the genetic isolation of Africa from Eurasia which enabled the Australopithecine hominids to evolve in isolation from the rest of the world. Be that as it may, further speculation would be unprofitable until the gap in the hominoid fossil record between 7 and 5 Ma has been finally closed, and until the time of divergence between pongid and hominid is more precisely known.

Homo habilis: the first stone toolmaker

Like the present-day great apes, the early Pliocene Australopithecine hominids were opportunistic users of twigs, sticks and stones. Once used, these temporary tools were forthwith discarded, never to be used again. There is no conclusive evidence that any of these creatures ever conceived of the idea of deliberately modifying stone fragments in order to make stone tools until the very late Pliocene, roughly 2.5 Ma ago. At this time in the Gona Valley near Hadar in the southern Afar Rift of Ethiopia, stone tools made to a recognisable and replicated pattern made their first appearance in the archaeological record (Harris 1980, 1983; Semaw *et al.* 1997). These tools were extremely simple and highly effective (Roche 1980; Gowlett 1984b). Their manufacture and use imply that their hands were capable of precision gripping, and not all of the late Pliocene

hominids were capable of such a grip (Susman 1994). At Hadar in Ethiopia, *Homo* fossils and stone tools occur together and are precisely dated to 2.33 ±0.07 Ma ago (Kimbel *et al.* 1996). In order to make these tools, several sharp flakes were dislodged from a large pebble by striking a single hard blow with a suitable hammerstone (Schick and Toth 1995). Armed with a handful of such flakes, a scavenging hominid could cut through the tough hide of some large and recently dead herbivore, detach sizeable portions of meat, and flee unscathed with this protein-rich food while the African carnivores were still dozing in the tropical heat of the day (J.D. Clark 1976a). The degree to which these early toolmakers were hunters or scavengers is still a moot point, and has prompted some excellent experimental studies of the factors controlling archaeological bone assemblages in the African savanna (Blumenschine 1988; Tappen 1995) as well as in limestone caves in southern Africa (Brain 1981b). The pebble tool or 'Oldowan tradition' persisted with minimal modification for a further million years, until about 1.5 Ma ago, in most parts of Africa (although persisting locally until 0.6 Ma in the Middle Awash Valley of Ethiopia: Clark *et al.* 1994), prompting us to ask why these early, small-brained, bipedal hominids first became toolmakers 2.5 Ma ago, rather than earlier or later.

Late Pliocene environmental changes

As we have seen in earlier chapters, a number of major environmental changes took place towards 2.5 Ma ago (for detailed discussions see Brain 1981a; Behrensmeyer 1982; Adamson and Williams 1987; Vrba 1988, 1995; Dupont and Leroy 1995). The most notable change, with far-flung repercussions for global climate, was the sudden accumulation of ice in North America and Europe. This rapid build-up of ice in the northern hemisphere is very evident in the oxygen isotope record of deep-sea cores, and is reasonably accurately dated to 2.5–2.4 Ma ago (Shackleton and Opdyke 1977; Shackleton *et al.* 1984).

There were several major side-effects of this rapid accumulation of ice at high and middle latitudes in the northern hemisphere. At Pliocene Lake Gadeb in the southeastern uplands of Ethiopia (Williams *et al.* 1979; Gasse 1980; Eberz *et al.* 1988), abundant pollen grains preserved in the lacustrine diatomites indicate that the montane climate in this equatorial region became significantly cooler and drier towards 2.5–2.35 Ma ago (Bonnefille 1983, 1995). In China, loess began to accumulate 2.4 Ma ago, when the late Pliocene climate became cold and dry (Heller and Liu 1982). In southern Europe, pollen evidence shows that the seasonally fluctuating Mediterranean type of climate began about 2.3–2.4 Ma ago (Suc 1984; Suc *et al.* 1997). In short, the very late

Pliocene was a time of relatively rapid global cooling and desiccation, with an increase in seasonality and progressive fragmentation of tropical forest and woodland habitats into a mosaic of more open woodland and savanna grassland. These habitat changes would undoubtedly have exerted selective pressures upon the hominids and other fauna, with responses ranging from minimal or subtle adaptations through to local and regional extinctions.

Prehistoric meat-eating and butchery sites

It is tempting to speculate that late Pliocene intertropical cooling and desiccation associated with global cooling and ice-cap growth at high latitudes may have displaced and otherwise fragmented the African vegetation belts and so initiated competition for food among the various robust and gracile Australopithecine hominids (Assefa *et al.* 1982). In any event, the possessors of sharp stone flakes would have been able to supplement their mostly vegetarian diet with meat cut from animals that they themselves may not have killed. Several days of decay are needed before the hides of the larger African herbivores can be tackled effectively by carnivores (Assefa *et al.* 1982), time enough for opportunistic tool-using hominids to scavenge some midday meat.

Although Binford (1981, 1983) has severely criticised some of the evidence relating to possible prehistoric big-game hunting and butchery sites, there is no doubt that as time went on, meat became an increasingly important part of the diet of such ancestral humans as *Homo habilis* and *Homo erectus*. Before a prehistoric butchery site can be accepted as such, four sets of criteria need to be met. The first and most obvious prerequisite is a single carcass in undisturbed or primary context (Clark and Haynes 1970). A second requirement is a recognisable spatial pattern of stone tools, also in primary context and in some form of functional association with the bones of the butchered animal (Clark and Kurashina 1979). Ideally, the bones should have visible cutmarks close to their extremities and roughly perpendicular to the long axis of the disarticulated bones (Jones 1980; Potts and Shipman 1981; Bunn 1981; Schick and Toth 1995). Pseudo-cutmarks can be produced by trampling animals and by various other natural processes (Behrensmeyer *et al.* 1986), so that a fourth and final prerequisite is micro-wear on the cutting edge of discarded flakes known from experimental work to be solely caused by cutting through hide, ligaments and flesh (Keeley 1980; Keeley and Toth 1981; Schick and Toth 1993).

HOMO ERECTUS, FIRE AND THE ACHEULIAN TRADITION

Homo habilis is widely believed to have been the first stone toolmaker, and the originator of the Oldowan (or pebble tool) tradition (Fig. 11.4). For a million years, from roughly 2.5 Ma to 1.5 Ma BP (and locally until 0.6 Ma: Clark *et al.* 1994), there was virtually no change in the stone tools, but a dramatic change took place in most but not all places at about 1.5 Ma BP, associated with the emergence of *Homo erectus*, a creature with a larger and more complex brain than its predecessor *Homo habilis* (Fig. 11.3). The new stone toolmaking technique consisted in detaching large flakes from a big lump of rock; the flakes were then fashioned into symmetrical handaxes and cleavers 15–30 cm long by removing smaller flakes from both sides ('bifacial flaking') all around the periphery to give a sharp, serrated edge (Fig. 11.4). These bifacially worked cleavers and handaxes were characteristic of the 'Acheulian tradition', which persisted, with progressive refinements, until about 150 ka ago, together with an improved version of the pebble tool tradition, termed the 'Developed Oldowan.' (Together, the Oldowan and Acheulian comprise the Early Stone Age or Lower Palaeolithic.)

A second major technological discovery attributable to *Homo erectus* is that of fire (Barbetti *et al.* 1980; Gowlett *et al.* 1981; Clark and Harris 1985; Brain and Sillen 1988). The use of fire not only had important implications for diet, ease of mastication, ease of digestion and the curing and storage of smoked meat (Clark and Harris 1985) but also provided greater security, warmth and the possibility of greater social interaction by all members of the group after sunset, including an increasing reliance on verbal communication. It is noteworthy that with the discovery of fire, small bands of *Homo erectus* moved out of the tropical savanna lowlands to occupy high altitude grasslands such as the Gadeb plains in the east central highlands of Ethiopia (Clark 1987), as well as moving out of Africa to occupy new sites at higher latitudes in Europe and Asia, (Fig. 11.5) by at least 1.5 Ma ago, as indicated by some recent and apparently well-dated evidence from two sites in Java for early hominid occupation some 1.8–1.6 Ma ago (Swisher *et al.* 1994). In addition, *Homo erectus* or some immediate predecessor appears to have been present in the Caucasus Mountains near Dmanisi in west Asia by 1.8 Ma ago (Gabunia and Vekua 1995), at Longgupo Cave in southern China by 1.9 Ma ago (Huang *et al.* 1995), and in the northern Dead Sea Rift of Israel by c. 2.0 Ma ago (Braun *et al.* 1991).

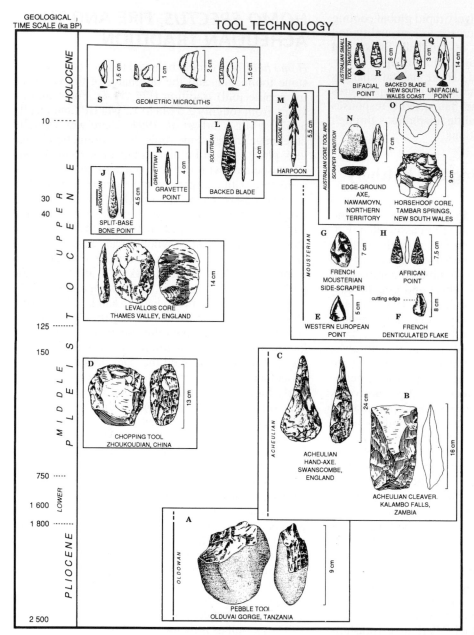

Fig. 11.4 The origin and development of Palaeolithic tool technology from 2.5 Ma to 10 ka. Neolithic polished stone tools are not shown here. (After Mulvaney 1975; Clark 1977; White and O'Connell 1982; Gowlett 1984a; Fagan 1989)

MIGRATION OF *HOMO ERECTUS* OR AN EARLIER *HOMO* FROM AFRICA TO EURASIA

The precise timing of the earliest migrations of *Homo erectus* out of Africa and into Eurasia is still a matter for debate, as is the widely accepted assumption that *Homo erectus* first evolved in Africa rather than in Asia. There is also a strong possibility that the first migrant was not *Homo erectus* at all, but some earlier and ancestral form of *Homo*, albeit still a tool-maker. What is the evidence?

One of the best known *Homo erectus* sites in Asia is the 'Peking Man' site of Choukoutien 50 km southwest of Beijing in northeastern China. Occupation of the limestone cave began towards 0.7 Ma ago and continued intermittently until roughly 230 ka BP, indicating a movement into cooler latitudes by at least 0.5 Ma ago (Liu 1983; Wu and Lin 1983; Clark 1992). Two older *Homo erectus* sites near Lantian in central China have recently been redated palaeomagnetically (see Appendix) to 0.65 Ma BP for the Chenjiawo mandible and to 1.15 Ma BP for the Gongwangling cranium (An and Ho 1989). More recent work suggests that a pre-*erectus* stone-toolmaking

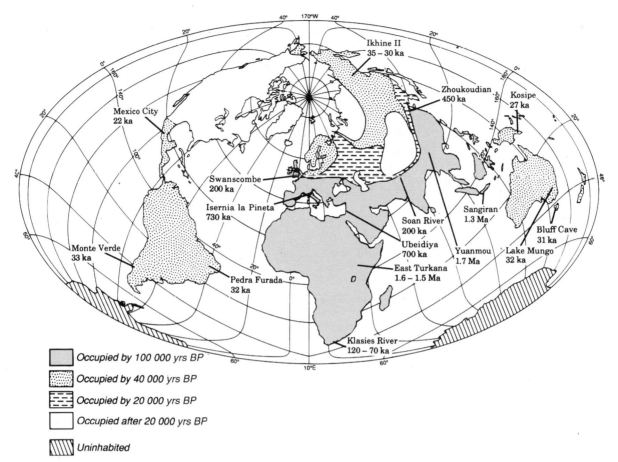

Occupied by 100 000 yrs BP

Occupied by 40 000 yrs BP

Occupied by 20 000 yrs BP

Occupied after 20 000 yrs BP

Uninhabited

Fig. 11.5 Successive stages in the prehistoric settlement of the world, showing the very late entry into Australia and the Americas. (After Fagan 1989; Barraclough 1982; White and O'Connell 1982; Davis 1986; Scarre *et al.* 1988; Dillehay and Collins 1988; Cosgrove 1989)

hominid may have been present in China at Long-gupo Cave south of the Yangtse River and near the eastern border of Sichuan Province as early as 1.9 Ma ago (Huang *et al.* 1995). Stable isotope analysis of loess–palaeosol deposits coeval with hominid-bearing sediments in central China indicate that *Homo erectus* migrated from subtropical southern China across the Qinling mountain barrier into the temperate Loess Plateau of central China towards 1.15 Ma ago (Wang *et al.* 1997), pre-dating the earliest reliable evidence of permanent occupation in temperate Europe by 150 ka. Although many of the *Homo erectus* sites from Java are still poorly dated, there is now some more persuasive radiometric evidence from two hominid sites in Java that *Homo erectus* may have been present there as early as 1.6–1.8 Ma ago (Swisher *et al.* 1994). On the basis of cranial morphology and brain size some of the other *Homo erectus* fossils from Java appear to be somewhat younger than the three oldest *Homo erectus* sites in China, but resolution of this matter must

await more rigorous dating of both the Chinese and the Javanese hominid fossils.

In Syria, Acheulian bifaces are associated with an extinct Middle Pleistocene fauna of *Elephas*, *Equus* and *Hippo*, yielding an approximate age range from c. 0.7 Ma to c. 125 ka BP. Upper Acheulian artefacts in the Golan Heights region of Syria/Israel are bracketed by lavas dated to 470 ka and 230 ka BP. In the Jordan Valley of Israel, the Ubeidiya Formation contains chopper tools and flakes comparable to those in Upper Bed II at Olduvai, suggesting a possible age of 1.4 Ma and a bracketing age range of 1.5 Ma to 0.7 Ma BP, consistent with the associated pollen and other fossil remains (Tchernov 1987). Later work has yielded a palaeomagnetically-derived age of c. 2 Ma for tool-bearing deposits which pre-date the Ubeidiya Formation (Braun *et al.* 1991). At Kuldera in Tajikistan in central Asia, stone tools similar to those excavated at Ubeidiya have recently been dated to 0.85 Ma (Ranov *et al.* 1995). In France, hearths in the Escale cave in the Massif Central date back to the start of the Middle Pleistocene (0.7 Ma BP), and in Hungary and Romania, flaked pebbles, flake tools and hearths occur in Middle Pleistocene formations dated between c. 0.7 Ma and c. 125 ka.

On the basis of the foregoing evidence, it seems fair to conclude that *Homo erectus* was certainly present in the warmer parts of western and central Europe, beyond the maximum limits of the ice, by at least 0.7 Ma, and possibly over a million years earlier in the Middle East, the Caucasus, Java and China. The cranial capacity of *Homo erectus* showed a steady increase during this time from around 775 cc at 1.6 Ma to 1300 cc by 0.2 Ma ago, especially during the Middle Pleistocene from about 600 ka to 150 ka BP (Ruff *et al.* 1997), but there was remarkably little change in the complexity of the Acheulian and Developed Oldowan tool-kits during this long interval of time (Fig. 11.4). To quote J. Desmond Clark: '*Homo erectus*, though he exploited a wide range of resources, did so only at a very low level of efficiency and with minimal ability to specialize' (Clark 1976a, p. 47). The paradox here, as Clark has noted, is that with the emergence of *Homo erectus* there was a drastic change in man's ability to adapt to a wide range of environments, but only a relatively small change in stone tool technology.

Despite the lack of any major change in stone tool technology between about 1.5 Ma and 0.2 Ma, the multipurpose Acheulian and Developed Oldowan tool-kits enabled the Early Stone Age peoples to acquire enough plant and animal foods to survive and proliferate. As noted earlier, an additional and very important item in the material repertoire of *Homo erectus* was fire, the use of which dates to 1.0–1.5 Ma at Swartkrans cave in South Africa (Brain and Sillen 1988), to perhaps 1.5 Ma at Chesowanja in Kenya (Gowlett *et al.* 1981) and possibly also at Gadeb in the Ethiopian uplands (Barbetti *et al.* 1980), and to c. 0.7 Ma at Escale cave in France and 0.5 Ma at Choukoutien cave in China (Gowlett 1984a). With the warmth and protection it afforded, fire enabled small bands of *Homo erectus* hunters to venture into hitherto unoccupied parts of Europe and Asia where the long cold winters required more effective shelters than were needed in the African savanna. No very early hearths have yet been recovered from the tropical and subtropical regions of Java and southern China, perhaps because they have not yet been sought from sites where they are most likely to have been preserved, or perhaps because fire was not much used in those areas at that time.

FROM *HOMO ERECTUS* TO *HOMO SAPIENS*

Apart from a gradual increase in cranial capacity between 1.5 Ma and roughly 0.6–0.3 Ma ago, there is very little evidence of any progressive morphological changes in the skeletal anatomy of *Homo erectus* until about 0.7–0.5 Ma ago when populations showing a combination of *Homo erectus* and *Homo sapiens* traits began to appear in Africa (Grün and Stringer 1991; Schwarcz and Grün 1992; Clark 1992; Aiello 1993; Clark *et al.* 1994; Stringer and McKie 1996; Bräuer *et al.* 1997). The timing of the transition from *Homo erectus* to *Homo sapiens* is less well dated in Europe and Asia. At such sites as Laetoli in Tanzania, Omo in southern Ethiopia/northern Kenya, and Broken Hill in Zambia, the skulls have a relatively large cranial capacity of c. 1200 cc but retain the large brow ridges, low sloping frontals and thick bones characteristic of *Homo erectus*. All of these sites are probably about 0.5 Ma old. There are still very few reliable dates for either the *H. erectus–H. sapiens* transition or for the archaic *Homo sapiens* fossils which postdate this transition. The application of electron spin resonance dating (see Appendix) to hominid tooth enamel is a significant breakthrough, and has recently yielded an age of 259 ± 35 ka for a late archaic *H. sapiens* fossil from Florisbad Spring near Bloemfontein in South Africa (Grün *et al.* 1996). Two recent uranium series dates (see Appendix) of 270 and 300 ka for a hominid cranium and femur from the Lake Turkana region in Kenya have likewise extended the antiquity of *Homo sapiens* (Brauer *et al.* 1997). From these and other new age estimates, it now appears that the transition from *Homo erectus* to archaic *Homo sapiens* dates back to at least 700–500 ka, and the transition from early to late archaic *Homo sapiens* to around 350–250 ka BP, with the emergence of modern *Homo sapiens* possibly extending back to 150 ka (Bräuer et al 1997). By about 100 ka BP, anatomically modern humans were present at Border Cave and Klasies River Mouth in South Africa, in the Omo valley of southern Ethiopia and at Qafzeh and Es Skhul in Israel, indicating that the transition from late archaic to modern *Homo sapiens* must have taken place between 250 and 100 ka ago (Stringer *et al.* 1989; Clark *et al.* 1994; Stringer and McKie 1996; Grün *et al.* 1996; Bräuer *et al.* 1997).

FROM EARLY STONE AGE TO MIDDLE STONE AGE

By 250–130 ka two distinct toolmaking traditions were widespread in Africa and Eurasia. The Acheulian tradition involved the use of large, standardised bifacial tools, typified by the cleavers and handaxes illustrated in Fig. 11.4. An assemblage of very variable choppers, scrapers and stone knives was characteristic of the Developed Oldowan tradition. Equipped with one or both of these Middle Pleistocene tool-kits, the *Homo erectus* populations were able to occupy every type of habitat apart from deserts and lowland and montane evergreen forests. The Early Stone Age (or Lower Palaeolithic) was characterised by a very slow rate of cultural change, a limited range of foraging activities, and an unspecialised use of natural resources. Flakes were used to obtain meat, wooden sticks to dig up roots, tubers and rhizomes, and hammer stones to crack nuts (Clark 1975, 1976a).

The Middle Stone Age (or Middle Palaeolithic), which in Africa dates back to about 200 ka and in Europe and the Middle East to about 160 ka (Aiello 1993), was characterised by a revolutionary new technique that involved prior flaking and preparation of the parent corestone to allow greater control over the form of the finished flake. These Levallois cores and flakes are diagnostic of the Middle Palaeolithic industries of Europe and Asia as well as of the Middle Stone Age assemblages throughout Africa (J.D. Clark 1988, 1992). The co-occurrence at many cave sites of small bifacial implements (Mousterian points) with fragmented bones of game animals suggests that the stone points were being used as spear tips – a good indication that composite tools of wood and stone were now being widely used. The recent discovery of a two-metre wooden hunting spear found in close association with the butchered remains of more than ten horses in deposits 380–400 ka in age at Schoningen in Germany indicates a high degree of hunting competence among these Middle Pleistocene hominids (Thieme 1997).

It would be misleading to claim a simple equivalence between the inception of the Middle Stone Age and the demise of *Homo erectus*, but it is worth noting that this second major innovation in stone toolmaking (the first being the large bifacially worked flake tools of the Acheulian) coincided very broadly with the emergence of recognisably modern humans – *Homo sapiens* – some 200–100 ka ago. The increase in brain complexity at this time strongly suggests a greater development of language skills, with an associated increase in the degree of social cohesion necessary in group hunting and sharing (Trinkhaus and Howells 1979). Whether the ability to articulate words was the same in all *Homo sapiens* populations is still a matter of debate (Trinkhaus 1986) but the recent discovery of a 60 ka Middle Palaeolithic hyoid bone associated with Neanderthal bones at Kebara cave on Mount Carmel in Israel shows that humans were certainly morphologically capable of fully modern speech at that time (Arensburg *et al.* 1989).

MIDDLE STONE AGE RITUAL, ART AND INNOVATION

Ritual and art did not originate in the Middle Stone Age, as the curious case of the defleshed Bodo skull indicates. Recovered in 1976 from Bodo catchment in the Middle Awash Valley of Ethiopia, this robust cranium belonged to a late Middle Pleistocene hominid transitional in appearance between *Homo erectus* and *Homo sapiens* (Conroy *et al.* 1978). When cleaned of its secondary calcium carbonate coating and examined in Addis Ababa by Dr Tim White in 1981, the skull revealed cut marks on the facial bones and inside the eye sockets, indicating deliberate defleshing of the face and removal of the scalp, presumably for ritual purposes (White 1986). Since the skull has recently been dated to c. 0.6 Ma (Clark *et al.* 1994), the ritual defleshing must be of similar antiquity.

The Middle Stone Age was a time of more intensive settlement and regional specialisation in Africa and Eurasia (Fig. 11.5) and a time when a number of distinctively human cultural traits first became apparent (Fig. 11.6). The excavations at Shanidar in the Zagros hills of northern Iraq strongly suggest that these Neanderthal hunters were not only caring for their sick and elderly, but were also burying their dead over 60 ka ago (Trinkhaus 1983), although Gargett (1989) has called into question the validity of some of the evidence for deliberate Middle Palaeolithic disposal of the dead.

Recent uranium–thorium dating of the cave travertine at the Mousterian site of Tata in Hungary shows that the cave was occupied towards the end of the last interglacial some 100 ka ago. Among the artefacts was a polished ivory plaque 11 cm long carved from a mammoth molar and rubbed with red ochre (Schwarcz and Skoflek 1982). The *Homo sapiens neanderthalensis* sculptor responsible for this plaque had a cranial capacity of 1400–1500 cc, which is fully equivalent to that of present-day humans.

The upper Semliki valley in Zaire has recently yielded Middle Stone Age artefacts dating to about 90 ka associated with a remarkable series of barbed and unbarbed bone points (Yellen *et al.* 1995; Brooks *et al.* 1995). Elsewhere in Africa such a bone point industry has hitherto been found only in Late Stone Age sites used by *Homo sapiens sapiens*, suggesting the possibility that certain Middle Stone Age societies displayed a level of technological and behavioural competence entirely consistent with fully

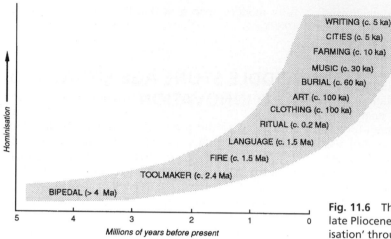

Fig. 11.6 The development of human culture during the late Pliocene and Quaternary, showing increasing 'hominisation' through time. (After Tobias 1979; Williams 1985b)

modern Upper Palaeolithic human societies (Yellen *et al.* 1995).

LATE STONE AGE DIVERSITY AND MIGRATIONS

In 1987 Cann, Stoneking and Wilson published the results of a detailed analysis of mitochondrial DNA and its possible bearing on human evolution (Cann *et al.* 1987; Stoneking and Cann 1989). Their major conclusion was that modern humans are descended from a common female ancestor who appears to have lived in Africa between about 0.5 Ma and 0.05 Ma ago, and most probably around 200 ka ago. Although vigorously disputed by some who argue for 'multire-gional evolution' or 'regional continuity' (Frayer *et al.* 1993) or questioned as to its details by others (Aiello 1993), the 'out-of-Africa' hypothesis now has many supporters among archaeologists, geneti-cists and human palaeontologists, and seems to be entirely consistent with the latest work on hominid dating in Africa and Eurasia (Stoneking *et al.* 1992; Clark 1992; Mellars 1992; Rogers 1995; Stringer and McKie 1996).

The sudden appearance throughout Europe of the Aurignacian industries associated with fully modern human skeletal remains and the relatively rapid demise of the European Neanderthals and their asso-ciated Middle Palaeolithic (Mousterian) industries is strong evidence that Europe was colonised between about 43 and 35 ka by fully modern people moving from the Middle East initially into eastern and cen-tral Europe (c. 43 ka) and finally into western Europe by c. 35 ka (Mellars 1992). The problem is slightly compounded by the recent thermoluminescence dat-ing of Neanderthal remains from Saint-Césaire in

France to 36.3 ± 2.7 ka, which makes them some-what younger than some of the Upper Palaeolithic (Aurignacian) sites in Spain which are believed to have been occupied by modern humans (Mercier *et al.* 1991; Stringer and Grün 1991). A possible expla-nation is that there was some degree of coexistence and perhaps cultural exchange between the two *Homo sapiens* groups present in western Europe at this time (Mellars 1992; Aiello 1993; Hublin *et al.* 1996), much as was evident in the Middle East con-siderably earlier (Shreeve 1996).

As time progressed, increasing use was made of materials other than stone or wood, including bone, antler, mammoth ivory and shell (Straus 1985; Mel-lars 1989). Many of these materials were fashioned into pendants or carved into animal and human fig-ures, particularly during Late Stone Age or Upper Palaeolithic times, when small, beautifully made stone engraving and cutting tools began to proliferate from about 40–35 ka onwards in Europe and western Siberia (Gowlett 1984a; Fagan 1995).

It was towards the close of the Middle Palaeolithic and the start of the Upper Palaeolithic in Europe and Asia that the last great prehistoric migrations took place (Fig. 11.5). Northern Australia was first occu-pied by anatomically modern humans some 50–60 ka ago (Roberts *et al.* 1990). By 35–40 ka the first Aus-tralians had reached as far south as Swan River in Western Australia (Pearce and Barbetti 1981) and Lake Mungo in semi-arid New South Wales (Barbetti and Allen 1972; Bowler *et al.* 1972), implying a series of prior sea journeys between mainland Asia and Northern Australia by a competent maritime people (Hallam 1977; White and O'Connell 1978; Jones 1979; Flood 1989).

Eastern Siberia was not occupied until about 30–35 ka (Klein 1975), putting a maximum age of c. 30 ka for the first entry into Alaska via the Bering land bridge during times of glacially lowered sea-level (Chapter 6), always assuming that the first Americans were tundra dwellers who preferred to follow the herds on foot rather than to voyage by boat. Since they were primarily hunters who depended upon the great herds of mammoths for their food, fuel and raw materials for shelter, it seems highly probable that they did walk across the Bering land bridge which would then have been a grassy plain nearly 1500 km wide from north to south.

Very few sites older than about 20–25 ka have yet been found in Central and South America, although there is growing evidence that some prehistoric sites in Brazil may be as old as 35–40 ka (Guidon and Delibrias 1985, 1986; Bednarik 1989). In North America the majority of prehistoric sites are younger than 12–14 ka, indicating a major influx of Palaeo-Indians after 12 ka only, once the MacKenzie Valley had become free of ice (Martin 1973).

PLEISTOCENE FAUNAL EXTINCTIONS

Ever since the well-publicised excavations of Boucher de Perthes (1847, 1857, 1864) in the Somme Valley of northern France in the 1840s and 1850s, geologists on both sides of the English Channel were alerted to the coexistence in Europe of Palaeolithic stone tools and a now-extinct fauna, including woolly rhinoceros, sabre-toothed tigers and mammoths (Evans 1860; Prestwich 1860; Lyell 1873). At about the same time, the fossil bones of large extinct marsupials were being recovered from now dry lake and swamp deposits in southeastern Australia (Owen 1870). The initial reaction of natural scientists in both Europe and Australia was to invoke climatic change as the sole cause of these extinctions. However, growing recognition of the efficiency of Late Stone Age hunting and trapping skills, and the apparent synchroneity between the demise of mammoths in North America and the first major influx of Palaeo-Indian mammoth hunters, has led many researchers to abandon a climatic explanation in favour of human predation and what Paul Martin so vividly refers to as 'Pleistocene overkill' (see Martin and Wright 1967; and its monumental successor, Martin and Klein 1984).

Controversy about the causes of the Pleistocene faunal extinctions continues unabated (Flannery 1994; Leakey and Lewin 1995), and although some of the debate is as polarised today as it was last century, a greater measure of consensus now appears to be emerging. We will comment briefly on the four major hypotheses proposed so far, for each may be valid in certain places at particular times.

The climatic change hypothesis has the merit that the geological, biological and isotopic evidence of repeated fluctuations in Quaternary temperature and precipitation is often of high quality and accurately dated. A recent variation on this theme is Mörner's (1978) suggestion that marine regressions will lead to a regional fall in groundwater levels as sea-level drops, resulting in the drying out of springs and lakes, causing widespread faunal extinctions. In Australia, Horton (1980, 1984) has drawn attention to the coincidence between formerly wooded areas and megafaunal fossil finds, arguing that late Pleistocene aridity destroyed the woodland habitat of the large browsing marsupials, tethering them to dwindling waterholes around which they soon consumed all the available forage. The site of Lancefield Swamp in Victoria strongly appears to support this hypothesis (Gillespie et al. 1978). A weakness of this hypothesis is that it does not convincingly account for why the fauna survived repeated previous Quaternary droughts, but succumbed to the latest major drought, by which time humans were also present in Australia.

The fact that the Pleistocene fauna of Australia and North America appeared to have survived unscathed the climatic vicissitudes of the Middle and early Late Pleistocene, only to disappear shortly after the arrival of humans in those two continents, gave rise to the hypothesis of 'Man the destroyer' espoused by Merrilees (1968) in Australia and by Martin (1967; and Martin and Klein (1984) in North America. Although on the face of it a logical and plausible explanation, the 'overkill' hypothesis has several flaws. First, as many Australian workers have been at pains to emphasise, there is a curious absence of kill sites, and there was a very long period of coexistence between humans and megafauna (well in excess of 20 000 years) so that Martin's (1984) 'Blitzkrieg' model of overkill is certainly not applicable to Australia (Gillespie et al. 1978; McIntyre and Hope 1978; Sanson et al. 1980; Gorecki et al. 1984). More recent work is beginning to show that some well-dated sites do in fact contain concentrations of broken megafauna bones and artefacts in primary context, such as, for instance, Cuddie Springs in New South Wales between 19 and 30 ka ago (Dodson et al. 1993).

Although mindful of Flint's dictum that 'absence of evidence is not evidence of absence', many researchers accepted the apparent lack of butchery sites and the long period of coexistence between megafauna and prehistoric hunters as convincing evidence that direct human predation was not a sufficient and necessary cause of late Pleistocene faunal extinctions. However, they were not convinced that climatic change was an adequate explanation either, for the reasons outlined earlier, and so sought a third explanation: human modification of the original vegetation through the use of fire (Jones 1968, 1969). There is certainly good evidence from northern Queensland that some species of tropical trees did become extinct in the late Pleistocene at a time when high charcoal concentrations indicate more intense and perhaps more frequent fires (Kershaw 1978, 1984). However, as Horton (1982) has pointed out, fires may promote new plant growth and may enhance the grazing potential, so that wallabies and kangaroos may benefit rather than suffer from the modification to their habitat caused by fire.

Haynes (1991) has reviewed the evidence for an interval of widespread drought in the western United States between 11.3 ka and 10.9 ka. This drought coincided with the time of the cold and dry Younger Dryas event in northern Europe, as well as with the very time when the clovis Palaeo-Indians were hunting and butchering mammoths and bisons concentrated around dwindling water-holes. The extinction of the mammoth (*Mammuthus columbi*) at this time may therefore reflect both climatic desiccation and human predation (Haynes 1991).

Unconvinced by the efficiency of either climatic change or hunting or burning as the major agents of Pleistocene faunal extinctions, an increasing number of natural scientists are now invoking a more complex and multicausal explanation. Long-term changes in plate movements cause changes in atmospheric and ocean circulation, in the distribution of land and sea, and in marine regressions and transgressions (see Chapters 2 and 6). The record of Cainozoic plant and animal extinctions reflects the interplay of physical and biological factors (Butzer 1982). The result is a repatterning of food supplies (which may favour grazers at the expense of browsers, as in Australia); changing competition between different groups of plants and animals; and cyclic swings from complex ecosystems with a diverse fauna and flora to simpler ecosystems dominated by fewer species. Against this background of progressive long-term changes in fauna and flora, it is easy to envisage that rapid Quaternary climatic fluctuations, human predation and fire, acting in concert, will selectively destroy those species already made vulnerable by the longer-term changes in climate and habitat caused by Cainozoic plate movements. In short, the arrival of humans in Australia towards 50 ka was but one of the factors responsible for the demise of the megafauna, but may have been the final and decisive factor accelerating the processes of extinction. Calaby's (1976) verdict is worth citing in this context:

> ... the weight of evidence favours climatic changes as the ultimate major cause of extinction and the most that Pleistocene men may have done was to hasten the

extinction of the remaining already doomed species that were still around when he arrived.

ISOTOPIC EVIDENCE OF PALAEO-DIET

One of the major reasons for the continuing debate over prehistoric faunal extinctions is the lack of precise quantitative information about what these creatures ate. Were they browsers or grazers or mixed feeders? It is difficult to argue convincingly that many of the now-extinct large Pleistocene marsupials of Australia were dominantly browsers when this hypothesis is simply asserted and never adequately tested. Fortunately, we now have the means to test whether the giant kangaroos like Procopton ate grass or browsed on shrubs and trees by analysing the $^{13}C/^{12}C$ ratios in unmineralised samples of bone.

Plants can fix atmospheric carbon by photosynthesis in one of three possible ways. All trees, most shrubs, and grasses growing in shaded forests or temperate climates follow the Calvin or C3 pathway of photosynthesis (Van der Merwe 1982). Grasses adapted to growing in strong sunlight, including most tropical grasses, follow the Hatch-Slack or C_4 pathway of photosynthesis, and most succulent plants follow the third pathway, which involves crassulacean acid metabolism (CAM pathway of photosynthesis). All three photosynthetic systems fractionate the carbon isotope ratio of atmospheric CO_2 in quite different ways (Vogel *et al.* 1978; Vogel 1978; van der Merwe 1982).

The outcome of this differential fractionation of the carbon isotopes during photosynthetic fixation of carbon is that C4 plants have $\delta^{13}C$ values of -9% to 16% (average -12.5%), C3 plants have $\delta^{13}C$ values of -20% to -35% (average -16.5%), and CAM plants have mean $\delta^{13}C$ values of roughly -16.5% (Van der Merwe 1982).

Further fractionation ensues when the plants are eaten by animals, including humans. Bone collagen is enriched by roughly 5% relative to the mean ratio of the plant food eaten. The $\delta^{13}C$ value of the bone collagen of browsing animals (such as kudu in South Africa) is about -21.5%, since they only eat leaves of trees and shrubs, i.e. C3 plants (Van der Merwe 1982). Grazing animals, which eat only C4 grasses, will have $\delta^{13}C$ values of -8% to -10%, and mixed feeders, such as sable antelope, values of -13% to -15%. Isotopic measurements on bone or teeth therefore provide a means of assessing the proportions of C3 and C4 plants eaten by prehistoric animals and people. Ambrose and De Niro (1986) have recently used both carbon and nitrogen isotope ratios in bone collagen to distinguish between human diets in

Africa, including groups eating marine foods, cereal grains, and pastoralists. In the latter case, they could clearly distinguish between camel pastoralists and capri-bovine pastoralists.

An additional factor needs to be considered when using plant carbon isotopes to infer animal and human diet. Recent research by Martinelli *et al.* (1991) has revealed that in certain environments such as the flood plain forests of the Amazon, much of the biogenic CO_2 may be recycled before it is mixed completely into the atmosphere. The intensity of this recycling is greater in the eastern than in the western Amazon basin, and increases systematically inland, demonstrating that carbon isotope gradients can vary across the same ecosystem, both between different species and within the same species (Martinelli *et al.* 1991).

One of the most intriguing aspects of human diet is the dramatic way in which it changed at the end of the Pleistocene. The inception of plant and animal domestication towards 10 ka denoted the end of the Palaeolithic and the start of the Neolithic tradition in every inhabited continent except Australia. With the Neolithic came a dramatic change in food production which ushered in some of the most revolutionary changes ever experienced by human societies during the long saga of prehistoric cultural evolution.

NEOLITHIC PLANT AND ANIMAL DOMESTICATION

For over two million years the genus *Homo* lived by gathering plant foods and hunting wild animals.

Homo sapiens sapiens remained a hunter–gatherer for a 100 000 years until about the start of the Holocene, 10 000 years ago. Within the next 5000 years or so, virtually the entire human population had become predominantly farmers or herders (see Fig. 11.7), although wild animals continued to be hunted and wild plants to be gathered long after the emergence of agriculture (Legge and Rowley-Conway 1987). What prompted the change?

Cohen (1977) has argued that agriculture does not provide a more secure, more palatable, more nutritious or more varied diet than that of most hunter–gatherers, but farming does ensure that more calories of food can be obtained from a given area of land in a given time than by simple collecting of wild foods. He went on to argue that growing pressure upon food resources toward the end of the Pleistocene, caused by a continually growing world population, was the primary stimulus which caused former hunters and gatherers to become pastoralists and farmers.

Cohen's views have been challenged by a number of archaeologists, some of whom have argued that Neolithic methods of food production were a cause rather than a consequence of human population growth (Hassan 1980). It is worth noting that world population 10 000 years ago is estimated at around 5 million, increasing by a factor of 20 to 100 million some 5000 years ago (May 1978). By 300 years ago, world population amounted to c. 500 million, reached 1000 million (1 billion) by 1850, 2 billion by 1930, over 4 billion by 1978 and over 5 billion by 1990. Such dramatic increase was only possible with the change in methods of food production initiated during the Neolithic.

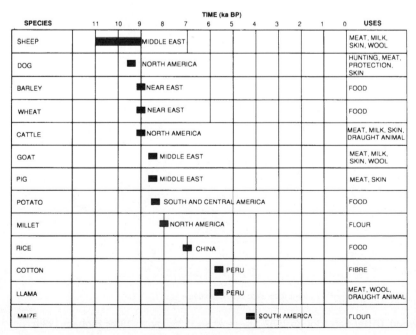

Fig. 11.7 Origins of domesticated plants and animals. (After Bray and Trump 1970; Barraclough 1982; Gowlett 1984a; Scarre *et al.* 1988; Fagan 1989)

SPECIES	USES
SHEEP (MIDDLE EAST)	MEAT, MILK, SKIN, WOOL
DOG (NORTH AMERICA)	HUNTING, MEAT, PROTECTION, SKIN
BARLEY (NEAR EAST)	FOOD
WHEAT (NEAR EAST)	FOOD
CATTLE (NORTH AMERICA)	MEAT, MILK, SKIN, DRAUGHT ANIMAL
GOAT (MIDDLE EAST)	MEAT, MILK, SKIN, WOOL
PIG (MIDDLE EAST)	MEAT, SKIN
POTATO (SOUTH AND CENTRAL AMERICA)	FOOD
MILLET (NORTH AMERICA)	FLOUR
RICE (CHINA)	FOOD
COTTON (PERU)	FIBRE
LLAMA (PERU)	MEAT, WOOL, DRAUGHT ANIMAL
MAIZE (SOUTH AMERICA)	FLOUR

It is highly unlikely that any one explanation will adequately account for the origins of agriculture (Smith 1995). Here, as elsewhere, we should be sceptical of unicausal explanations, and reject over-facile correlations between different events as proof of a causal explanation. Whether the onset of farming followed a wet or dry climate is not, in itself, proof that the change in climate initiated agriculture in that locality. After all, many previous similar climatic fluctuations did not. Far too many attempts to explain agricultural origins succumb to the *post hoc ergo propter hoc fallacy* by claiming that because B followed A, A must have caused B.

The ultimate causes of early farming and herding were probably a variable combination of social, economic, technological and environmental factors.

The actual process of domestication is illuminating. Stemler (1980) has emphasised that the two major prerequisites for cereal domestication were the harvesting of whole grain-heads of mutant plants with a tough rachis, and the later sowing of grains from such mutant plants. With tough-stemmed tropical panicoid grasses like sorghum and millet, an efficient harvesting tool was essential (Stemler 1980). J.D. Clark (1971, 1980) has pointed out that early farming in the Nile Valley was preceded by a long interval of pre-adaptation to agriculture. Certainly, efficient sickles were already widely used for harvesting wild grasses in the Nile Valley and the Near

East and at least during the late Upper Palaeolithic at which time grindstones also became abundant (Close 1989). Where environmental stress was great, as at Dhar Tichitt in Mauritania, the transition from collecting wild grasses to harvesting domesticated cereals was remarkably rapid, in this instance a few centuries only (Munson 1976; Stemler 1980; Williams 1985b).

In Africa and many parts of the Middle East and Asia, archaeologically visible evidence of early farming and herding coincides with a change from a cold, dry and windy terminal Pleistocene climate to a warm and wet early Holocene climate (Adamson *et al.* 1980; Wendorf and Schild 1980; Williams 1984b). Careful scrutiny of the onset of plant and animal domestication in North Africa shows that the beginning of agriculture cannot be directly linked to climatic change, since it was strongly time-transgressive (Williams 1985b). Furthermore, plant and animal domestication often started at quite different times at the same site, and the onset of agriculture was often quite different at localities in relatively close proximity to one another (Williams 1985b).

Regardless of whether the inception of early farming and herding was associated with demographic pressure upon resources (Cohen 1977), or to a technical capacity to harvest, store and sow appropriate cereal grains (Stemler 1980), or to a combination of climatic, ecological and technological factors (J.D.

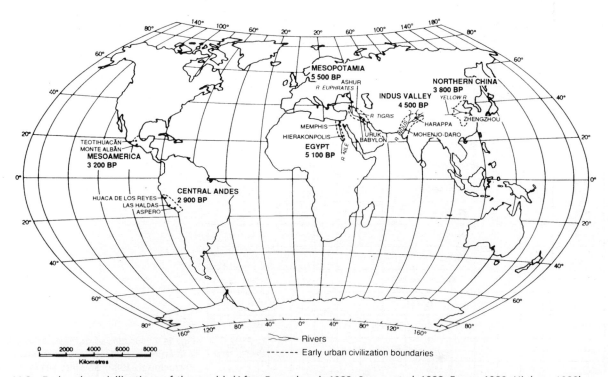

Fig. 11.8 Early urban civilizations of the world. (After Barraclough 1982; Scarre *et al.* 1988; Fagan 1989; Higham 1989)

Clark 1976b, 1980, 1984), the net effects were generally the same.

As noted earlier, relative to Upper Palaeolithic hunting and gathering, Neolithic agriculture would have provided more food calories per unit area of land per unit time (Cohen 1977). In addition, the combination of milk and a porridge of cooked cereal grains was ideal food for young children, and easier for them to digest than meat, especially if the latter was inadequately cooked (Stemler 1980). Cultivation, to be effective, requires a more sedentary lifestyle than herding or hunting, hence allowing a far higher population density than non-agricultural societies would find viable. The consequences of the 'Neolithic Revolution' were an increase in sedentary living and in population, leading to the emergence of urban civilizations during the mid to late Holocene in widely separated parts of the world, including South America, the Nile Valley, Mesopotamia, the Indus Valley and China (Fig. 11.8).

From seasonal lakeshore dwellings to settled villages is a reflection of a profound change in social structure baldly summarized by the adjectives Mesolithic and Neolithic. Evolutionary changes in stone tool technology brought about revolutionary changes in food production. From Enkidu to Gilgamesh, from individualistic to hierarchical, from ephemeral camp to enduring city, the Neolithic foundations for our present urban civilization had now been laid.

(Williams 1985b, p. 183)

ATMOSPHERIC CIRCULATION DURING THE QUATERNARY

Blow, winds, and crack your cheeks!
rage! blow!
You cataracts and hurricanoes, spout
Till you have drench'd our steeples,
drown'd the cocks!
William Shakespeare (1564–1616),
King Lear

Climatic change was the hallmark of the Quaternary. Relative to the previous 60 million or so years of Tertiary geological history, the climatic fluctuations of the Quaternary were unprecedented in terms of the speed and amplitude of global temperature oscillations (Kukla and Cilek 1996). The mean annual temperature difference between the last glacial maximum (18±3 ka) and the early Holocene 'climatic optimum' or 'hypsithermal' (9±3 ka) amounted to about 10°C in many temperate land areas of the world, which is roughly equivalent to the inferred long-term drop in land and sea temperatures between the early Eocene and early Pliocene in these same regions. The surface of the Southern Ocean cooled by about 10–15°C over a time-span of roughly 40 million years (Shackleton and Kennett 1975). Temperature drops of similar magnitude were evident in eastern Australia over the much shorter time-span of about 100 000 years which separated the last interglacial (125±5 ka) from the last glacial maximum in this region. The Vostok ice core from Antarctica displays a similar pattern of rapid fluctuations in temperature over the past 160 ka, as well as dramatic contrasts in the concentration of atmospheric carbon dioxide and methane, with much lower amounts of both gases being present during times of lower global temperature (Barnola *et al.* 1987, 1991; Jouzel *et al.* 1989; Chappellaz *et al.* 1990; H.J. Smith *et al.* 1997). More recent work suggests that some care is needed when interpreting the carbon dioxide content of air bubbles trapped in ice, since some carbon dioxide may be produced after bubble formation (H.J. Smith *et al.* 1997). In the last few years ice cores from Antarctica, Greenland, Peru and central Asia have yielded an impressive array of palaeoclimatic data, including variations in rates of aeolian dust deposition, the incidence of ENSO events, and the impact of recent warming trends on ice-cap ablation (Thompson and Mosley-Thompson 1992; Brecher and Thompson 1993; Thompson *et al.* 1993).

Temperature and precipitation are the two climatic parameters of prime concern in most reconstructions of Quaternary climates. Ever since the classic work of Emiliani (1955) in which he used variations in the oxygen isotopic composition of the calcareous tests of marine planktonic foraminifera collected from 11 deep-sea sediment cores in the Atlantic and Pacific oceans to deduce a series of alternating cold and warm stages, the marine record has yielded a wealth of palaeoclimatic information relating to changes in global ice volume as well as more local changes in ocean temperature and salinity. Some ingenious techniques have been used of late to reconstruct changes in Quaternary temperature, including analyses of atmospheric noble gases dissolved in groundwater (Stute *et al.* 1992) and terrestrial heat flow measurements (Huang *et al.* 1997).

In their attempts to understand the nature and causes of Quaternary climatic fluctuations, research workers have adopted two different approaches, which have too often been poorly integrated with one another despite being essentially complementary. The palaeoclimatic modellers use a variety of global circulation models (GCMs) to simulate likely patterns of past climatic change. Some of these models incorporate quite realistic past geographies; some are coupled ocean–atmosphere models; but few are capable of precise spatial resolution and none can yet cope adequately with cloud dynamics. The palaeo-geographers and palaeoecologists, on the other hand, within the time resolution of the dating methods used and the spatial resolution of the evidence, whether geological, biological, archaeological or geochemical, can usually contrive to reconstruct past changes in local, regional or global marine and terrestrial environments reasonably well, albeit with varying degrees of quantitative accuracy (Feng and Epstein 1994; Nowak *et al.* 1994; Van Devender *et al.* 1994; Anderson and Van Devender 1995; Wright 1996; Benson *et al.* 1997). Earlier chapters in this book have already discussed the scope and limitations of the evidence used to reconstruct former environments. A criticism frequently levelled at the producers of palaeoclimatic reconstructions based on such proxy data is that they proceed from the primary data (for example, pollen evidence) directly to an inference about climate, when it might be more appropriate to ignore climate and simply deduce what one can about that aspect of the environment (in this case, vegetation) with which the evidence is most directly concerned. Such criticism is often justified, and just as often forgotten.

A notable exception to many of the above strictures are the attempts by the CLIMAP project members to reconstruct seasonal changes in global geography at the last glacial maximum, taken to be 18 ± 3 ka (CLIMAP Project Members 1976, 1981). Much of this work was based on the great improvements in our understanding of past ocean behaviour made possible by advances in marine biostratigraphy and in oxygen isotope analyses of suitable deep-sea cores (Hays *et al.* 1976a; McIntyre *et al.* 1976). The CLIMAP palaeogeographical reconstructions, especially of winter and summer sea-surface temperatures, provided useful constraints for GCM models (Gates 1976), stimulating fresh modelling experiments, some of them devoted to testing the role of past variations in the obliquity and precession of the Earth in controlling or influencing Quaternary climatic changes (Hays *et al.* 1976a; Imbrie and Imbrie 1979; Kerr 1986; Kutzbach and Guetter 1986; Ruddiman *et al.* 1986; COHMAP 1988). There have been some very fruitful by-products from the CLIMAP and COHMAP projects and related efforts to test the models against the observational data, and vice versa. One very useful outcome has been to expose possible discrepancies between palaeotemperature estimates from land and sea (Rind and Peteet 1985; Lézine and Hooghiemstra 1990; Guilderson *et al.* 1994), thereby helping to improve future models. However, the ultimate test of any global circulation model is how well it can simulate present-day cli-

matic patterns, so we begin with a brief account of our present climate.

PRESENT-DAY GLOBAL ATMOSPHERIC AND OCEANIC CIRCULATION PATTERNS

Global atmospheric circulation is a function of the amount of solar energy received at the surface of the Earth. Insolation is at a maximum in intertropical latitudes, but the incidence of incoming short-wave solar radiation varies seasonally as a result of the tilt in the Earth's axis, and the annual apparent migration of the overhead sun between the two tropics. Variations in the distribution of land, sea and ice, as well as in cloud cover, will also modify the heat budget of the Earth, so that in some latitudes or regions there is a net heat loss from the surface of the Earth by outgoing long-wave radiation. Feedbacks are shown in Fig. 1.6, and average global climate in Figs 1.3 and 1.4 of Chapter 1. Our concern here is with seasonal variation. Figure 12.1 shows the mean temperature distribution over land and sea during January and July, reflecting the impact of the northern winter/southern summer and northern summer/southern winter respectively. The near-surface changes in wind patterns associated with these seasonal temperature (and pressure) patterns are illustrated in Fig. 12.2. The rotation of the Earth and the variation in speed of surface rotation with latitude (fast at the equator, slow at the poles) modifies the simple Hadley circulation model outlined in Chapter 9 and is responsible for the oblique orientation of westerlies and trade winds. Figure 12.3 is a more detailed portrayal of sea-surface temperatures, shown here for February and August to allow for the slower heating and cooling of the oceans relative to the land. Between them, Figs 1.3, 1.4, 12.1, 12.2 and 12.3 give a highly generalised picture of the geographical and seasonal variations in temperature and precipitation, and of the global distribution of presently wet, dry and cold climatic zones in relation to atmospheric and oceanic circulation. Earlier chapters have described changes in the late Quaternary distribution of land, sea and ice, showing how the areas then occupied by deserts and by humid regions show a significant contrast with the present. We now enlarge on this contrast, focusing initially upon the CLIMAP reconstructions of sea-surface temperatures.

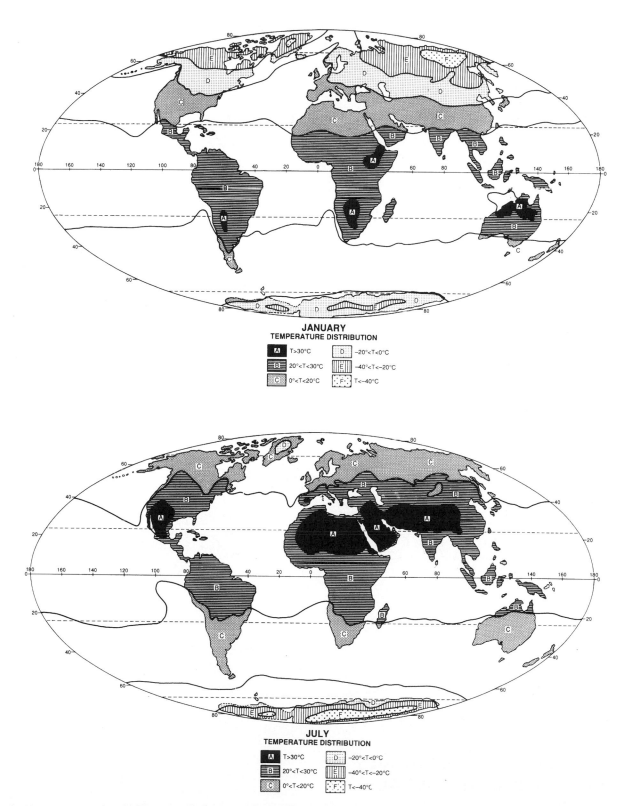

JANUARY
TEMPERATURE DISTRIBUTION

A	T>30°C	D	−20°<T<0°C
B	20°<T<30°C	E	−40°<T<−20°C
C	0°<T<20°C	F	T<−40°C

JULY
TEMPERATURE DISTRIBUTION

A	T>30°C	D	−20°<T<0°C
B	20°<T<30°C	E	−40°<T<−20°C
C	0°<T<20°C	F	T<−40°C

Fig. 12.1 Present-day temperature distribution in January and July. (After Tanke and Gulik 1989)

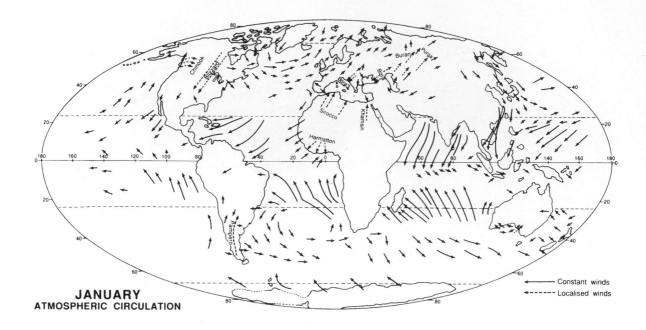

JANUARY
ATMOSPHERIC CIRCULATION

Constant winds
Localised winds

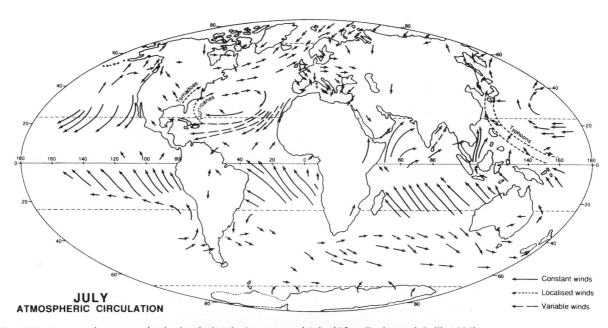

JULY
ATMOSPHERIC CIRCULATION

Constant winds
Localised winds
Variable winds

Fig. 12.2   Present-day atmospheric circulation in January and July. (After Tanke and Gulik 1989)

SEA-SURFACE TEMPERATURES DURING THE LAST GLACIAL MAXIMUM

Figure 12.4 is a simplified version of the CLIMAP (1976) reconstruction of sea-surface temperatures during the northern summer (August) 18 000 years ago, but replotted on the map projection used throughout much of this book. It was prepared by converting numerical estimates of three planktonic groups (foraminifera, radiolaria and coccoliths), the fossil remains of which were preserved in deep-sea sediments laid down around 18 ka BP, into estimates of

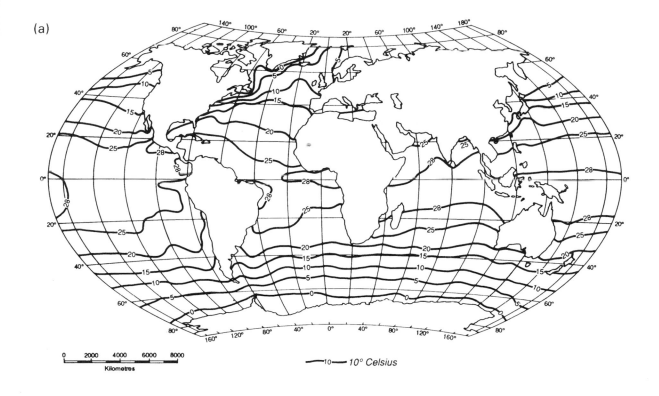

(a)

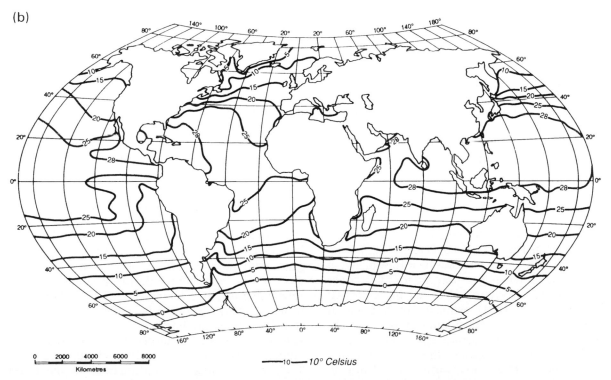

(b)

Fig. 12.3 Present-day sea surface temperatures (a) in February and (b) in August. (After Bartholomew *et al*. 1980; Gorshkov 1978)

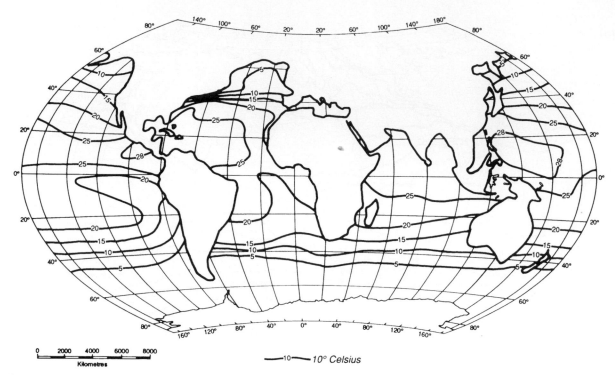

Fig. 12.4 Reconstructed August sea-surface temperatures during the last glacial maximum (18 ka). (After CLIMAP Project Members 1976)

sea-surface temperatures at last glacial maximum using the transfer function statistical technique described in Chapter 7. Only in the Pacific were all three planktonic groups used, and there were occasional major discrepancies ($\geq 1.6°C$) between temperatures in the eastern equatorial Pacific estimated from the zooplankton and phytoplankton groups, in which case the coccolith values were used. These and other sources of error were discussed in detail by the CLIMAP Project Members (1976).

The next step is to compare the differences between measured present-day sea-surface temperature values and those deduced for the last glacial maximum. Figure 12.5 does this for February and August, using the CLIMAP Project Members (1981) data as evaluated by Rind and Peteet (1985), who found that the 18 ka temperatures near Hawaii and in the Pacific subtropical gyre were up to 2°C warmer then than now. They concluded that the CLIMAP Project Members (1981) estimates of last glacial maximum sea-surface temperatures at low and subtropical latitudes, were inconsistent with estimates of 18 ka land temperatures in those latitudes, and were probably up to 2°C too warm. A problem with this conclusion is that the comparison between 18 ka sea-surface temperatures and 18 ka land temperatures estimated from tropical upland pollen spectra and snowline depression on tropical mountains

(Hawaii, Colombia, East Africa and New Guinea) may not be appropriate. Indeed, a more relevant comparison is with lowland pollen taxa blown offshore and preserved in 18 ka marine sediments, since the glacial lowering of temperature was apparently less (-2 to $-4°C$) in certain tropical lowlands than in the adjacent high tropical mountains (-5 to $-15°C$) and much more in line with the adjacent marine record (Bonnefille *et al.* 1990; Van Campo *et al.* 1990; Lézine and Hooghiemstra 1990; Van der Hammen and Absy 1994; Colinvaux *et al.* 1996b).

The reconstruction of former changes in land and sea temperatures is a prerequisite for reconstructing the associated changes in global atmospheric circulation patterns, for only thus can we really come to grips with possible causes of Quaternary climate change. The comparative dearth of southern hemisphere marine palaeotemperature records, especially in high latitude seas like the Southern Ocean, has stimulated some exciting new research relating to the seas around Australia during the last glacial maximum (LGM) (Barrows *et al.* 1996). Figures 12.6 and 12.7 are preliminary reconstructions of the LGM February and August sea-surface temperatures and temperature anomalies relative to the present-day around Australia. These four maps provide a very useful refinement of the CLIMAP reconstructions of

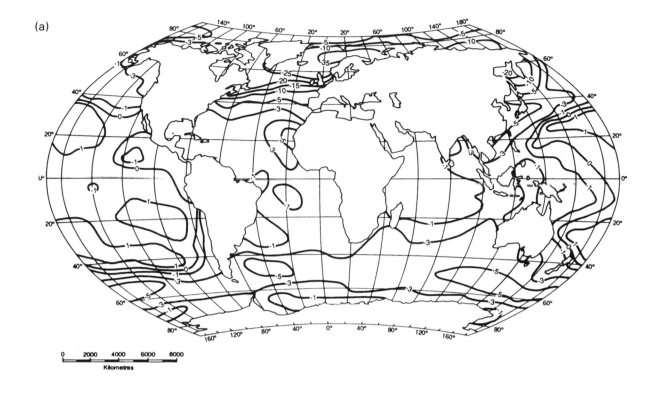

(a)

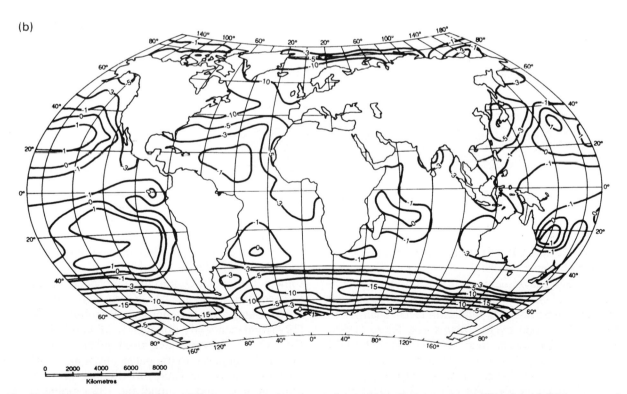

(b)

Fig. 12.5 Sea-surface temperature differences between present-day and last glacial maximum (18 ka) for (a) February and (b) August. (After Rind and Peteet 1985)

February sea-surface temperature map (LGM)

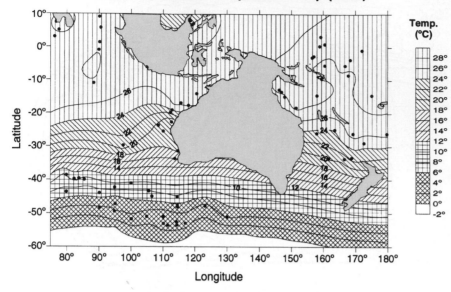

August sea-surface temperature map (LGM)

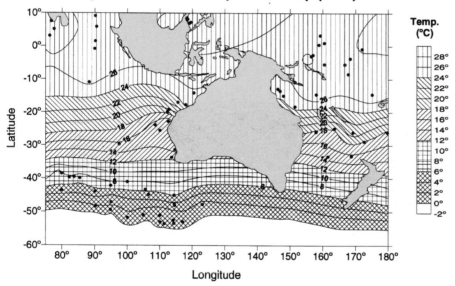

Fig. 12.6 Reconstructed February and August sea-surface temperatures during the last glacial maximum (18 ka). (After Barrows *et al.* 1996)

15 years earlier, and will continue to be improved as new data become available.

A further reason behind current efforts to reconstruct global climates during the LGM has to do with concerns over possible human-induced changes to the global carbon cycle. Adams and Faure (1996) have compiled world maps of the inferred LGM (18 ka) and early Holocene (8 ka) vegetation, and a major international project is now under way to prepare continental-scale maps of the 18 ka plant cover (Adams and Faure 1997). Using present-day estimates of the above- and below-ground carbon storage in major biomes, they concluded that about 1000 gigatonnes of carbon (GtC) were released from land

ecosystems as the earth shifted from interglacial to full glacial conditions, equivalent to a release of some 476 ppm of CO_2 into the atmosphere (1 ppm of CO_2 equates to about 2.1 GtC) (Adams and Faure 1996). Using a similar approach, Friedlingstein *et al.* (1995) independently obtained a value of 612 ± 105 GtC. Other workers have used global circulation models (as opposed to the palaeoenvironmental data used by Adams and Faure) to simulate present and LGM climates and to model the 18 ka global vegetation cover and terrestrial carbon storage. Using this approach, Prentice *et al.* (1993) inferred a 300–700 GtC increase in terrestrial carbon storage following the last glacial maximum. The estimates

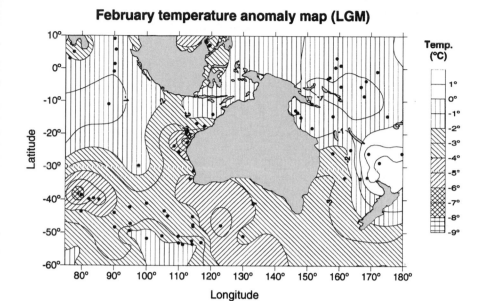

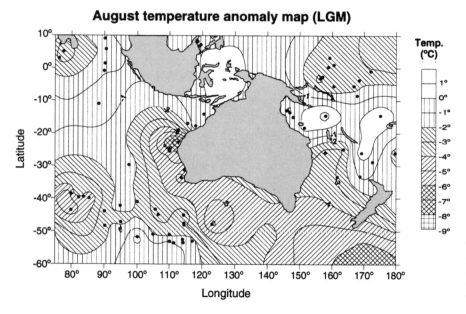

Fig. 12.7 Reconstructed February and August sea-surface temperature differences between present-day and the last glacial maximum (18 ka). (After Barrows *et al.* 1996)

from both approaches accord with changes in oceanic ^{13}C inferred from measurements on benthic foraminifera, which suggest increases in carbon storage on land amounting to 500–1000 GtC (Prentice *et al.* 1993; Adams and Faure 1996).

Three key questions emerge from these exercises. Did changes in atmospheric CO_2 precede or follow global changes in ice volume, sea-surface temperature, and terrestrial plant cover? What combination of factors and feedback loops was responsible for the contrast in atmospheric composition between glacials and interglacials? To what extent have human activities during the late Holocene decreased the amount of carbon stored in terrestrial ecosystems?

These questions are considered elsewhere in this volume, but we can at present only offer qualified answers to all three.

ATMOSPHERIC CIRCULATION PATTERNS DURING THE LATE QUATERNARY

Attempts to model Quaternary global atmospheric circulation patterns are necessarily hypothetical, and are usually limited to the last 20 000 years or so by the lack of well-dated marine and terrestrial evidence

(Davis and Sellers 1994; Wright 1996). One approach much favoured by palaeoclimatologists is to compare the LGM (18±3 ka) with the early Holocene (9±2 ka), which is often considered the equivalent of an interglacial.

An example of this approach is the attempt by Nicholson and Flohn (1980) to identify changes in the position of the intertropical convergence zone (ITCZ) over Africa during the northern summer and winter at three contrasted times (18 ka, 10–8 ka, 6.5–4.5 ka; Fig. 12.8). The merit of such an attempt, however much one may dispute the detail, is that it is inherently testable and refutable (Rognon 1987; Coetzee and Van Zinderen Bakker 1989; Littman 1989). The presence or absence of warm or cold oceanic water offshore at particular times, and of aridity on land, can be tested against the evidence of ocean cores (Sarnthein *et al.* 1982; Pokras and Mix 1985; Diester-Haass *et al.* 1988; Rognon and Coudé-Gaussen 1996). Pollen spectra from marine and terrestrial sediments and lake-level histories can be used to check estimates of the efficacy and areal extent of monsoon rains and the ITCZ (Butzer *et al.* 1972; Street and Grove 1979; Hastenrath and Kutzbach 1983; Williams *et al.* 1987; Bonnefille *et al.* 1990; Gasse *et al.* 1990; Lézine 1991; Lézine and Casanova 1991; Maley 1997; Bergonzini *et al.* 1997). Several workers have also noted that the late Quaternary African monsoons were highly sensitive to changes in the Earth's orbital geometry (Kutzbach *et al.* 1996), and this is reflected in fluctuations in Nile discharge over the past 464 000 years (Rossignol-Strick 1983) as well as in fluctuations in lake levels throughout the tropics during the past 18 000 years (Kutzbach and Street-Perrott 1985).

The strong contrast between the mainly warm and wet early Holocene climate of Africa and the mainly cold, dry and windy terminal Pleistocene climate

(Williams 1985b; Kadomura 1995; Baker *et al.* 1995; Thomas and Thorp 1995; Stokes *et al.* 1997) has its counterpart in peninsular India (Duplessy 1982; Williams and Clarke 1984, 1995) as well as in Australia (Fig. 12.9). As a result of lower sea-level during the last glacial maximum, conservatively estimated at −135 m (although some estimates opt for a lowering of −175 m for northern Australia), the land area of Australia was increased by a fifth and what is now the Arafura Sea and the Gulf of Carpentaria was a land bridge over 1200 km wide linking the much-enlarged mainland of Australia to a slightly bigger Papua New Guinea. Both summer and winter rainfall were well below present levels in the tropical northeast (Kershaw 1978) and the temperate southeast (Bowler 1978; Dodson and Ono 1997), and many of the now-vegetated dune systems were then active as shown in Fig. 12.9a. Pollen evidence from Papua New Guinea, northeastern and southeastern Australia, supplemented by geomorphic evidence from elsewhere in Australia, and by geochemical, microfossil, trace element and isotopic analyses of certain volcanic lakes in Victoria, all point to warmer and wetter early Holocene climates in much of Australia. Times of maximum precipitation may not have been synchronous throughout Australia, since the peak in tropical summer rainfall towards the middle of the Holocene seems to have coincided with a phase of reduced winter rainfall in the Victorian maar lakes of the far southeast (De Deckker *et al.* 1988; Williams 1994c).

Figure 12.9 illustrates another aspect of Quaternary environmental change alluded to in Chapter 6. The late Pleistocene land bridge connecting Australia with Papua New Guinea diverted the warm south equatorial current from its path between these two large islands, thereby depriving northern Australia of a major source of moist maritime air, further accen-

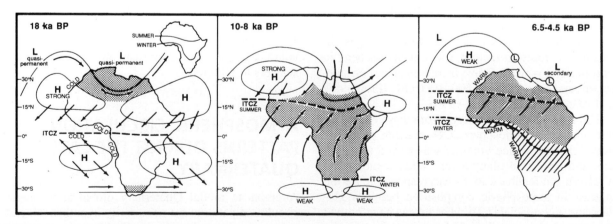

Fig 12.8 Putative atmospheric circulation scheme for Africa during the very late Pleistocene (20–12 ka), the early Holocene (10–8 ka) and the mid-Holocene (6.5–4.5 ka). (After Nicholson and Flohn 1980)

tuating late Pleistocene aridity in that region. Sub-aerial exposure of the Great Barrier Reef had a similar effect, and it was not until the very early Holocene that the postglacial rise in sea-level finally submerged both reef and land bridge, causing a sudden influx of moist air to what are now the coastal lowlands of tropical northern Australia. The early Holocene increase in summer rainfall was reflected in the sudden expansion of rainforest in northeastern Australia into localities previously covered in eucalypt forest or woodland (Kershaw 1978).

A final example will suffice to show the value of using a combination of field evidence and theoretical

models to reconstruct former Quaternary environments in different parts of the world. The modelled reconstruction of likely wind directions over North America, the North Atlantic and Europe during the last glacial maximum winter (18 ka, January) by Kutzbach and Wright (1985; and Fig. 12.10) is based on prior knowledge of the probable extent and thickness of ice caps and sea-ice in this region at this time. Unlike the 18 ka climate reconstruction for Africa (Nicholson and Flohn 1980; and Fig. 12.8), this modelled reconstruction shows both high- and low-altitude winds, and identifies the probable winter and summer temperature and moisture status of a

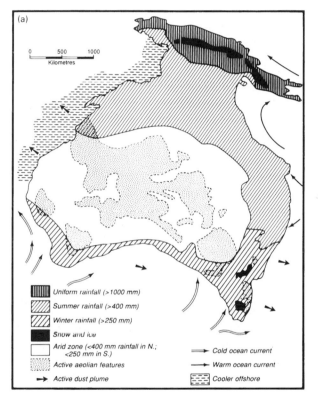

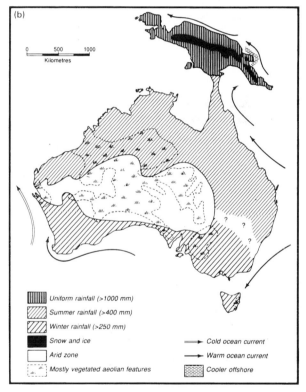

Fig. 12.9 Morphoclimatic map of Australia–Papua New Guinea, during (a) the last glacial maximum (18 ka) and (b) the early Holocene (9 ka). (After Williams 1984c)

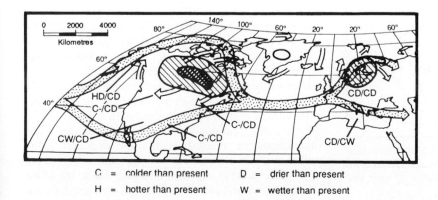

Fig. 12.10 Modelled reconstruction of high-altitude winds (continuous dotted arrows) and low-altitude winds (short arrows) for January during the last glacial maximum (18 ka) over North America, the North Atlantic and Europe. Also shown are the January (before backslash) and July (after backslash) temperature and precipitation anomalies. (After Kutzbach and Wright 1985)

number of localities relative to the present day. It is interesting to note that even during the mostly colder and drier LGM, certain localities are inferred to have been seasonally warmer then than now.

Predictions of this sort are testable, and can be a useful guide for further field research. The 9.5 ka wind directions shown in Fig. 12.11 differ from those shown in Fig. 12.10 in being based on early Holocene dune alignments around the margins of a much-shrunken Laurentide ice sheet. Nevertheless, the anticyclonic wind directions associated with the ice cap-induced high pressure system are still very much in evidence, enhancing the credibility of the 18 ka reconstruction of surface wind directions. Later work has confirmed the validity of this approach (Wright 1996; Yu *et al.* 1997), while giving some emphasis to the regional influence upon climate of large glacial lakes such as Lake Agassiz.

GLOBAL PALAEOHYDROLOGY AND LINKS BETWEEN OCEANIC AND ATMOSPHERIC CIRCULATION

Mean global precipitation on land and sea amounts to 857 mm/a, equivalent to roughly 400×10^3 km³/a over the oceans and 100×10^3 km³/a over the land. Although 80% of all precipitation falls over the oceans, evaporation from the oceans yields 440×10^3 km³/a of water equivalent, or about 86% of the land precipitation. In other words, the oceans supply an extra 6% of the land precipitation, amounting to 36×10^3 km³/a, which is returned to the oceans as a similar volume of annual runoff (Bloom 1978). Of the total precipitation over land, roughly a third is

from maritime sources and two-thirds from continental evapotranspiration from plants, lakes, swamps and rivers. During interglacial (and present) times, some 97.6% of the world's water supply is stored in the oceans, and about 2% in ice caps. During glacial maxima, about 10% is stored in ice sheets and 89.5% in the oceans. The present ice sheets contain 26×10^6 km³ of ice, equivalent to a sea-level change of 65 m for a world ocean surface area of 362×10^6 km². During the last glacial maximum the total ice volume was about 77×10^6 km³, equivalent to 197 m of sea-level change (Flint 1971). The sea-level fell about 135 m below present level at 18 ka (see Chapter 4) which is remarkably close to Flint's (1971) estimate of around 132 m (197–65 m).

The global water budget summarised above has oscillated between interglacial and glacial modes at orbitally controlled intervals throughout the Quaternary. Within these two extremes there were countless smaller fluctuations (Petit-Maire *et al.* 1991; Williamson and Oeschger 1993; Thompson *et al.* 1995; Ditlevsen *et al.* 1996; Marsh and Ditlevsen 1997), many of them linked to changes in oceanic circulation patterns which in turn influenced the atmospheric circulation over land and sea (Broecker and Denton 1989, 1990; Porter and An 1995; An and Porter 1997; Rousseau and Wu 1997). Although some of the attempts to relate sudden climatic fluctuations to changes in the North Atlantic deep-water flux during the late Quaternary perforce remain speculative (Broecker *et al.* 1985; Street-Perrott and Perrott 1990; Charles and Fairbanks 1992), they are based on an impressive and growing body of well-dated palaeoceanographic evidence (Prell 1984; Boyle 1990; Hasselmann 1991).

Several of these attempts to interpret possible late Quaternary interactions between ocean, atmosphere and land deserve more detailed comment. In Chapter 8 we discussed the historic influence of El Niño–Southern Oscillation (ENSO) events upon droughts and floods in Africa, Australia, India and China as a classic example of how ocean–atmosphere interactions can control extreme climatic events on land at scales of 10^{-1} to 10^2 years. Very precise high-resolution measurements of the strontium/calcium ratios in early Holocene corals from the Huon Peninsula in Papua New Guinea have revealed that the equatorial western Pacific ocean was about 2–3°C cooler towards 7.3–8.9 ka, and that stronger ENSO events may have been more common at that time (McCulloch *et al.* 1996). In the Santa Barbara basin off the southwest coast of California, century-scale fluctuations in coastal vegetation deduced from pollen analysis of laminated marine sediments reveal a possible ENSO signal extending back to 24 ka (Heusser and Sirocko 1997). In the next section, we

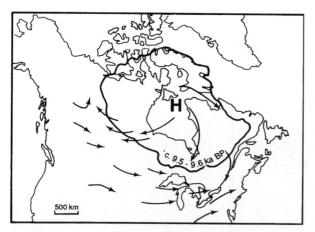

Fig. 12.11 Map of the early Holocene (9.5 ka) Laurentide ice sheet with anticyclonic wind directions as inferred from alignments of early Holocene dunes. (After Kutzbach and Wright 1985)

focus on late Quaternary climatic changes at scales of 10^2 to 10^4 years.

Paradoxically, the best records of late Quaternary climates in peninsular India are those furnished by deep-sea cores from the northern Indian Ocean and Arabian Sea (Prell *et al.* 1986; Duplessy 1982; Prell 1984). Duplessy (1982) used differences in the oxygen isotopic composition of planktonic foraminifera (see Chapter 7) from the northern Indian Ocean to reconstruct the probable Holocene and late Pleistocene climates on land. He concluded that the southwest summer monsoon was weaker at 18 ka than it is today, since the upwelling along the southern coast of Arabia had disappeared at that time, implying much weaker southwesterly winds, and so much reduced summer rainfall. He also deduced that the Ganges and Brahmaputra contributed much less water to the ocean at 18 ka, since the salinity gradient in the Bay of Bengal was very much steeper than today. The inference of glacial aridity in India drawn from the ocean cores is entirely consistent with the aeolian and pollen evidence from northwest India (Goudie *et al.* 1973; Singh *et al.* 1974) and with the alluvial and fossil evidence from the Son and Belan rivers in north-central India, which suggests that they were more seasonal and had more sparsely vegetated catchments at 18 ka than during the Holocene (Williams and Royce 1982; Williams and Clarke 1984, 1995). Prell (1984) used the presence or absence of cold upwelling along the southern coast of Arabia, as deduced from deep-sea cores, to reconstruct a longer history of monsoons in this region. Phases of weakened summer monsoons appear to coincide with the 21 ka orbital precession cycle (Prell 1984).

With improvements in the resolution with which late Quaternary marine cores can now be dated, many workers now accept that the most recent deglaciation (c. 16–7 ka) consisted of several distinct climatic oscillations, with rapid ice melting at 14–12 ka and 10–7 ka separated by an interval of no or very little melting (Duplessy *et al.* 1981; Berger *et al.* 1985;

Fairbanks 1989; Jansen and Veum 1990). Particular interest has been focused on the Younger Dryas event, long recognised by Scandinavian palynologists as a return to a cold, dry, near-glacial climate in northwest Europe for a few centuries between 11 and 10 ka (Blikra and Longva 1995; Hajdas *et al.* 1995). The Younger Dryas cold phase is also evident in North Atlantic deep-sea cores, in the melting history of the Laurentide ice sheet, in Greenland ice cores, in African and Tibetan lake histories, in the alluvial history of the White Nile, and even in the isotopic composition of freshwater mollusca from southern Africa (Leventer *et al.* 1982; Dansgaard *et al.* 1989; Broecker *et al.* 1989; Fontes *et al.* 1993; Denton and Hendy 1994; Goslar *et al.* 1995; J.E. Smith *et al.* 1997; Anderson 1997).

An ingenious attempt to explain the abrupt climatic oscillations that characterise the last deglaciation is that of Broecker and colleagues, who argue that the global oceanic thermohaline circulation is controlled by fluctuations in Atlantic salinity (Broecker *et al.* 1985; Broecker and Denton 1989). How far such Atlantic salinity changes reflect changes in glacial meltwater influx is still a moot point (Rahmstorf 1995; Manabe and Stouffer 1995; Adkins *et al.* 1997), but there is no longer any argument about the importance of Quaternary ocean–atmosphere interactions in controlling climatic fluctuations on land, whether modulated by Milankovitch forcing or not (Kutzbach and Street-Perrott 1985; Gallimore and Kutzbach 1996; de Noblet *et al.* 1996; Wright 1996; Pollock 1997).

As scientific understanding of the actual and potential impact of human actions upon world climate improves, there is likely to be considerably more research focused upon elucidating past and present human interactions with our environment. For examples drawn from India, Japan and China, see Agrawal (1995), Yasuda (1995) and Zhang (1995), respectively. Present and possible future human impacts are the subject of the following chapter.

CHAPTER
13

ENVIRONMENTAL CHANGES: PAST, PRESENT, FUTURE

The evidence suggests that agriculture has been practised for 7000 years, and the consequent reclamation and cultivation of land has completely destroyed the natural vegetation in many areas.

Zhao Ji, Zheng Guangmei, Wang Huadong, Xu Jialin
The Natural History of China (1990)

We have now reviewed the characteristics of the major global environments that existed during the Quaternary. The inescapable conclusion is that during the Quaternary, change was essentially ubiquitous. Environments globally, regionally and locally were unstable, from time to time displaying increasing dryness or more widespread pluvial conditions; falling or rising sea-level; changes in windiness, continentality, seasonal solar radiation, and dustiness of the atmosphere. On land, snowlines fluctuated through a height range of more than 1 km; vegetation communities adapted repeatedly to altered conditions; lakes dried out or overflowed; rivers changed in character from stable meandering systems to aggrading braided ones and vice versa; dunes were periodically active or stable; and soils were affected by deflation, glacial scour, erosion by meltwater floods, and ground ice effects. In the oceans, temperatures, carbonate balance, circulation, oxygenation, nutrient availability and biological productivity were all subject to significant variation.

In considering the natural archives, where information on much of this variability resides, we have seen evidence for both slow and very rapid change. In view of the long periods that are needed for evolutionary change in organisms, for the evolution of new landforms, and for the Earth to adjust to changing ice and water loads, we can safely conclude that many aspects of the Earth's environments must, in the Quaternary, have been essentially permanently prevented from reaching a final equilibrium. Rather, there must at all times have been many readjustments underway simultaneously. This seems to be the case with sea-level today: adjustments from the last deglaciation are still taking place (Chapter 6) while the climate may already be changing in such a way as to destabilise sea-level once more.

This makes clear to us one of the great benefits to be had from a knowledge of the Quaternary and its environments: it allows us to perceive our present circumstances in their proper context, and dispels any lingering ideas that we may have entertained that the global environment is a stable backdrop to human affairs or biological evolution. Furthermore, from the better-known parts of the late Quaternary, at least, it is clear that no previous time has been quite like the present (Howard 1997). The previous interglacial (the Eemian), which we might have expected to be a time at least broadly similar to the present, had higher sea-levels and may have been quite unstable climatically. In contrast, the Holocene seems only to have experienced changes that were relatively slow and of lower amplitude than those inferred to have occurred in the Eemian. Therefore, we cannot rely on finding Quaternary records that will enable us, by analogy, to know exactly how the environment will change in the future, as the present interglacial comes to an end. Nor, unfortunately, can we expect to find strict Quaternary analogues of a world affected by human activity, to see what the effects of, for example, global warming might be on the land, oceans and the biosphere in coming decades. Bearing these caveats in mind, we can none the less identify many additional benefits that can be derived from refining our knowledge of the Quaternary.

Indeed, such benefits are manifold and important. The study of Quaternary environments has required investigation of the processes of readjustment to changes that are exhibited at various temporal and spatial scales. We have had to investigate a diversity of linkages through which changes in one global system affect others. In seeking explanations for the events recorded in the natural archives, researchers have had to pose, test and refine many key ideas about the operation of these global environmental mechanisms. The climatic role of the global thermohaline system of the oceans (the oceanic 'conveyor belt') and the mechanisms that affect it has formed one such study; the conditions required to support ice growth and decay on land has been another. In combination, the environmental records derived through the investigation of the Quaternary, and the mechanisms that have been proposed to account for the observed changes, provide a vital resource for modelling and understanding the global environment. Thus, although there may be no strict analogues of the present or the near future, the study of Quaternary environments has provided a good deal of what is required in order to understand, model and predict future changes. The mechanisms that operate within global environments, and which must be understood in order to forecast future environmental changes have, in many cases, been elucidated as a result of investigating the Quaternary. As has been made clear in earlier chapters, there are still very significant gaps in our knowledge, so that ongoing research is required to fill these and to refine and extend what is already known.

Among the reasons why strict Quaternary analogues of the contemporary environment, and equally of future environments, cannot be found, we must highlight the diverse activities of people as agents of environmental change. We are responsible for a diversity of activities affecting the environment, and in this chapter we will consider a selection of these. These few examples should serve to make the point that we are now capable of influencing the future course of environmental changes, and must employ our developing knowledge of the mechanisms lying behind the global environment and its behaviour in order to forsee and manage our potential impacts. As we shall see, wrought carelessly and in ignorance, changes to environmental processes caused by people may seriously upset mechanisms which previously regulated aspects of the environment, many of which we are yet to understand fully. Our present environment, in the global sense, is governed by the kinds of conditions that existed in earlier brief, warm interglacial periods: the sea-level is high, the atmosphere is moist and warm (and hence relatively enriched in the natural greenhouse gases), and land and sea ice are restricted largely to the high northern and southern latitudes. Under the continuing influence of the Milankovitch solar forcing, we may expect that in due course these conditions will give way gradually to increasing cold and eventually to the renewed development of glacial conditions. Indeed, much of the evidence reviewed earlier in this book offers the possibility that the peak of the present interglacial interval may have been reached about 4 ka BP, and that we are already experiencing descent into the next glacial stage.

THE HUMAN POPULATION IN THE CONTEXT OF THE LATE QUATERNARY

The Quaternary saw the development of humankind and the diversification of our culture. Exceptionally rapid growth in the human population and its cultural richness has occurred during the last glacial and the Holocene (Chapter 11). These developments have certainly been fostered by climatic amelioration, and the concomitant development of the organised Neolithic agricultural food production that this has permitted. We now exist as a technologically powerful, numerous, and resource-consuming group of organisms such as has not been present in any earlier interglacial. Growth of the human population up until the present displays an ever-increasing acceleration which is more than exponential. The present net rate of global population increase is nearly 100×10^6 per annum (i.e. about 2 million additional people every week!). Simultaneously, as technological and cultural development continues, our demands for energy and resources are also accelerating (Gilland 1988; Simmons 1989). Should global population growth continue at its present rate, the world population (presently 5.7 billion) would double in only 43 years (Cohen 1995).

What are the implications of this special circumstance? An obvious one arises from our generation of greenhouse gases, and the inadvertent climate change that this may induce, which we consider briefly below. The height of an interglacial is the least desirable time to force climatic warming: the global climate is already at or near the peak of its natural temperature variation (at least on the Quaternary time-scale), and the added heat arising from the anthropogenic greenhouse effect thus amounts to a gigantic global experiment. Fortunately, our awareness of our relationship to the environment, and our dependence upon it, have also evolved to a level not seen before. We possess significant knowledge which can be brought to bear on many issues now facing us as a species. The perspective of the Quaternary is a useful one here, and permits us to understand the history of our environment and to see many of the ways in which it has responded to previous stimuli. With the aid of this knowledge, we are able to evaluate the possible consequences of deliberate or inadvertent human impacts. In addition, many of our landscapes, and the soils which support us, our animals and our crops, are legacies of events occurring during the Quaternary, and our growing understanding of these events will be important in aiding our stewardship of the Earth's resources for the future.

BIOTA IN THE QUATERNARY

As described in earlier chapters, the 8–10 ka period of warming that marked the end of the last glacial brought global temperatures back to levels that had not been experienced since the preceding interglacial 100 ka before. This was thus a rapid temperature swing to which biota had to adjust. Plant and animal communities in glacial or interglacial times survive ensuing climatic changes only in areas which provide a habitat where they are able to reproduce. Temperate forests of the interglacials survive glacial conditions only at low altitudes and latitudes; cold-climate communities survive the interglacials only at high altitudes and latitudes. Such areas where organisms may continue to exist despite shifts in their habitat are termed *refugia*. There are organisms which have survived the Tertiary climatic cooling and desiccation in such refugia, and certainly remnants of glacial communities also exist today through this mechanism. During glacial conditions, the oak, pistachio and olive woodlands of the Mediterranean contracted to refugia which probably included Israel and Jordan (Roberts 1989) and possibly other areas of the southeastern Mediterranean. During postglacial times, tree species recolonised in a westerly direction, replacing the glacial herb–steppe communities. Low-latitude rainforests in Africa, South America and Australia were similarly much more restricted in their extent during the dry full-glacial conditions, retreating to as yet unknown refugia (Nicholson 1989). Refugia from glacial cold and altered food availability must also have been exploited by many animal groups. The rates at which forest recolonisation from refugia takes place have been estimated in several areas, and range up to several kilometres per annum; for some species, such as the North American beech, the rates are only a tenth of this (Roberts 1988). Cold-tolerant species exist today at higher altitudes in interglacial refugia, much as they must have done at 125 ka BP. This poses particular problems for conservation, owing to the limited extent of some of these communities and their fragmentation among multiple isolated sites. Mapping and identification of the flora and fauna are necessary precursors to informed management of these remaining communities. However, the realisation that continual change is what really characterises the environment should guide attitudes which suggest that we must attempt to preserve (e.g. in reserves) environments in their present, transitory configurations. Now that many ecosystems are isolated by settlements, croplands or plantation forests, opportunities for migration are greatly restricted. Future environmental changes, perhaps anthropogenically caused, will thus pose a new hazard for biota, and human intervention (perhaps through the creation of migration corridors,

or the artificial relocation of organisms to more suitable environments) may be required to preserve communities and genetic diversity. We must recognise too that the composition of many ecosystems has been deliberately modified already through human intervention. The introduction of foreign organisms for cultivation, or for the biological control of pest organisms deliberately or accidentally transported, has been widespread during the last two centuries especially.

DROUGHT, OVERGRAZING AND DESERTIFICATION

During the Quaternary, climatic, hydrological and vegetation changes altered the characteristics of the land-surface, and albedo and moisture feedbacks arose to add to the variation.

The intensifying human use of the landscape has, during the Holocene, wrought new kinds of changes in the landscape. Widespread effects may be identified under the general heading of land degradation (Blaikie and Brookfield 1987). The Mediterranean region provides abundant examples of deforestation, soil erosion, and loss of ecosystem productivity resulting from human occupation. The forests of this region have long been exploited for fuel and building materials, and cleared for agriculture. Destruction of the vegetation has been exacerbated by the numerous flocks of grazing animals. In many areas it is only possible to speculate about the pre-disturbance condition of the vegetation.

In the extensive dry regions across North Africa and into the Middle East, the remains of once-productive agricultural land remind us of these changes. The general deterioration of the soil and vegetation cover caused by human occupation, and associated with falling crop productivity which has caused several infamous famines, has been termed *desertification* (Dregne 1983; Thomas & Middleton 1994; Williams & Balling 1996; Xue 1996). It is a phenomenon now widespread in and around the margins of the world's drylands (Fig. 13.1). However, it is necessary to recall our Quaternary perspective; in parallel with the effects of human occupation of the landscape are the ongoing climatic shifts associated with external causes such as the Milankovitch mechanism. It is thus necessary to isolate the separate effects of these agencies before the true human impact on the landscape can be judged.

Let us consider as an example the infamous famine-prone Sahel region, which runs east–west across Africa, including parts of Senegal, Mauritania, Mali, Niger, Chad, Burkina Faso, Nigeria, Sudan and Ethiopia. This region probably gets its name from an

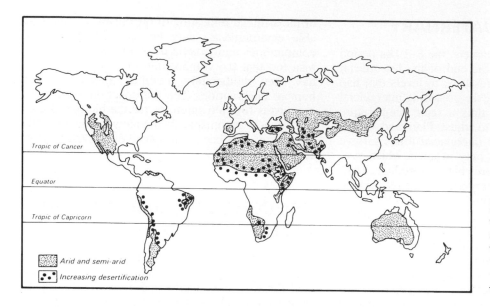

Tropic of Cancer

Equator

Tropic of Capricorn

Arid and semi-arid
Increasing desertification

Fig. 13.1 Regions of the world where there has been a marked increase in desertification between 1977 and 1984. (After M.A.J. Williams 1986, based mainly on data tabulated by Mabbutt 1985)

Arabic word meaning shore or coast, and is indeed the southern margin of the hyper-arid Sahara which dominates Algeria, Libya and Egypt to the north. Across the Sahel, the annual average rainfall declines northward at a mean rate of 1 mm/km; annual totals in the south are around 600 mm, and in the north, around 100 mm (Le Houérou 1989). People have had an impact on the environments of the Sahel for millennia, but this is now increasing because of high population growth rates which have developed especially in the last few decades. Cropping is progressively replacing low-density rangeland grazing, and fallow periods are being reduced to allow more frequent harvests (Le Houérou 1989). In addition, land in the drier northern regions (where the coefficient of variability of the annual total rainfall – inversely proportional to its reliability – is higher) is being pressed into use. For example, in Niger cultivation 40 years ago was restricted to areas south of 15° N (annual rainfall about 400 mm or more) but has now spread north to 16°20′ N, where the annual rainfall is less than 250 mm; high incidence of crop failure has resulted, together with the loss of the previously useful grassland (Le Houérou 1989). Sharply increasing stock numbers, when combined with drought, have left the soil surface unprotected against wind, rain splash, rill and sheet erosion. The soil surface becomes compacted and impervious to water; soil moisture levels fall and large areas become bare in the typical sequence of events involved in desertification. Through events of this kind, the southern margin of the Sahara in some places gives the appearance of having moved 80–100 km southward in the Sudan between 1958 and 1975; similar shifts are also suggested in Mali, Chad and elsewhere.

It is necessary to examine the possible role of external climate change as a fundamental cause of the desert expansion, since ongoing change is, as noted earlier, the hallmark of the Quaternary. Petit-Maire (1990a,b, 1994) has done this in the context of what is known of the Quaternary environments of north Africa. At 130 ka BP, during interglacial conditions, lakes existed throughout the Sahara; however, these dried up progressively as the Wisconsin–Weichsel glacial developed, and at 18 ka BP the arid Sahara extended as far south as 14° N, where relict dunefields have been mapped (Petit-Maire 1990a). Forests retreated to the upland refugia referred to earlier. Human activity in the landscape cannot have been responsible for this environmental change – population numbers were too low, and in any case the same sequence of climatic events is essentially global in its occurrence (see Chapter 9). During the late glacial and early Holocene, rainfall across the Sahel once again increased, lakes refilled, the hyper-arid Sahara retreated northwards, and conditions improved until about 9 ka BP. However, the interval of benevolent climate was shortlived: by about 6 ka BP, drought was repeatedly affecting northern Mali, and the southern margin of the Saharan environment has today extended to about 17° N (Petit-Maire 1990a). It seems unlikely that the Neolithic peoples of this area, who were cattle herders, and who only populated the area after about 7 ka BP (Petit-Maire 1990b), can have been responsible for the late Holocene environmental deterioration (see Chapter 11). Rather, if indeed the peak of the present interglacial has passed, a trend towards aridification is what must be expected, as conditions slowly cool and become more like those of the last glacial. However, overgrazing, soil deterioration, and resultant loss of grasses and woody species act to reinforce the externally-driven tendency. It has been hypothesised that the higher albedo of the desertified landscapes acts through altered fluxes of heat and moisture into the atmosphere, as well as through reduced surface roughness. Atmospheric circulation models predict declining rainfall (especially over the Sahel) and southwards displacement of wetter areas if surface albedo is increased (Nicholson 1989), but empirical verification of these ideas is still required. A considerably more significant role may be played by fluctuations in sea-surface temperatures, and links with global-scale phenomena, such as the El Niño–Southern Oscillation (ENSO) phenomenon, may in due course provide us with some ability to forecast periods of drought in areas like the Sahel (Rasmusson 1987). Thus a working conclusion at present is that there are probably natural external factors promoting drought in areas like the Sahel, but that the conditions are almost certainly made worse by the anthropogenic desertification.

Globally, the significance of the albedo effect and related changes involved in desertification continue to be investigated. Certainly, desertification has affected large areas, including the Sahara and its margins, the Rajasthan desert in India, and smaller famous instances such as the Lebanon, which has suffered massive deforestation. Many of these cases were discussed by Sagan *et al.* (1979), who concluded that globally, in excess of $9 \times 10^6 \text{km}^2$ has been affected; they estimated the rate of growth of desertified land as about 0.1% of the land surface of the globe per decade. The albedo of such surfaces increases from perhaps 0.16 to 0.35 (Sagan *et al.* 1979); the effect of this is to lower the atmospheric temperature, and, on the basis of Quaternary analogues, this must be expected to result in turn in lower rainfall. A similar role for dust deflated from the overgrazed surfaces has been hypothesised: that it may block incoming solar radiation, hence cooling the surface and resulting in descending air and reduced opportunity for precipitation (Hansen and Lacis 1990). Sagan *et al.* (1979) speculated that the long history of anthropogenic desertification, which together with salinisation and deforestation, has now in total affected about 15% of the global land area, may indeed have contributed to global cooling since the relative warmth of the earlier Holocene. They estimated the anthropogenic component of this cooling to amount to $1°C$.

Large-scale deforestation is another concern in several areas of the globe, not least because of the risk that atmospheric moisture fluxes, and hence the atmospheric heat engine and global rainfall patterns, may be adversely affected. In particular, Amazonian and southeast Asian forest areas, where rates of for-

est cover decline are significant, have been investigated from this perspective (e.g. Gash *et al.* 1996). It appears that albedo may be raised by up to 0.05 after deforestation, altering the surface energy balance (Bastable *et al.* 1993; Culf *et al.* 1995). Furthermore, precipitation recycling, which relies on evapotranspiration pumping to return water vapour to the atmosphere, and which accounts for up to 25% of Amazonian rainfall, may be diminished significantly by the removal of forest (Eltahir and Bras 1994; Zeng *et al.* 1996). Forest decline related to Quaternary environmental change in the areas presently affected by clearing would have differed in character from that caused by modern human intervention, not least in terms of the rapidity of the change in vegetation cover.

IRRIGATION AND SALINISATION

Hydrological changes were marked in the Quaternary, as illustrated by the early Holocene pluvial conditions recorded across the Sahara, described in Chapter 9, which arose because of changes in the monsoon system. These changes have been successfully modelled using GCM methods (e.g. Hall and Valdes 1997). But hydrological change is now in many cases caused by people, either deliberately or as an unintended side-effect of human land-use, so that modelling of future changes will require additional kinds of economic and sociological data.

To allow cropping in dry regions, people have long used irrigation. In the Negev desert of Israel, Nabatean peoples 2 ka ago appropriated and directed runoff water on to their fields by major earthworks and stone embankments which 'harvested' water from extensive areas in the surrounding hills. They were successfully able to grow grain crops and fruits, and some of their techniques are being reintroduced at the present day, together with newer techniques, to permit the growth of trees in what is termed the 'savannisation' of the desert (Evenari *et al.* 1971; Israel Land Development Authority 1990). Larger-scale irrigation, employing water diverted from rivers and dams, is, however, associated with the problem of *salinisation*. The irrigation water brings with it dissolved materials which accumulate in the soil to increase the salt content there. Rising water tables produced by the irrigation, combined with strong evaporation at the surface, progressively result in salts being set down in the upper layers of the soils where they produce conditions hostile to continued plant growth. More than 50% of the world's irrigated land is salinised to some degree (Rice and Vandermeer 1990). Techniques to cope with the salt accumulation, including the use of evaporation ponds where the salts are harvested, become essential where the salinisation is significantly reducing crop yields. Severe salinisation, which converts productive land to increasingly bare saline flats, produces a similar albedo increase to that of desertification already described, and undoubtedly also contributes to global environmental change. The effects of salinisation on various civilizations have been documented by Hillel (1994).

Irrigation also represents a disturbance to the hydrology of the landscape surface, in which water is diverted from one area to another, and lost to evaporation and transpiration. Consequences are often felt in the ecosystems that suffer water loss. However, in areas where irrigation water, or water for stock or domestic use, is derived from groundwater, additional problems may arise. In many areas, such as the Great Artesian Basin of Australia, the groundwater that is harvested from boreholes has been moving slowly through the aquifers for tens or hundreds of ka. Much of it must have entered through groundwater recharge areas during wet intervals of the Quaternary. The water thus in reality is a store of 'fossil' groundwater that may not be replaced in the contemporary climatic environment at a rate which can sustain the rate of extraction. If this is so, then extensive use of groundwater from such aquifers will progressively deplete the groundwater store, and will be unsustainable. In Australia, where much of the pastoral country relies heavily on groundwater, there are 400 000 boreholes in use (Habermehl 1985); in some aquifers, extraction is in approximate equilibrium with recharge, but in others, falling yields imply that extraction is proceeding too rapidly. In the Great Artesian Basin, dating of the water by isotopic methods has shown that much of it dates from recharge occurring since the middle Quaternary (Bentley et al. 1986).

HUMAN EFFECTS ON THE ATMOSPHERE

Earlier chapters of this book presented ice core and other data showing that the composition of the atmosphere underwent significant changes during the Quaternary, both in terms of gaseous constituents such as methane, carbon dioxide and water vapour, and in loadings of dust, sulphate particles and other particulates. In this area, modern human activities have once again created a series of environmental forcings unlike those of the earlier Quaternary, in both character and rate of impact.

The human influence on the composition of the global atmosphere may be the most far-reaching of our environmental disturbances. Sulphate particles are generated in industrial processes, fossil fuel combustion and biomass burning. In the atmosphere, these act to reduce incoming solar radiation (Taylor and Penner 1994). On the other hand, many of the activities already mentioned result in the release of infra-red absorbing 'greenhouse' gases, which reduce heat loss from the Earth–atmosphere system (Table 13.1; Gribbin 1986; Pearman 1988; Schneider 1989; Leggett 1990; Bouma et al. 1996).

The metabolism of cattle, for example, releases methane, a very effective greenhouse gas. Modern plastics production and refrigeration plants release the chlorofluorocarbon gases (CFCs), which are potent greenhouse gases whose atmospheric abundance has roughly doubled since observations began in 1977 (Smil 1990). Deforestation and the combustion of fossil fuels all release CO_2, a less potent but none the less important greenhouse gas, as well as nitrous oxide (Cofer et al. 1991). Ice core records (see Chapter 3; Fig. 13.2) together with analysis of relatively young Antarctic firn (Battle et al. 1996) and contemporary observation programmes (Figs 13.3, 13.4, 13.5) clearly indicate that human activities are increasing the concentration of CO_2 and other gases in the atmosphere, and a tendency towards

Table 13.1 Summary of key greenhouse gases affected by human activities

	CO_2 (ppmv)	CH_4 (ppbv)	N_2O (ppbv)	CFC-11 (pptv)	HCFC-22 (a CFC substitute) (pptv)	CF_4 (a perfluorocarbon) (pptv)
Pre-industrial concentration	~280	~700	~275	0	0	0
Concentration in 1994	358	1720	312	268	110	72
Rate of concentration change (% per annum)	0.4	0.6	0.25	0	5	2
Atmospheric lifetime (years)	50–200	12	120	50	12	50 000

Source: Houghton et al. 1996.

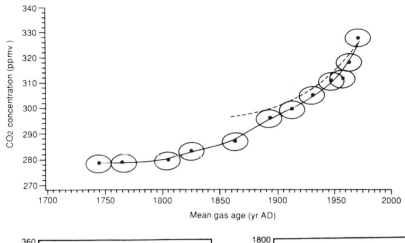

Fig. 13.2 Measured mean CO_2 concentration plotted against estimated age, based on analysis of air occluded in West Antarctic ice. (After Neftel *et al.* 1985)

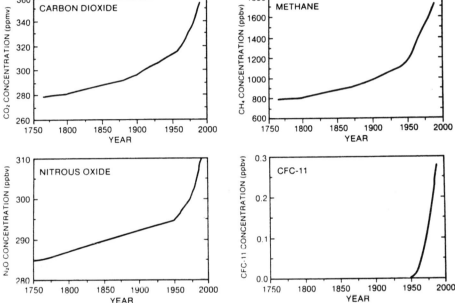

Fig. 13.3 Patterns of increasing concentration of CO_2, methane, nitrous oxide and CFCs over the last 250 years. (From Houghton *et al.* 1990)

global warming is the almost certain consequence. This tendency is reinforced by the increasing absolute humidity of warmer air, since water vapour is also a significant greenhouse gas (Del Genio *et al.* 1991; Rind *et al.* 1991). The pre-industrial CO_2 concentration was not absolutely stable, with levels falling by about 6 ppmv during the Little Ice Age, for example (Etheridge *et al.* 1996). However, the concentration of around 280 ppmv that characterised the period 1000–1600 AD clearly contrasts with the rapid rise toward the present value of nearly 350 ppmv that began around 1800 AD.

Release of CO_2 has resulted from forest clearance and the resulting oxidation of litter and soil organic matter (Woodwell *et al.* 1983). Clear signals of increased global biomass burning over the last century or two, and reflecting the massive expansion of settle-

ment, pastoralism and agriculture, have been recorded in ice-core records (e.g. Holdsworth *et al.* 1996; Legrand and De Angelis 1996). Faure *et al.* (1990) have estimated that 33–47% of the Holocene phytomass equal to 275–490 GT of carbon (1 GT equals 10^{15} g) has been passed into or through the atmosphere by this mechanism. This leaves a present phytomass carbon store of about 580 GT of carbon, compared with the estimated glacial phytomass store of 300 GT and the interglacial value of 900 GT (Faure 1990). There have also been changes in the soil carbon store and in peat deposits, assessed by J.M. Adams *et al.* (1990). In addition to phytomass loss, the combustion of the fossil fuels coal, oil and gas has released about 20×10^9 t of CO_2. Only about 40% of the released CO_2 has evidently remained in the atmosphere, increasing the level in the atmosphere from the pre-Industrial Revolution value of 280 ppmv (parts per million by volume) to the present value of 350 ppmv. The remaining CO_2 has evidently been absorbed by the oceans or other parts of the biosphere in ways that are incompletely understood (Smil 1990).

The effects of the greenhouse gases on the global climate are estimated from numerical models of the global atmosphere. These are still developing and do not incorporate all of the processes which must be modelled, nor do they have the spatial resolution to forecast climatic outcomes reliably at continental or regional scales. In addition, the prediction of future climatic outcomes depends upon the adoption of an appropriate scenario describing the future generation of the gases involved. This is complex because it involves predictions of the growth of the global population, together with its demands for energy and resources. One scenario commonly employed is the 'business as usual' scenario which involves extrapolation of contemporary trends. Other scenarios involve the foreshadowed adoption of controls on both population and the generation of greenhouse gases; the rate of adoption of such controls varies from scenario to scenario. There is no way at present to know which (if any) scenario will approximate the final outcome.

The general conclusion from the 'business as usual' scenario is that atmospheric CO_2 levels may double in the next 50 years or so. While the climate models yield quite variable results for the climatic consequences of this, a general conclusion is that the mean atmospheric temperature will rise by $3.5 \pm 1.5°C$ (Fig. 13.6). In consequence of this change, many other effects follow. There will be retreat of valley glaciers, contributing water to the oceans. The sea-level will tend to rise in response, but largely because of the thermal expansion of the surface water (Wigley and Raper 1987; Mikolajewicz *et al*. 1990). However, it has been suggested that the warmer oceans will yield higher rainfalls in many areas, possibly including the Antarctic ice cap, so that water may also be removed from the oceans. On the basis of the effects presently understood, the predicted overall sea-level response is for a rise of 8–29 cm, most probably 18 cm, by 2030, and of 21–71 cm, most probably 71 cm, by 2070 (Warrick and Oerlemans 1990). Even larger effects, including a sea-level rise of up to 2 m, have been foreshadowed over the coming several centuries (Manabe and Stouffer 1993). Sea level is presently rising, judged against stable coasts, by 1–2 mm per annum (Warrick and Oerlemans 1990), and global temperatures are confirmed to be rising (Jones *et al*. 1986; Jones and Wigley 1990; see Fig. 13.7). Increased coastal erosion and greater storm-generated marine inundation of low-lying areas are among a

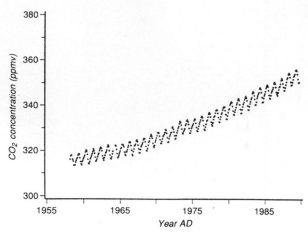

Fig. 13.4 Monthly atmospheric CO_2 concentrations for Mauna Loa, Hawaii. (After Boden *et al*. 1990)

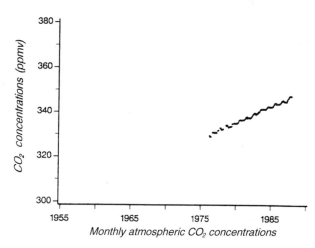

Fig. 13.5 Monthly atmospheric CO_2 concentrations for Cape Grim, Tasmania. (After Boden *et al*. 1990)

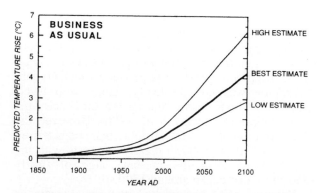

Fig. 13.6 Simulation of increases in global mean temperature from 1850–1990 produced by increased concentration of greenhouse gases, and predictions of the continuing rise for 1990–2100 based upon 'business-as-usual' emissions. (After Houghton *et al*. 1990)

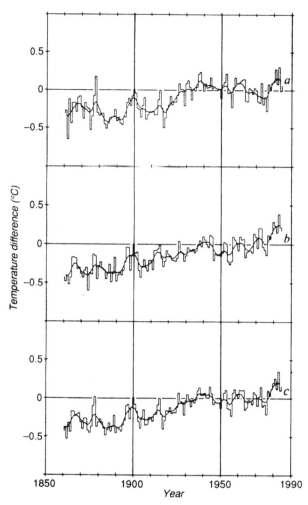

Fig. 13.7 Annual temperature variation since 1861, based on sea-surface temperature data. Smooth curves are 10-year Gaussian filtered values. (a) Northern hemisphere; (b) southern hemisphere; (c) global. (After Jones *et al.* 1986)

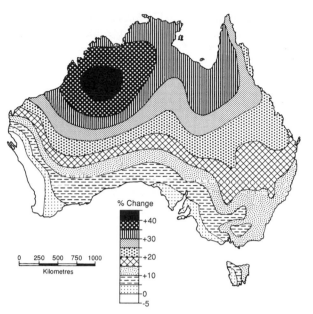

Fig. 13.8 Pattern of percentage change in net primary productivity in Australia relative to the present for a scenario of approximately doubled atmospheric CO_2 concentration. (After Pittock and Nix 1986)

multitude of potentially costly and damaging consequences of the predicted sea-level rise (Gornitz 1991).

Ecosystem changes will also follow any greenhouse warming, and are likely to involve a mixture of undesirable and beneficial effects. Agriculture may benefit from increased rainfall in some areas (Fig. 13.8). In locations that do not experience significantly increased rainfall, soil moisture levels will fall, and the suitability of land for crop production will change. Altered rainfall seasonality or storm intensity will have implications for surface erosional processes, particularly in agricultural and rangelands. Plant productivity increases in an atmosphere enriched in CO_2, and the effect increases in warmer conditions, so that there may be beneficial consequences for crop production; yields are estimated

to increase in some areas by 50–60% (Smil 1990; R.M. Adams *et al.* 1990). Also, frost damage may be reduced. However, microbial respiration also increases in warmer conditions, so that there may be a feedback mechanism here to release additional CO_2 (Melillo *et al.* 1990). This may be reinforced by the death of temperate forests and their replacement by grassland as a result of moisture stress. A particular concern with the prospective greenhouse climatic warming is that, unlike the warming that led from the last glacial phase into the Holocene, it will be very rapid. The anticipated rates of up to 0.3°C per decade (Melillo *et al.* 1990) may be too fast for an accommodating migration of ecosystems to occur, especially in view of the fact that potential migration pathways may not be available because of urban or agricultural development.

Many people have considered that the demand for food and for improved standards of living will ensure that fossil fuel consumption and deforestation will continue well into the future. Whether or not this proves, in the end, to be so, we would be foolhardy as a species to ignore the risk involved. A search for increasingly efficient forms of transportation and power generation, and plans to control global population, will be beneficial in extending the lifetime of available global resources, whether or not such steps are strictly required for the protection of the global climate. However, such action as we take may prove to be too little, too late. Therefore consideration is being given to ways in which the global environment might be deliberately modified to counteract undesirable changes – a somewhat anthropocentric view of the environment (Fyfe 1990b).

WHAT KINDS OF ACTIONS COULD WE TAKE?

In order to control the CO_2 content of the atmosphere, Faure *et al.* (1990) have envisaged a number of strategies. By diverting water to dry regions and cultivating forests, it would be possible to store 10 GT of carbon in each $10^6 km^2$ of forest. Alternatively, artificial peatlands could be created in continental depressions. The stimulation of corals and algae through the supply of extra nutrients to the water could result in the storage of additional carbonates in the marine environment, or vegetation could be submerged and buried offshore to simulate the fossil storage of carbon: 4 km3 of vegetation buried would store about 1 GT of carbon. Finally, CO_2 could be stored in seawater by the deliberate pumping of deep cold water to the surface or by the deliberate breakup of sea ice to produce sinking cold water.

Continuing sea-level rise will pose enormous problems for coastal cities and in the other areas already mentioned. Deliberate manipulation of global sea-level has been suggested by Newman and Fairbridge (1986), who argued that the construction of reservoirs during the past few decades has stored water equivalent to a 0.75 mm/a sea-level rise. This has only occurred, however, as a result of the construction of vast numbers of dams, estimated by Milliman (1997) to have totalled nearly 30 000 between 1950 and 1982. Much larger systems of water storage could continue to be employed in this way; Newman and Fairbridge (1986) contemplated the pumping of seawater into major continental depressions such as the Imperial Valley of California, the Dead Sea rift between Israel and Jordan, and the Qattara depression in Egypt. Additionally, the level of the Aral–Caspian sea, presently 28 m below sea-level, could be lifted. Raising this by 10 m could store sufficient water to stabilise sea-level for about a decade. There are many difficulties with these ideas, including the environmental cost of inundating the areas mentioned, many of which are scenically or culturally important. Furthermore, evaporation from the proposed sites could lead to the deposition of salt, and continued delivery of seawater could produce long-term changes in ocean chemistry, with possible subsequent climatic consequences. The enormous additional mass loaded on to the crust may induce brittle failure there and result in increased earthquake activity, as has already been associated with some large water-supply reservoirs. Finally, there is evidence that dams restrict the supply of key nutrients to the oceans. While human influence has increased river inputs of phosphorus and nitrogen to the oceans by about 400% (Humborg *et al.* 1997), dams may curtail the movement of silica. This may induce a decline in the abundance of siliceous diatoms in the marine environment. We saw earlier (in Chapter 4) that such changes may have played a significant role in Quaternary climate change. Here, then, is another means by which human activity may in some respects mimic, yet go well beyond, processes identified in the Quaternary record.

Human modification of the global climate may trigger additional feedbacks in other ways, some of which can be anticipated from a knowledge of Quaternary events. A very significant decline in the stability of permafrost may follow from global warming. Presently, about 25% of land in the northern hemisphere is underlain by permafrost (Koster 1995). This area may decline by up to 20% in the next 50 years, since the anticipated mean global warming of a few degrees may be associated with a warming of up to 9°C in winter at high latitudes (Anisimov and Nelson 1996). Resulting oxidation of organic matter stored in permafrost may provide a source of additional CO_2 beyond that responsible for the initial warming. It has also been suggested that the rate at which the atmospheric CO_2 concentration rises may be critical to the stability of the oceanic thermohaline circulation system (Stocker and Schmittner 1997). Modelling indicates that very rapid rises may result in a shutdown of the thermohaline circulation, which is known to be susceptible to perturbation (Tziperman 1997). If this occurred, the ocean surface would become warmer, so releasing more CO_2. (Indeed, a 20th-century decline in the areal extent of Antarctic summer sea ice, by up to 25%, has recently been suggested on the basis of whaling records (de la Mare 1997).) This, and the permafrost decline just mentioned, provide examples of positive feedback effects, many important instances of which were noted in our earlier discussions of Quaternary change. In coming decades, these may serve to magnify, perhaps greatly, the direct climatic effects wrought by human activities.

CONCLUSION

We have seen that people are now so numerous and have such widespread effects on the global environment that we are partly responsible for its behaviour. We are directly causing change in the composition of the atmosphere, in the water balance and albedo of large areas of the land-surface, and in the cycling of carbon through the ecosystems of the planet. The environmental changes which we are continuing to generate have analogues in the environmental changes of the Quaternary, but only in certain respects. Our knowledge of the events of the Quaternary is insufficient to identify and understand the multitude of interconnections and feedback linkages that were involved in the environmental instability which was its hallmark. However, we do have knowledge of the environmental changes and of the rates at which they occurred. This reveals that the anthropogenically-induced environmental changes which are presently underway are fundamentally different, at least in terms of their rate of occurrence, to those of the Quaternary. Consequently, greater stress will be placed on organisms required to adapt to changing habitats. Our knowledge of the events of the Quaternary period also allows us to judge the magnitude of contemporary environmental changes against those of the past, and by analogy with the nature of past environments, to gauge something of their probable consequences. However, decisions about the future of human impacts have to be made within the contemporary social and economic framework (e.g. see Parry and Duncan 1995; Lamb 1995). Some changes may be beneficial, like the increases in Australian wheat yield attributed by Nicholls (1997) to recent warming, or the lengthening of the growing season in high northern latitudes shown from satellite data by Myneni *et al.* (1997). Other consequences, like rising sea-level, may be enormously costly environmentally and economically (Wind 1987), or in terms of human health (McMichael *et al.* 1996). However, the systems being affected are so complex that proof of the outcome is often elusive, and this has led some workers (e.g. Wigley *et al.* 1996) to argue in favour of a 'do-nothing' approach to avoid any unnecessary impost on the economy. However, few aspects of the environment seem likely to escape the growing human impact (Samuels and Prasad 1994; Moore *et al.* 1996). Warming will affect the hydrological cycle via altered rainfall, plant water use, and evaporative losses. River flows will thus be affected, with possible effects on flooding, sustainability of dry-weather flows, urban water supply, and aquatic and terrestrial biodiversity (Ely *et al.* 1996; Arnell and Reynard 1996; Szilder and Lozowski 1996; Leavesley *et al.* 1997; Panagoulia and Dimou 1997; Sefton and Boorman 1997; Sellers *et al.* 1997). The consequences for agriculture are potentially serious, as agriculture currently accounts for 87% of global fresh water consumption (Pimentel *et al.* 1997), and there are limits to how much of the available supply humanity can appropriate for its own use (Postel *et al.* 1996). The changing extent of marsh and swamp which might follow hydrological change could act as a feedback, via altered fluxes of methane (Petit-Maire *et al.* 1991). Changed river flows in turn will affect patterns of erosion and sedimentation (Arbogast and Johnson 1994), and the delivery of nutrients to lakes and to the sea.

These instances of the growing human impact upon the environment and its regulatory mechanisms make it clear that there are clear and compelling reasons to continue to refine our understanding of the Quaternary. We must understand its legacy, appreciate the ongoing global change that seems inevitable when viewed from a Quaternary perspective, and apply this knowledge to the development of a more complete understanding of our own place within the global environment, and of our responsibility for its management. As Karl *et al.* (1997) pointed out, the future of the global climate and of environments generally now depends in part upon a series of choices that people must make, concerning activities which generate aerosol particles, consume or divert fresh water, release greenhouse gases, or alter the albedo and other characteristics of parts of the land area of the globe. To the natural mechanisms responsible for the Quaternary events described in this book we must now clearly add our own effects. This makes it even more clear that further development in our understanding of Quaternary events can provide an important key to an informed assessment of our own potential impacts on global environments. This is an exciting challenge, and we hope that readers of this book will be motivated to become involved in the continuing study of the Quaternary and of our place within it.

APPENDIX

DATING METHODS IN QUATERNARY RESEARCH

Since geological time is not salami, slicing it up has no particular virtue. But if it is to be sliced there is no need to botch the job, and chronometric dating provides the guidelines

Claudio Vita-Finzi
Recent Earth History (1973)

PRELIMINARY EXPLANATION

Most methods for measuring the age of Quaternary geological and archaeological materials are based on orderly and progressive change of some constituent of the specimen being dated. Several major methods, such as radiocarbon dating, are based on decay of a radioactive element, present in trace quantities; others depend on the accumulation over time of trapped electrons (e.g. luminescence dating) and some are based on slow chemical reactions (e.g. amino acid racemisation).

The radioactive element, chemical compound or other dating ingredient changes through time in its own characteristic way but most follow one of four pathways illustrated in Fig. A1. In some cases, the relative concentration of the dating ingredient decreases exponentially as time passes (curve A); in others, it is initially zero and increases with time (curve B) or it may be initially non-zero and increase with

time (curve C). In a few cases, the concentration of the dating ingredient declines from a high initial value towards a lower value, non-zero asymptote (curve D). For a given curve, an age can be calculated if the present and the initial concentrations of the dating ingredient are known.

Only where nature has recorded the passage of years in annual layers such as tree rings or ice layers can the age of a past event be established with calendar-like precision. Other age measurements carry some uncertainty. This is explained more fully later, but to illustrate the concept, Fig. A1 shows a small vertical bar labelled M, which represents the measured concentration of an ingredient (radiocarbon, say) that changes through time according to curve A. No measurement is exactly reproduceable and the uncertainty, or measurement 'error', is shown by the length of M. Although the true age of the sample might lie at the intersection of M and curve A, one cannot know this but can only know that it lies

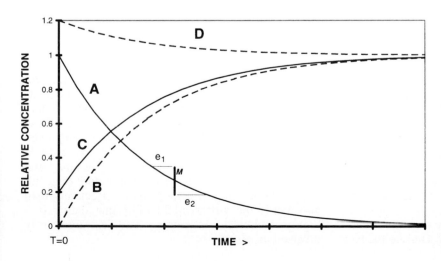

Fig. A1 Representative changes through time of relative concentrations of isotopes or compounds used for dating. Curve A, relative concentration decreases exponentially with time; curve B, concentration initially is zero and increases with time; curve D, concentration declines from a high initial value towards a lower value, non-zero asymptote. M represents the measured concentration of an ingredient (radiocarbon, say) at a point on curve A; measurement 'error' is shown by the length of M. The age lies between the points where lines e_1 and e_2 intersect A

between the points where lines e_1 and e_2 intersect curve A, owing to the uncertainty in M. Compared with real measurements, the length of M is rather exaggerated in Fig. A1, but it is clear that the smaller the uncertainty or error associated with measurement M, the greater the precision of the age estimate. It can also be seen that the lengths of e_1 and e_2, and hence the size of the age-error, depends not only on the length of M but also on where M intersects the age curve, i.e. it depends on the age itself.

There is a major difference between *precision* and *accuracy* in dating. Precision refers to the measurements themselves; accuracy refers to the degree of agreement between an age measurement and the real age of a sample, which may differ for any of several reasons. The age curve may be mistaken (the age calculated using curve B would differ from that based on curve C, for example), or the rate of change of the age curve may not be known correctly, or the dating ingredient may have leaked out of the specimen being dated. Some dating techniques are both precise and accurate, while others are neither. Dating specialists strive to maximise both the precision and accuracy of their methods; field scientists need to know how to interpret the stated precision (usually the age error) and need to know how to assess whether a date is likely to be accurate or not. To have confidence that an age measurement is accurate within its stated uncertainty is essential, and high precision is obviously desirable when the ordering of closely-spaced events is in question.

Dating is technically complex – much more so than might be imagined by non-specialists – and there are perils in using mail-order dates, even though the laboratory analysis may be of the highest quality. This Appendix is a partial introduction for those Quaternary scientists who have no direct access to dating laboratories. Laboratory technicalities are barely touched upon. The reader is introduced to basic principles of dating and some of the vernacular (which is loaded with terms such as radioisotope, disequilibrium, Gaussian error and so forth), and reasons why age measurements are occasionally way off the mark are also discussed. The space devoted to each method varies considerably; radiocarbon, for example, is reviewed at some length because it is highly popular, its technicalities are well understood, and it illustrates many practical aspects of dating. Luminescence methods are also described at moderate length, because both the principles and the techniques are more complex than is popularly conceived to be the case. The specialist will decently bypass this Appendix, but the writer recommends that field scientists whose research calls for dating read it in full *and* consult appropriate texts.

For best results, field scientists are urged to collaborate closely with dating specialists. Despite the fact that dating laboratories strive for accuracy and take care not to overstate precision, conflict within a set of dates sometimes does occur. The results may come from different laboratories that do not regularly measure cross-check samples, which can be revealing. For example, results from cross-check samples distributed amongst $^{230}\text{Th}/^{234}\text{U}$ laboratories varied by more than $\pm 30\%$ of the average age (Ivanovich *et al.* 1984). Most commonly, however, the dated samples themselves are faulty, owing to diagenesis and chemical exchange with their surroundings before they were exhumed. Conflict amongst dates sometimes arises because stratigraphy has been misinterpreted; occasionally, samples are simply mislabelled or muddled. Conflicting dates sometimes are rationalised rather than investigated; outlying dates may be rejected, which is dangerous if the outlier is the only uncontaminated sample, or all results may be averaged (also dangerous: the average of good and bad is bad), or the specialist declares that the laboratory cannot be wrong. Rather than rationalise after the event, it is better that the field and laboratory scientists design an experiment to resolve the matter, which might originate with processes outside their previous experience.

Radiocarbon, selected uranium-series and luminescence dating are the principal methods reviewed here; potassium–argon, electron spin resonance (ESR) and amino-acid methods are outlined more briefly (Table A1). Time series based on orbital variations and geomagnetic reversals, which underpin Quaternary time stratigraphy, are reviewed briefly. Elements of the relevant physical or chemical principles are explained, but the reader whose research calls for dates is strongly advised to consult textbooks devoted to dating, such as Aitken (1990) or Geyh and Schleicher (1990), before contacting a laboratory. Some other methods used in Quaternary research, such as fission track, obsidian hydration and various chemical dating techniques are not discussed here but are described in the texts referred to above.

Perhaps the most conspicuous omission in this Appendix is accelerator mass spectronomy (AMS) dating of rock surfaces using very low-level cosmogenic radioisotopes, including those generated *in situ*, applications of which are expanding rapidly in Quaternary and geomorphological research (Geyh and Schleicher 1990). Compared with radiocarbon, these methods are not widely available and are deployed by specialist groups, largely centred on particle accelerators. However, the interested reader may find it stimulating to dip into the AMS literature (e.g. Jull *et al.* 1997), which illustrates more effectively than any words the writer might offer that dating is a highly technical, interesting and extraordinarily diverse business.

Table A1 Quaternary dating methods described in this Appendix

Method	Basis; (curve), technique[1]	Range[2]	Best materials	Precision	Cautions
I. Radioisotope parent – stable daughter					
Radiocarbon (conventional)	$^{14}C \rightarrow {}^{14}N$ (A), radiometry	0–40 ka	Wood, resin, charcoal, peat, shell, coral	0–6 ka, 60 years; 6–30 ka, 1%; > 30 ka, > 1%	Calibration is required for real-year ages. Method is problematic near the dating limit
Radiocarbon (AMS)	Ditto; (A), AMS	0–40 ka	Ditto (milligram samples); bone	Ditto	
Potassium–argon; argon–argon	$^{40}K \rightarrow {}^{40}Ar$ (B), MS	0.5 Ma to 5 Ga	Igneous and metamorphic rocks and minerals	0.5%	Falsified by original Ar in young lavas
II. Disequilibrium between parent and daughter radioisotopes					
U-series: I, zero initial ^{230}Th	$^{238}U \rightarrow {}^{234}U \rightarrow {}^{230}Th \rightarrow \ldots$ (B), α-spec. or TIMS	0–250 ka (α-spec) 0–500 ka (TIMS)	Coral, speleotherm, eggshell	1% (α-spec) < 0.5% (TIMS)	Thorough checks for diagenesis are essential
U-series: II, excess initial ^{230}Th	Ditto, (D), ditto	0–250 ka	Deep-sea sediments	10–20%	Affected by bioturbation and sedimentation rate
III. Trapped electrons					
TL (thermo-luminescence)	Palaeodose and dose rate; (B; C) TL reader	0 to 100–500 ka	Pottery, hearths, tephras; Quartz or feldspar sediments; loess	10%; 10–15%	Dosimetry affected by non-systematic factors. Resetting may be doubtful
OSL (optical dating)	Ditto; (B, C), OSL reader	Ditto	Quartz or feldspar sediments	7–10%	Ditto
ESR (electron spin resonance)	Ditto; (B, C), ESR spec.	0–1 Ma	Coral, teeth, calcite, gypsum	10–20%	Dosimetry problems; open systems
IV. Chemical methods					
Amino-acid racemisation	Racemisation; (B, C), chromatography	0 to 100–500 ka	Eggshell, shells, forams, wood	> 15%	Strongly affected by temperature
V. Correlation methods: orbital ('Milankovitch') variations and geomagnetic reversals					

Notes: [1] Basis = basic principle (i.e. radioactive decay process); curve = trend of process (curves A, B, C, D in Fig. A1); techniques include radiometry, MS (mass spectrometry), AMS (accelerator mass spectrometry), and others mentioned in the text.
[2] ka = 1000 years, Ma = million years and Ga = million years.

OVERVIEW OF METHODS BASED ON RADIOACTIVE DECAY

Isotopes and radioisotopes

To introduce methods based on radioactive decay, the reader is reminded of some terminology and basic concepts of atomic structure and radioactivity. Atoms are comprised of a nucleus and a set of orbiting electrons, which on an atomic scale circulate around the nucleus roughly like planets around the sun. In a normal substance, atomic nuclei are isolated from one another by electrons and play no part in its chemistry, which is determined by the orbital structure of the electrons. Atomic nuclei contain protons (which have a positive electrical charge) and neutrons (with no electrical charge) except for hydrogen atoms, which have a single proton as their nucleus. When atoms are in an electrically neutral state, the numbers of electrons and protons are equal.

The nucleus of each element has its own characteristic number of protons (hydrogen has one, helium two, and so on to uranium with 92) but the number of neutrons in any given element can vary, giving rise to different *isotopes*. Oxygen nuclei, for example, have eight protons; in addition, ordinary oxygen has eight neutrons but a less common isotope of oxygen has 10 neutrons. These two isotopes of oxygen are usually written as ^{16}O and ^{18}O, respectively, where the superscripts ^{16} and ^{18} indicate the total number of protons plus neutrons in each isotope. Most isotopes are stable but some, known as *radioisotopes*, transform spontaneously into other elements, accompanied by radioactive emission of subatomic particles. The rate at which a given parent radioisotope transforms or *decays* to its daughter is not affected by temperature, pressure or its surroundings.

The most powerful forms of geological dating are based on measurements of the relative proportions of radioisotopes and their daughters, which may be produced by one of the following three radioactive processes:

a. *Alpha decay*: a nucleus emits an alpha (α) particle comprising of two neutrons plus two protons and becomes an isotope two steps lower in the elemental series, as in the decay of ^{234}U to ^{230}Th:

$$^{234}U \rightarrow {}^{230}Th + \alpha$$

b. *Beta emission*: a neutron emits a beta-particle (β) and becomes a proton, and the nucleus is transformed into an isotope of the next-higher element, illustrated by the decay of radiocarbon to common nitrogen:

$$^{14}C \rightarrow {}^{14}N + \beta$$

c. *Electron capture*: a proton captures an electron, becomes a neutron and emits a gamma photon (γ), and the atom becomes an isotope of the next-lower element, as in the decay of potassium-40 to argon-40:

$$^{40}K \rightarrow {}^{40}Ar + \gamma$$

Radioisotope dating: the basic concept

The moment at which a radioisotope atom decays to an atom of its daughter element cannot be predicted, but has a fixed probability which is unaffected by external factors. Thus, although the moment at which a given atom will decay is unknown, all atoms of a given radioisotope have the same *decay constant* (λ); the probability of decay is high for short-lived species and is low for long-lived species. The *mean lifetime* (τ) of a radioisotope is the inverse of its decay constant. When the number of radioactive atoms is large, the overall decay is a continuous series of random events and the net rate of decay can be written thus:

$$dn/dt = -n/\tau \qquad (A1)$$

where n is the number of atoms of the radioisotope at any time t. Integrating Equation (A1) gives the number of atoms at time t:

$$n = N \exp(-t/\tau) \qquad (A2)$$

where N is the initial number of atoms at $t = 0$ (note: the decay constant $\lambda = 1/\tau$ equally may be used in the decay equation). Equation (A2), represented by curve A in Fig. A1, is the basis of dating methods based on the simple decrease of a radioisotope such as radiocarbon.

The time it takes for any given quantity of a radioisotope to reduce by half is equal to $\tau/1.443$, which is referred to as the *half-life*. Rearranging Equation (A2) gives the following equation:

$$t = -\tau.\log(n/N) = -\tau.\log(n/(n+d)) \qquad (A3)$$

where d is the number of daughter atoms. The age of a specimen containing a radioisotope of known half-life can be calculated from Equation (A3) if n can be measured and N is known. If N is not known, it may be derived by measuring the numbers of both the parent (n) and its daughter (d) because $N = n+d$, provided that the daughter itself is not a radioisotope. The dating equation is more complex when parent and daughter are both radioisotopes (as in uranium-series dating, described later) but the principles are similar.

The time range of a dating method depends on the half-life of the radioisotope on which it is based. For example, the half-life of ^{14}C is 5730 years and radiocarbon dating extends to about 40 000 years or about seven half-lives, whereas ^{40}K has a half-life of 1.19×10^{10} years and can be used for rocks ranging back to the age of the Earth. The time range of a dating method is also affected by measurability of the radioisotope and other factors.

Terrigenous and cosmogenic radioisotopes

Radioisotopes used for dating have two origins; some have been constituents of the Earth since it condensed from the primordial matter that encircled the early sun, and others are produced by cosmic rays bombarding nuclei of elements in the atmosphere and Earth's surface. Radioisotopes with long half-lives are used for dating ancient rocks, such as ^{238}U (half-life = 4.47×10^9 years), ^{232}Th (1.4×10^{10} years) and ^{40}K (1.19×10^{10} years). Together with other elements, long-lived radioisotopes are recycled by plate tectonic processes through the crust and mantle; their concentrations in different rocks and reservoirs is dictated by their chemistry.

Although some long-lived radioisotopes are used in Quaternary dating, notably ^{40}K, methods based on shorter-lived types are more widely used, including those generated by cosmic rays entering the atmosphere such as ^{10}Be, ^{14}C and ^{36}Cl, which are either adsorbed by soils or enter a biogeochemical cycle such as the carbon cycle. Cosmogenic radionuclides generated in surface rocks remain *in situ* until they either weather out or decay radioactively. Cosmogenic production depends on cosmic ray flux, which varies with altitude and latitude, and is strongly influenced by the Earth's magnetic field, which has changed in the past. Hence, chronologies based on cosmogenics do not correspond exactly with those based on terrigenous radioisotopes but can be calibrated through cross-dating programmes (see radiocarbon dating below).

Measurement: radiometry, mass spectrometry and AMS

The abundance of most radioisotopes used in dating is very low. Their measurement involves highly specialised procedures, too technical to usefully describe here, but the different methods and their relative advantages are outlined briefly. In all cases, the radioisotope of interest is chemically extracted and converted to a form suitable for the measurement instrument, taking extreme care to avoid contamination because many radioisotopes, such as ^{14}C, occur in our environment. The concentration of the radioisotope is then measured by *radiometry*, *mass spectrometry*, or *accelerator mass spectrometry* (AMS).

Radiometric measurement is based on counting the rate at which α, β, or γ particles are produced by radioactive decay in the sample. The energy of every particle is recorded, to distinguish between particles from the sample and from extraneous sources such as cosmic rays. Counting instruments typically include sophisticated screening arrangements to eliminate external radiation. To determine the concentration accurately, a reasonably large number of events must be counted because the error is inversely proportional to the square root of the number of counts. Thus, a measurement based on 1000 counts has an error of $\pm 3.3\%$ and one based on 10 000 counts has an error of $\pm 1\%$.

Radioactivity plays no part in mass spectrometry, which is a technique for separating atoms or molecules of different masses and measuring their proportions. A chemically purified sample is ionised, the ions are accelerated electrically, the beam is deflected through a magnetic field and ions of different masses are separated because they are deflected differently. The separate ion currents are measured and the proportion of each isotope is calculated. Mass spectrometric techniques sometimes are labelled by the technical basis of the ion source (e.g. TIMS = thermal ion (source) mass spectrometer).

Conventional mass spectrometry is limited by the relative abundance of the isotope of interest; many factors affect this, but generally the abundance of a rare isotope relative to a more common sister isotope should be no smaller than about 10^{-6}. Radioisotopes that occur at lower levels, however, beyond the capabilities of conventional mass spectrometry, can be measured by accelerator mass spectrometry (AMS), which differs from conventional mass spectrometry in that the ions are accelerated to many millions of volts and are detected by methods that do not work at the lower voltages of mass spectrometry. Isotopes with relative abundances as low as 10^{-14} can be measured by AMS.

A radiometry laboratory can be established for less than a few hundred thousand dollars; a mass spectrometer laboratory for precise measurements of isotope ratios can cost more than five times as much, and to establish an AMS facility costs many millions of dollars. Factors other than cost dictate the choice of measurement technique, however. When the initial concentration of a radioisotope is not known precisely, its daughter must be measured (Equation A4) and mass spectrometry is unavoidable if the daughter is a stable isotope, as in potassium–argon dating, for example.

Although it is the least expensive, the practicability of radiometry depends on whether the element can be prepared in a form suitable for counting. It also depends on the radioactivity of the sample, and effectiveness decreases fairly sharply if the count rate is much less than about 1 particle every few minutes. For a specimen of a given size the maximum count rate is

$$n = 6 \times 10^{23} RW/A\tau \qquad (A4)$$

where n is the mean count rate in counts per minute, R is abundance of the radioisotope relative to a stable sister isotope in the prepared sample, W is the sample weight, A is atomic weight of the radioisotope and τ is the mean lifetime expressed in minutes (note that the actual rate depends on the sample age and is less than that given by Equation A4). For example, ^{14}C is better suited than ^{36}Cl for measurement by radiometry. For ^{14}C, $R = 1.2 \times 10^{-12}$ and $\tau = 8268$ years; substituting in Equation (A4) gives about 11 counts per minute (cpm) per gram of carbon, well above the usefulness threshold. A similar calculation for ^{36}Cl ($R = 2 \times 10^{-11}$, $\tau = 432\,900$ years) gives about 1.3 cpm. Hence, ^{14}C can be usefully measured by radiometry and AMS; ^{36}Cl is best measured by AMS.

Summarising, some radioisotopes are suitable for radiometry but not mass spectrometry and some require mass spectrometry for measurement of stable daughter isotopes; either technique can be used for some radioisotopes, but there are others that can only be measured by AMS. The appropriate technique for each Quaternary dating method is indicated in Table A1.

PRACTICAL DATING WITH RADIOISOTOPES

Radiocarbon, uranium-series and K–Ar are well proven and widely established dating methods in Quaternary research, whereas ^{36}Cl and ^{10}Be are representatives of the new set of cosmogenic radioisotopes measured by AMS at relatively few laboratories. They differ widely in terms of time range, applications and measurement techniques (Table A1). The basis of each method is summarised below, beginning with radiocarbon because it illustrates many practical aspects of dating.

Radiocarbon

OVERVIEW

Carbon, the basis of life, has two stable isotopes, ^{12}C and ^{13}C, of which ^{12}C (common carbon) is the more abundant by about 500 times. Radiocarbon or ^{14}C is a cosmogenic radioisotope that forms in the upper atmosphere by cosmic rays bombarding nitrogen nuclei, and which then mixes throughout the atmos-

phere, biosphere and ocean, where its concentration relative to ^{12}C is extremely low, about 10^{-12}. Radiocarbon decays by β emission to ordinary nitrogen with a half-life of 5730 years, although the original estimate by W. Libby (1950) of 5568 years is retained in radiocarbon dating (see 'Radiocarbon dating conventions', below).

Radiocarbon enters the tissues, bones or shells of all living organisms, as well as carbonate deposits such as travertine and speleothems. In any remains of a newly dead organism, the concentration of ^{14}C relative to ^{12}C is initially similar to that in the atmosphere or the water in which the specimen lived, but decreases thereafter, following curve A of Fig. A1 (Equation A3 above). Thus, the age of an ancient specimen can be determined if the initial concentration of ^{14}C is known and its present concentration is measured, provided that no carbon has entered or left the specimen since it formed. The initial concentration of ^{14}C cannot be estimated from its daughter, which is a minute amount of ordinary nitrogen, but is assumed to have been the same as it would be in a modern specimen of the same type of material. Cross-dating studies using ^{14}C and other methods have shown that this assumption is not exactly correct (see 'Calibration' below).

Until the mid-1980s, radiocarbon was routinely measured radiometrically by gas-proportional or liquid scintillation counting, which requires a few grams of carbon to be most effective (see the discussion following Equation A4 above). The AMS technique, now used increasingly, permits ^{14}C to be measured with samples smaller than a milligram. ^{14}C dates of the highest precision have been achieved with gas proportional radiometry, but liquid scintillation radiometry, which is operationally more versatile, is favoured by many laboratories including large commercial services such as Beta Analytic. AMS has a considerable advantage owing to the very small sample size required, where original carbon is sparse or difficult to extract, but is more expensive.

Radiocarbon is a powerful and versatile dating method. To take advantage of what the method has to offer, the user should have an understanding of systematic errors, problems near the dating limit, conventions adopted in radiocarbon dating, reservoir effects and calibration, which are now reviewed.

SYSTEMATIC ERRORS

Radiometric measurement of ^{14}C entails counting of β particles produced by ^{14}C decay. As described earlier, decay is a random process and the measured rate is unavoidably subject to uncertainty, which is represented as: count rate $\pm$ error, i.e. $R \pm \sigma$. The error-term σ is one standard deviation of a Gaussian error distribution and is inversely proportional to the number of counts recorded. For example, suppose that one sample yields 2500 decay events in a period of 1000 minutes and that another yields 10 000 events; the count rates are 2.5 ± 0.05 (i.e. error = $\pm 2\%$ of the count rate) and 10 ± 0.1 (error = $\pm 1\%$) counts per minute, respectively.

Other factors contribute to the error term for a radiocarbon date. Equation (A3) indicates that the age should be calculated from the measured ^{14}C count divided by the initial count rate. Earlier it was explained that the initial count rate cannot be measured; hence, in practice, the equation is

$$t = -1.443\lambda \log ((SAM-BKG) / (STD-BKG)) \quad (A5)$$

where SAM, BKG and STD are count rates for the sample, instrument background and a modern standard, respectively. BKG is the recorded count rate for a sample with zero ^{14}C, and the modern standard is an internationally distributed substance, calibrated to the ambient ^{14}C level in 1950 AD (STD is discussed further under 'Radiocarbon dating conventions', below). SAM, BKG and STD count rates have error terms, which are combined in the age error.

The age error conventionally is one standard deviation of the age-error distribution, i.e. there is a 68% probability that the age lies within the stated error range. However, many scientific publications nowadays report the age error as $\pm 2\sigma$, in which case the probability that the age lies within the stated error range is 95%. Radiocarbon age measurements that lie within 2σ of background (BKG) are usually reported as 'greater than' the calculated age.

Radiocarbon ages determined by AMS are not based on counts of radioactive decay events, but have age error terms, none the less, because ion counting is subject to random fluctuations, similar to radiometric counting.

CLOSED SYSTEMS AND PROBLEMS NEAR THE DATING LIMIT

It would be impossible to determine a meaningful age if ^{14}C atoms wandered through a specimen while it lay in a geological or archaeological deposit. Very few carbon compounds can safely be regarded as chemically inert, particularly when buried for thousands of years in near-surface deposits inhabited by microbes and percolated by groundwater. Charcoal, for example, can adsorb dissolved organic compounds from soil or shallow groundwater, and the same compounds can contaminate cellulose in wood when it becomes degraded. In the case of shells, carbon is exchanged with bicarbonate ions in percolating water. A radiocarbon determination of a sample so affected can provide useful information about contamination processes, if the sample age is known beforehand, but cannot yield a reliable age.

The likelihood of chemical exchanges and contamination increases with sample age. Furthermore, the effect of a trace of modern contamination becomes acute as the dating limit is approached, where the quantity of residual ^{14}C is very small (curve A, Fig. A1). Both effects make reliable dating difficult close to the dating limit, where even the best screening protocols sometimes break down (Chappell *et al.* 1996).

Various procedures are applied routinely to eliminate contamination, such as successive acid–alkali–acid leaching of wood samples or use of a dental drill to remove discoloured parts of a shell. Diagenesis, which is commonly accompanied by chemical exchange with surrounding fluids, can be detected by petrographic and electron microscopy, and contamination can often be identified and eliminated by radiocarbon assay of different chemical fractions of a sample. Some materials are more susceptible to contamination than others; resins, for example, are chemically less reactive than shells. The possibility of contamination should always be considered and, where it is suspected, dating results are likely to be most reliable when the field and laboratory scientists collaborate closely.

RADIOCARBON DATING CONVENTIONS

Radiocarbon dates from different laboratories are placed on a common footing through the adoption of conventions to do with half-life, modern standard and other factors that affect the calculated age result. These are summarised as a guide for the non-specialist; however, a specialist should be consulted whenever there is doubt about which conventions have been or should be followed.

Half-life

A date can be compared with others only when the half-life on which it is based is known to all concerned. Systematic measurements of ^{14}C show the half-life to be 5730 years, but the original estimate by W. Libby (1950) of 5568 years has been the basis of so many radiocarbon dates that to change would cause confusion. Dates based on the 5568 half-life are usually referred to as *conventional radiocarbon ages*; to convert to the 5730-year base, multiply a conventional age by 1.03.

Isotopic fractionation

The initial concentration of ^{14}C in a carbon-based substance not only depends on the medium in which it formed (see reservoir effects, below) but is also affected by *isotopic fractionation*. Consider the following generalised reactions:

$$CO_2 \text{ (atmosphere)} \rightarrow \text{(photosynthesis)} \rightarrow \text{plant tissue} \quad (A6a)$$

$$CO_2 \text{ (atmosphere)} \rightarrow CO_2 \text{ (aqueous)} \rightarrow CaCO_3 \text{ (travertine)} \quad \text{(A6b)}$$

The rate of each reaction varies with the isotopic species involved; for example (A6a) proceeds slower with $^{13}CO_2$ than with $^{12}CO_2$ and is slowest with $^{14}CO_2$. This *isotopic fractionation* effect is greater for reaction (A6a) than for (A6b); hence, plant tissue is depleted in ^{13}C and ^{14}C relative to travertine even though the carbon in both originated in the atmosphere. This effect is taken into account in ^{14}C dating; the fractionation coefficient for a sample is derived from its $^{13}C/^{12}C$ isotope ratio and the ^{14}C count rate (SAM in Equation A5) is adjusted accordingly.

Reservoir effects

The concentration of ^{14}C differs between major carbon reservoirs such as the atmosphere, ocean and biosphere, because the residence time of carbon atoms is different in each. The effect is greatest for the ocean, so that until recently the ^{14}C concentrations in ocean surface water and deep water were about 5% and 10% less than in atmospheric CO_2, respectively (^{14}C levels in all reservoirs has changed since atmospheric bomb testing commenced: see next section). Hence, the ^{14}C dates of shallow-water marine shells are about 400–500 years older than the age of a contemporary tree. Adjustments made to radiocarbon ages to allow for *reservoir effects* are usually based on the ^{14}C activity of specimens that formed this century, before 1950 AD, in the same reservoir as the sample being dated.

Bomb effects

Atmospheric ^{14}C rose sharply after atmospheric testing of thermonuclear bombs commenced, and by the year 1963 was about twice the pre-1950 level, but has decreased slowly since. The bomb 'spike' is useful for tracing mixing processes in the global carbon cycle, particularly in the oceans, but because of it, organisms living today cannot be used as the basis for the ^{14}C modern standard.

Modern standard

The initial concentration of ^{14}C in a sample is equated to the count rate for a selected substance known as a *modern standard* (STD in Equation A5), which has the same count rate as wood that grew in 1950 AD would have (1950 AD is the zero age point on the radiocarbon time-scale). In reality, the initial concentration varied in the past because production of ^{14}C depends on the flux of cosmic rays entering the upper atmosphere, which has changed owing to past variations of the Earth's magnetic field. Furthermore, the distribution of carbon between the atmosphere, biosphere and oceans has varied and has affected local ^{14}C concentrations, also. None the less, radiocarbon dates are calculated as if the initial concentration at all times in the past was the same as the modern standard value.

Calibration

Past variations in the initial concentration of ^{14}C have been determined using tree ring sequences covering the last 10 000 years, which have been painstakingly

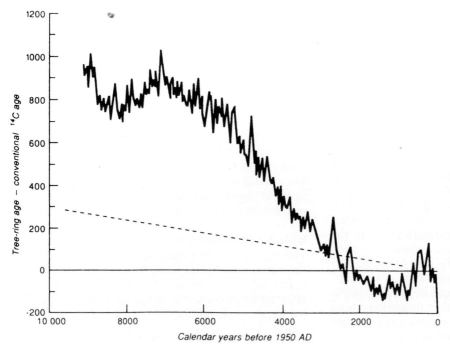

Fig. A2 Differences between tree-ring ('calendar' ages) and conventional radiocarbon years for the last 9000 years, based on close ^{14}C dating of tree-ring sequences (solid wiggly line) (Stuiver *et al.* 1993). The dashed line represents the difference between conventional ages and ^{14}C ages calculated using the measured half-life of 5730 years; differences between the solid and dashed lines reflect past variations of ^{14}C production rate and, possibly, changes of CO_2 exchange between the atmosphere and ocean

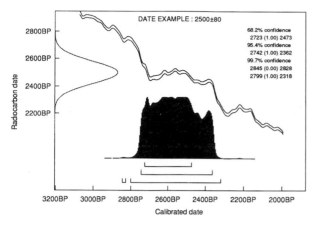

Fig. A3 Calibration of a conventional ^{14}C date, from the OxCal package (Stuiver *et al.* 1993). The bell curve attached to the vertical axis represents the conventional ^{14}C age (2500 ± 80 yr BP) and the black irregular curve is the corresponding calibrated age probability distribution. The wiggled narrow band represents the ^{14}C – tree-ring age relationship for 2000–3000 calibrated years BP. The calibrated age lies in the range 2473–2723 BP with 68% probability, and in the range 2362–2742 with 95% probability (Bronk Ramsey 1995). Bars below the calibrated age probability distribution represent the 68.2% and 95.4% confidence limits of this distribution. When the calibrated distribution is polymodal, these bars may be discontinuous, in which case the calibrated result is expressed as more than one possible age, each with its own probability value

assembled and carefully cross-dated by several radiocarbon laboratories. Radiocarbon ages do not match annual ring-count ages and divergences tend to increase with distance into the past, but the pattern is complex (Fig. A2).

Inspection of Fig. A2 suggests that the graph could be used to calibrate a radiocarbon age calculated by Equation (A5), to get the equivalent age in true years. This is not simple, however. Owing to the many significant wiggles in the relationship between ^{14}C and tree-ring ages, a given ^{14}C age may correspond to more than one calendar age; furthermore, the error term of a ^{14}C age conforms to a Gaussian error distribution, but the equivalent calibrated age has an irregular error distribution (Fig. A3). The statistical problems involved in calibration of ^{14}C ages are not simple but have been extensively analysed and, as shown in Fig. A3, calibration can now conveniently be done for ^{14}C ages up to about 10 000 years BP with the OxCal package (Stuiver *et al.* 1993).

Calibration programs such as OxCal may be applied to radiocarbon dates obtained from wood and other terrestrial organic substances, but should not be applied indiscriminately to all ^{14}C dates, particularly those where the reservoir correction may have varied in the past or is unknown. In the latter cases, results

from different sites are probably best compared in terms of their conventional radiocarbon ages.

Uranium-series disequilibrium dating

The decay chains that start with ^{238}U, ^{235}U and ^{232}Th and finish with stable isotopes of lead contain many members, with half-lives ranging from seconds to gigayears, and offer the means for measuring ages across the full sweep of geological time. At one end of the scale, U–Th–Pb dating is applied to some of the most ancient igneous and metamorphic rocks; at the other, the decay of ^{210}Pb to ^{206}Pb is used for dating sediments less than 100 years old. In between lie the U-series disequilibrium dating methods, based on one or more pairs of radioisotopes at the upper end of the ^{238}U and ^{235}U decay chains, which range to several hundred thousand years and are applied to a wide range of materials including corals, speleothems, terrestrial and marine sediments, ancient peats, evaporites and bones.

Ages determined by these methods are based on the degree of disequilibrium between higher members of the two uranium decay series:

$$^{238}\text{U (4.47 Ga)} \rightarrow {}^{234}\text{U (245 ka)} \rightarrow {}^{230}\text{Th (75.4 ka)} \rightarrow \ldots \quad \text{(A7a)}$$

$$^{235}\text{U (0.7 Ga)} \rightarrow {}^{231}\text{Pa (33 ka)} \rightarrow \ldots \quad \text{(A7b)}$$

(numbers in brackets are half-lives). In a closed system, equilibrium exists when every member of the chain decays to its daughter at the same rate as it is produced by the decay of its parent; hence, if parent and daughter isotopes become separated owing to differences in their chemistry, the equality between rates of production and decay is disrupted and the system is in disequilibrium. For example, ^{234}U is present in seawater and is available for uptake by marine organisms but its daughter ^{230}Th is virtually absent, because its solubility is very low and it is precipitated with sediment to the sea floor. Marine organisms such as corals, which take up uranium in their skeletons, thus contain no initial thorium but ^{230}Th is generated in the skeleton by ^{234}U decay, after death.

When disequilibrium exists between a parent and a daughter radioisotope, the *activity* (the rate of emission of radioactive particles) of the daughter is given by the following equation:

$$[\text{Ad}] = (\lambda_d/(\lambda_d - \lambda_p)) \cdot [\text{Ap}]_o (\exp(-\lambda_p t) - \exp(-\lambda_d t)) + [\text{Ad}]_o \exp(-\lambda_d t) \quad \text{(A8)}$$

where [Ad] is the activity of the daughter at time t, $[\text{Ad}]_o$ and $[\text{Ap}]_o$ are activities of daughter and parent at time zero, and λ_d and λ_p are their decay constants (decay constant = 1/mean lifetime, defined earlier). This equation is obviously more complicated than that for simple decay with a stable daughter isotope

(Equation A2) and the derivation of an age is also a little more complex.

Disequilibrium can occur in two ways: either the activity of the daughter is below the equilibrium level at time zero and equilibrium is subsequently approached from below (curves B or C in Fig. A1), or it exceeds the equilibrium level at time zero and equilibrium is approached from above (similar to curve D, Fig. A1). Both can be used for dating. Furthermore, the two uranium decay chains (A7a and A7b) offer a means of cross-dating, because half-lives in the two series are different. Not only can the age of a uranium-bearing specimen be calculated from ^{230}Th/^{234}U and ^{231}Pa/^{235}U, and from ^{234}U/^{238}U if this pair is initially in disequilibrium, but also it can be calculated from ^{230}Th/^{231}Pa, because the natural abundance ratio ^{238}U/^{235}U = 137.88 is constant. Changes through time of these ratios resemble the curves in Fig. A1 but are not exactly the same, owing to there being several decay terms in the relevant equations.

As with all other methods, U-series dating is compromised unless a specimen has been a closed system since it formed, and it is very difficult to establish whether the system leaked early, late, continuously or at a fluctuating rate. Dating with various open-system U-series models has usually met with indifferent success but, overall, U-series methods have been applied successfully to more types of materials than are described below, from marine phosporites to arid-zone evaporites, as reviewed in specialised texts such as that by Geyh and Schleicher (1990).

SIMPLE CLOSED SYSTEMS: CORALS AND SPELEOTHEMS

Probably the greatest impact of U-series disequilibrium dating in Quaternary research has been through its application to corals. U-series ages of raised coral terraces in Barbados and Papua New Guinea were the first radiometric dates to support the Milankovitch theory of climatic change (U-series measurements of deep-sea sediments gave similar indications but the dating precision was too low for conclusions to be drawn confidently). Corals ranging up to several hundred thousand years old have been dated with high precision, contributing to studies of reef growth, sea-level changes and tectonic movements. The method has now achieved such accuracy that coral less than 25 years old has been dated to within a few years, and it has been used to extend ^{14}C calibration through cross-dating late Quaternary corals by ^{230}Th/^{234}U and ^{14}C (Bard *et al.* 1990b). The method is also routinely used for dating speleothems.

Ages of upper Quaternary corals are calculated from their ^{230}Th/^{234}U ratio, which is zero in living corals and approaches equilibrium after 400–500 ka. The initial ^{234}U/^{238}U ratio in corals is also in disequi-librium, because ^{234}U is relatively enriched in both fresh water and seawater owing to ^{234}U in weathered rocks being less tightly bound than ^{238}U. The ^{234}U/^{238}U ratio varies very little in modern corals (1.145) and takes more than a million years to approach equilibrium. Hence, the age of an upper Quaternary coral can be calculated from both ^{230}Th/^{234}U and ^{234}U/^{238}U, which provides a good cross-check on the results. Alternatively, the initial ^{234}U/^{238}U ratio is calculated from the measured ratio and the ^{230}Th/^{234}U age; a result different from the value in modern corals indicates that the age is likely to be unreliable. A similar means of cross-checking is useful for speleothems but is less secure, because ^{234}U/^{238}U can vary in rivers and groundwater.

The isotopes ^{230}Th, ^{234}U and ^{238}U are measured either radiometrically (typically by α spectrometry) or by thermal ionisation mass spectrometry (TIMS). The dating precision achieved by spectrometry has improved from about ± 8% in the 1960s to about 1.5% nowadays, and the precision achieved by TIMS can be better than ± 0.5% of the age.

Acceptability criteria

Four criteria must be met for a ^{230}Th/^{234}U age from a coral specimen to be accepted: the original coralline fabric and mineralogy of the specimen must be preserved; there must be no measurable ^{232}Th; and both the uranium content and the initial ^{234}U/^{238}U ratio must be similar to values found in modern corals. A specimen that fails any of these is unlikely to have been a closed system since its formation. The fabric should be examined petrographically and by X-ray diffraction for diagenesis (XRD examination alone is insufficient because it will not detect diagenetic aragonite). The ^{232}Th test is useful for reefs founded on continental or volcanic basement, where detrital thorium may be included in the coral, including detrital ^{230}Th.

Similar tests arc used to evaluate dates from speleothems, although the tests are not as sharp as for corals because the uranium concentration and ^{234}U/^{238}U ratio vary through time. Furthermore, speleothems sometimes contain ^{232}Th, signalling that detrital thorium, including ^{230}Th, was present initially (usually with clay inclusions), but although several stratagems have been tried, age-adjustments to allow for this often fail to convince (cf. Geyh and Schleicher 1990).

CLOSED SYSTEMS WITH INITIAL EXCESS OF A RADIOISOTOPE

In some systems, such as ^{234}U/^{238}U in seawater, the activity of a daughter radioisotope initially exceeds that of its parent. Deep-sea sediments and manganese nodules contain ^{230}Th in excess over that produced by decay of *in situ* ^{234}U, because ^{230}Th from decay of ^{234}U in seawater is precipitated to the sea floor. Once trapped in the sediment, the precipitated or *unsupported* ^{230}Th slowly decays while that which is produced *in situ* increases.

The initial-excess model has been applied to ^{230}Th/^{234}U dating of deep-sea sediments and manganese nodules, but accuracy is compromised by uncertainties about the level of unsupported initial ^{230}Th, partly caused by variations of sedimentation rate, and the method has been superseded by the orbitally-based chronology for deep-sea cores, described later.

OPEN SYSTEM MODELS

Many materials form with little or no initial uranium, but accumulate it afterwards. Some of these either occur widely in Quaternary deposits (e.g. molluscs) or are extraordinarily significant (e.g. hominid fossils) so that the urge to attempt to make age measurements is strong, despite the fact that the materials are obviously not closed to uranium. Cross-dating shows that ages derived from mollusc shells are realistic only 50% of the time (Ivanovitch *et al*. 1983), which clearly is not satisfactory. When the uranium content is sufficiently high, the reliability of ^{230}Th/^{234}U ages for such materials can be checked by determining the ^{231}Pa/^{235}U age (cf. Equation A7b), but the accuracy remains compromised because ^{235}U, like ^{238}U, was absent when the shell or bone was formed.

Ages have been calculated using various open system models, including early versus linear uptake of uranium, but the degree to which these represent actual conditions can rarely be gauged and uncertainties are too great for chronologies to be based on these methods. However, information gained from U-series measurements from open-system materials, particularly in sites with important fossils, is often valuable when a battery of dating methods including TL and ESR are applied to a problematic site.

Potassium–argon and argon–argon methods

These methods, both based on the decay of ^{40}K to ^{40}Ar, which has a very long half-life (1.19×10^{10} years), are not only applied across the full sweep of geological time but have contributed significantly to Quaternary chronology, particularly through the dating of geomagnetic reversals, which are key time-markers for most continuous sequences of Quaternary deposits. Other notable applications include the precise dating of tuffs and tephras within Lower Pleistocene sedimentary series, including hominid fossil-bearing deposits in north-east Africa.

Potassium occurs in many minerals, mostly as ^{39}K (93.258%) and ^{41}K (6.730%). The radioactive isotope ^{40}K (0.01167% of natural potassium) decays to ^{40}Ca by β emission (89%), and also to ^{40}Ar by electron capture (11%) with a half-life of 1.19×10^{10} years, which is the basis of K–Ar dating (^{40}K $-$ ^{40}Ca is rarely used because radiogenic ^{40}Ca is very difficult to measure against natural ^{40}Ca, common in most K-bearing minerals). In K–Ar dating, radiogenic ^{40}Ar, which increases with sample age (following curve B, Fig. A1), must be distinguished from naturally-occurring ^{40}Ar, which comprises 99.6% of atmospheric argon (argon is the third most abundant atmospheric gas, at 0.93%). In conventional K–Ar dating, the ^{40}K concentration is determined from a measurement of total potassium. Argon isotopes are measured mass spectrometrically, in gas extracted by heating the sample. In order to distinguish radiogenic ^{40}Ar from atmospheric argon that may have diffused into a specimen, radiogenic ^{40}Ar is gauged from the degree to which the measured ^{40}Ar/^{36}Ar exceeds the atmospheric ratio of 295.5.

The argon–argon method measures potassium indirectly by first irradiating the sample with fast neutrons to convert ^{39}K to ^{39}Ar; the argon isotopes, including ^{39}Ar, are then measured mass spectrometrically. The measurement of ^{40}Ar, required for the age calculation, is derived from ^{39}Ar by irradiating a standard sample to calibrate the ^{39}K $\rightarrow$ ^{39}Ar reaction and multiplying by the ^{40}K/^{39}K abundance ratio.

Age measurements at the very young end of the scale are more likely to be compromised by traces of contamination by unidentified atmospheric argon or when primordial ^{40}Ar is retained in the sample. The ^{39}Ar/^{40}Ar method, although technically demanding, has the advantage that the ^{40}K/^{40}Ar ratio can be measured (*via* the ^{39}Ar surrogate) as the argon is extracted through stepwise heating of the sample. Being derived from ^{39}K atoms in the same minerals that contain ^{40}K, the ^{39}Ar is released from lattice sites close to the ^{40}Ar; hence, a sample that had retained original ^{40}Ar, or had gained or lost it (all of which falsify a conventional K–Ar age), would be detected because during progressive heating it would fail to produce a stable plateau of age versus temperature.

For details of these powerful methods, the reader is referred to McDougall and Harrison (1988).

METHODS BASED ON ACCUMULATION OF TRAPPED ELECTRONS

Many minerals can be dated by taking a measure of electrons that have lodged in 'traps' in their crystal lattices. The electrons in most crystalline minerals are bound into the lattice but they can be dislodged by α, β or γ particles, emitted by radioactive elements within the crystal or its neighbourhood. The dislodged electrons may come to rest in lattice traps, which typically occur where the lattices are deformed by impurities. The number of trapped electrons increases with time, in proportion to the number of traps available and the rate at which electrons are dislodged by α, β or γ particles. Thus, the age of a sample might be calculated if the initial and final numbers of trapped electrons were measured, and if the rates at which free electrons are produced and trapped were known.

Concealed beneath the 'might' in the last sentence are more issues in atomic and mineral physics than underlie methods based on radioactive decay. Trapped-electron methods have become widely used in Quaternary studies but many of their applications are still experimental; indeed, there is disagreement among specialists as to the reliability of these methods with certain materials and over certain time ranges. None the less, the range of materials to which these dating methods can be applied is impressive and includes sediments of quartz, feldspar and certain other minerals; loess, pottery, volcanic tephra, ancient hearths, burnt flint, teeth, coral and limestone cave deposits.

Trapped electron dating includes thermoluminescence or TL, optically stimulated luminescence or OSL, and electron spin resonance (ESR) techniques, which differ in terms of the techniques used to measure trapped electrons. These methods are applied to an extraordinary range of topics, not all to do with dating; their versatility and technicalities are described by Aitken (1985), Grün (1989) and Ikeya (1993).

Practical considerations

In principle, the age of a sample could be calculated if the initial and final numbers of trapped electrons were measured and if the rates at which free electrons are produced and trapped were known. Absolute values of these variables are difficult to measure but can be evaluated indirectly. Conventionally, the sample to be dated is divided into a number of similar aliquots, which are irradiated as a stepped series with β or γ rays, and the proportion of trapped electrons in each aliquot is then assessed by TL, OSL or ESR.

Graphs of applied radiation dose versus TL, OSL or ESR measurements from a series of aliquots commonly are used to illustrate simplified explanations of luminescence and ESR dating, but such graphs camouflage several factors which are better explained by a few elementary equations.

The general equation that describes measurements from a series of aliquots can be written

$$Z = f(D) \qquad (A9)$$

where Z is the amplitude of the TL, OSL or ESR signal and D is the applied radiation dose. In the simplest model, the rate at which traps are filled is proportional to the number of unfilled traps and to the rate (k) at which radiation is applied ('dose rate'); i.e. $dZ/dt = k(Z_o - Z)$ where Z = the measured signal and Z_o = the signal when all traps are full (saturated). The solution in terms of time is

$$Z = Z_o(1 - \exp(-kt)) \qquad (A9a)$$

Equation (9a) is represented by curve B, Fig. A1. As $D = kt$, $f(D) = Z_o(1 - \exp(-D))$. Hence, Z as a function of total radiation dose can be defined by statistically fitting a model such as Equation (A9a) to measurements from a series of aliquots. Once fitted, the function $f(D)$ is used to estimate the radiation dose that generated the natural signal Z(nat) in the sample. The age of a sample can then be calculated, provided that the natural dose rate is known.

In its usual form, the age equation is deceptively simple:

$$\text{Age} = \text{ED}/(\text{natural dose rate}) \qquad (A10)$$

where ED (equivalent dose) is the dose of radiation required to generate Z(nat). Equation (A10) is the only one to be found in many elementary accounts of TL and ESR dating, and may falsely give the impression that the methods themselves are simple.

Measurement of ED and natural dose rate is not straightforward. The simple age equation (A10) camouflages several factors that affect ED: there may be more than one model for $f(D)$; Z may differ from zero when t = zero; and the growth of trapped electrons when a sample is artificially irradiated may differ from growth over a long period. Equation (A10) gives no hint of phenomena such as supralinear response at low doses, incomplete bleaching and dose sensitisation, which must be known before ED can be evaluated. Furthermore, Equation (A10) conceals the fact that accurate measurement of natural dose rate is not simple, and assumes that the natural dose rate was constant throughout the history of the sample.

TL and OSL dating

Thermoluminescence (TL) and optically-stimulated luminescence (OSL) or 'optical' dating are considered together because both use luminescence to measure the proportion of trapped electrons in a sample, although the measurements are made differently. Both have their merits, but optical dating has significant advantages for dating sediments.

MEASUREMENT PRINCIPLES

Photon emission or luminescence occurs when an electron is displaced from an electron trap and is captured by a 'hole trap' or luminescence centre, which is created whenever an electron is displaced by α, β or γ radiation and lodges in an electron trap. Traps may be emptied and luminescence stimulated by heating the sample (TL) or, in some cases, by illumination with light of a suitable wavelength (OSL). Broadly, the intensity of luminescence indicates the proportion of trapped electrons in the specimen. More accurately, most minerals contain several species of traps which differ in 'depth', such that the deeper the trap the greater the energy required to displace electrons from it. Hence, traps of different depths are emptied at different temperatures and a graph of luminescence versus temperature (known as a 'glow curve') typically has several peaks, each of which represents electrons displaced from a particular species of trap. Dating is normally based on deeper traps which retain electrons for millions of years at normal temperatures.

Some traps in certain minerals can be emptied by exposing a sample to intense light of an appropriate wavelength. For example, the 325°C TL peak in quartz represents a trap that can be emptied by exposure to light with a wavelength of 500 nm. By alternately exposing a sample to the light source and to a sensitive light detector, the number of trapped electrons can be gauged. OSL dating uses this principle to measure a series of aliquots, and ED is estimated as for TL.

RESETTING AT TIME ZERO, AND THE MERITS OF OSL VERSUS TL

To estimate ED, not only must the data from a series of irradiated aliquots be fitted to the appropriate growth curve, but also the initial value must be known. The initial value can be determined experimentally and is usually zero for materials exposed to a high temperature at time zero, such as pottery or volcanic tephra, but sediments are more problematic. Sedimentary grains are commonly recycled many times before the moment of burial. Thus, dating must be based on traps that were emptied (or 'reset') when the grains were exposed to sunlight, immediately before burial, but the initial value is not necessarily zero and is determined by simulating field exposure in the laboratory.

TL dating is applied routinely where sample materials were strongly heated at or soon before they were deposited, such as pottery, hearths and volcanic tephras. OSL is preferable for sediments because it allows the light-sensitive traps, which were reset by sunlight before burial, to be measured selectively and with greater precision than by TL. Quartz, the most ubiquitous sedimentary mineral, illustrates both advantages of OSL. The trap species in quartz which have TL peaks at 325°C and 375°C can be reset by sunlight, but to reset the 375°C trap takes up to sev-

eral hours of exposure whereas the 325°C trap is reset within a few tens of seconds. Furthermore, signals from the two traps overlap strongly when measured by TL, but the sensitive 325°C trap can be measured without interference, with higher precision, by OSL.

ESR dating

ESR dating differs from TL and OSL in that the proportion of electrons in each trap species is measured directly, rather than from the luminescence produced when the electron is displaced from its trap to a luminescence centre. Otherwise, the principles are the same. Samples are divided into aliquots, irradiated in the laboratory, and the ED is calculated from the signal–dose growth curve in the same manner.

An ESR spectrometer subjects a sample to microwave excitation in a magnetic field; unpaired electrons in lattice-defect traps resonate at specific combinations of microwave frequency and magnetic field strength. The resonance signal-strength is proportional to the number of trapped electrons in a given trap species and is measured by the spectrometer. In practice, the microwave frequency is set and magnetic field intensity is varied; each trap species resonates at its own characteristic field strength, allowing different traps to be separated.

With appropriate traps, the precision of ESR-based estimates of ED can be comparable to those of TL or OSL, but ESR has not overtaken luminescence as a means of dating quartz or feldspar sediments, because critical traps such as that represented by the 325°C TL peak are measured more accurately by OSL or TL. There are other materials where ESR has the advantage. For example, ESR dating has been applied to carbonates such as coral, shells and speleothems; cross-checks against U-series results show that the ages obtained from well-preserved fossil corals by ESR are dependable, although less precise than obtained by U-series. However, the greatest successes of ESR dating have been with fossil teeth, enabling the ages of important fossils to be established directly, rather than by indirect association with deposits dated by other methods (Grün and Stringer 1991).

Dosimetry and age-errors

Dates based on TL, OSL or ESR measurements commonly have lower precision than those based on radioactive decay, partly because the precision of the ED estimate is usually about ± 4–8%. Furthermore, measurements of the dose rate (known as *dosimetry*) have systematic errors and are liable to other uncertainties.

After burial, a sample receives radiation from radioactive decay of uranium (U), thorium (Th) and potassium (K) in its immediate environment. Samples preserved within a few metres of the Earth's suface also receive cosmic radiation. In principle the environmental dose rate could be estimated from measurements of the concentrations of U, Th and K in a specimen and its surrounding matrix. In practice, daughter elements of the U and Th decay chains should be assayed by gamma spectrometry because they may not be in equilibrium with their parents, in which case the age calculation is modified. Furthermore, the dosimetry effect of γ radiation is affected by the water content of the sample and its matrix. A net dose rate error is usually compiled from standard errors in all the contributing measurements, coupled with uncertainty about the water content; commonly, systematic errors in dose rate can exceed $\pm$ 10%.

Accuracy is compromised by past changes of water content of a specimen and its matrix, and more serious effects can arise from past changes in the quantities of U or K owing to chemical migration, which cannot be identified after the event. Where U or K migration is suspected, ages can be calculated for certain limiting models. In the case of fossil teeth, for example, ESR ages are calculated on the basis of both early uptake and continuous uptake of U, and the age of a specimen is assumed to lie between extremes given by these alternatives (Grün and Stringer 1991).

Validation

Dating based on trapped electrons is at an earlier stage of development and is intrinsically more complex than well-established methods based on radioactive decay. The precision and accuracy of TL, OSL and ESR dates are significantly poorer than ^{230}Th/^{234}U, K–Ar and calibrated ^{14}C age measurements of unaltered, closed-system specimens, but they can be applied to important classes of materials which cannot be dated by other methods. Potential sources of error in evaluating ED can be circumvented by using combinations of additive and regenerative growth curves and, in the case of TL, by confirming that the sample gives a uniform plateau of age versus TL temperature (cf. Aitken 1985). Other sources of error, including uncertainties about dosimetry history, can only be evaluated by making parallel measurements of age or dosimetry variables, by as many means as possible.

There are unsolved problems in trapped electron-based dating, such as the matter of anomalous fading in feldspars (cf. Aitken 1985). There is more experimentation and less agreement about detailed laboratory procedures between laboratories using trapped electron-based methods, than for the major radioactive decay methods. A high degree of consistency has been achieved in studies based on relatively large

numbers of dates produced by different methods, but chronologies based on a handful of TL or ESR dates are often disputed. In short, even though individual age measurements pass internal checks, collaboration in the field and laboratory, multiple dating and usage of as many methods as possible is likely to produce the most enduring results.

DATING BASED ON SLOW CHEMICAL REACTIONS

Much endeavour in Quaternary research is aimed to bring the chronology of climatic changes and other events ever more sharply into focus. None the less, certain dating methods based on slow chemical reactions, which inherently have low precision, are surprisingly useful. Perhaps no single result from measurements of this type has been more celebrated than the demonstration that the Piltdown hominid was a hoax, which was based on traces of fluorine and uranium in the bones themselves (Oakley 1980). Although affected by all sorts of environmental vagaries, chemical methods often provide a ready means for selecting the best hypothesis when the age of a deposit is highly uncertain. Furthermore, owing to the effect of temperature on reaction rate, chemical methods can be used to estimate palaeotemperature, when the age of a specimen is determined by other means. Two notable examples of chemical methods are those based on amino-acid racemisation and obsidian hydration.

Amino-acid racemisation

Many molecules can be structured in alternative ways, just as a motor car can have the controls either on the left or the right, or apartments can be identical except that their floor plans are mirror images. Owing to the existence of asymmetric carbon bonds in most basic molecules of life, such as amino acids, alternative molecular patterns or *stereoisomers* can exist. Stereoisomers of a given compound are subtiled by the direction in which plane-polarised light is rotated when it is passed through a solution of the compound: left and right rotations distinguish the L (levo) and D (dextro) forms, respectively.

Each molecular species in living matter exists dominantly in the L isomeric form, for reasons to do with the way molecules fit together. A life molecule, like a door key, can in principle exist as either of two mirror-images, but only one of the two will fit perfectly and its inverted image fails to do so. Thus, natural selection prevails; for any given life molecule, a vital isomer exists and its alternative form, being non-functional, is rare, at least until the parent organism dies.

Life molecules such as amino acids can transform from the naturally selected L form into the D alternative and back again, but D forms will be discarded metabolically. When the organism dies, transformations proceed back and forth until there are about the same number of each. The progressive change from L dominance to an equilibrium mixture of L and D is termed *racemisation* for isomers with one asymmetric carbon atom, and *epimerisation* for those with two. At equilibrium, the exact proportions of D and L forms in the *racemic mixture* depends on their relative stabilities. For most amino acids, the L–D transformation is a reversible first-order chemical reaction and the following equation describes the change through time of the D/L ratio:

$$\ln ((1 + R)/(1 - K.R)) = E + (k_L + k_D).t \qquad (A11)$$

where R is the D/L ratio at time t; k_L and k_D are the foward and reverse rate constants, respectively, and $K = k_D/k_L$. E is the initial D/L ratio, which is not zero in some cases. If $E = 0$, the change through time of the D/L ratio is similar to curve B in Fig. A1, and is similar to curve C if $E > 0$.

Racemisation dating is based on amino acids extracted from all types of fossils in which tissue, proteins or other life molecules are preserved. The rate of racemisation differs between molecular types and is very strongly affected by temperature. Chemical environment also affects the rate, more so for some amino acids than others, and the same amino acids behave differently according to their type of fossil host. Thus, for a given type of material, rate constants k_L and k_D should be calibrated in laboratory tests at different controlled temperatures.

The temperature sensitivity is such that the age uncertainty can be as much as 10% if the mean temperature during the history of the specimen is uncertain by only 1°C. Hence, dating studies based on amino-acid racemisation (AAR) are typically calibrated with dates obtained by other methods. Although precision is not high, AAR methods have proven to be very useful for assigning various types of deposits to specific glacial or interglacial intervals within the Quaternary series (e.g. Miller and Mangerud 1986). When the age is well controlled by other dating methods and optimum sample material such as eggshell is used, AAR has been used to establish mean palaeotemperatures to within a degree or two by AAR measurements (Miller *et al.* 1997).

ORBITAL AND GEOMAGNETIC REVERSAL CHRONOLOGIES

Perhaps no method has had a greater impact on Quaternary chronology than correlation with time series

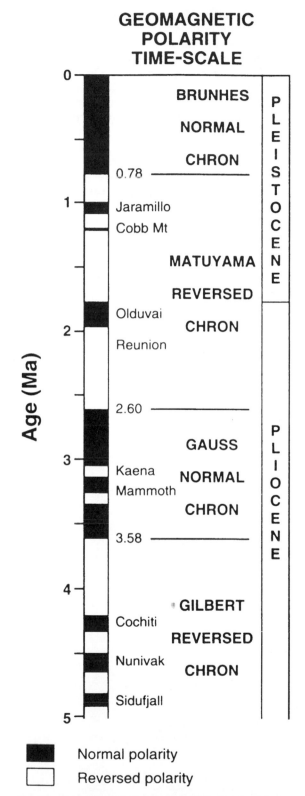

Fig. A4 Geomagnetic time-scale for the last 5 million years, after Shackleton *et al.* (1990) and Hilgen (1991). Løvlie (1989) provides a useful review of the use of the geomagnetic reversal series for Quaternary correlation and chronology

of orbital variations. Correlation with geomagnetic reversals has also been important. The orbitally-based chronology ultimately is based on the Milankovitch theory of climate change, reviewed in Chapter 5, which proposes that the Quaternary ice ages were regulated by orbitally-driven variations of seasonal insolation. As well as supported by U-series ages of raised coral reefs which represent high sea-levels within the last glacial cycle, the orbital chronology is precisely tied to the K–Ar Brunhes–Matuyama (B/M) geomagnetic reversal at 778 000 years BP (Tauxe *et al.* 1996). The geomagnetic reversal series for the last 5 million years, which has been modestly revised in the light of orbital time series and is the key for chronology and correlation between a large number of Pliocene and Quaternary stratigraphic sites (Løvlie 1989), is summarised in Fig. A4.

The strongest support for the Milankovitch theory and its attendant orbital chronology derives from deep-sea core records. To begin with, geomagnetic reversals, which had been identified in sequences of lava flows dated by K–Ar, were also identified in deep-sea cores, and approximate time-scales were interpolated between these fixed points. Using this chronology, periodic variations of oxygen isotope ratios in the cores were shown to be similar to the orbital variations. It is now widely accepted that orbital variations are the pacemaker of the Pleistocene ice ages (although the relationship between the amplitudes of orbital forcing and climatic response is far from regular), and the chronology now is tuned to orbital time series (Martinson *et al.* 1987).

The orbital chronology is almost universally applied to continuous sedimentary sequences that record many Quaternary glacial cycles, of which deep-sea sediment cores are by far the most important. Correlation between deep-sea records is generally based on oxygen isotope variations, which are aligned with the orbital 'clock', because major changes in the size of the continental ice sheets cause changes of $\delta^{18}O$ throughout the world's oceans. Changes of other variables recorded in deep-sea cores, such as sea-surface temperatures indicated by microfossils, are not synchronous throughout the oceans, however (Imbrie *et al.* 1989). Hence, for a given Quaternary record, exact correlations should not be made between it and the orbital chronology until phase relationships are established by some independent means.

REFERENCES

Aaby, B. and Berglund, B.E. 1986. Characterization of peat and lake deposits. In Berglund, B.E. (ed.), *Handbook of Holocene Palaeoecology and Palaeohydrology*. Wiley, Chichester, 231–246.

Abell, P.I. and Williams, M.A.J. 1989. Oxygen and carbon isotope ratios in gastropod shells as indicators of palaeoenvironments in the Afar region of Ethiopia. *Palaeogeography, Palaeoclimatology, Palaeoecology* 74, 265–278.

Adam, D.P. 1975. Ice ages and the thermal equilibrium of the earth, II. *Quaternary Research* 5, 161–171.

Adams, J. 1995. Weathering and glacial cycles. *Nature* 373, 110.

Adams, J.M. and Faure, H. 1996. Changes in moisture balance between glacial and interglacial conditions: influence on carbon cycle processes. In Branson, J., Brown, A.G. and Gregory, K.J. (eds), *Global Continental Changes: the Context of Palaeohydrology*. Geological Society Special Publications 115, 27–42.

Adams, J.M. and Faure, H. 1997. Preliminary vegetation maps of the world since the last glacial maximum: an aid to archeological understanding. *Journal of Archaeological Science* 24, 623–647.

Adams, J.M., Faure, H., Faure-Denard, L., McGlade, J.M. and Woodward, F.I. 1990. Increases in terrestrial carbon storage from the last glacial maximum to the present. *Nature* 348, 711–714.

Adams, R.M., Rosenzweig, C., Peart, R.M., Ritchie, J.T., McCarl, B.A., Glyer, J.D., Curry, R.B., Jones, J.W., Boote, K.J. and Allen, L.H. 1990. Global climate change and US agriculture. *Nature* 345, 219–224.

Adamson, D.A. and Fox, M.D. 1982. Change in Australian vegetation since European settlement. In Smith, J.M.B. (ed.), *A History of Australasian Vegetation*. McGraw-Hill, Sydney, 109–146.

Adamson, D.A., Gasse, F., Street, F.A. and Williams, M.A.J. 1980. Late Quaternary history of the Nile. *Nature* 288, 50–55.

Adamson, D.A. and Pickard, J. 1986. Cainozoic history of the Vestfold Hills. In Pickard, J. (ed.), *Antarctic Oasis: Terrestrial environments and history of the Vestfold Hills*. Academic Press, Sydney, 63–97.

Adamson, D.A. and Williams, M.A.J. 1987. Geological setting of Pliocene rifting in the Afar Depression of Ethiopia. *Journal of Human Evolution* 16, 597–610.

Adkins, J.F., Boyle, E.A., Keigwin, L. and Cotijo, E. 1997. Variability of the North Atlantic thermohaline circulation during the last interglacial period. *Nature* 390, 154–156.

Agrawal, D.P. 1995. Environmental and civilizational processes in India: their global relevance. In Ito, S. and Yasuda, Y. (eds), *Nature and Humankind in the Age of Environmental Crisis*. International Research Centre for Japanese Studies, Kyoto, 45–55.

Aharon, P. 1984. Implications of the coral-reef record from New Guinea concerning the astronomical theory of ice ages. In Berger, A., Imbrie, J., Hays, J., Kukla, G. and Saltzman, B. (eds), *Milankovitch and Climate. Understanding the Response to Astronomical Forcing*. Reidel, Dordrecht, 379–389.

Aharon, P. and Chappell, J. 1986. Oxygen isotopes, sea level changes and the temperature history of a coral reef environment in New Guinea over the last 10^5 years. *Palaeogeography, Palaeoclimatology, Palaeoecology* 56, 337–379.

Aiello, L.C. 1993. The fossil evidence for modern human origins in Africa: a revised view. *American Anthropologist* 95, 73–96.

Aitken, M.J. 1985. *Thermoluminescence Dating*. Academic Press, London.

Aitken, M.J. 1990. *Science-Based Dating in Archaeology*. Longman, London.

Allan, R., Lindesay, J. and Parker, D. 1996. *El Niño Southern Oscillation and Climatic Variability*. CSIRO Publishing, Collingwood, Melbourne.

Alley, R.B. and MacAyeal, D.R. 1994. Ice-rafted debris associated with binge/purge oscillations of the Laurentide ice sheet. *Paleoceanography* 9, 503–511.

Alpagut, B., Andrews, P., Fortelius, M., Kappelman, J., Temizsoy, I., Çelebi, H. and Lindsay, W. 1996. A new specimen of *Ankarapithecus meteai* from the Sinap formation of central Anatolia. *Nature* 382, 349–351.

Altabet, M.A., François, R., Murray, D.W. and Prell, W.L. 1995. Climate-related variations in denitrification in the Arabian Sea from sediment $^{15}N/^{14}N$ ratios. *Nature* 373, 506–509.

Ambrose, S.H. and De Niro, M.J. 1986. Reconstruction of African human diet using bone collagen carbon and nitrogen isotope ratios. *Nature* 318, 321–324.

An, Z. and Ho, C.K. 1989. New magnetostratigraphic dates of Lantian *Homo erectus*. *Quaternary Research* 32, 213–221.

An, Z. and Porter, S.C. 1997. Millennial-scale climatic oscillations during the last interglacial in central China. *Geology* 25, 603–606.

Andersen, B.G. and Borns, H.W. Jr. 1994. *The Ice Age World. An Introduction to Quaternary History and Research with Emphasis on North America and Northern Europe during the Last 2.5 Million Years*. Scandinavian University Press, Oslo.

Anderson, B.G. 1981. Late Weichselian ice sheets in Eurasia and Greenland. In Denton, G.H. and Hughes, T.J. (eds), *The Last Great Ice Sheets*. Wiley, New York, 1–65.

Anderson, D.E. 1997. Younger Dryas research and its implications for understanding abrupt climatic change. *Progress in Physical Geography* 21, 230–249.

Anderson, N.R. and Malahoff, A. (eds) 1977. *The Fate of Fossil Fuel CO₂ in the Oceans.* Plenum Press, New York.

Anderson, R.S. and van Devender, T.R. 1995. Vegetation history and palaeoclimates of the coastal lowlands of Sonora, Mexico – pollen records from packrat middens. *Journal of Arid Environments* 30, 295–306.

Andreas, E.L. and Ackley, S.F. 1982. On the differences in ablation seasons of Arctic and Antarctic sea ice. *Journal of Atmospheric Science* 39, 440–447.

Andrews, J.T. 1987. The late Wisconsin glaciation and deglaciation of the Laurentide ice sheet. In Ruddiman, W.F. and Wright, H.E. Jr. (eds), *North America and Adjacent Oceans During the Last Deglaciation.* The Geology of North America Vol. K-3. Geological Society of America, Boulder, CO, 13–37.

Andrews, J.T., Erlenkeuser, H., Tedesco, K., Aksu, A.E. and Jull, A.J.T. 1994. Late Quaternary (stage 2 and 3) meltwater and Heinrich events, northwest Labrador Sea. *Quaternary Research* 41, 26–34.

Andrews, J.T. and Miller, G.H. 1985. Holocene sea level variations within Frobisher Bay. In Andrews, J.T. (ed.), *Quaternary Environments: Eastern Canadian Arctic, Baffin Bay and Western Greenland.* Allen and Unwin, Boston, 585–607.

Andrews, J.T. and Tedesco, K. 1992. Detrital carbonate-rich sediments, northwestern Labrador Sea: implications for ice-sheet dynamics and iceberg rafting (Heinrich) events in the North Atlantic. *Geology* 20, 1087–1090.

Anisimov, O.A. and Nelson, F.E. 1996. Permafrost distribution in the Northern Hemisphere under scenarios of climatic change. *Global and Planetary Change* 14, 59–72.

Arbogast, A.F. and Johnson, W.C. 1994. Climatic implications of the late Quaternary alluvial record of a small drainage basin in the central Great Plains. *Quaternary Research* 41, 298–305.

Archer, D. and Maier-Reimer, E. 1994. Effect of deep-sea sedimentary calcite preservation on atmospheric CO₂ concentration. *Nature* 367, 260–263.

Archer, M., Hand, S.J. and Godthelp, H. 1995. Tertiary environmental and biotic change in Australia. In Vrba, E.S., Denton, G.H., Partridge, T.C. and Burckle, L.H. (eds), *Paleoclimate and Evolution, with Emphasis on Human Origins.* Yale University Press, New Haven, 77–90.

Arensburg, B., Tillier, A.M., Vandermeersch, B., Duday, H., Shepartz, L.A. and Rak, Y. 1989. A middle Palaeolithic human hyoid bone. *Nature* 338, 758–760.

Arnell, N.W. and Reynard, N.S. 1996. The effects of climate change due to global warming on river flows in Great Britain. *Journal of Hydrology* 183, 397–424.

Ash, J.E. and Wasson, R.J. 1983. Vegetation and sand mobility in the Australian desert dunefield. *Zeitschrift für Geomorphologie, Supplementband* 45, 7–25.

Assefa, G., Clark, J.D. and Williams, M.A.J. 1982. Late Cenozoic history and archaeology of the Upper Webi Shebele Basin, east central Ethiopia. *Sinet: Ethiopian Journal of Science* 5, 27–46.

Augustinus, P.G.E.F. 1989. Cheniers and chenier plains: a general introduction. *Marine Geology* 90, 219–229.

Australian Academy of Science 1976. *Report of a Committee on Climatic Change.* Australian Academy of Science, Canberra.

Axelrod, D.I. and Raven, P.H. 1978. Late Cretaceous and Tertiary vegetation history of Africa. In Werger, M.A.J. (ed.), *Biogeography and Ecology of Southern Africa.* Junk, The Hague, 77–130.

Ayliffe, D., Williams, M.A.J. and Sheldon, F. 1996. Stable carbon and oxygen isotopic composition of early-Holocene gastropods from Wadi Mansurab, north-central Sudan. *The Holocene* 6, 157–169.

Bach, W. 1984. *Our Threatened Climate: Ways of Averting the CO₂ Problem through Rational Energy Use.* Reidel, Boston.

Bagnold, R.A. 1941. *The Physics of Blown Sand and Desert Dunes.* Methuen, London.

Baird, R.F. 1991. The taphonomy of late Quaternary cave localities yielding vertebrate remains in Australia. In Vickers-Rich, P., Monaghan, J.M., Baird, R.F. and Rich, T.H. (eds), *Vertebrate Palaeontology of Australasia.* Pioneer Design Studio, Melbourne, 267–310.

Baker, V.R. 1983. Late Pleistocene fluvial systems. In Wright, H.E. and Porter, S.C. (eds), *Late Quaternary Environments of the United States. Volume 1. The Late Pleistocene.* London, Longman, 115–129.

Baker, V.R. 1988. Overview. In Baker, V.R., Kochel, R.C. and Patton, P.C. (eds), *Flood Geomorphology.* Wiley, New York, 1–8.

Baker, V.R., Bowler, J.M., Enzel, Y. and Lancaster, N. 1995. Late Quaternary palaeohydrology of arid and semi-arid regions. In Gregory, K.J., Starkel, L. and Baker, V.R. (eds), *Global continental palaeohydrology.* Wiley, Chichester, 203–231.

Balling, N. 1980. The land uplift in Fennoscandia, gravity field anomalies and isostasy. In Mörner, N.-A. (ed.), *Earth Rheology, Isostasy and Eustasy.* Wiley, Chichester, 297–321.

Balout, L. 1955. *Préhistoire de L'Afrique du Nord.* Arts et Métiers Graphiques, Paris.

Barbetti, M. and Allen, H. 1972. Prehistoric man at Lake Mungo, Australia, by 32,000 years BP. *Nature* 240, 46–48.

Barbetti, M., Clark, J.D., Williams, F.M. and Williams, M.A.J. 1980. Palaeomagnetism and the search for very ancient fireplaces in Africa. *Anthropologie* 18, 229–304.

Bard, E., Arnold, M., Duprat, J., Moyes, J. and Duplessy, J.-C. 1987. Reconstruction of the last deglaciation: deconvolved records of δ¹⁸O profiles, micropaleontological variations and accelerator mass spectrometric ¹⁴C dating. *Climate Dynamics* 1, 101–112.

Bard, E., Fairbanks, R., Arnold, M., Maurice, P., Duprat, J., Moyes, J. and Duplessy, J.-C. 1989. Sea-level estimates during the last deglaciation based on δ¹⁸O and accelerator mass spectrometry ¹⁴C ages measured in *Globigerina bulloides. Quaternary Research* 31, 381–391.

Bard, E., Hamelin, B. and Fairbanks, R.G. 1990a. U–Th ages obtained by mass spectrometry in corals from Barbados: sea level during the past 130 000 years. *Nature* 346, 456–458.

Bard, E., Hamelin, B., Fairbanks, R.G. and Zindler, A. 1990b. Calibration of the ¹⁴C timescale over the past

30,000 years using mass spectrometric U–Th ages from Barbados corals. *Nature* 345, 405–410.

Bard, E., Jouannic, C., Hamelin, B., Pirazzoli, P., Arnold, M., Faure, G., Sumosusastro, P. and Syaefudin 1996. Pleistocene sea levels and tectonic uplift based on dating corals from Sumba Island, Indonesia. *Geophysical Research Letters* 23, 1473–1476.

Bard, E., Rostek, F. and Sonzogni, C. 1997. Interhemispheric synchrony of the last deglaciation inferred from alkenone palaeothermometry. *Nature* 385, 707–710.

Barker, W.R. and Greenslade, P.J.M. (eds) 1982. *Evolution of the Flora and Fauna of Arid Australia.* Peacock, Adelaide.

Barnett, T.P. 1983. Recent changes in sea level and their possible causes. *Climatic Change* 5, 15–38.

Barnett, T.P. 1984. The estimation of 'global' sea level change: a problem of uniqueness. *Journal of Geophysical Research* 89(C5), 7980–7988.

Barnola, J.-M., Pimienta, P., Raynaud, D. and Korotkevich, Y.S. 1991. CO_2–climate relationship as deduced from the Vostok ice core: a re-examination based on new measurements and on a re-evaluation of the air dating. *Tellus* 43B, 83–90.

Barnola, J.-M., Raynaud, D., Korotkevich, Y.S. and Lorius, C. 1987. Vostok ice core provides 160,000 year record of atmospheric CO_2. *Nature* 329, 408–414.

Barraclough, G. (ed.), 1982. *The Times Concise Atlas of World History.* Angus and Robertson, London.

Barrows, T.T., Ayress, M.A. and Hunt, G.R. 1996. A reconstruction of last glacial maximum sea-surface temperatures in the Australasian region. *Quaternary Australasia* 14, 27–31.

Bartholomew, J.C., Geelan, P.J.M., Lewis, H.A.C., Middleton, P. and Winkleman, B. (eds) 1980. *The Times Atlas of the World (Comprehensive edn).* Times Books, London.

Bartlein, P.J. and Whitlock, C. 1993. Palaeoclimatic interpretation of the Elk Lake pollen record. In Bradbury, J.P. and Dean, W.E. (eds), *Elk Lake, Minnesota: Evidence for Rapid Climatic Change in the North-Central United States.* Geological Society of America Special Paper 276, 275–293.

Bastable, H.G., Shuttleworth, W.J., Dallarosa, R.L.G., Fisch, G. and Nobre, C.A. 1993. Observations of climate, albedo, and surface radiation over cleared and undisturbed Amazonian forest. *International Journal of Climatology* 13, 783–796.

Battle, M., Bender, M., Sowers, T., Tans, P.P., Butler, J.H., Elkins, J.W., Ellis, J.T., Conway, T., Zhang, N., Lang, P. and Clarke, A.D. 1996. Atmospheric gas concentrations over the past century measured in air from firn at the South Pole. *Nature* 383, 231–235.

Baulin, V.V. and Danilova, N.S. 1984. Dynamics of late Quaternary permafrost in Siberia. In Velichko, A.A. (ed.), English language version: Wright, H.E. and Burnosky, C.W. (eds), *Late Quaternary Environments of the Soviet Union.* Longman, Essex, 69–77.

Bé, A.W.H. and Tolderlund, D.S. 1971. Distribution and ecology of living planktonic foraminifera in surface waters of the Atlantic and Indian Oceans. In Funnell, B.M. and Riedel, W.R. (eds), *The Micropalaeontology of the Oceans.* Cambridge University Press, Cambridge, 105–149.

Beck, J.W., Récy, J., Taylor, F., Edwards, R.L. and Cabioch, G. 1997. Abrupt changes in early Holocene tropical sea surface temperature derived from coral records. *Nature* 385, 705–707.

Bednarik, R.G. 1989. On the Pleistocene settlement of South America. *Antiquity* 63, 101–111.

Begét, J.E. and Hawkins, D.B. 1989. Influence of orbital parameters on Pleistocene loess deposition in central Alaska. *Nature* 337, 151–153.

Behl, R.J. and Kennett, J.P. 1996. Brief interstadial events in the Santa Barbara basin, NE Pacific, during the past 60 kyr. *Nature* 379, 243–246.

Behrenfeld, M.J., Bale, A.J., Kolber, Z.S., Aiken, J. and Falkowski, P.G. 1996. Confirmation of iron limitation of phytoplankton photosynthesis in the equatorial Pacific Ocean. *Nature* 383, 508–511.

Behrensmeyer, A.K. 1982. The geological context of human evolution. *Annual Review of Earth and Planetary Science* 10, 39–60.

Behrensmeyer, A.K., Gordon, K.D. and Yanagi, G.T. 1986. Trampling as a cause of bone surface damage and pseudo-cutmarks. *Nature* 319, 768–771.

Behrensmeyer, A.K. and Kidwell, S.M. 1985. Taphonomy's contributions to paleobiology. *Paleobiology* 11, 105–119.

Bell, M. and Laine, E.P. 1985. Erosion of the Laurentide region of North America by glacial and glaciofluvial processes. *Quaternary Research* 23, 154–174.

Bell, M. and Walker, M.J.C. 1992. *Late Quaternary Environmental Change. Physical and Human Perspectives.* Longman, London.

Benda, L. 1995. *Das Quartär Deutschlands.* Gebrüder Borntraeger, Stuttgart.

Bender, M., Sowers, T., Dickson, M.-L., Orchardo, J., Grootes, P., Mayewski, P.A. and Meese, D.A. 1994. Climate correlations between Greenland and Antarctica during the past 100,000 years. *Nature* 372, 663–666.

Bender, M.L., Fairbanks, R.G., Taylor, F.W., Matthews, R.K., Goddard, J.G. and Broecker, W.S. 1979. Uranium-series dating of the Pleistocene reef tracts of Barbados, West Indies. *Bulletin of the Geological Society of America* 90, 577–594.

Bennett, K.D. 1983a: Devensian late-glacial and Flandrian vegetational history at Hockham Mere, Norfolk, England. *New Phytologist* 95, 457–487.

Bennett, K.D. 1983b: Postglacial population expansion of forest trees in Norfolk, UK. *Nature* 303, 164–167.

Bennett, K.D. 1997. *Evolution and Ecology: the Pace of Life.* Cambridge University Press, Cambridge.

Benson, L. 1994. Carbonate deposition, Pyramid Lake Subbasin, Nevada: 1. Sequence of formation and elevational distribution of carbonate deposits (tufas). *Palaeogeography, Palaeoclimatology, Palaeoecology* 109, 55–87.

Benson, L. and Thompson, R.S. 1987. The physical record of lakes in the Great Basin. In Ruddiman, W.F. and Wright, H.E. (eds), *North America and Adjacent Oceans During the Last Deglaciation.* The Geology of North America, Volume K-3, Geological Society of America, 241–260.

Benson, L., Burdett, J., Lund, S., Kashgarian, M. and Mensing, S. 1997. Nearly synchronous climate change in the northern hemisphere during the last glacial termination. *Nature* 388, 263–265.

Benson, L., Kashgarian, M. and Rubin, M. 1995. Carbonate deposition, Pyramid Lake subbasin, Nevada: 2. Lake levels and polar jet stream positions reconstructed from radiocarbon ages and elevations of carbonates (tufas) deposited in the Lahontan basin. *Palaeogeography, Palaeoclimatology, Palaeoecology* 117, 1–30.

Benson, R.H. 1984. The Phanerozoic 'Crisis' as viewed from the Miocene. In Berggren, W.A. and van Couvering, J.A. (eds), *Catastrophes and Earth History*. Princeton University Press, Princeton, NJ, 437–446

Bentley, H.W., Phillips, F.M., Davis, S.N., Habermehl, M.A., Airey, P.L., Calf, G.E., Elmore, D., Grove, H.E. and Torgerson, T. 1986. Chlorine 36 dating of very old groundwater 1. The Great Artesian Basin, Australia. *Water Resources Research* 20, 1991–2001.

Berger, A.L. 1977. Support for the astronomical theory of climatic change. *Nature* 269, 44–45.

Berger, A.L. 1978a: A simple algorithm to compute long term variations of daily or monthly insolation. Contribution No. 18, *Institut d'Astronomie et de Geophysique Georges Lemaître*, Université Catholique de Louvain.

Berger, A.L. 1978b: Long-term variations of caloric insolation resulting from the earth's orbital elements. *Quaternary Research* 9, 139–167.

Berger, A.L. 1978c. Long-term variations of daily insolation and Quaternary climatic changes. *Journal of the Atmospheric Sciences* 35, 2362–2367.

Berger, A.L. 1979. Insolation signatures of Quaternary climatic changes. *Il Nuovo Cimento* 2C, 63–87.

Berger, A.L. 1980. The Milankovitch astronomical theory of paleoclimates: a modern review. *Vistas in Astronomy* 24, 103–122.

Berger, A.L. 1981a: The astronomical theory of paleoclimates. In Berger, A. (ed.), *Climatic Variations and Variability: Facts and Theories*. NATO ISI Series C: 72. Reidel, Dordrect, 501–525.

Berger, A.L. 1981b: Spectrum of climatic variations and possible causes. In Berger, A. (ed.), *Climatic Variations and Variability: Facts and Theories*. NATO ISI Series C Vol. 72. Reidel, Dordrecht, 411–432.

Berger, A. (ed.) 1981c. *Climatic Variations and Variability: Facts and Theories*. Reidel, Dordrecht.

Berger, A.L. 1984. Accuracy and frequency stability of the earth's orbital clements during the Quaternary. In Berger, A., Imbrie, J., Hays, J., Kukla, G. and Saltzman, B. (eds), *Milankovitch and Climate. Understanding the Response to Astronomical Forcing*. NATO ISI Series C, Vol. 126. Reidel, Dordrecht, 3–39.

Berger, A. 1988. Milankovitch theory and climate. *Reviews of Geophysics* 26, 624–657.

Berger, A. 1989a. Pleistocene climatic variability at astronomical frequencies. *Quaternary International* 2, 1–14.

Berger, A. 1989b. The spectral characteristics of pre-Quaternary climatic records, an example of the relationship between the astronomical theory and geosciences. In Berger, A., Schneider, S. and Duplessy, J.-C. (eds), *Climate and Geo-Sciences*. Kluwer, Dordrecht, 47–76.

Berger, A. 1991. Long-term history of climate ice ages and Milankovitch periodicity. In Sonett, C.P., Giampapa, M.S. and Matthews, M.S. (eds), *The Sun in Time*. University of Arizona, Tucson, 498–510.

Berger, A., Guiot, J., Kukla, G. and Pestiaux, P. 1981. Long-term variations of monthly insolation as related to climatic changes. *Sonderdruck aus der Geologischen Rundschau* 70, 748–758.

Berger, A. and Loutre, M.F. 1991. Insolation values for the climate of the last 10 million years. *Quaternary Science Reviews* 10, 291–317.

Berger, A. and Loutre, M.F. 1994a. Astronomical forcing through geological time. In De Boer, P.L. and Smith, D.G. (eds), *Orbital Forcing and Cyclic Sequences*. International Association of Sedimentologists, Special Pub. No. 19, 15–24.

Berger, A. and Loutre, M.F. 1994b. Precession, eccentricity, obliquity, insolation and paleoclimates. In Duplessy, J.-C. and Spyridakis, M.-T. (eds), *Long-term Climatic Variations. Data and Modelling*. NATO ISI Series I, Vol. 22. Springer Verlag, Berlin, 107–151.

Berger, A., Loutre, M.-F. and Tricot, C. 1993a: Insolation and earth's orbital periods. *Journal of Geophysical Research* 98(D6), 10341–10362.

Berger, A.L. and Pestiaux, P. 1984. Accuracy and stability of the Quaternary terrestrial insolation. In Berger, A., Imbrie, J., Hays, J., Kukla, G. and Saltzman, B. (eds), *Milankovitch and Climate. Understanding the Response to Astronomical Forcing*. NATO ISI Series C, Vol. 126. Reidel, Dordrecht, 83–111.

Berger, A., Tricot, C., Gallée, H. and Loutre, M.F. 1993b: Water vapour, CO_2 and insolation over the last glacial–interglacial cycles. *Philosophical Transactions of the Royal Society of London B*, 341, 253–261.

Berger, W.H., Burke, S. and Vincent, E. 1987. Glacial–Holocene transition: climate pulsation and sporadic shutdown of NADW production. In Berger, W.H. and Labeyrie, L.D. (eds), *Abrupt Climatic Change*. NATO ASI Series C: 216. Reidel, Dordrecht, 279–297.

Berger, W.H., Killingley, J.S. and Vincent, E. 1985. Timing of deglaciation from an oxygen isotope curve for Atlantic deep-sea sediments. *Nature* 314, 156–158.

Berger, W.H. and Vincent, E. 1986. Sporadic shutdown of North Atlantic deep water production during the glacial–Holocene transition? *Nature* 324, 53–55.

Bergonzini, L., Chalié, F. and Gasse, F. 1997. Paleoevaporation and paleoprecipitation in the Tanganyika Basin at 18,000 years B.P. inferred from hydrologic and vegetation proxies. *Quaternary Research* 47, 295–305.

Berkey, C.P. and Morris, F.K. 1927. *Geology of Mongolia. Natural history of Central Asia, Volume II*. American Museum of Natural History, New York.

Betancourt, J.L., van Devender, T.R. and Martin, P.S. 1990. *Packrat Middens: the Last 40,000 Years of Biotic Change*. University of Arizona Press, Tucson.

Bettis, E.A. and Autin, W.J. 1997. Complex response of a midcontinent North America drainage system to Late Wisconsinan sedimentation. *Journal of Sedimentary Research* 67, 740–748.

Bigarella, J.J. and Ferreira, A.M.M. 1985. Amazonian geology and the Pleistocene and the Cenozoic environments and paleoclimates. In Prance, G.T and Lovejoy, T.E. (eds), *Amazonia*. Pergamon, Oxford, 49–71.

Billard, A. and Orombelli, G. 1986. Quaternary glaciations in the French and Italian piedmonts of the Alps. *Quaternary Science Reviews* 5, 407–419.

Binford, L.R. 1981. *Bones: Ancient Men and Modern Myths*. Academic Press, New York.

Binford, L.R. 1983. *In Pursuit of the Past: Decoding the Archaeological Record*. Thames and Hutton, New York.

Birchfield, G.E. 1987. Changes in deep-ocean water $\delta^{18}O$ and temperature from the last glacial maximum to the present. *Paleoceanography* 2, 431–442.

Birchfield, G.E. and Ghil, M. 1993. Climate evolution in the Pliocene and Pleistocene from marine-sediment records and simulations: internal variability versus orbital forcing. *Journal of Geophysical Research* 98(D6), 10385–10339.

Birchfield, G.E. and Grumbine, R.W. 1985. 'Slow' physics of large continental ice sheets and underlying bedrock and its relation to the Pleistocene ice ages. *Journal of Geophysical Research* 90(B13), 11294–11302.

Birchfield, G.E. and Weertman, J. 1982. A model study of the role of variable ice albedo in the climate response of the earth to orbital variations. *Icarus* 50, 462–472.

Bird, E.C.F. 1993. *Submerging Coasts. The Effect of Rising Sea Level on Coastal Environments*. Wiley, Chichester.

Bird, E.C.F. 1996. Coastal erosion and rising sea levels. In Milliman, J.D. and Haq, B.U. (eds), *Sea-Level Rise and Coastal Subsidence. Causes, Consequences, and Strategies*. Kluwer, Dordrecht, 87–103.

Birkeland, P.W. 1974. *Pedology, Weathering, and Geomorphological Research*. Oxford University Press, Oxford.

Birkeland, P.W. 1984. *Soils and Geomorphology*. Oxford University Press, Oxford.

Birks, H.J.B. 1981. The use of pollen analysis in the reconstruction of past climates: a review. In Wigley, T.M.L., Ingram, M.J. and Farmer, G. (eds), *Climate and History*. Cambridge University Press, Cambridge, 111–138.

Birks, H.J.B. and Birks, H.H. 1981. *Quaternary Palaeoecology*. Edward Arnold, London.

Björck, S., Kromer, B., Johnsen, S., Bennike, O., Hammarlund, D., Lemdahl, G., Possnert, G., Rasmussen, T.L., Wohlfarth, B., Hammer, C.U. and Spurk, M. 1996. Synchronized terrestrial–atmospheric deglacial records around the North Atlantic. *Science* 274, 1155–1160.

Blaikie, P. and Brookfield, H. 1987. *Land Degradation and Society*. Methuen, London.

Blanchon, P. and Shaw, J. 1995. Reef drowning during the last deglaciation: evidence for catastrophic sea-level rise and ice-sheet collapse. *Geology* 23, 4–8.

Bleil, U. and Thiede, J. (eds) 1990. *Geological History of the Polar Oceans: Arctic versus Antarctic*. Kluwer, Dordrecht.

Blikra, L.H. and Longva, O. 1995. Frost-shattered debris facies of Younger Dryas age in the coastal sedimentary secessions in western Norway: palaeoenvironmental implications. *Palaeogeography, Palaeoclimatology, Palaeoecology* 118, 89–110.

Bloemendal, J. and deMenocal, P. 1989. Evidence for a change in the periodicity of tropical climate cycles at 2.4 Myr from whole-core magnetic susceptibility measurements. *Nature* 342, 897–900.

Bloom, A.L. 1967. Pleistocene shorelines: a new test of isostasy. *Bulletin of the Geological Society of America* 78, 1477–1494.

Bloom, A.L. 1978. *Geomorphology: a Systematic Analysis of Late Cenozoic Landforms*. Prentice Hall, New Jersey.

Bloom, A.L. 1983. Sea level and coastal morphology of the United States through the Late Wisconsin glacial maximum. In Wright, H.E. and Porter S.C. (eds), *Late Quaternary Environments of the United States*. University of Minnesota Press, Minneapolis, 215–236.

Bloom, A.L., Broecker, W.S., Chappell, J.M.A., Matthews, R.K. and Mesolella, K.J. 1974. Quaternary sea level fluctuations on a tectonic coast: new $^{230}Th/^{234}U$ dates from the Huon Peninsula, New Guinea. *Quaternary Research* 4, 185–205.

Blumenschine, R.J. 1988. An experimental model of the timing of hominid and carnivore influence on archaeological bone assemblages. *Journal of Archaeological Science* 15, 483–502.

Board on Global Change (Commission on Geosciences, Environment, and Resources, National Research Council) 1994. *Solar Influences on Global Change*. National Academy Press, Washington, DC.

Boden, T.A, Kanciruk, P. and Farrell, M.P. 1990. *Trends '90*. Information Analysis Center, Oak Ridge National Laboratory, Tennessee.

Bond, G.C., Broecker, W.S., Johnsen, S., McManus, J., Labeyrie, L., Jouzel, J. and Bonani, G. 1993. Correlations between climate records from North Atlantic sediments and Greenland Ice. *Nature* 365, 143–147.

Bond, G., Heinrich, H., Broecker, W., Labeyrie, L., McManus, J., Andrews, J., Huon, S., Jantschik, R., Clasen, S., Simet, C., Tedesco, K., Klas, M., Bonani, G. and Ivy, S. 1992. Evidence for massive discharges of icebergs into the North Atlantic Ocean during the last glacial period. *Nature* 360, 245–249.

Bond, G.C. and Lotti, R. 1995. Iceberg discharges into the North Atlantic on millennial time scales during the last glaciation. *Science* 267, 1005–1010.

Bonnefille, R. 1983. Evidence for a cooler and drier climate in the Ethiopian uplands towards 2.5 Myr ago. *Nature* 303, 487–491.

Bonnefille, R. 1995. A reassessment of the Plio-Pleistocene pollen record of East Africa. In Vrba, E.S., Denton, G.H., Partridge, T.C. and Burckle, L.H. (eds), *Paleoclimate and Evolution*. Yale University Press, New Haven, 1131–1137.

Bonnefille, R., Roeland, J.C. and Guiot, J. 1990. Temperature and rainfall estimates for the past 40,000 years in equatorial Africa. *Nature* 346, 347–349.

Bonnefille, R., Vincens, A. and Buchet, G. 1987. Palynology, stratigraphy and palaeoenvironment of a Pliocene hominid site (2.9–3.3 MY) at Hadar, Ethiopia. *Palaeogeography, Palaeoclimatology, Palaeoecology* 60, 249–281.

Boucher de Perthes, J. 1847. *Antiquités Celtiques et Antédiluviennes. Mémoire sur l'industrie primitive et les arts à leur origine*. Volume 1. Treuttel and Wertz, Paris.

Boucher de Perthes, J. 1857. *Antiquités Celtiques et Antédiluviennes. Mémoire sur l'industrie primitive et les arts à leur origine*. Volume 2. Treuttel and Wertz, Paris.

Boucher de Perthes, J. 1864. *Antiquités Celtiques et Antédiluviennes. Mémoire sur l'industrie primitive et les arts à leur origine.* Volume 3. Treuttel and Wertz, Paris.

Bouma, W.J., Pearman, G.I. and Manning, M.R. (eds) 1996. *Greenhouse: Coping with Climate Change.* CSIRO, Melbourne.

Bowen, D.Q., 1978. *Quaternary Geology. A Stratigraphic Framework for Multi-disciplinary Work.* Pergamon Press, Oxford.

Bowen, D.Q., Rose, J., McCabe, A.M. and Sutherland, D.G. 1986. Correlation of Quaternary glaciations in England, Ireland, Scotland and Wales. *Quaternary Science Reviews* 5, 299–340.

Bowler, J.M. 1978. Glacial age aeolian events at high and low altitudes: a Southern Hemisphere perspective. In Van Zinderen Bakker, E.M. (ed.), *Antarctic Glacial History and World Paleoenvironments.* Balkema, Rotterdam, 149–172.

Bowler, J.M. 1981. Australian salt lakes: a palaeohydrological approach. *Hydrobiology* 82, 431–444.

Bowler, J.M., Thorne, A.G. and Polach, H.A. 1972. Pleistocene man in Australia: age and significance of the Mungo skeleton. *Nature* 240, 48–50.

Bowler, J.M. and Wasson, R.J. 1984. Glacial age environments of inland Australia. In Vogel, J.C. (ed.), *Late Cainozoic Environments of the Southern Hemisphere.* Balkema, Rotterdam, 183–208.

Bowles, F.A. 1975. Paleoclimatic significance of quartz/illite variations in cores from the eastern equatorial North Atlantic. *Quaternary Research* 5, 225–235.

Boyd, R., Huang, Z. and O'Connell, S. 1994. Milankovitch cyclicity in Late Cretaceous sediments from Exmouth Plateau off northwest Australia. In De Boer, P.L. and Smith, D.G. (eds), *Orbital Forcing and Cyclic Sequences.* Int. Assoc. Sedimentologists, *Special Publication* No. 19, 145–166.

Boyd, R.F., Clark, D.L., Jones, G., Ruddiman, W.F., McIntyre, A. and Pisias, N.G. 1984. Central Arctic Ocean response to Pleistocene earth-orbital variations. *Quaternary Research* 22, 121–128.

Boyle, E.A. 1990. Quaternary deepwater paleoceanography. *Science* 249, 863–870.

Boyle, E.A. 1992. Cadmium and $d^{13}C$ paleochemical ocean distributions during the stage 2 glacial maximum. *Annual Reviews of Earth and Planetary Sciences* 20, 247–287.

Boyle, E.A. and Keigwin, L. 1987. North Atlantic thermohaline circulation during the past 20,000 years linked to high-latitude surface temperature. *Nature* 330, 35–40.

Boyle, E.A. and Rosener, P. 1990. Further evidence for a link between late Pleistocene North Atlantic surface temperatures and North Atlantic deep-water production. *Palaeogeography, Palaeoclimatology, Palaeoecology (Global and Planetary Change Section)* 89, 113–124.

Boyle, E. and Weaver, A. 1994. Conveying past climates. *Nature* 372, 41–42.

Bradbury, J.P. and Dean, W.E. (eds) 1993. *Elk Lake, Minnesota: Evidence for Rapid Climate Change in the North-Central United States.* Geological Society of America, Special Paper, 276.

Bradley, R.S. 1985. *Quaternary Paleoclimatology.* Allen and Unwin, London.

Bradley, R.S. and Jones, P.D. (eds) 1995. *Climate Since A.D. 1500* (revised edn). Routledge, London.

Brain, C.K. 1981a: The evolution of man in Africa: was it a consequence of Cainozoic cooling? *Alex L. du Toit Memorial Lectures No.17, Geological Society of South Africa,* Annexure to Volume 84, 1–19.

Brain, C.K. 1981b: *The Hunters or the Hunted. An Introduction to African Cave Taphonomy.* University of Chicago Press, Chicago.

Brain, C.K. and Sillen, A. 1988. Evidence from the Swartkrans cave for the earliest use of fire. *Nature* 336, 464–466.

Bräuer, G., Yokoyama, Y., Falguères, C. and Mbua, E. 1997. Modern human origins backdated. *Nature* 386, 337–338.

Braun, D., Ron, H. and Marco, S. 1991. Magnetostratigraphy of the hominid tool-bearing Erk el Ahmar Formation in the northern Dead Sea Rift. *Israel Journal of Earth Sciences* 40, 191–197.

Bravard, J.-P. and Petts, G.E. 1996. Human impacts on fluvial hydrosystems. In Petts, G.E. and Amoros, C. (eds), *Fluvial Hydrosystems.* Chapman and Hall, London, 242–262.

Bray, J.R. 1977. Pleistocene volcanism and glacial initiation. *Science* 197, 251–254.

Bray, J.R. 1979. Surface albedo increase following massive Pleistocene explosive eruptions in western North America. *Quaternary Research* 12, 204–211.

Bray, J.R. and Trump, D. 1970. *The Penguin Dictionary of Archaeology.* Penguin Books, Harmondsworth.

Brecher, H.H. and Thompson, L.G. 1993. Measurement of the retreat of Qori Kalis Glacier in the tropical Andes of Peru by terrestrial photogrammetry. *Photogrammetric Engineering and Remote Sensing* 59, 1017–1022.

Broecker, W., Bond, G., Klas, M., Clark, E. and McManus, J. 1992. Origin of the northern Atlantic's Heinrich events. *Climate Dynamics* 6, 265–273.

Broecker, W.S. 1987a. The biggest chill. *Natural History Magazine* 97, 79–89.

Broecker, W.S. 1987b. Unpleasant surprises in the greenhouse? *Nature* 328, 123–126.

Broecker, W.S. 1990. Salinity history of the Northern Atlantic during the last deglaciation. *Paleoceanography* 5, 459–467.

Broecker, W.S. 1992. The great ocean conveyor. In Levi, B.G., Hafemeister, D. and Scribner, R. (eds), *Global Warming: Physics and Facts. Conference Proceeding 247.* American Institute of Physics, New York, 129–161.

Broecker, W.S. and Denton, G.H. 1989. The role of ocean–atmosphere reorganisations in glacial cycles. *Geochimica et Cosmochimica Acta* 53, 2465–2501.

Broecker, W.S. and Denton, G.H. 1990. The role of ocean–atmosphere reorganizations in glacial cycles. *Quaternary Science Reviews* 9, 305–341.

Broecker, W.S. and Peng, T.-H. 1989. The cause of the glacial to interglacial atmospheric CO_2 change: a polar alkalinity hypothesis. *Paleoceanography* 3, 215–239.

Broecker, W.S., Kennett, J.P., Flower, B.P., Teller, J.T., Trumbore, S., Bonani, G. and Wolfli, W. 1989. Routing of meltwater from the Laurentide ice sheet during the Younger Dryas cold episode. *Nature* 341, 318–321.

Broecker, W.S., Peteet, D.M. and Rind, D. 1985. Does the ocean–atmosphere system have more than one stable mode of operation? *Nature* 315, 21–25.

Broecker, W.S., Thurber, D.L., Goddard, J., Ku, T.-L.,

Matthews, R.K. and Mesolella, K.J. 1968. Milankovitch hypothesis supported by precise dating of coral reefs and deep-sea sediments. *Science* 159, 297–300.

Bromage, T.G. and Schrenk, F. 1995. Biogeographic and climatic basis for a narrative of early hominid evolution. *Journal of Human Evolution* 28, 109–114.

Bronger, A., Winter, R., Derevjanko, O. and Aldag, S. 1995. Loess–palaeosol sequences in Tadjikistan as a palaeoclimatic record of the Quaternary in Central Asia. In Derbyshire, E. (ed.), *Wind Blown Sediments in the Quaternary record*. Quaternary Proceedings No. 4. John Wiley, Chichester, 69–81.

Bronk, Ramsey C. 1995. Oxcal v. 2.0 r:4sd: 123prob [chron]. Oxford Radiocarbon Accelerator Unit Publication 18, Oxford.

Brook, E.J., Sowers, T. and Orchardo, J. 1996. Rapid variations in atmospheric methane concentration during the past 110,000 years. *Science* 273, 1087–1091.

Brooks, A.S., Helgren, D.M., Cramer, J.S., Franklin, A., Hornyak, W., Keating, J.M., Klein, R.G., Rink, W.J., Schwarcz, H., Leith Smith, J.N., Stewart, K., Todd, N.E., Verniers, J. and Yellen, J.E. 1995. Dating and context of three Middle Stone Age sites with bone points in the Upper Semliki Valley, Zaire. *Science* 268, 548–553.

Brown, L. (ed.) 1997. *State of the World 1997*. W.W. Norton, New York.

Brown, R.J.E. 1970. *Permafrost in Canada: Its Influence on Northern Development*. University of Toronto Press, Toronto.

Brubaker, K.L. and Entekhabi, D. 1996. Analysis of feedback mechanisms in land–atmosphere interaction. *Water Resources Research* 32, 1343–1357.

Brunet, M., Beauvilain, A., Coppens, Y., Heintz, E., Moutaye, A.H.E. and Pilbeam, D. 1995. The first australopithecine 2,500 kilometres west of the Rift Valley (Chad). *Nature* 378, 273–275.

Bunn, H.T. 1981. Archaeological evidence for meat-eating by Plio-Pleistocene hominids from Koobi Fora and Olduvai Gorge. *Nature* 291, 574–577.

Burbank, D.W. and Li, J. 1985. Age and palaeoclimatic significance of the loess of Lanzhou, north China. *Nature* 316, 429–431.

Burke, K., Francis, P. and Wells, G. 1990. Importance of the geological record in understanding global change. *Palaeogeography, Palaeoclimatology, Palaeoecology, Global and Planetary Change Section*, 89, 193–204.

Burton, K.W., Ling, H.-F. and O'Nions, R.K. 1997. Closure of the Central American isthmus and its effect on deep-water formation in the North Atlantic. *Nature* 386, 382–385.

Bush, M.B. and Colinvaux, P.A. 1990. A pollen record of a complete glacial cycle from lowland Panama. *Journal of Vegetation Science* 1, 105–118.

Bush, M.B., Piperno, D.R. and Colinvaux, P.A. 1989. A 6,000 year history of Amazon maize cultivation. *Nature* 340, 303–305.

Butler, B.E. 1956. Parna – an aeolian clay. *Australian Journal of Science* 18, 145–151.

Butzer, K.W. 1974. *Environment and Archaeology. An Ecological Approach to Prehistory* (3rd edn). Chicago, Aldine.

Butzer, K.W. 1982. *Archaeology as Human Ecology*. Cambridge University Press, Cambridge.

Butzer, K.W. and Hansen, C.L. 1968. *Desert and river in Nubia. Geomorphology and Prehistoric Environments at the Aswan Reservoir*. University of Wisconsin Press, Madison.

Butzer, K.W., Isaac, G.L., Richardson, J.L. and Washbourn-Kamau, C. 1972. Radiocarbon dating of East African lake levels. *Science* 175, 1069–1076.

Calaby, J.H. 1976. Some biogeographical factors relevant to the Pleistocene movement of man in Australasia. In Kirk, R.L. and Thorne, A.G. (eds), *The Origin of the Australians*. Australian Institute of Aboriginal Studies, Canberra, 23–28.

Cann, R.L., Stoncking, M. and Wilson, A.C. 1987. Mitochondrial DNA and human evolution. *Nature* 325, 31–36.

Carter, R.M. and Johnson, D.P. 1986. Sea-level controls on the post-glacial development of the Great Barrier Reef, Queensland. *Marine Geology* 71, 137–164.

Catt, J.A. 1995. Soils in aeolian sequences as evidence of Quaternary climatic change: problems and possible solutions. In Derbyshire, E. (ed.), *Wind Blown Sediments in the Quaternary Record*. Quaternary Proceedings No. 4. John Wiley, Chichester, 59–68.

Cerling, T.E., Harris, J.M., MacFadden, B.J., Leakey, M.G., Quade, J., Eisenmann, V. and Ehleringer, J.R. 1997. Global vegetation change through the Miocene/Pliocene boundary. *Nature* 389, 153–158.

Cerveny, R.S. 1991. Orbital signals in the diurnal cycle of radiation. *Journal of Geophysical Research* 96(D9), 17209–17215.

Cess, R.D. 1992. Comparison of general circulation models. In Levi, B.G., Hafemeister, D. and Scribner, R. (eds), *Global Warming: Physics and Facts*. Conference Proceeding 247. American Institute of Physics, New York, 46–54.

Cess, R.D. and Wronka, J.C. 1979. Ice ages and the Milankovitch theory: a study of interactive climate feedback mechanisms. *Tellus* 31, 185–192.

Chao, B.F. 1996. 'Concrete' testimony to Milankovitch cycle in earth's changing obliquity. *EOS* 77, 433.

Chappell, J.M.A. 1973. Astronomical theory of climatic change: status and problem. *Quaternary Research* 3, 221–236.

Chappell, J. 1974a: Late Quaternary glacio- and hydro-isostasy, on a layered earth. *Quaternary Research* 4, 429–440.

Chappell, J. 1974b: Geology of coral terraces, Huon Peninsula, New Guinea: a study of Quaternary tectonic movements and sea-level changes. *Bulletin of the Geological Society of America* 85, 553–570.

Chappell, J.M.A. 1974c. Relationships between sealevels, ^{18}O variations and orbital perturbations, during the past 250,000 years. *Nature* 252, 199–202.

Chappell, J. 1983. Aspects of sea levels, tectonics, and isostasy since the Cretaceous. In Gardner, R. and Scoging, H. (eds), *Mega-geomorphology*. Clarendon, Oxford, 56–72.

Chappell, J. 1987. Ocean volume change and the history of sea water. In Devoy, R.J.N. (ed.), *Sea Surface Studies: a Global View*. Croom-Helm, London, 33–56.

Chappell, J., Head, M.J. and Magee, J. 1996. Beyond the radiocarbon limit in Australian archaeology and Quaternary research. *Antiquity* 70, 543–552.

Chappell, J. and Polach, H. 1991. Post-glacial sea-level rise from a coral record at Huon Peninsula, Papua New Guinea. *Nature* 349, 147–149.

Chappell, J., Rhodes, E.G., Thom, B.G. and Wallensky, E. 1982. Hydro-isostasy and the sea-level isobase of 5500 B.P. in north Queensland, Australia. *Marine Geology* 49, 81–90.

Chappell, J. and Shackleton, N.J. 1986. Oxygen isotopes and sea level. *Nature* 324, 137–140.

Chappell, J. and Veeh, H.H. 1978. Late Quaternary tectonic movements and sea-level changes in Timor and Atauro Island. *Bulletin of the Geological Society of America* 89, 356–368.

Chappellaz, J., Barnola, J.M., Raynaud, D., Korotkevich, Y.S. and Lorius, C. 1990. Ice-core record of atmospheric methane over the past 160,000 years. *Nature* 345, 127–131.

Chappellaz, J., Blunier, T., Raynaud, D., Barnola, J.M., Schwander, J. and Stauffer, B. 1993. Synchronous changes in atmospheric CH_4 and Greenland climate between 40 and 8 kyr BP. *Nature* 366, 443–445.

Charles, C. 1997. Cool tropical punch of the ice ages. *Nature* 385, 681–683.

Charles, C.D. and Fairbanks, R.G. 1992. Evidence from Southern Ocean sediments for the effect of North Atlantic deep-water flux on climate. *Nature* 355, 416–419.

Charles, C.D., Rind, D., Jouzel, J., Koster, R.D. and Fairbanks, R.G. 1994. Glacial-interglacial changes in moisture sources for Greenland: influences on the ice core record of climate. *Science* 263, 508–511.

Chen, F.H., Bloemendal, J., Wang, J.M., Li, J.J. and Oldfield, F. 1997. High-resolution multi-proxy climate records from Chinese loess: evidence for rapid climatic changes over the last 75 kyr. *Palaeogeography, Palaeoclimatology, Palaeoecology* 130, 323–335.

Chen, J.H., Curran, H.A., White, B. and Wasserburg, G.J. 1991. Precise chronology of the last interglacial period: ^{234}U–^{230}Th data from fossil coral reefs in the Bahamas. *Bulletin of the Geological Society of America* 103, 82–97.

Chinn, T.J. 1996. The Southern Hemisphere glacial record – Antarctica and New Zealand. *Papers and Proceedings of the Royal Society of Tasmania* 130, 17–24.

Chivas, A.R., De Deckker, P. and Shelley, J.M.G. 1986. Magnesium and strontium in non-marine ostracod shells as indicators of palaeosalinity and palaeotemperature. *Hydrobiologia* 143, 135–142.

Chivas, A.R., De Deckker, P., Cali, J.A., Chapman, A., Kiss, E. and Shelley, J.M.G. 1993. Coupled stable-isotope and trace-element measurements of lacustrine carbonates as paleoclimatic indicators. *American Geophysical Monographs* 78, 113–121.

Christopher, S.A., Kliche, D.V., Chou, J. and Welch, R.M. 1996. First estimates of the radiative forcing of aerosols from biomass burning using satellite data. *Journal of Geophysical Research* 101(D16), 21265–21273.

Clapperton, C.M., Hall, M., Mothes, P., Hole, M.J., Still, J.W., Helmens, K.F., Kuhry, P. and Gemmell, A.M.D. 1997. A Younger Dryas icecap in the equatorial Andes. *Quaternary Research* 47, 13–28.

Clark, G. 1977. *World Prehistory in New Perspective.* Cambridge University Press, Cambridge.

Clark, J.A. 1976. Greenland's rapid postglacial emergence: a result of ice-water gravitational attraction. *Geology* 4, 310–312.

Clark, J.A. 1980. A numerical model of worldwide sea level changes on a viscoelastic earth. In Mörner N.-A. (ed.), *Earth Rheology, Isostasy and Eustasy.* Wiley, Chichester, 525–534.

Clark, J.A., Farrell, W.E. and Peltier, W.R. 1978. Global changes in postglacial sea level: a numerical calculation. *Quaternary Research* 9, 265–287.

Clark, J.A. and Lingle, C.S. 1979. Predicted relative sea-level changes (18,000 years BP to present) caused by late-glacial retreat of the Antarctic ice sheet. *Quaternary Research* 11, 279–298.

Clark, J.D. 1971. A re-examination of the evidence for agricultural origins in the Nile Valley. *Proceedings of the Prehistoric Society* 37, 34–79.

Clark, J.D. 1975. Africa in prehistory: peripheral or paramount? *Man (N.S.)* 10, 175–198.

Clark, J.D. 1976a. African origins of man the toolmaker. In Isaac, G.L. and McCown, E.R. (eds), *Human Origins: Louis Leakey and the East African Evidence.* Benjamin, Menlo Park, California, 1–53.

Clark, J.D. 1976b. Prehistoric populations and pressures favouring plant domestication in Africa. In Harlan, J.R., de Wet, J.M.J. and Stemler, A.B.L. (eds), *Origins of African Plant Domestication.* Mouton, The Hague, 67–105.

Clark, J.D. 1980. Human populations and cultural adaptations in the Sahara and Nile during prehistoric times. In Williams, M.A.J. and Faure, H. (eds), *The Sahara and the Nile.* Balkema, Rotterdam, 527–582.

Clark, J.D. 1984. The domestication process in Northeast Africa: ecological change and adaptive strategies. In Krzyzaniak, L. and Kobusiewicz, M. (eds), *Origin and Early Development of Food-Producing Culture in North-Eastern Africa.* Polish Academy of Sciences, Poznan, 25–41.

Clark, J.D. 1987. Transitions: *Homo erectus* and the Acheulian: the Ethiopian sites of Gadeb and the Middle Awash. *Journal of Human Evolution* 16, 809–826.

Clark, J.D 1988. The Middle Stone Age of East Africa and the beginnings of regional identity. *Journal of World Prehistory* 2, 235–305.

Clark, J.D. 1992. African and Asian perspectives on the origins of modern humans. *Philosophical Transactions of the Royal Society of London* 337, 201–217.

Clark, J.D., Asfaw, B., Assefa, G., Harris, J.W.K., Hurashina, H., Walter, R.C., White, T.D. and Williams, M.A.J. 1984. Palaeoanthropological discoveries in the Middle Awash Valley, Ethiopia. *Nature* 307, 423–428.

Clark, J.D., de Heinzelin, J., Schick, K.D., Hart, W.K., White, T.D., Woldegabriel, G., Walter, R.C., Suwa, G., Asfaw, B., Vrba, E. and Selassie, Y.H. 1994. African *Home erectus*: old radiometric ages and young Oldowan assemblages in the Middle Awash Valley, Ethiopia. *Science* 264, 1907–1910.

Clark J.D. and Harris, J.W.K. 1985. Fire and its roles in early hominid lifeways. *African Archaeological Review* 3, 3–27.

Clark, J.D. and Haynes, C.V. 1970. An elephant butchery site at Mwanganda's village, Karonga, Malawi and its relevance for Palaeolithic archaeology. *World Archaeology* 1, 390–411.

Clark, J.D. and Kurashina, H. 1979. Hominid occupation of the East-Central Highlands of Ethiopia in the Plio-Pleistocene. *Nature* 282, 33–39.

Clark, J.S. 1988. Charcoal-stratigraphic analysis on petrographic thin sections: recent fire history in northwest Minnesota. *Quaternary Research* 30, 67–80.

Clark, J.S., Cachier, H., Goldammer, J.G. and Stocks, B. (eds) 1997. *Sediment Records of Biomass Burning and Global Change.* Springer, Berlin.

Clark, J.S. and Royall, P.D. 1995. Transformation of a northern hardwood forest by aboriginal (Iroquois) fire: charcoal evidence from Crawford Lake, Ontario, Canada. *The Holocene* 5, 1–9.

Clark, M.J. (ed.), 1988. *Advances in Periglacial Geomorphology.* Wiley, New York.

Clark, P.U. 1994. Unstable behaviour of the Laurentide ice sheet over deforming sediment and its implications for climate change. *Quaternary Research* 41, 19–25.

Clark, P.U., Alley, R.B., Keigwin, L.D., Licciardi, J.M., Johnsen, S.J. and Wang, H. 1996. Origin of the first global meltwater pulse following the last glacial maximum. *Paleoceanography* 11, 563–577.

Clark, P.U. and Bartlein, P.J. 1995. Correlation of late Pleistocene glaciation in the western United States with North Atlantic Heinrich events. *Geology* 23(6), 483–486.

Clark, P.U., Clague, J.J., Curry, B.B., Dreimanis, A., Hicock, S.R., Miller, G.H., Berger, W.H., Eyles, N., Lamothe, M., Miller, B.B., Mott, R.J., Oldale, R.N., Stea, R.R., Szabo J.P., Thorleifson, L.H. and Vincent, J.-S. 1993. Initiation and development of the Laurentide and Cordilleran ice sheets following the last interglaciation. *Quaternary Science Reviews* 12, 79–114.

Clark, R.L. 1981. The prehistory of bushfires. In Stanbury, P. (ed.), *Bushfires: Their Effect on Australian Life and Landscape.* The Macleay Museum, University of Sydney, 61–74.

Clark, R.L. 1983. Pollen and charcoal evidence for the effects of Aboriginal burning on the vegetation of Australia. *Archaeology in Oceania* 18, 32–37.

Clemens, S.C., Murray, D.W. and Prell, W.L. 1996. Nonstationary phase of the Plio–Pleistocene Asian monsoon. *Science* 274, 943–948.

Clemens, S.C. and Tiedemann, R. 1997. Eccentricity forcing of Pliocene–Early Pleistocene climate revealed in a marine oxygen-isotope record. *Nature* 385, 801–804.

Clemmensen, L.B., Øxnevad, I.E.I. and De Boer, P.L. 1994. Climatic controls on ancient desert sedimentation: some late Palaeozoic and Mesozoic examples from NW Europe and the western interior of the USA. In De Boer, P.L. and Smith, D.G. (eds), *Orbital Forcing and Cyclic Sequences.* International Association of Sedimentologists, Special Publication No. 19, 439–457.

CLIMAP Project Members 1976. The surface of the ice-age Earth. *Science* 191 (4232),

CLIMAP Project Members 1981. Seasonal reconstructions of the earth's surface at the Last Glacial Maximum. *Geological Society of America Map and Chart Series* MC-36, 1–18 and 9 maps.

Close, A.E. 1989. Lithic development in the Kubbaniyan (Upper Egypt). In Krzyzaniak, L. and Kobusiewicz, M. (eds), *Late Prehistory of the Nile Basin and the Sahara.* Poznan Archaeological Museum, Poznan, 117–125.

Coe, M.T. 1995. The hydrologic cycle of major continental drainage and ocean basins: a simulation of the modern and mid-Holocene conditions and a comparison with observations. *Journal of Climate* 8, 535–543.

Coetzee, J.A. and Van Zinderen Bakker, E.M. 1989. Palaeoclimatology of east Africa during the last glacial maximum: a review of changing theories. In Mahaney, W.C. (ed.), *Quaternary and Environmental Research on East African Mountains.* Balkema, Rotterdam, 189–198.

Cofer, W.R., Levine, J.S., Winstead, E.L. and Stocks, B.J. 1991. New estimates of nitrous oxide emissions from biomass burning. *Nature* 349, 689–691.

Cohen, J.E. 1995. Population growth and Earth's carrying capacity. *Science* 269, 341–346.

Cohen, M.N. 1977. *The Food Crisis in Prehistory. Overpopulation and the Origins of Agriculture.* Yale University Press, New Haven.

Cohmap 1988. Climatic changes of the last 18,000 years: observations and model simulations. *Science* 241, 1043–1052.

Colinvaux, P.A. 1989. The past and future Amazon. *Scientific American* 260, 68–74.

Colinvaux, P.A., de Oliviera, P.E., Moreno, J.E., Miller, M.C. and Bush, M.B. 1996a. A long pollen record from lowland Amazonia: forest and cooling in glacial times. *Science* 274, 85–88.

Colinvaux, P.A., Liu, K.-B., de Oliveira, P., Bush, M.B., Miller, M.C. and Steinitz Kannan, M.S. 1996b. Temperature depression in the lowland tropics in glacial times. *Climate Change* 32, 19–33.

Collier, M., Webb, R.H. and Schmidt, J.C. 1996. *Dams and Rivers. A Primer on the Downstream Effects of Dams.* United States Geological Survey Circular 1126.

Collier, M.P., Webb, R.H. and Andrews, E.D. 1997. Experimental flooding of the Grand Canyon. *Scientific American* 276, 66–73.

Conroy, G.C., Jolly, C.J., Cramer, D. and Kalb, J.E. 1978. Newly discovered fossil hominid skull from the Afar depression, Ethiopia. *Nature* 276, 67–70.

Cooke, R. and Warren, A. 1973. *Geomorphology in Deserts.* Batsford, London.

Cooke, R., Warren, A. and Goudie, A. 1993. *Desert Geomorphology.* UCL Press, London.

Coope, G.R. 1970. Interpretations of Quaternary insect fossils. *Annual Review of Entomology* 15, 97–120.

Coope, G.R. 1975. Mid-Weichselian climate changes in Western Europe, reinterpreted from coleopteran assemblages. In Suggate, R.P. and Cresswell, M.M. (eds), *Quaternary Studies.* Royal Society of New Zealand, Wellington, 101–108.

Cooper, D.J., Watson, A.J. and Nightingale, P.D. 1996. Large decrease in ocean-surface CO_2 fugacity in response to *in situ* iron fertilization. *Nature* 383, 511–513.

Coppens, Y. 1989. Hominid evolution and the evolution of the environment. *Ossa* 14, 157–163.

Corliss, B.H. 1983. Quaternary circulation of the Antarctic Circumpolar Current. *Deep-Sea Research* 30, 47–61.

Corliss, B.H. and Fois, E. 1990. Morphotype analysis of deep-sea benthic foraminifera from the northwest Gulf of Mexico. *Palaios* 5, 589–605.

Cortijo, E., Duplessy, J.C., Labeyrie, L., Leclaire, H., Duprat, J. and van Weering, T.C.E. 1994. Eemian cooling in the Norwegian Sea and North Atlantic Ocean preceding continental ice-sheet growth. *Nature* 372, 446–449.

Cosgrove, R. 1989. Thirty thousand years of human colonisation in Tasmania: new Pleistocene dates. *Science* 243, 1706–1708.

Costa, J.E., Miller, A.J., Potter, K.W. and Wilcock, P.R. (eds) 1995. *Natural and Anthropogenic Influences in Fluvial Geomorphology.* The Wolman Volume. Geophysical Monograph 89. American Geophysical Union, Washington, DC.

Cowie, J.W. and Bassett, M.G. 1989. 1989 global stratigraphic chart, with geochronometric and magnetostratigraphic calibration. Supplement to *Episodes* 12(2).

Croke, J., Magee, J. and Price, D. 1996. Major episodes of Quaternary activity in the lower Neales River, northwest of Lake Eyre, central Australia. *Palaeogeography, Palaeoclimatology, Palaeoecology* 124, 1–15.

Crombie, M.K., Arvidson, R.E., Sturchio, N.C., El Alfy, Z. and Abu Zeid, K. 1997. Age and isotopic constraints on Pleistocene pluvial episodes in the Western Desert, Egypt. *Palaeogeography, Palaeoclimatology, Palaeoecology* 130, 337–355.

Cronin, T.M. and Raymo, M.E. 1997. Orbital forcing of deep-sea benthic species diversity. *Nature* 385, 624–627.

Crowell, J.C. and Frakes, L.A. 1970. Phanerozoic glaciation and the causes of ice ages. *American Journal of Science* 268, 193–224.

Crowley, T.J. 1992. North Atlantic deep water cools the Southern Hemisphere. *Paleoceanography* 7, 489–497.

Cuffey, K.M., Clow, G.D., Alley, R.B., Stuiver, M., Waddington, E.D. and Saltus, R.W. 1995. Large Arctic temperature change at the Wisconsin–Holocene glacial transition. *Science* 270, 455–458.

Culf, A.D., Fisch, G. and Hodnett, M.G. 1995. The albedo of Amazonian forest and ranch land. *Journal of Climate* 8, 1544–1554.

Dahm, C.N. and Molles, M.C. Jr. 1992. Streams in semi-arid regions as sensitive indicators of global climate change. In Firth, P. and Fisher, S.G. (eds), *Global Climate Change and Freshwater Ecosystems.* Springer-Verlag, New York, 250–260.

Damuth, J.E. and Fairbridge, R.W. 1970. Equatorial Atlantic deep-sea arkosic sands and ice-age aridity in tropical South America. *Bulletin of the Geological Society of America* 81, 189–206.

Damuth, J.E. and Kumar, N. 1975. Amazon cone: morphology, sediments, age, and growth pattern. *Bulletin of the Geological Society of America* 86, 863–878.

Dansgaard, W., Clausen, H.B., Gundestrup, N., Johnsen, S.J. and Rygner, C. 1985. Dating and climatic interpretation of two deep Greenland ice cores. In Langway, C.C., Oeschger, H. and Dansgaard, W. (eds), *Greenland Ice Core: Geophysics, Geochemistry, and the Environment.* American Geophysical Union, Washington, DC, 71–76.

Dansgaard, W., Johnsen, S.J., Clausen, H.B., Dahl-Jensen, D., Gundestrup, N., Hammer, C.U. and Oeschger, H. 1984. North Atlantic climatic oscillations revealed by deep Greenland ice cores. In Hansen, J.E. and Takahashi, T. (eds), *Climate Processes and Climate Sensitivity.* Geophysical Monograph 29. American Geophysical Union, Washington, DC, 288–298.

Dansgaard, W., Johnsen, S.J., Clausen, H.B., Dahl-Jensen D., Gundestrup, N.S., Hammer, C.U., Hvidberg, C.S., Steffensen, J.P., Sveinbjörnsdottir, A.E., Jouzel, J. and Bond, G. 1993. Evidence for general instability of past climate from a 250–kyr ice-core record. *Nature* 364, 218–220.

Dansgaard, W., White, J.W.C. and Johnsen, S.J. 1989. The abrupt termination of the Younger Dryas climate event. *Nature* 339, 532–534.

Darwin, C. 1871. *The Descent of Man.* Random House, New York.

Davies, J.L., 1969. *Landforms of Cold Climates.* Australian National University Press, Canberra.

Davis, J.L. and Mitrovica, J.X. 1996. Glacial isostatic adjustment and the anomalous tide gauge record of eastern North America. *Nature* 379, 331–333.

Davis, M.B. 1967. Pollen accumulation rates at Rogers Lake, Conn., during late and postglacial time. *Review of Palaeobotany and Palaeoecology* 2, 219–230.

Davis, M.B. 1976. Pleistocene biogeography of temperate deciduous forests. *Geoscience and Man* 13, 13–26.

Davis, M.B., Woods, K.D., Webb, S.L. and Futyma, R.P. 1986. Dispersal versus climate: expansion of Fagus and Tsuga into the Upper Great Lakes region. *Vegetatio* 67, 93–103.

Davis, O.K. and Sellers, W.D. 1994. Orbital history and seasonality of regional precipitation. *Human Ecology* 22, 97–113.

Davis, R.S. 1986. The Soan in Central Asia? Problems in Lower Palaeolithic cultural history. In Jacobsen, J. (ed.), *Studies in the Archaeology of India and Pakistan.* Oxford and IBH Publishing Company, New Delhi, 1–17.

Deblonde, G. and Peltier, W.R. 1993. Late Pleistocene ice age scenarios based on observational evidence. *Journal of Climate* 6, 709–727.

De Cisneros, C. and Vera, J.A. 1993. Milankovitch cyclicity in Purbeck peritidal limestones of the Prebetic (Berriasian, southern Spain). *Sedimentology* 40, 513–537.

De Deckker, P. 1988. Large Australian lakes during the last 20 million years: sites for petroleum source rocks or metal ore deposition or both? In Fleet, A.J., Kelts, K.R. and Talbot, R. (eds), *Lacustrine Petroleum Source Rocks.* Special Publication 40, Geological Society of London, 45–58.

De Deckker, P. 1997. The significance of the oceans in the Australasian region with respect to global palaeoclimates: future directions. *Palaeogeography, Palaeoclimatology, Palaeoecology* 131, 511–515.

De Deckker, P., Corrège, T. and Head J. 1991. Late Pleistocene record of eolian activity from tropical northeastern Australia suggesting that the Younger Dryas is not an unusual climatic event. *Geology* 19, 602–605.

De Deckker, P., Kershaw, A.P. and Williams, M.A.J. 1988. Past environmental analogues. In Pearman, G.I. (ed.), *Greenhouse: Planning for Climate Change.* CSIRO, Melbourne, 473–488.

Del Genio, A.D., Lacis, A.A. and Reudy, R.A. 1991. Simulations of the effect of a warmer climate on atmospheric humidity. *Nature* 351, 382–385.

Deloison, Y. 1992. Empreintes de pas à Laetoli (Tanzanie). Leur apport à une meilleure connaissance de la locomotion des Hominidés fossiles. *Comptes Rendus de l'Académie des Sciences (Paris)* 315, 103–109.

DeMenocal, P.B. 1995. Plio-Pleistocene African climate. *Science* 270, 53–59.

DeMenocal, P.B. and Bloemendal, J. 1995. Plio-Pleistocene climatic variability in subtropical Africa and the paleoenvironment of hominid evolution: A combined data-model approach. In Vrba, E.S., Denton, G.H., Partridge, T.C. and Burckle, L.H. (eds), *Paleoclimate and Evolution, with Emphasis on Human Origins*. Yale University Press, New Haven, 262–288.

De Noblet, N.I., Prentice, I.C., Joussaume, S., Texier, D., Botta, D. and Haxeltine, A. 1996. Possible role of atmosphere–biosphere interactions in triggering the last glaciation. *Geophysical Research Letters* 23, 3191–3194.

Denton, G.H. and Hendy, C.H. 1994. Younger Dryas age advance of Franz Josef glacier in the southern Alps of New Zealand. *Science* 264, 1434–1437.

Denton, G.H. and Hughes, T.J. (eds), 1981. *The Last Great Ice Sheets*. Wiley, New York.

Denton, G.H. and Hughes, T.J. 1981. The Arctic Ice Sheet: an outrageous hypothesis. In Denton, G.H. and Hughes, T.J. (eds), *The Last Great Ice Sheets*. Wiley, New York, 440–467.

Denton, G.H. and Karlén, W. 1973. Holocene climatic variations – their pattern and possible cause. *Quaternary Research* 3, 155–205.

Derbyshire, E. 1996. Quaternary glacial sediments, glaciation style, climate and uplift in the Karakoram and northwest Himalaya: review and speculations. *Palaeogeography, Palaeoclimatology, Palaeoecology* 120, 147–157.

Derbyshire, E., Keen, D.H., Kemp, R.A., Rolph, T.A., Shaw, J. and Meng, X.M. 1995. Loess–palaeosol sequences as recorders of palaeoclimatic variations during the last Glacial–Interglacial cycle: some problems of correlation in north-central China. In Derbyshire, E. (ed.), *Wind blown sediments in the Quaternary Record*. Quaternary Proceedings No.4. Wiley, Chichester, 7–18.

Deuser, W.G., Ross, E.H. and Waterman, L.S. 1976. Glacial and pluvial periods: their relationship revealed by Pleistocene sediments of the Red Sea and Gulf of Aden. *Science* 191, 1168–1170.

Devoy, R.J.N. (ed.), 1987a: *Sea Surface Studies: A Global View*. Croom Helm, London.

Devoy, R.J.N. 1987b: Sea-level changes during the Holocene: the North Atlantic and Arctic Oceans. In Devoy, R.J.N. (ed.), *Sea Surface Studies: A Global View*. Croom Helm, London, 294–347.

Dickson, R.R. and Brown, J. 1994. The production of North Atlantic deep water: sources, rates and pathways. *Journal of Geophysical Research* 99(C6), 12319–12341.

Diester-Haas, L., Heine, K., Rothe, P. and Schrader, H. 1988. Late Quaternary history of continental climate and the Benguela Current off South West Africa.

Palaeogeography, Palaeoclimatology, Palaeoecology 65, 81–91.

Dillehay, T.D. and Collins, M.D. 1988. Early cultural evidence from Monte Verde in Chile. *Nature* 332, 150–152.

Ding, Z.L., Rutter, N. and Liu, T. 1993. Pedostratigraphy of Chinese loess deposits and climatic cycles in the last 2.5 Myr. *Catena* 20, 73–91.

Dingle, R.V., Siesser, W.G. and Newton, A.R. 1983. *Mesozoic and Tertiary Geology of Southern Africa*. Balkema, Rotterdam.

Ditlevsen, P.D., Svensmark, H. and Johnsen, S. 1996. Contrasting atmospheric and climate dynamics of the last glacial and Holocene periods. *Nature* 379, 810–812.

Dockal, J.A. and Worsley, T.R. 1991. Modeling sea level changes as the Atlantic replaces the Pacific: submergent versus emergent observers. *Journal of Geophysical Research* 96(B4), 6805–6810.

Dodge, R.E., Fairbanks, R.G., Benninger, L.K. and Maurrasse, F. 1983. Pleistocene sea levels from raised coral reefs of Haiti. *Science* 210, 1423–1425.

Dodson, J.R. and Ono, Y. 1997. Timing of late Quaternary vegetation response in the 30–50° latitude bands in southeastern Australia and northeastern Asia. *Quaternary International* 37, 89–104.

Dodson, J., Fullagar, R., Furby, J., Jones, R. and Prosser, I. 1993. Humans and megafauna in a late Pleistocene environment from Cuddie Springs, north western New South Wales. *Archaeology in Oceania* 28, 94–99.

Dokken, T.M. and Hald, M. 1996. Rapid climatic shifts during isotope stages 2–4 in the polar North Atlantic. *Geology* 24, 599–602.

Domack, E.W., Jull, A.J.T. and Nakao, S. 1991. Advance of East Antarctic outlet glaciers during the Hypsithermal: implications for the volume state of the Antarctic ice sheet under global warming. *Geology* 19, 1059–1062.

Dong Guangrong (ed.), 1991. *Quaternary Environmental Research on the Deserts in China*. Institute of Desert Research, Academia Sinica, Lanzhou.

Douglas, B.C. 1991. Global sea level rise. *Journal of Geophysical Research* 96(C4), 6981–6992.

Dregne, H.L.F. 1983. *Desertification of Arid Lands*. Harwood, London.

Drever, J.I. 1994. The effect of land plants on weathering rates of silicate minerals. *Geochimica et Cosmochimica Acta* 58, 2325–2332.

Dunbar, R.B. and Cole, J.E. 1993. *Coral Records of Ocean-Atmosphere Variability. Report from the Workshop on Coral Paleoclimate Reconstruction*. NOAA Climate and Global Change Program, Special Report 10, September 1993.

Dunkerley, D.L. and Brown, K.J. 1997. Desert soils. In Thomas, D.S.G. (ed.), *Arid Zone Geomorphology: Process, Form and Change in Drylands* (2nd edn). Wiley, New York, 55–68.

Dunwiddie, P.W. 1979. Dendrochronological studies of indigenous New Zealand trees. *New Zealand Journal of Botany* 17, 251–266.

Duplessy, J.-C. 1982. Glacial to interglacial contrasts in the Northern Indian Ocean. *Nature* 295, 494–498.

Duplessy, J.-C. 1996. *Quand l'Océan se Fâche. Histoire Naturelle du Climat*. Editions Odile Jacob, Paris.

Duplessy, J.-C., Delibrias, G., Turon, J.L., Pujol, C. and Duprat, J. 1981. Deglacial warming of the northeastern Atlantic Ocean: correlation with the paleoclimatic evolution of the European continent. *Palaeogeography, Palaeoclimatology, Palaeoecology* 35, 121–144.

Duplessy, J.-C. and Labeyrie, L.D. 1994. Surface and deep water circulation changes during the last climatic cycle. In Duplessy, J.-C. and Spyridakis, M.-T. (eds), *Long-term Climatic Variations. Data and Modelling.* NATO ASI Series I: 22, Springer, Berlin, 277–298.

Dupont, L.M. and Leroy, S.A.G. 1995. Steps towards drier climatic conditions in northwestern Africa during the Upper Pliocene. In Vrba, E.S., Denton, G.H., Partridge, T.C. and Burckle, L.H. (eds), *Paleoclimate and Evolution, with Emphasis on Human Origins.* Yale University Press, New Haven, 289–298.

Dutton, J.F. and Barron, E.J. 1997. Miocene to present vegetation changes: a possible piece of the Cenozoic cooling puzzle. *Geology* 25, 39–41.

Dwyer, G.S., Cronin T.M., Baker, P.A., Raymo M.E., Buzas, J.S. and Corrège, T. 1995. North Atlantic deep-water temperature change during late Pliocene and late Quaternary climatic cycles. *Science* 270, 1347–1351.

Dyke, A.S. and Prest, V.K. 1987. Late Wisconsinan and Holocene history of the Laurentide ice sheet. *Géographie physique et Quaternaire* 41, 237–263.

Eberz, G.W., Williams, F.M. and Williams, M.A.J. 1988. Plio-Pleistocene volcanism and sedimentary facies changes at Gadeb prehistoric site, Ethiopia. *Geologische Rundschau* 77, 513–527.

Ehlers, J. 1996. *Quaternary and Glacial Geology.* Wiley, Chichester.

Eltahir, E.A.B. and Bras, R.L. 1994. Precipitation recycling in the Amazon basin. *Quarterly Journal of the Royal Meteorological Society* 120, 861–880.

Elverhøi, A., Andersen, E.S., Dokken, T., Hebbeln, D., Spielhagen, R., Svendsen, J.I., Sørflaten, M., Rørnes, A., Hald, M. and Forsberg, C.F. 1995. The growth and decay of the Late Weichselian ice sheet in western Svalbard and adjacent areas based on provenance studies of marine sediments. *Quaternary Research* 44, 303–316.

Elverhøi, A., Fjeldskaar, W., Solheim, A., Nyland-berg, M. and Russwurm, L. 1993. The Barents Sea ice sheet – a model of its growth and decay during the last ice maximum. *Quaternary Science Reviews* 12, 863–873.

Ely, L.L., Enzel, Y., Baker, V.R., Kale, V.S. and Mishra, S. 1996. Changes in the magnitude and frequency of late Holocene monsoon floods on the Narmada River, central India. *Bulletin of the Geological Society of America* 108, 1134–1148.

Emiliani, C. 1955. Pleistocene temperatures. *Journal of Geology* 63, 538–578.

Emiliani, C. (ed.) 1981. The oceanic lithosphere. In *The Sea (*Vol. 7). Wiley, New York, 347–352.

Etheridge, D.M., Steele, L.P., Langenfelds, R.L., Francey, R.J., Barnola, J.-M. and Morgan, V.I. 1996. Natural and anthropogenic changes in atmospheric CO_2 over the last 1000 years from air in Antarctic ice and firn. *Journal of Geophysical Research* 101(D2), 4115–4128.

Etkins, R. and Epstein, E.S. 1982. The rise of global mean sea level as an indication of climatic change. *Science* 215, 287–289.

Evans, D.A., Beuker, N.J. and Kirschvink, J.L. 1997. Low-latitude glaciation in the Palaeoproterozoic era. *Nature* 386, 262–266.

Evans, J. 1860. On the occurrence of flint implements in undisturbed beds of gravel, sand, and clay. *Archaeologia* 38, 280–307.

Evenari, M. Shanan, L. and Tadmor, N. 1971. *The Negev: the Challenge of a Desert.* Harvard University Press, Cambridge, MA.

Fagan, B.M. 1989. *People of the Earth.* (6th edn). Scott, Foresman, Illinois.

Fagan, B.M. 1995. *People of the Earth: An Introduction to World Prehistory.* (8th edn). Harper Collins College Publishers, New York.

Fairbanks, R.G. 1989. A 17,000 year glacio-eustatic sea level record: influence of glacial melting rates on Younger Dryas event and deep-ocean circulation. *Nature* 342, 637–642.

Fairbridge, R.W. 1961. Eustatic changes in sea level. *Physics and Chemistry of the Earth* 4, 99–185.

Fairbridge, R.W. 1970. World palaeoclimatology of the Quaternary. *Revue de Géographie Physique et de Géologie Dynamique* 12, 97–104.

Fairbridge, R.W. and Finkl, C.W. Jr. 1984. Tropical stone lines and podzolized sand plains as paleoclimatic indicators for weathered cratons. *Quaternary Science Reviews* 3, 41–72.

Farley, K.A. and Patterson, D.B. 1995. A 100–kyr periodicity in the flux of extraterrestrial ^{3}He to the sea floor. *Nature* 378, 600–603.

Farrell, J.W., Pedersen, T.F., Calvert, S.E. and Nielsen, B. 1995. Glacial–interglacial changes in nutrient utilization in the equatorial Pacific Ocean. *Nature* 377, 514–517.

Faure, H. 1966. Évolution des grand lacs sahariens à l'Holocène. *Quaternaria* 8, 167–175.

Faure, H. 1990. Changes in the global continental reservoir of carbon. *Palaeogeography, Palaeoclimatology, Palaeecology, Global and Planetary Change Section* 82, 47–52.

Faure, H.F., Faure-Denard, L. and Fairbridge, R.W. 1990. Possible effects of man on the carbon cycle in the past and in the future. In Paepe, R., Fairbridge, R.W. and Jelgersma, S. (eds), *Greenhouse Effect, Sea Level and Drought.* Kluwer, Dordrecht, 459–462.

Feibel, C.S. 1997. Debating the environmental factors in hominid evolution. *GSA Today* 7, 1–7.

Feibel, C.S., Brown, F.H. and McDougall, I. 1989. Stratigraphic context of fossil hominids from the Omo Group deposits: northern Turkana Basin, Kenya and Ethiopia. *American Journal of Physical Anthropology* 78, 595–622.

Feng, X. and Epstein, S. 1994. Climatic implications of an 8,000–year hydrogen isotope time series from Bristlecone pine trees. *Science* 265, 1079–1081.

Ferland, M.A., Roy, P.S. and Murray-Wallace, C.V. 1995. Glacial lowstand deposits on the outer continental shelf of southeastern Australia. *Quaternary Research* 44, 294–299.

Fernandez-Jalvo, Y. 1996. Small mammal taphonomy and the Middle Pleistocene environments of Dolina, northern Spain. *Quaternary International* 33, 21–34.

Fichefet, T., Hovine, S. and Duplessy, J.-C. 1994. A model

study of the Atlantic thermohaline circulation during the last glacial maximum. *Nature* 372, 252–255.

Filippelli, G.M. 1997. Intensification of the Asian monsoon and a chemical weathering event in the late Miocene–early Pliocene: implications for late Neogene climate change. *Geology* 25, 27–30.

Fisher, A.G. 1984. The two Phanerozoic supercycles. In Berggren W.A. and Van Couvering, J.A. (eds), *Catastrophes and Earth History*. Princeton University Press, Princeton, 129–150.

Fisher, T.G. and Smith, D.G. 1994. Glacial Lake Agassiz: its northwest maximum extent and outlet in Saskatchewan (Emerson phase). *Quaternary Science Reviews* 13, 845–858.

Flannery, T. 1994. *The Future Eaters. An Ecological History of the Australasian Lands and People*. Reed Books, Chatswood, NSW.

Flenley, J.R. 1978. *The Equatorial Rainforest: A Geological History*. Butterworth, London.

Flint, R.F. 1971. *Glacial and Quaternary Geology*. Wiley, New York.

Flood, J. 1989. *Archaeology of the Dreamtime. The Story of Prehistoric Australia and its People* (2nd edn). Collins, Sydney.

Foley, R.A. 1994. Speciation, extinction and climatic change in hominid evolution. *Journal of Human Evolution* 26, 275–289.

Foley, R.A. and Lee, P.C. 1989. Finite social space, evolutionary pathways, and reconstructing hominid behavior. *Science* 243, 901–906.

Fontes, J.-C. and Gasse, F. 1991. Palhydaf (Palaeohydrology in Africa) program: objectives, methods, major results. *Palaeogeography, Palaeoclimatology, Palaeoecology* 84, 191–125.

Fontes, J.-C., Gasse, F., Callot, Y., Plaziat, J.-C., Carbonnel, P., Dupeuple, P.A. and Kaczmarska, I. 1985. Freshwater to marine-like environments from Holocene lakes in Northern Sahara. *Nature* 317, 608–610.

Fontes, J.-C., Mélières, F., Gibert, E., Qing, L. and Gasse, F. 1993. Stable isotope and radiocarbon balances of two Tibetan lakes (Sumxi Co, Longmu Co) from 13,000 BP. *Quaternary Science Reviews* 12, 875–887.

Forester, R.M. 1987. Late Quaternary paleoclimate records from lacustrine ostracodes. In Ruddiman, W.F. and Wright, H.E. (eds), *North America and Adjacent Oceans During the Last Deglaciation*. The Geology of North America. Vol. K-3. Geological Society of America, Boulder, 261–276.

Forman, S.L. 1990. Post-glacial relative sea-level history of northwestern Spitsbergen, Svalbard. *Bulletin of the Geological Society of America* 102, 1580–1590.

Foucault, A. and Stanley, D.J. 1989. Late Quaternary palaeoclimatic oscillations in East Africa recorded by heavy minerals in the Nile delta. *Nature* 339, 44–46.

François, R., Bacon, M.P., Altabet, M.A. and Labeyrie, L.D. 1993. Glacial/interglacial changes in sediment rain in the SW Indian sector of subantarctic waters as recorded by ^{230}Th, ^{231}Pa, U and d^{15}N. *Paleoceanography* 8, 611–629.

Franzén, L.G. 1994. Are wetlands the key to the ice-age cycle enigma? *Ambio* 23, 300–308.

Frayer, D.W., Wolpoff, M.H., Thorne, A.G., Smith, F.H.

and Pope, G.G. 1993. Theories of modern human origins: the paleontological test. *American Anthropologist* 95, 14–50.

Frey, D.G. (ed.) 1969. Symposium on paleolimnology. *Internationale Vereinigung für theoretische und angewandte Limnologie, Mitteilungen* 17, 1–448.

Friedlingstein, P., Prentice, K.C., Fung, I.Y., John, J.G. and Brassueur, G.P. 1995. Carbon–biosphere–climate interactions in the last glacial maximum climate. *Journal of Geophysical Research* 100, 7203–7221.

Fritts, H.C. 1976. *Tree Rings and Climate*. Academic Press, London.

Fronval, T., Jansen, E., Bloemendal, J. and Johnsen, S. 1995. Oceanic evidence for coherent fluctuations in Fennoscandian and Laurentide ice sheets on millenium timescales. *Nature* 374, 443–446.

Fuhrer, K., Neftel, A., Anklin, M., Staffelbach, T. and Legrand, M. 1996. High-resolution ammonium ice core record covering a complete glacial–interglacial cycle. *Journal of Geophysical Research* 101(D2), 4147–4164.

Fuji, N. and Horowitz, A. 1989. Brunhes epoch palaeoclimates of Japan and Israel. *Palaeogeography, Palaeoclimatology, Palaeoecology* 72, 79–88.

Fulton, R.J. (ed.), 1989. *Quaternary Geology of Canada and Greenland*. Geological Survey of Canada: Geology of Canada No. 1; and Geological Society of North America: Geology of North America Vol. K-1.

Funnell, B.M. and Riedel, L.W.R. (eds) 1971. *The Micropalaeontology of the Oceans*. Cambridge University Press, Cambridge.

Fyfe, W.S. 1990a. Geosphere forcing: plate tectonics and the biosphere. *Palaeogeography, Palaeoclimatology, Palaeoecology, Global and Planetary Change Section* 89, 185–191.

Fyfe, W.S. 1990b. The International Geosphere/Biosphere Programme and global change: an anthropocentric or an ecocentric future? A personal view. *Episodes* 13, 100–102.

Gabunia, L. and Vekua, A. 1995. A Plio-Pleistocene hominid from Dmanisi, East Georgia, Caucasus. *Nature* 373, 509–512.

Gallée, H., van Ypersele, J.P., Fichefet, T., Marsiat, I., Tricot, C. and Berger, A. 1992. Simulation of the last glacial cycle by a coupled, sectorally averaged climate-ice sheet model. 2: Response to insolation and CO_2 variations. *Journal of Geophysical Research* 97(D14), 15713–15740.

Gallimore, R.G. and Kutzbach, J.E. 1996. Role of orbitally induced changes in tundra area in the onset of glaciation. *Nature* 381, 503–505.

Ganeshram, R.S., Pedersen, T.F., Calvert, S.E. and Murray, J.W. 1995. Large changes in oceanic nutrient inventories from glacial to interglacial periods. *Nature* 376, 755–758.

Gargett, R.H. 1989. Grave shortcomings. The evidence for Neanderthal burial. *Current Anthropology* 30, 157–190.

Gash, J.H.C., Nobre, C.A., Roberts, J.M. and Victoria, R.L. 1996. *Amazonian Deforestation and Climate*. Wiley, Chichester.

Gasse, F. 1975. *L'évolution des Lacs de l'Afar Central (Ethiopie et TFAI) du Plio-Pléistocèn à l'Actuel: Reconstitution des Paléomilieux Lacustres à Partir de l'Étude des Diatomées*. DSc thesis, University of Paris VI, 406.

Gasse, F. 1980. Les diatomées lacustres plio-pléistocènes du Gadeb (Éthiopie) systématique, paléoécologie, biostratigraphie. *Revue Algologique*, Mémoire hors-série 3.

Gasse, F. and Fontes, J.-C. 1992. Climatic changes in northwest Africa during the last deglaciation (16–7 ka BP). In Bard, E. and Broecker, W.S. (eds), *The Last Deglaciation: Absolute and Radiocarbon Chronologies*. Springer-Verlag, Berlin, 295–325.

Gasse, F., Rognon, P. and Street, F.A. 1980. Quaternary history of the Afar and Ethiopian Rift Lakes. In Williams, M.A.J. and Faure, H. (eds), *The Sahara and the Nile*. Balkema, Rotterdam, 361–400.

Gasse, F., Tehet, R., Durand, A., Gibert, E. and Fontes, J.-C. 1990. The arid–humid transition in the Sahara and the Sahel during the last deglaciation. *Nature* 346, 141–146.

Gates, W.L. 1976. Modelling the Ice-Age climate. *Science* 191, 1138–1144.

Gautier, A. 1988. The final demise of *Bos ibericus? Sahara* 1, 37–48.

Geyh, M.A. and Schleicher, H.S. 1990. *Absolute Age Determination*. Springer-Verlag, Berlin.

Ghil, M. 1981. Internal climatic mechanisms participating in glaciation cycles. In Berger, A. (ed.), *Climatic Variations and Variability: Facts and Theories*. NATO ISI Series C: 72. Reidel, Dordrecht, 539–557.

Ghil, M. 1989. Deceptively-simple models of climatic change. In Berger, A., Schneider, S. and Duplessy, J.-C. (eds), *Climate and Geo-sciences. A Challenge for Science and society in the 21st Century*. NATO ISI Series C Volume 285. Kluwer, Dordrecht, 211–240.

Ghil, M. 1991. Quaternary glaciations: theory and observations. In Sonett, C.P. Giampapa, M.S. and Matthews, M.S. (eds), *The Sun in Time*. University of Arizona, Tucson, 511–542.

Gilland, B. 1988. Population, economic growth, and energy demand 1985–2020. *Population and Development Review* 14, 233–244.

Gillespie, R., Horton, D.R., Ladd, P., Macumber, P.G., Thorne, R. and Wright, R.V.S. 1978. Lancefield Swamp and the extinction of the Australian megafauna. *Science* 200, 1044–1048.

Godwin, H. and Tallantire, P.A. 1951. Studies in the post-glacial history of British vegetation. XII. Hockham Mere, Norfolk. *Journal of Ecology* 39, 285–307.

Gorecki, P.P., Horton D.R., Stern, N. and Wright, R.V.S. 1984. Co-existence of humans and megafauna in Australia: improved stratified evidence. *Archaeology in Oceania* 19, 117–119.

Gornitz, V. 1991. Global coastal hazards from future sea level rise. *Palaeogeography, Palaeoclimatology, Palaeoecology, Global and Planetary Change Section* 89, 379–398.

Gornitz, V. 1993. Mean sea level changes in the recent past. In Warrick, R.A., Barrow, E.M. and Wigley, T.M.L. (eds), *Climate and Sea Level Change: Observations, Projections and Implications*. Cambridge University Press, Cambridge, 25–44.

Gornitz, V., Lebedeff, S. and Hansen, J. 1982. Global sea level trend in the past century. *Science* 215, 1611–1614.

Gornitz, V., Rosenzweig, C. and Hillel, D. 1997. Effects of anthropogenic intervention on the land hydrologic cycle and global sea level rise. *Global and Planetary Change* 14, 147–161.

Gorshkov, S.G. (ed.), 1978. *World Ocean Atlas Volume 2. Atlantic and Indian Oceans*. Pergamon Press, Australia.

Goslar, T., Arnold, M., Bard, E., Kuc, T., Pazdur, M.F., Ralska-Jasiewiczowa, M., Rózanski, K., Tisnerat, N., Walanus, A., Wicik, B. and Wickowski, K. 1995. High concentration of atmospheric ^{14}C during the Younger Dryas cold episode. *Nature* 377, 414–417.

Goudie, A. 1983. *Environmental Change*. (2nd edn). Clarendon Press, Oxford.

Goudie, A.S., Allchin, B. and Hegde, K.T.M. 1973. The former extensions of the Great Indian Sand Desert. *Geographical Journal* 134, 243–257.

Gowlett, J.A.J. 1984a: Ascent to Civilization. *The Archaeology of Early Man*. Collins, London.

Gowlett, J.A.J. 1984b: Mental abilities of Early Man: a look at some hard evidence. In Foley, R. (ed.), *Hominid Evolution and Community Ecology*. Academic Press, London, 167–192.

Gowlett, J.A.J., Harris, J.W.K., Walton, D.A. and Wood, B.A. 1981. Early archaeological sites, further hominid remains and traces of fire from Chesowanja, Kenya. *Nature* 294, 125–129.

Graedel, T.E. and Crutzen, P.J. 1995. *Atmosphere, Climate, and Change*. Scientific American Library, New York.

Gray, J. (ed.) 1988. Aspects of freshwater paleoecology and biogeography. *Palaeogeography, Palaeoclimatology, Palaeoecology* 62, 1–623.

Green, D.G. 1981. Time series and postglacial forest ecology. *Quaternary Research* 15, 265–277.

Green, D.G. and Dolman, G.S. 1988. Fine resolution pollen analysis. *Journal of Biogeography* 15, 685–701.

Greenland Ice-Core Project (GRIP) Members 1993. Climate instability during the last interglacial period recorded in the GRIP ice core. *Nature* 364, 203–207.

Gregory, K.J., Starkel, L. and Baker, V.R. (eds) 1995. *Global Continental Palaeohydrology*. Wiley, Chichester.

Gribbin, J. 1976. Mason develops Milankovitch ice age theory. *Nature* 260, 396.

Gribbin, J. (ed.) 1986. *The Breathing Planet*. Blackwell, Oxford.

Grimm, E. 1988. Data analysis and display. In Huntley, B. and Webb, T. III (eds), *Vegetation History*. Kluwer, Dordrecht, 43–76.

Grimm, E.C., Jacobson, G.L., Watts, W.A., Hansen, B.C.S. and Maasch, K.A. 1993. A 50,000-year record of climate oscillations from Florida and its temporal correlation with the Heinrich events. *Science* 261, 198–200.

Grimm, N.B. 1993. Implications of climate change for stream communities. In Kareiva, P.M., Kingsolver, J.G. and Huey, R.B. (eds), *Biotic Interactions and Global Change*. Sinauer, Sunderland, MA, 293–314.

Gross, M.G. 1982. *Oceanography: A View of the Earth*. Prentice Hall, New Jersey.

Grove, A.T. 1980. Geomorphic evolution of the Sahara and the Nile. In Williams, M.A.J. and Faure, H. (eds), *The Sahara and the Nile*. Balkema, Rotterdam, 7–16.

Grove, A.T. and Warren, A. 1968. Quaternary landforms and climate on the south side of the Sahara. *Geographical Journal* 134, 194–208.

Grove, J.M. 1988. *The Little Ice Age*. Methuen, London.

Grün, R. 1989. Electron spin resonance (ESR) dating. *Quaternary International* 1, 65–109.

Grün, R., Brink, J.S., Spooner, N.A., Taylor, L., Stringer, C.B., Franciscus, R.G. and Murray, A.S. 1996. Direct dating of Florisbad hominid. *Nature* 382, 500–501.

Grün, R. and Stringer, C.B. 1991. Electron spin resonance dating and the evolution of modern humans. *Archaeometry* 33, 153–199.

Guidon, N. and Delibrias, G. 1985. Inventaire des sites sud-américains antérieurs à 12,000 ans. *L'Anthropologie* 89, 385–407.

Guidon N. and Delibrias, G. 1986. Carbon-14 dates point to man in the Americas 32,000 years ago. *Nature* 321, 769–771.

Guilderson, T.P., Fairbanks, R.G. and Rubenstone, J.L. 1994. Tropical temperature variations since 20,000 years ago: modulating interhemispheric climate change. *Science* 263, 663–664.

Guiot, J. 1990. Methodology of the last climatic cycle reconstruction in France estimated from pollen data. *Palaeogeography, Palaeoclimatology, Palaeoecology* 80, 49–69.

Habermehl, M.A. 1985. Groundwater in Australia. In *Hydrogeology in the Service of Man*. Memoirs of the 18th International Association of Hydrogeologists Congress. International Association of Hydrological Sciences Publication No. 154, 31–52.

Haffer, J. 1969. Speciation in Amazonian forest birds. *Science* 165, 131–137.

Hajdas, I., Zolitschka, B., Ivy-Ochs, S.D., Beer, J., Bonani, G., Leroy, S.A.G., Negendank, J.W., Ramrath, M. and Suter, M. 1995. AMS radiocarbon dating of annually laminated sediments from Lake Holzmaar, Germany. *Quaternary Science Reviews* 14, 137–143.

Hall, C.M., Walter, R.C., Westgate, J.A. and York D. 1984. Geochronology, stratigraphy and geochemistry of Cindery Tuff in Pliocene hominid-bearing sediments of the Middle Awash, Ethiopia. *Nature* 308, 26–31.

Hall, N.M.J. and Valdes, P.J. 1997. A GCM simulation of the climate 6000 years ago. *Journal of Climate* 10, 3–17.

Hallam, S.J. 1977. The relevance of Old World archaeology to the first entry of man into New Worlds: colonization seen from the Antipodes. *Quaternary Research* 8, 128–148.

Hamilton, T.D. and Thorson, R.M. 1983. The Cordilleran ice sheet in Alaska. In Wright, H.E. and Porter, S.C. (eds), *Late Quaternary environments of the United States. Volume 1. The Late Pleistocene*. Longman, London, 38–52.

Hanna, E. 1996. Have long-term solar minima, such as the Maunder Minimum, any recognisable climatic effect? *Weather* 51, 304–312.

Hansen, J.E. and Lacis, A.A. 1990. Sun and dust versus greenhouse gases: an assessment of their relative roles in global climatic change. *Nature* 346, 713–719.

Haq, B.U., Hardenbol, J. and Vail, P.R. 1987. Chronology of the fluctuating sea levels since the Triassic. *Science* 235, 1156–1167.

Haq, B.U. 1995. Growth and decay of gas hydrates: a forcing mechanism for abrupt climate change and sediment wasting on ocean margins. In Troelstra, S.R., van Hinte, J.E. and Ganssen, G.M. (eds), *The Younger Dryas*. North-Holland, Amsterdam, 191–203.

Harmon, R.S., Schwarcz, H.P. and Ford, D.C. 1978. Late Pleistocene sea level history of Bermuda. *Quaternary Research* 9, 205–218.

Harmon, R.S., Mitterer, R.M., Kriausakul, N., Land, L.S., Schwarcz, H.P., Garrett, P., Larson, G.J., Vacher, H.L. and Rowe, M. 1983. U-series and amino-acid racemization geochronology of Bermuda: implications for eustatic sea level fluctuation over the past 250,000 years. *Palaeogeography, Palaeoclimatology, Palaeoecology* 44, 41–70.

Harris, J.M., Brown, F.H., Leakey, M.G., Walker, A.C. and Leakey, R.E. 1988. Pliocene and Pleistocene hominid-bearing sites from west of Lake Turkana, Kenya. *Science* 239, 27–33.

Harris, J.W.K. 1980. Early Man. In Sherratt, A. (ed.), *The Cambridge encyclopedia of archaeology*. Cambridge University Press, Cambridge, 62–70.

Harris, J.W.K. 1983. Cultural beginnings: Plio-Pleistocene archaeological occurrences from the Afar, Ethiopia. *African Archaeological Review* 1, 2–31.

Harris, S.A. 1985. Distribution and zonation of permafrost along the eastern ranges of the Cordillera of North America. *Biuletyn Peryglacjalny* 30, 107–118.

Harrison, S.P., Metcalfe, S.E., Street-Perrott, F.A., Pittock, A.B., Roberts, C.N. and Salinger, M.J. 1984. A climatic model of the last glacial/interglacial transition based on palaeotemperature and palaeohydrological evidence. In Vogel, J. (ed.), *Late Cainozoic Environments of the Southern Hemisphere*. Balkema, Rotterdam, 21–34.

Hartman, D.L. 1994. *Global Physical Climatology*. Academic Press, San Diego.

Harvey, L.D.D. 1988. Climatic impact of ice-age aerosols. *Nature* 334, 333–335.

Hassan, F.A. 1980. Prehistoric settlement along the Main Nile. In Williams, M.A.J. and Faure, H. (eds), *The Sahara and the Nile*. Balkema, Rotterdam, 421–450.

Hasselmann, K. 1981. Construction and verification of stochastic climate models. In Berger, A. (ed.), *Climatic Variations and Variability: Facts and Theories*. NATO ISI Series C: 72. Reidel, Dordrecht, 481–497.

Hasselmann, K. 1991. Ocean circulation and climate change. *Tellus* 43AB, 82–103.

Hastenrath, S. and Kutzbach, J.E. 1983. Palaeoclimatic estimates from water and energy budgets of East African lakes. *Quaternary Research* 19, 141–153.

Hay, W.W. 1993. The role of polar deep water formation in global climate change. *Annual Reviews of Earth and Planetary Science* 21, 227–254.

Haynes, C.V. 1991. Geoarchaeological and palaeohydrological evidence for a Clovis-age drought in North America and its bearing on extinction. *Quaternary Research* 35, 438–450.

Haynes, C.V., Jr. 1989. Bagnold's barchan: A 57–yr record of dune movement in the eastern Sahara and implications for dune origin and paleoclimate since Neolithic times. *Quaternary Research* 32, 153–167.

Hays, J.D. 1978. A review of the Late Quaternary climatic history of Antarctic Seas. In Van Zinderen Bakker, E.M.

(ed.), *Antarctic Glacial History and World Palaeoenvironments*. Balkema, Rotterdam, 57–71.

Hays, J.D., Imbrie, J. and Shackleton, N.J. 1976a. Variations in the earth's orbit: pacemaker of the Ice Ages. *Science* 194, 1121–1132.

Hays J.D., Lozano, J.A., Shackleton, N. and Irving, G. 1976b. Reconstruction of the Atlantic and Western Indian Ocean sectors of the 18,000 B.P. Antarctic Ocean. *Geological Society of America Memoir* 145, 337–372.

Hedgpeth, J.W. 1969. Introduction to Antarctic zoogeography. *Antarctic map folio series*, Folio 11. 1. American Geographical Society, New York.

Heinrich, H. 1988. Origin and consequences of cyclic ice-rafting in the northeast Atlantic Ocean during the past 130,000 years. *Quaternary Research* 29, 142–152.

Held, I. 1982. Climate models and the astronomical theory of the Ice Ages. *Icarus* 50, 449–461.

Heller, F. and Liu, T.-S. 1982. Magnetostratigraphical dating of loess deposits in China. *Nature* 300, 431–433.

Hennenberg, M. 1989. Morphological and geological dating of early hominid fossils compared. *Current Anthropology* 30, 527–529.

Hennenberg, M. and Thackeray, J.F. 1995. A single-lineage hypothesis of hominid evolution. *Evolutionary Theory* 11, 31–38.

Herbert, T.D. and Fischer, A.G. 1986. Milankovitch climatic origin of mid-Cretaceous black shale rhythms in central Italy. *Nature* 321, 739–743.

Hereford, R. 1993. Entrenchment and widening of the Upper San Pedro River, Arizona. *Geological Society of America Special Paper 282*.

Hesse, P.P. 1994. The record of continental dust from Australia in Tasman Sea sediments. *Quaternary Science Reviews* 13, 257–272.

Heusser, L.E. and Sirocko, F. 1997. Millennial pulsing of environmental change in southern California from the past 24 ky: a record of Indo-Pacific ENSO events? *Geology* 25, 243–246.

Higham, C. 1989. *The Archaeology of Mainland Southeast Asia: from 10,000 BC to the Fall of Angkor*. Cambridge University Press, Cambridge.

Hilgen, F.J. 1991. Extension of the astronomically calibrated (polarity) time scale to the Miocene–Pliocene boundary. *Earth and Planetary Science Letters* 107, 349–368.

Hill, A. 1987. Causes of perceived faunal change in the later Neogene of East Africa. *Journal of Human Evolution* 16, 583–596.

Hill, A. and Ward, S. 1988. Origin of the Hominidae: the record of African large hominoid evolution between 14 My and 4 My. *Yearbook for Physical Anthropology* 31, 49–83.

Hillaire-Marcel, C. and Occhietti, S. 1980. Chronology, paleogeography, and paleo-climatic significance of the late and post-glacial events in eastern Canada. *Zeitschrift für Geomorphologie* N.F. 24, 373–392.

Hillel, D. 1994. *Rivers of Eden: the struggle for water and the quest for peace in the Middle East*. Oxford University Press, New York.

Hoelzmann, P. 1993. Palaeoecology of Holocene lacustrine sediments in Western Nubia, SE Sahara. In Thorweihe, U. and Schandelmeier, H. (eds), *Geoscientific Research in Northeast Africa*. Proceedings of the International Conference on Geoscientific Research in Northeast Africa, Berlin, Germany, 17–19 June 1993. Balkema, Rotterdam, 569–574.

Hofmann, C., Courtillot, V., Féraud, G., Rochette, P., Yirgu, G., Ketefo, E. and Pik, R. 1997. Timing of the Ethiopian flood basalt event and implications for plume birth and global change. *Nature* 389, 838–841.

Holdsworth, G., Higuchi, K., Zielinski, G.A., Mayewski, P.A., Wahlen, M., Deck, B., Chylek, P., Johnson, B. and Damiano, P. 1996. Historical biomass burning: late 19th century pioneer agriculture revolution in northern hemisphere ice core data and its atmospheric interpretation. *Journal of Geophysical Research* 101(D18), 23317–23334.

Holland, H.D. and Petersen, U. 1995. *Living Dangerously. The Earth, its Resources, and the Environment*. Princeton University Press, Princeton, NJ.

Hollin, J.T. and Schilling, D.H. 1981. Late Wisconsin–Weichselian mountain glaciers and small ice caps. In Denton, G.H. and Hughes, T.J. (eds), *The Last Great Ice Sheets*. Wiley, New York, 179–206.

Hooghiemstra, H. 1988. The orbital-tuned marine oxygen isotope record applied to the Middle and Late Pleistocene pollen record of Funza (Colombian Andes). *Palaeogeography, Palaeoclimatology, Palaeoecology* 66, 9–17.

Hooghiemstra, H. 1989. Quaternary and Upper-Pliocene glaciations and forest development in the tropical Andes: evidence from a long high-resolution pollen record from the sedimentary basin of Bogota, Colombia. *Palaeogeography, Palaeoclimatology, Palaeoecology* 72, 11–26.

Hooghiemstra H. 1995. Environmental and paleoclimatic evolution in Late Pliocene–Quaternary Colombia. In Vrba, E.S., Denton, G.H., Partridge, T.C. and Burckle, L.H. (eds), *Paleoclimate and Evolution*. Yale University Press, New Haven, 249–261.

Hooghiemstra, H., Bechler, A. and Beug, H.-J. 1987. Isopollen maps for 18,000 years b.p. of the Atlantic offshore of northwest Africa: evidence for paleowind circulation. *Paleoceanography* 2, 561–582.

Hope, G.S. 1984. Papua New Guinea at 18,000 BP. In Chappell, J.M.A. and Grindrod, A. (eds), *Proceedings of the first CLIMANZ Conference, held at Howman's Gap, Victoria, Australia, February 8–13, 1981*. Department of Biogeography and Geomorphology, Research School of Pacific Studies, Australian National University, Canberra, 55.

Hopkins, M.S., Ash, J., Graham, A.W., Head, J. and Hewett, R.K. 1993. Charcoal evidence of the spatial extent of the *Eucalyptus* woodland expansions and rainforest contractions in north Queensland during the Late Pleistocene. *Journal of Biogeography* 20, 357–372.

Horton, D.R. 1980. A review of the extinction question: man, climate and megafauna. *Archaeology and Physical Anthropology in Oceania* 15, 86–97.

Horton, D.R. 1982. The burning question: Aborigines, fire and Australian ecosystems. *Mankind* 13, 237–251.

Horton, D.R. 1984. Red kangaroos: last of the megafauna. In Martin, P.S. and Klein, R.G. (eds), *Quaternary Extinctions*. University of Arizona Press, Tuscon, 639–680.

Houghton, J.T., Jenkins, G.J. and Ephraums, J.J. (eds) 1990. *Climate Change: The Intergovernmental Panel on Climate Change Scientific Assessment.* Cambridge University Press, Cambridge.

Houghton, J.T., Meira Filho, L.G., Bruce, J., Hoesung Lee, Callander, B.A., Haites, E., Harris, N. and Maskell, K. (eds) 1995. *Climate Change 1994. Radiative Forcing of Climate Change and an Evaluation of the IPCC IS92 Emission Scenarios.* Cambridge University Press, Cambridge.

Houghton, J.T., Meira Filho, L.G., Callander, B.A., Harris, N., Kattenberg, A. and Maskell, K. (eds) 1996. *Climate Change 1995. The Science of Climate Change.* Cambridge University Press, Cambridge.

Howard, W.R. 1997. A warm future in the past. *Nature* 388, 418–419.

Hsieh, J.C.C. and Murray, B. 1996. A ~ 24000 year period climate signal in 1.7 - 2.0 million year old Death Valley strata. *Earth and Planetary Science Letters* 141, 11–19.

Hsü, K.J. 1983. *The Mediterranean was a Desert. A Voyage of the Glomar Challenger.* Princeton University Press, Princeton, NJ.

Hsü, K.J., Montaderts, L., Bernoulli, D., Cita, M.B., Erikson, A., Garrison, R.E., Kidd, R.B., Mélières, F., Müller, C. and Wright, R. 1977. History of the Mediterranean salinity crisis. *Nature* 267, 399–403.

Huang, S., Pollock, H.N. and Shen, P.V. 1997. Late Quaternary temperature changes seen in world-wide continental heat flow measurements. *Geophysical Research Letters* 24, 1947–1950.

Huang, W.P., Ciochon, R., Gu, Y.M., Larick, R., Fang, Q.R., Schwarcz, H., Yonge, C., de Vos, J. and Rink, W. 1995. Early *Homo* and associated artefacts from Asia. *Nature* 378, 275–278.

Hublin, J.-J., Spoor, F., Braun, M., Zonneveld, F. and Condemi, S. 1996. A late Neanderthal associated with Upper Palaeolithic artefacts. *Nature* 381, 224–226.

Huggett, R.J. 1991. *Climate, Earth Processes and Earth History.* Springer-Verlag, Berlin.

Hughen, K.A., Overpeck, J.T., Peterson, L.C. and Trumbore, S. 1996. Rapid climate changes in the tropical Atlantic region during the last deglaciation. *Nature* 380, 51–54.

Hughes, T. 1987. Ice dynamics and deglaciation models when ice sheets collapsed. In Ruddiman, W.F. and Wright, H.E. Jr. (eds), *North America and Adjacent Oceans During the Last Deglaciation.* The Geology of North America Vol. K-3, Geological Society of America, 183–220.

Hulton, N., Sugden, D., Payne, A. and Clapperton, C. 1994. Glacier modeling and the climate of Patagonia during the last glacial maximum. *Quaternary Research* 42, 1–19.

Humborg, C., Ittekkot, V., Cociasu, A. and Bodungen, B.V. 1997. Effect of Danube River dam on Black Sea biogeochemistry and ecosystem structure. *Nature* 386, 385–388.

Hunt, C.B. 1972. *Geology of Soils, their Evolution, Classification, and Uses.* W.H. Freeman, San Francisco.

Hunt, K.D. 1994. The evolution of human bipedality: ecology and functional morphology. *Journal of Human Evolution* 26, 183–202.

Hunt, K.D. 1996. The postural feeding hypothesis: an ecological model for the evolution of bipedalism. *South African Journal of Science* 92, 77–90.

Hunten, D.M., Gérard, J.-C. and François, L.M. 1991. The atmosphere's response to solar irradiation. In Sonett, C.P., Giampapa, M.S. and Matthews, M.S. (eds), *The Sun in Time.* University of Arizona, Tucson, 463–497.

Huntley, B. 1990. European post-glacial forests: compositional changes in response to climatic change. *Journal of Vegetation Science* 1, 507–518.

Husen, D. van 1989. The last interglacial–glacial cycle in the eastern Alps. *Quaternary International* 3/4, 115–121.

Hutchinson, D.R., Golmshtok, A.J., Zonenshain, L.P., Moore, T.C., Scholz, C.A. and Klitgord, K.D. 1992. Depositional and tectonic framework of the rift basins of Lake Baikal from multichannel seismic data. *Geology* 20, 589–592.

Hyde, W.T. and Peltier, W.R. 1985. Sensitivity experiments with a model of the ice age cycle: the response to harmonic forcing. *Journal of the Atmospheric Sciences* 42, 2170–2188.

Ikeya, M. 1993. *New applications of electron spin resonance.* World Scientific, Singapore.

Imbrie, J. 1981. Time-dependent models of the climatic response to orbital variations. In Berger, A. (ed.), *Climatic Variations and Variability: Facts and Theories.* NATO ISI Series C: 72. Reidel, Dordrecht, 527–538.

Imbrie, J. 1982. Astronomical theory of the Pleistocene ice ages: a brief historical review. *Icarus* 50, 408–422.

Imbrie, J. 1994. Measuring the gain of the climate system's response to Milankovitch forcing in the precession and obliquity bands. In Duplessy, J.-C. and Spyridakis, M.-T. (eds), *Long-term Climatic Variations. Data and Modelling.* Springer, Berlin, 403–410 .

Imbrie, J. and Imbrie, K.P. 1979. *Ice Ages: Solving the Mystery.* Macmillan, London.

Imbrie, J. and Kipp, N.G. 1971. A new micropaleontological method for quantitative paleoclimatology: Application to a late Pleistocene Carribean core. In Turekian, K.K. (ed.), *The Late Cenozoic Glacial Ages.* Yale University Press, New Haven, CT, 71–181.

Imbrie, J., Berger, A., Boyle, E.A., Clemens, S.C., Duffy, A., Howard, W.R., Kukla, G., Kutzbach, J., Martinson, D.G., McIntyre, A., Mix, A.C., Molfino, B., Morley, J.J., Peterson, L.C., Pisias, N.G., Prell, W.L., Raymo, M.E., Shackleton, N.J. and Toggweiler, J.R. 1993a. On the structure and origin of major glaciation cycles. 2. The 100,000–year cycle. *Paleoceanography* 8, 699–735.

Imbrie, J., Berger, A. and Shackleton, N.J. 1993b. Role of orbital forcing: a two-million-year perspective. In Eddy, J.A. and Oeschger, H. (eds), *Global Changes in the Perspective of the Past.* Wiley, Chichester, 263–277.

Imbrie, J., Boyle, E.A., Clemens, S.C., Duffy, A., Howard, W.R., Kukla, G., Kutzbach, J., Martinson, D.G., McIntyre, A., Mix, A.C., Molfino, B., Morley, J.J., Peterson, L.C., Pisias, N.G., Prell, W.L., Raymo, M.E., Shackleton, N.J. and Toggweiler, J.R. 1992. On the structure and origin of major glaciation cycles. 1. Linear responses to Milankovitch forcing. *Paleoceanography* 7, 701–738.

Imbrie, J., Hays, J.D., Martinson, D.G., McIntyre, A., Mix, A.C., Morley, J.J., Pisias, N.G., Prell, W.L. and

Shackleton, N.J. 1984. The orbital theory of Pleistocene climate: support from a revised chronology of the marine ^{18}O record. In Berger, A.L., Imbrie, J., Hays, J., Kukla, G. and Saltzman, B. (eds), *Milankovitch and Climate: Understanding the Response to Astronomical Forcing*. Reidel, Dordrecht, 269–305.

Imbrie, J., McIntyre, A. and Mix A. 1989. Oceanic response to orbital forcing in the late Quaternary: observational and experimental strategies. In Berger, A., Schneider, S. and Duplessy, J.-C.(eds), *Climate and Geo-sciences. A Challenge for Science and Society in the 21st Century*. NATO ISI Series C Volume 285. Kluwer, Dordrecht, 121–164.

Iqbal, M. 1983. *An Introduction to Solar Radiation*. Academic Press, Toronto.

Irion, G. 1984. Sedimentation and sediments of Amazonian rivers and evolution of the Amazonian landscape since Pliocene times. In Sioli, H. (ed.), *The Amazon*. Junk, Dordrecht, 201–214.

Isdale, P. 1984. Fluorescent bands in massive corals record centuries of coastal rainfall. *Nature* 310, 378–379.

Israel Land Development Authority 1990. Redeeming the Negev Desert in Israel. In *United Nations Environment Programme, Exchange of Environment Experience Series*, Book 3. Infoterra Programme Activity Centre, Nairobi, 1–16.

Issawi, B. 1983. Ancient rivers of the eastern Egyptian desert. *Episodes* 1983(2), 3–6.

Ivanovich, M., Ku, T.-L., Harmon, R.S. and Smart, P.L. 1984. Uranium series intercomparison project (USIP). *Nuclear Instrumental Methods in Physics Research* 223, 466–471.

Ivanovich, M., Vita-Finzi, C. and Hennig, G.J. 1983. Uranium series dating of molluscs of uplifted Holocene beaches in the Persian Gulf. *Nature* 302, 408–410.

Jaanusson, V. 1991. Morphological changes leading to hominid bipedalism. *Lethaia* 24, 443–457.

Jablonski, N.G. and Chaplin, G. 1993. Origin of habitual terrestrial bipedalism in the ancestor of the Hominidae. *Journal of Human Evolution* 24, 259–280.

Jansen, E. 1989. The use of stable oxygen and carbon isotope stratigraphy as a dating tool. *Quaternary International* 1, 151–166.

Jansen E. and Veum, T. 1990. Evidence of two-step deglaciation and its impact on North Atlantic deep-water circulation. *Nature* 343, 612–616.

Jansson, C.R. 1966. Recent pollen spectra from the deciduous and coniferous–deciduous forests of northeastern Minnesota: a study in pollen dispersal. *Ecology* 47, 804–825.

Jelgersma, S. 1961. Holocene sea level changes in the Netherlands. *Mededelingen Geologische Stichting* Serie C-VI, No. 7 13–14.

Jiang, X. and Peltier, W.R. 1996. Ten million year histories of obliquity and precession: the influence of the ice-age cycle. *Earth and Planetary Science Letters* 139, 17–32.

Johanson, D.C. 1989. The current status of Australopithecus. In Giacobini, G. (ed.), *Hominidae: Proceedings of the 2nd International Congress of Human Paleontology*. Jaca, Milan, 77–96.

Johanson, D.C. and Edey, M.A. 1981. *Lucy: The Beginnings of Humankind*. Simon and Schuster, New York.

Johanson, D.C., Masao, F.T., Eck, G.G., White, T.D., Walter, R.C., Kimbel, W.H., Asfaw, B., Manega, P., Ndessokia, P. and Suwa, G. 1987. New partial skeleton of *Homo habilis* from Olduvai Gorge, Tanzania. *Nature* 372, 205–209.

Johanson, D.C. and White, T. 1979. A systematic assessment of early African Hominids. *Science* 202, 321–330.

Johnsen, S.J., Clausen, H.B., Dansgaard, W., Fuhrer, K., Gundestrup, N., Hammer, C.U., Iversen, P., Jouzel, J., Stauffer, B. and Steffensen, J.P. 1992. Irregular glacial interstadials recorded in a new Greenland ice core. *Nature* 359, 311–313.

Johnsen, S.J., Clausen, H.B., Dansgaard, W., Gundestrup, N.S., Hammer, C.U. and Tauber, H. 1995. The Eem stable isotope record along the GRIP ice core and its interpretation. *Quaternary Research* 43, 117–124.

Johnson, D.L. and Watson-Stegner, D. 1987. Evolution model of pedogenesis. *Soil Science* 143, 349–366.

Jones, A.T. 1993. Review of the chronology of marine terraces in the Hawaiian archipelago. *Quaternary Science Reviews* 12, 811–823.

Jones, P. 1980. Experimental butchery with modern stone tools and its relevance for Palaeolithic archaeology. *World Archaeology* 12(2), 153–165.

Jones, P.D. and Wigley, T.M.L. 1990. Global warming trends. *Scientific American* 263, 66–73.

Jones, P.D., Wigley, T.M.C. and Wright, P.B. 1986. Global temperature variations between 1861 and 1984. *Nature* 322, 430–434.

Jones, R. 1968. The geographical background to the arrival of man in Australia and Tasmania. *Archaeology and Physical Anthropology in Oceania* 3, 186–215.

Jones, R. 1969. Fire-stick farming. *Australian Natural History* 16, 224–228.

Jones, R. 1979. The fifth continent: problems concerning the human colonization of Australia. *Annual Review of Anthropology* 8, 445–466.

Jouzel, J. 1994. Ice cores north and south. *Nature* 372, 612–613.

Jouzel, J., Barkov, N.I., Barnola, J.M., Bender, M., Chappellaz, J., Genthon, C., Kotlyakov, V.M., Lipenkov, V., Lorius, C., Petit, J.R., Raynaud, D., Raisbeck, G., Ritz, C., Sowers, T., Stievenard, M., Yiou, F. and Yiou, P. 1993. Extending the Vostok ice-core record of palaeoclimate to the penultimate glacial period. *Nature* 364, 407–412.

Jouzel, J., Barkov, N.I., Barnola, J.M., Genthon, C., Korotkevich, Y.S., Kotlyakov, V.M., Legrand, M., Lorius, C., Petit, J.P., Petrov, V.N., Raisbeck, G., Raunaud, D., Ritz, C. and Yiou, F. 1989. Global change over the last climatic cycle from the Vostok ice core record (Antarctica). *Quaternary International* 2, 15–24.

Jouzel, J., Lorius, C., Petit, J.R., Genthon, C., Barkov, N.I., Kotlyakov, V.M. and Petrov, V.M. 1987. Vostok ice core: a continuous isotope temperature record over the last climatic cycle (160,000 years). *Nature* 329, 403–408.

Jouzel, J., Lorius, C., Petit, J.R., Ritz, C., Stievenard, M., Yiou, P., Barkov, N.I., Kotlyakov, V.M. and Lipenkov, V. 1994. The climatic record from Antarctic ice now extends back to 220 kyr BP. In Duplessy, J.-C. and Spyridakis, M.-T. (eds), *Long-term Climatic Variations. Data and Modelling*. NATO ASI Series I: 22. Springer, Berlin, 213–237.

Jull, A.J.T., Beck, J.W. and Burr, G.S. (eds) 1997. *Accelerator Mass Spectrometry*. Elsevier, Amsterdam.

Jutson, J.T. 1934. The physiography (geomorphology) of Western Australia. *Western Australia Geological Survey Bulletin* 95.

Kadomura, H. 1995. Palaeoecological and palaeohydrological changes in the humid tropics during the last 20,000 years, with reference to equatorial Africa. In Gregory, K.J., Starkel, L. and Baker, V.R. (eds), *Global Continental Palaeohydrology*. Wiley, Chichester, 177–202.

Kalb, J.E. 1993. Refined stratigraphy of the hominid-bearing Awash Group, Middle Awash Valley, Afar Depression, Ethiopia. *Newsletters on Stratigraphy* 29, 21–62.

Kapsner, W.R., Alley, R.B., Shuman, C.A., Anadakrishnan, S. and Grootes, P.M. 1995. Dominant influence of atmospheric circulation on snow accumulation in Greenland over the past 18,000 years. *Nature* 373, 52–55.

Karl, T.R., Nicholls, N. and Gregory, J. 1997. The coming climate. *Scientific American* 276, 54–59.

Karlén, W. and Kuylenstierna, J. 1996. On solar forcing of Holocene climate: evidence from Scandinavia. *The Holocene* 6, 359–365.

Kashiwaya, K., Atkinson, T.C. and Smart, P.L. 1991. Periodic variations in Late Pleistocene speleothem abundance in Britain. *Quaternary Research* 35, 190–196.

Kaufman, A.J. 1997. An ice age in the tropics. *Nature* 386, 227–228.

Keany, J. 1976. Diachronous deposition of ice-rafted debris in sub-Antarctic deep-sea sediments. *Geological Society of America Bulletin* 87, 873–882.

Keeley, L. 1980. *Experimental Determination of Stone Tool Use*. University of Chicago Press, Chicago.

Keeley, L.H. and Toth, N. 1981. Microwear polishes on early stone tools from Koobi Fora, Kenya. *Nature* 293, 464–465.

Kehew, A.E. and Teller, J.T. 1994. History of late glacial runoff along the southwestern margin of the Laurentide ice sheet. *Quaternary Science Reviews* 13, 859–877.

Keigwin, L.D., Curry, W.B., Lehman, S.J. and Johnsen, S. 1994. The role of the deep ocean in North Atlantic climate change between 70 and 130 kyr ago. *Nature* 371, 323–326.

Keller, C.K. and Wood, B.D. 1993. Possibility of chemical weathering before the advent of vascular land plants. *Nature* 364, 223–225.

Kennedy-Sutherland, E. 1983. The effects of fire exclusion on growth in mature Ponderosa Pine in Northern Arizona. Unpublished, University of Arizona.

Kennett, J.P. 1982. *Marine Geology*. Prentice-Hall, New Jersey.

Kennett, J.P. 1995. A review of polar climatic evolution during the Neogene, based on the marine sediment record. In Vrba, E.S., Denton, G.H., Partridge, T.C. and Burckle, L.H. (eds), *Paleoclimate and Evolution*. Yale University Press, New Haven, 49–64.

Keown, M.P., Dardeau, E.A. and Causey, E.M. 1986. Historic trends in the sediment flow regime of the Mississippi River. *Water Resources Research* 22, 1555–1564.

Kerr, R.A. 1986. Mapping orbital effects on climate. *Science* 234, 283–284.

Kerr, R.A. 1987. Milankovitch climate cycles through the ages. *Science* 235, 973–974.

Kerr, R.A. 1997. Upstart ice age theory gets attentive but chilly hearing. *Science* 277, 183–184.

Kershaw, A.P. 1978. Record of last interglacial cycle from north-eastern Queensland. *Nature* 272, 159–162.

Kershaw, A.P. 1984. Late Cenozoic plant extinctions in Australia. In Martin, P.S. and Klein, R.G. (eds), *Quaternary Extinctions: a Prehistoric Revolution*. University of Arizona Press, Tuscon, 601–709.

Kershaw, A.P. 1986. The last two glacial–interglacial cycles from northeastern Queensland: implications for climatic change and Aboriginal burning. *Nature* 322, 47–49.

Kershaw, A.P. and Hyland, B.P.M. 1975. Pollen transport and periodicity in a marginal rainforest situation. *Review of Palaeobotany and Palynology* 19, 129–138.

Kilden, P. 1991. The Lateglacial and Holocene evolution of the middle and lower River Scheldt, Belgium. In Starkel, L., Gregory, K.J. and Thornes, J.B. (eds), *Temperate Palaeohydrology: Fluvial Processes in the Temperate Zone during the Last 15000 years*. Wiley, Chichester, 283–299.

Kimbel, W.H., Johanson, D.C. and Rak, Y. 1994. The first skull and other new discoveries of *Australopithecus afarensis* at Hadar, Ethiopia. *Nature* 368, 449–451.

Kimbel, W.H., Walter, R.C., Johanson, D.C., Reed, K.E., Aronson, J.L., Assefa, Z., Marean, C.W., Eck, G.G., Bobe, R., Hovers, E., Rak, Y., Vondra, C., Yemane, T., York, D., Chen, Y., Evensen, N.M. and Smith, P.E. 1996. Late Pliocene *Homo* and Oldowan tools from the Hadar formation (Kada Hadar Member), Ethiopia. *Journal of Human Evolution* 31, 549–561.

Kingston, J.D., Marino, B.D. and Hill, A. 1994. Isotopic evidence for Neogene hominid paleoenvironments in the Kenya Rift Valley. *Science* 264, 955–959.

Klammer, G. 1984. The relief of the extra-Andean Amazon basin. In Sioli, H. (ed.), *The Amazon*. Junk, Dordrecht, 47–83.

Klein, R. 1975. First entry of man into the New World. *Quaternary Research* 5, 391–394.

Klein, R.G. 1986. Carnivore size and Quaternary climatic change in southern Africa. *Quaternary Research* 26, 153–170.

Klein, R.G. 1989. *The Human Career. Human Biological and Cultural Origins*. University of Chicago Press, Chicago.

Knox, J.C. 1984. Responses of river systems to Holocene climates. In Wright, H.E. (ed.), *Late Quaternary Environments of the United States. The Holocene*. Vol. 2. Longman, London, 26–41.

Koerner, R.M. and Fisher, D.A. 1990. A record of Holocene summer climate from a Canadian high-Arctic ice core. *Nature* 343, 630–631.

Kohl, H. 1986. Pleistocene glaciations in Austria. *Quaternary Science Reviews* 5, 421–427.

Kolber, Z.S., Barber, R.T., Coale, K.H., Fitzwater, S.E., Greene, R.M., Johnson, K.S., Lindley, S. and Falkowski, P.G. 1994. Iron limitation of phytoplankton photosynthesis in the equatorial Pacific Ocean. *Nature* 371, 145–149.

Kominz, M.A. and Pisias, N.G. 1979. Pleistocene climate: deterministic or stochastic? *Science* 204, 171–173.

Koster, E.A. 1995. The response of permafrost ecosystems to climate change. In Zwerver, S., van Rompaey, R.S.A.R., Kok, M.T.J. and Berk, M.M. (eds), *Climate Change Research: Evaluation and Policy Implications*. Elsevier, Amsterdam, 385–388.

Kramere, A. 1994. A critical analysis of claims for the existence of Southeast Asian australopithecines. *Journal of Human Evolution* 26, 3–21.

Kröpelin, S. 1993. Environmental change in the southeastern Sahara and the proposal of a Geo-Biosphere Reserve in the Wadi Howar area (NW Sudan). In Thorweihe, U. and Schandelmeier, H.(eds), *Geoscientific Research in Northeast Africa*. Proceedings of the International Conference on Geoscientific Research in Northeast Africa, Berlin, Germany, 17–19 June 1993. Balkema, Rotterdam, 561–568.

Ku, T.-L., Kimmel, M.A., Easton, W.H. and O'Neil, T.J. 1974. Eustatic sea level 120,000 years ago on Oahu, Hawaii. *Science* 183, 959–962.

Kukla, G. 1989. Long continental records of climate – an introduction. *Palaeogeography, Palaeoclimatology, Palaeoecology* 72, 1–9.

Kukla, G. and An, Z. 1989. Loess stratigraphy in central China. *Palaeogeography, Palaeoclimatology, Palaeoecology* 72, 203–225.

Kukla, G., Berger, A., Lotti, R. and Brown, J. 1981. Orbital signature of interglacials. *Nature* 290, 295–300.

Kukla, G. and Cilek, V. 1996. Plio-Pleistocene megacycles: record of climate and tectonics. *Palaeogeography, Palaeoclimatology, Palaeoecology* 120, 171–194.

Kukla, G. and Gavin, J. 1992. Insolation regimes of the warm to cold transitions. In Kukla, G.J. and Went, E. (eds), *Start of a Glacial*. NATO ISI Series 13. Springer, Berlin, 307–339.

Kutzbach, J.E. 1981. Monsoon climate of the early Holocene: climate experiment with the Earth's orbital parameters for 9000 years ago. *Science* 214, 59–61.

Kutzbach, J., Bonan, G., Foley, J. and Harrison, S.P. 1996. Vegetation and soil feedbacks on the response of the African monsoon to orbital forcing in the early to middle Holocene. *Nature* 384, 623–626.

Kutzbach, J.E. and Guetter, P.J. 1986. The influence of changing orbital parameters and surface boundary conditions on climate simulations for the past 18,000 years. *Journal of the Atmospheric Sciences* 43, 1726–1759.

Kutzbach, J.E., Prell, W.L. and Ruddiman, W.F. 1993. Sensitivity of Eurasian climate to surface uplift of the Tibetan Plateau. *Journal of Geology* 101, 177–190.

Kutzbach, J.E. and Street-Perrott, F.A. 1985. Milankovitch forcing of fluctuations in the level of tropical lakes from 18 to 0 kyr BP. *Nature* 317, 130–134.

Kutzbach, J.E. and Wright, H.E. 1985. Simulation of the climate of 18,000 years BP. Results for the North American/North Atlantic/European sector and comparison with the geologic record of North America. *Quaternary Science Reviews* 4, 147–187.

Labeyrie, L.D., Duplessy, J.C. and Blanc, P.L. 1987. Variations in mode of formation and temperature of oceanic deep waters over the past 125,000 years. *Nature* 327, 477–482.

Lamb, H.H. 1972. *Climate. Present, Past and Future. Vol. 1. Fundamentals and Climate Now*. Methuen, London.

Lamb, H.H. 1995. *Climate, History and the Modern World* (2nd edn). Routledge, London.

Lambeck, K. 1990. Late Pleistocene, Holocene and present sea-levels: constraints of future change. *Palaeogeography, Palaeoclimatology, Palaeoecology, Global and Planetary Change Section* 89, 205–217.

Lambeck, K. 1993a: Glacial rebound of the British Isles – I. Preliminary model results. *Geophysical Journal International* 115, 941–959.

Lambeck, K. 1993b: Glacial rebound of the British Isles – II. A high-resolution, high-precision model. *Geophysical Journal International* 115, 960–990.

Lambeck, K. 1995a: Late Devensian and Holocene shorelines of the British Isles and North Sea from models of glacio-hydro-isostatic rebound. *Journal of the Geological Society* 152, 437–448.

Lambeck, K. 1995b: Constraints on the Late Weichselian ice sheet over the Barents Sea from observation of raised shorelines. *Quaternary Science Reviews* 14, 1–16.

Lambeck, K. 1996. Shoreline reconstructions from the Persian Gulf since the last glacial maximum. *Earth and Planetary Science Letters* 142, 43–57.

Lambeck, K. and Nakada, M. 1990. Late Pleistocene and Holocene sea-level change along the Australian coast. *Palaeogeography, Palaeoclimatology, Palaeoecology, Global and Planetary Change Section* 89, 143–176.

Lambert, M.R.K. 1984. Amphibians and reptiles. In Cloudsley-Thompson, J.L. (ed.), *Sahara Desert*. Pergamon, Oxford, 205–227.

Lamplugh, G.W. 1902. Calcrete. *Geological Magazine* 9, 575.

Lancaster, N. 1995. *Geomorphology of Desert Dunes*. Routledge, London.

Laporte, L.F. and Zihlman, A.L. 1983. Plates, climate and hominoid evolution. *South African Journal of Science* 79, 96–110.

Larsen, E., Sejrup, H.P., Johnsen, S.J. and Knudsen, K.L. 1995. Do Greenland ice cores reflect NW European interglacial climate variations? *Quaternary Research* 43, 125–132.

Larsen, H.C., Saunders, A.D., Clift, P.D., Beget, J., Wei, W. and Spezzaferri, S. 1994. Seven million years of glaciation in Greenland. *Science* 264, 952–955.

Lautenschlager, M. and Herterich, K. 1990. Atmospheric response to ice age conditions: climatology near the earth's surface. *Journal of Geophysical Research* 95(D13), 22547–22557.

Lea, D.W., Shen, G.T. and Boyle, E.A. 1989. Coralline barium records temporal variability in equatorial Pacific upwelling. *Nature* 340, 373–376.

Leakey, M.D. and Hay, R.L. 1979. Pliocene footprints in the Laetoli Beds at Laetoli, Northern Tanzania. *Nature* 278, 317–323.

Leakey, M.D. and Harris, J.M. (eds) 1987. *Laetoli. A Pliocene Site in Northern Tanzania*. Clarendon Press, Oxford.

Leakey, M.G., Feibel, C.S., McDougall, I. and Walker, A. 1995. New four-million-year-old hominid species from Kanapoi and Allia Bay, Kenya. *Nature* 376, 565–571.

Leakey, R. and Lewin, R. 1992. *Origins Reconsidered. In Search of What Makes us Human*. Little, Brown and Company, London.

Leakey, R. and Lewin, R. 1995. *The Sixth Extinction. Biodiversity and its Survival*. Weidenfeld and Nicolson, London.

Leavesley, G.H., Turner, K., D'Agnese, F.A. and McKnight, D. 1997. Regional delineation of North America for the assessment of freshwater ecosystems and climate change. *Hydrological Processes* 11, 819–824.

Lee, K.E. and Wood, T.G. 1971. *Termites and Soils*. Academic Press, London.

Legge, A.J. and Rowley-Conway, P.A. 1987. Gazelle killing in Stone Age Syria. *Scientific American* 257, 76–83.

Leggett, J. (ed.) 1990. *Global Warming: the Greenpeace Report*. Oxford University Press, Oxford.

Legrand, M. and de Angelis, M. 1996. Light carboxylic acids in Greenland ice: a record of past forest fires and vegetation emissions from the boreal zone. *Journal of Geophysical Research* 101(D2), 4129–4145.

Legrand, M.R., Delmas, R.J. and Charlston, R.J. 1988. Climate forcing implications from Vostok ice-core sulphate data. *Nature* 334, 418–420.

Lehman, S.J., Jones, G.A., Keigwin, L.D., Andersen, E.S., Butenko, G. and Ostmo, S.-R. 1991. Initiation of Fennoscandian ice-sheet retreat during the last deglaciation. *Nature* 349, 513–516.

Le Houérou, H.N. 1989. *The Grazing Land Ecosystems of the African Sahel*. Springer-Verlag, Berlin.

Leroy, S. and Dupont, L. 1994. Development of vegetation and continental aridity in northwestern Africa during the Late Pliocene: the pollen record of ODP Site 658. *Palaeogeography, Palaeoclimatology, Palaeoecology* 109, 295–316.

Leroy, S.A.G. and Dupont, L.M. 1997. Marine palynology of the ODP 658 (NW Africa) and its contribution to the stratigraphy of Late Pliocene. *Geobios* 30, 351–359.

Leroy Ladurie, E. 1972. *Times of Feast, Times of Famine. A History of Climate Since the Year 1000*. Allen and Unwin, London.

Le Treut, H. and Ghil, M. 1983. Orbital forcing, climatic interactions, and glaciation cycles. *Journal of Geophysical Research* 88(C9), 5167–5190.

Leuenberger, M. and Siegenthaler, U. 1992. Ice-age atmospheric concentration of nitrous oxide from an Antarctice ice core. *Nature* 360, 449–451.

Leuenberger, M., Siegenthaler, U. and Langway, C.C. 1992. Carbon isotope composition of atmospheric CO_2 during the last ice age from an Antarctic ice core. *Nature* 357, 488–490.

Leventer, A., Williams, D.F. and Kennett, J.P. 1982. Dynamics of the Laurentide ice sheet during the last glaciation: evidence from the Gulf of Mexico. *Earth and Planetary Science Letters* 59, 11–17.

Lézine, A.-M. 1991. West African paleoclimates during the last climatic cycle inferred from an Atlantic deep-sea pollen record. *Quaternary Research* 35, 456–463.

Lézine, A.M. and Casanova, J. 1991. Correlated oceanic and continental records demonstrate past climate and hydrology of north Africa (0–140 ka). *Geology* 19, 307–310.

Lézine, A.M. and Hooghiemstra, H. 1990. Land–sea comparisons during the last glacial–interglacial transition: pollen records from west tropical Africa. *Palaeogeography, Palaeoclimatology, Palaeoecology* 79, 313–331.

Lhote, H. 1959. *The Search for the Tassili Frescoes*. Hutchinson and Co, London.

Libby, W.F. 1950: *Radiocarbon dating* (1st edn). University of Chicago Press, Chicago.

Lindstrom, D.R. and MacAyeal, D.R. 1993. Death of an ice sheet. *Nature* 365, 214–215.

Linsley, B.K. and Thunell, R.C. 1990. The record of deglaciation in the Sulu Sea: evidence for the Younger Dryas event in the tropical western Pacific. *Paleoceanography* 5, 1025–1039.

Lister, G.S. 1988. Stable isotopes from lacustrine Ostracoda as tracers for continental palaeoenvironments. In De Deckker, P., Colin, J.P. and Peypouquet, J.P. (eds), *Ostracoda in the Earth Sciences*. Elsevier, Amsterdam, 201–218.

Littmann, T. 1989. Climatic changes in Africa during the last glacial: facts and problems. *Palaeoecology of Africa and the Surrounding Islands* 20, 163–179.

Liu, H.-S. 1992. Frequency variations of the Earth's obliquity and the 100–kyr ice-age cycles. *Nature* 358, 397–399.

Liu, T.S. (ed.) 1985. *Loess and the Environment*. China Ocean Press, Beijing.

Liu, T.S. (ed.) 1987. *Aspects of Loess Research*. China Ocean Press, Beijing.

Liu, T.S. (ed.) 1991. *Loess, Environment and Global Change*. Science Press, Beijing.

Liu, T.S, Ding, Z., Chen, M. and An, Z. 1989. The global surface energy system and the geological role of wind stress. *Quaternary International* 2, 43–54.

Liu, T.S., Ding, M. and Derbyshire, E. 1996. Gravel deposits on the margins of the Qinghai-Xizang Plateau, and their environmental significance. *Palaeogeography, Palaeoclimatology, Palaeoecology* 120, 159–170.

Liu, T.S., Fei, G.X., Sheng, A.Z. and Xiang, F.Y. 1981. The dust fall on Beijing, China, on April 18, 1980. *Geological Society of America Special Paper* 186, 149–158.

Liu, T.S. and Yuan, B.Y. 1987. Paleoclimatic cycles in northern China (Luochuan loess section and its environmental implications). In Liu, T.S. (ed.), *Aspects of Loess Research*. China Ocean Press, Beijing, 3–26.

Liu, X.M., Shaw, J., Liu, T. and Heller, F. 1993. Magnetic susceptibility of the Chinese loess–palaeosol sequence: environmental change and pedogenesis. *Journal of the Geological Society* 150, 583–588.

Liu, Z. 1983. Peking Man's cave yields new finds. *Geographical Magazine* 55, 297–300.

Livingstone, I. and Warren, A. 1996. *Aeolian Geomorphology: An Introduction*. Addison Wesley Longman, Harlow.

Löffler, E. 1972. Pleistocene glaciation in Papua and New Guinea. *Zeitschrift für Geomorphologie N.F. Supplementband* 13, 32–58.

Löffler, H. (ed.) 1987. Paleolimnology IV. *Developments in Hydrobiology* 37, 1–431.

Lord, M.L. 1991. Depositional record of a glacial-lake outburst: glacial Lake Souris, North Dakota. *Bulletin of the Geological Society of America* 103, 290–299.

Lorius, C.J. 1989. Polar ice cores and climate. In Berger, A., Schneider, S. and Duplessy, J.-C. (eds), *Climate and*

Geo-Sciences. A Challenge for Science and Society in the 21st Century. NATO ASI Series C: 285. Kluwer, Dordrecht, 77–103.

Lorius, C., Jouzel, J., Ritz, C., Merlivat, L., Barkov, N.I., Korotkevich, Y.S. and Kotlyakov, V.M. 1985. A 150,000–year climatic record from Antarctic ice. *Nature* 316, 591–596.

Lourens, L.J., Antonarakou, A., Hilgen, F.J., van Hoof, A.A.M., Vergnaud-Grazzini, C. and Zachariasse, W.J. 1996. Evaluation of the Plio-Pleistocene astronomical timescale. *Paleoceanography* 11, 391–413.

Lovelock, J.E. and Kump, L.R. 1994. Failure of climate regulation in a geophysiological model. *Nature* 369, 732–734.

Løvlie, R. 1989. Paleomagnetic stratigraphy: a correlation method. *Quaternary International* 1, 129–149.

Lowe, J.J. and Walker, M.J.C. 1984. *Reconstructing Quaternary Environments.* Longman, London.

Lundqvist, J. 1986. Late Weichselian glaciation and deglaciation in Scandinavia. *Quaternary Science Reviews* 5, 269–292.

Lyell, C. 1873. *Geological Evidences of the Antiquity of Man, with an Outline of Glacial and Post-Tertiary Geology and Remarks on the Origin of Species, with Special Reference to Man's First Appearance on Earth.* (4th edn). John Murray, London.

Maarleveld, G.C. 1976. Periglacial phenomena and the mean annual temperature during the last glacial time in the Netherlands. *Biuletyn Peryglacjalny* 26, 57–78.

Mabbutt, J.A. 1977. *Desert Landforms.* Australian National University Press, Canberra.

Mabbutt, J.A. 1985. Desertification of the world's rangelands. *Desertification Control Bulletin*, UNEP, Nairobi, 12, 1–11.

MacAyeal, D.R. 1992. Irregular oscillations of the West Antarctic ice sheet. *Nature* 359, 29–32.

MacAyeal, D.R. 1993a: A low-order model of the Heinrich event cycle. *Paleoceanography* 8, 767–773.

MacAyeal, D.R. 1993b: Binge/purge oscillations of the Laurentide ice sheet as a cause of the North Atlantic's Heinrich events. *Paleoceanography* 8, 775–784.

MacAyeal D.R. 1995. Challenging an ice-core paleothermometer. *Science* 270, 444–445.

MacDonald, G.J. 1990. Role of methane clathrates in past and future climates. *Climatic Change* 16, 247–281.

Mack, G.H. and James, W.C. 1994. Paleoclimate and the global distribution of paleosols. *Journal of Geology* 102, 360–366.

Magee, J.W., Bowler, J.M., Miller, G H. and Williams, D.L.G. 1995. Stratigraphy, sedimentology, chronology and palaeohydrology of Quaternary lacustrine deposits at Madigan Gulf, Lake Eyre, South Australia. *Palaeogeography, Palaeoclimatology, Palaeoecology* 113, 3–42.

Magny, M. 1993. Solar influences on Holocene climatic changes illustrated by correlations between past lake-level fluctuations and the atmospheric ^{14}C record. *Quaternary Research* 40, 1–9.

Maher, B.A. and Thompson, R. 1992. Paleoclimatic significance of the mineral magnetic record of the Chinese loess and paleosols. *Quaternary Research* 37, 155–170.

Maher, B.A. and Thompson, R. 1994. Pedogenesis and paleoclimate: interpretation of the magnetic susceptibil-ity record of Chinese loess–paleosol sequences. Comments and reply. *Geology* 22, 857–860.

Mainguet, M., Canon, L. and Chemin, M.C. 1980. Le Sahara: géomorphologie et paléogeomorphologie éoliennes. In Williams, M.A.J. and Faure, H. (eds), *The Sahara and the Nile*. Balkema, Rotterdam, 17–35.

Mainguet, M. and Cossus, L. 1980. Sand circulation in the Sahara: geomorphological relations between the Sahara desert and its margins. *Palaeoecology of Africa* 12, 69–78.

Maley, J. 1980. Les changements climatiques de la fin du Tertiaire en Afrique: leur conséquence sur l'apparition du Sahara et de sa végétation. In Williams, M.A.J. and Faure, H. (eds), *The Sahara and the Nile*. Balkema, Rotterdam, 63–86.

Maley, J. 1982. Dust, clouds, rain types, and climatic variations in tropical north Africa. *Quaternary Research* 18, 1–16.

Maley, J. 1996. The African rain forest – main characteristics of changes in vegetation and climate from the Upper Cretaceous to the Quaternary. *Proceedings of the Royal Society of Edinburgh* 104B, 31–73.

Maley, J. 1997. Middle to late Holocene changes in tropical Africa and other continents: paleomonsoon and sea surface temperature variations In Nüzhet Dalfes, H., Kukla, G. and Weiss, H. (eds), *Third Millenium BC Climate Change and Old World Collapses*. Springer-Verlag, Berlin, 611–640.

Manabe, S. and Broccoli, A.J. 1985. The influence of continental ice sheets on the climate of an ice age. *Journal of Geophysical Research* 90, 2167–2190.

Manabe, S. and Stouffer, R.J. 1993. Century-scale effects of increased atmospheric CO_2 on the ocean–atmosphere system. *Nature* 364, 215–218.

Manabe, S. and Stouffer, R.J. 1995. Simulation of abrupt climate change induced by freshwater input to the North Atlantic ocean. *Nature* 378, 165–167.

Mangerud, J., Jansen, E. and Landvik, J.Y. 1996. Late Cenozoic history of the Scandinavian and Barents Sea ice sheets. *Global and Planetary Change* 12, 11–26.

Mangerud, J., Sønstegaard, E. and Sejrup, H.-P. 1979. Correlation of the Eemian (interglacial) stage and the deep sea oxygen-isotope stratigraphy. *Nature* 277, 189–192.

Mangini, A., Eisenhauer, A. and Walter, P. 1991. A spike of CO_2 in the atmosphere at glacial–interglacial boundaries induced by rapid deposition of manganese in the oceans. *Tellus* 43B, 97–105.

Mann, D.H. and Peteet, D.M. 1994. Extent and timing of the last glacial maximum in southwestern Alaska. *Quaternary Research* 42, 136–148.

Mannion, A.M. and Bowlby, S.R. (eds) 1992. *Environmental Issues in the 1990s*. Wiley, Chichester.

de la Mare, W.K. 1997. Abrupt mid-twentieth-century decline in Antarctic sea-ice extent from whaling records. *Nature* 389, 57–60.

Marsh, N.D. and Ditlevsen, P.D. 1997. Climate during glaciation and deglaciation identified through chemical tracers in ice-cores. *Geophysical Research Letters* 24, 1319–1322.

Marsiat, I.M. and Berger, A. 1990. On the relationship between ice volume and sea level over the last glacial cycle. *Climate Dynamics* 4, 81–84.

Martin, J.H. and Fitzwater, S.E. 1988. Iron deficiency limits phytoplankton growth in the north-east Pacific sub-arctic. *Nature* 331, 341–343.

Martin, P.S. 1967. Prehistoric overkill. In Martin, P.S. and Wright, H.E. (eds), *Pleistocene Extinctions: the Search for a Cause.* Yale University Press, New Haven, 75–120.

Martin, P.S. 1973. The discovery of America. *Science* 179, 969–974.

Martin, P.S. 1984. Prehistoric overkill: the global model. In Martin, P.S. and Klein, R.G. (eds), *Quaternary Extinctions: A Prehistoric Revolution.* University of Arizona Press, Tucson, 354–403.

Martin, P.S. and Klein, R. (eds) 1984. *Quaternary Extinctions: A Prehistoric Revolution.* University of Arizona Press, Tuscon.

Martin, P.S. and Wright, H.E. (eds) 1967. *Pleistocene Extinctions: the Search for a Cause.* Yale University Press, New Haven.

Martinelli, L.A., Devol, A.H., Victoria, R.L. and Ritchey, J.E. 1991. Stable carbon isotope variation in C_3 and C_4 plants along the Amazon River. *Nature* 353, 55–61.

Martinson, D.G. 1990. Evolution of the Southern Ocean winter mixed layer and sea ice: open ocean deepwater formation and ventilation. *Journal of Geophysical Research* 95 (C7), 641–654.

Martinson, D.G., Pisias, N.G., Hays, J.D., Imbrie, J., Moore, T.C. Jr and Shackleton, N.J. 1987. Age dating and the orbital theory of the ice ages: development of a high resolution 0 to 300,000 year chronology. *Quaternary Research* 21, 1–29.

Mason, B.J. 1976. Towards the understanding and prediction of climatic variations. *Quarterly Journal of the Royal Meteorological Society* 102, 473–498.

May, R.M. 1978. Human reproduction reconsidered. *Nature* 272, 491–495.

McAndrews, J.H. and Boyko-Diakonow 1989. Pollen analysis of varved sediments at Crawford Lake, Ontario: evidence of Indian and European farming. In Fulton, R.J. (ed.), *Quaternary Geology of Canada and Greenland. Geology of Canada.* No. 1. Canadian Government Publishing Centre, Ottawa, 528–530.

McCarthy, L., Head, L. and Quade, J. 1996. Holocene palaeoecology of the northern Flinders Ranges, South Australia, based on stick-nest rat (*Lepotillus* spp.) middens: a preliminary overview. *Palaeogeography, Palaeoclimatology, Palaeoecology* 123, 205–218.

McClure, H.A. 1976. Radiocarbon chronology of late Quaternary lakes in the Arabian desert. *Nature* 263, 755–756.

McCulloch, M.T., Gagan, M.K., Mortimer, G.E., Chivas, A.R. and Isdale, P.J. 1994. A high-resolution Sr/Ca and $d^{18}O$ coral record from the Great Barrier Reef, Australia, and the 1982–1983 El Niño. *Geochimica Cosmochimica Acta* 58, 2747–2754.

McCulloch, M., Mortimer, G., Esat, T., Xianhua, L., Pillans, B. and Chappell, J. 1996. High resolution windows into early Holocene climate: Sr/Ca coral records from the Huon Peninsula. *Earth and Planetary Science Letters* 138, 169–178.

McDonough, K.J. and Cross, T.A. 1991. Late Cretaceous sea level from a paleoshoreline. *Journal of Geophysical Research* 96(B4), 6591–6607.

McDougall, I. and Harrison, T.M. 1988. *Geochronology and Thermochronology by the $^{40}Ar/^{39}Ar$ Method.* Oxford Monographs on Geology and Geophysics 9. Oxford University Press, Oxford.

McEwen-Mason, J.R.C. 1991. The late Cainozoic magnetostratigraphy and preliminary palynology of Lake George, New South Wales. In Williams, M.A.J., De Deckker, P. and Kershaw, A.P. (eds), *The Cainozoic in Australia: A Re-appraisal of the Evidence.* Geological Society of Australia Special Publication No. 18, 195–209.

McIntyre, A. 1981. Seasonal reconstructions of the Earth's surface at the last Glacial Maximum by CLIMAP Project Members. *Geological Society of America Map and Chart Series* MC-36.

McIntyre, A. 1989. Surface water response of the equatorial Atlantic Ocean to orbital forcing. *Paleoceanography* 4, 19–55.

McIntyre, A. and Kipp, N.G., with Bé, A.W.H., Crowley, T., Kellogg, T., Gardner, J.V., Prell, W. and Ruddiman, W.F. 1976. Glacial North Atlantic 18,000 years ago: a CLIMAP reconstruction. *Geological Society of America Memoir* 145, 43–75.

McIntyre, A. and Molfino, B. 1996. Forcing of Atlantic equatorial and subpolar millenial cycles by precession. *Science* 274, 1867–1870.

McIntyre, A., Ruddiman, W.F., Karlin, K. and Mix, A.C. 1989. Surface water response of the equatorial Atlantic Ocean to orbital forcing. *Paleoceanography* 4, 19–55.

McIntyre, M.L. and Hope, J.H. 1978. Procoptodon fossils from the Willandra Lakes, western New South Wales. *The Artefact* 3, 117–132.

McManus, J.F., Bond, G.C., Broecker, W.S., Johnsen, S., Labeyrie, L. and Higgins, S. 1994. High-resolution climate records from the North Atlantic during the last interglacial. *Nature* 371, 326–329.

McMichael, A.J. 1993. *Planetary Overload. Global Environmental Change and the Health of the Human Species.* Cambridge University Press, Cambridge.

McMichael, A.J., Haines, A., Slooff, R. and Kovats, S. (eds) 1996. *Climate Change and Human Health.* World Health Organization, Geneva.

McTainsh, G. 1980. Harmattan dust deposition in northern Nigeria. *Nature* 286, 587–588.

McTainsh, G. 1985. Dust processes in Australia and West Africa: a comparison. *Search* 16, 104–106.

McTainsh, G.H. and Lynch, A.W. 1996. Quantitative estimates of the effect of climate change on dust storm activity in Australia during the Last Glacial Maximum. *Geomorphology* 17, 263–271.

Meese, D.A., Gow, A.J., Grootes, P., Mayewski, P.A., Ram, M., Stuiver, M., Taylor, K.C., Waddington, E.D. and Zielinski, G.A. 1994. The accumulation record from the GISP2 core as an indicator of climate change throughout the Holocene. *Science* 266, 1680–1682.

Meier, M.F. 1984. Contribution of small glaciers to global sea level. *Science* 226, 1418–1421.

Melillo, J.M., Callaghan, T.V., Woodward, F.I., Salati, E. and Sinha, S.K. 1990. Effects on ecosystems. In Houghton, J.T., Jenkins, G.J. and Ephraums, J.J. (eds), *Climatic Change: The IPCC Assessment.* Cambridge University Press, Cambridge, for the Intergovernmental Panel on Climatic Change, 283–310.

Mellars, P.A. 1989. Major issues in the emergence of modern humans. *Current Anthropology* 30, 349–385.

Mellars, P.A. 1992. Archaeology and the population-dispersal hypothesis of modern human origins in Europe. *Philosophical Transactions of the Royal Society of London* 337, 225–235.

Mercer, J.H. 1978. West Antarctic ice sheet and CO_2 greenhouse effect: a threat of disaster. *Nature* 271, 321–325.

Mercier, N., Valladhas, H., Joron, J.-L., Reyss, J.-L., Léveque, F. and Vandermeersch, B. 1991. Thermoluminescence dating of the late Neanderthal remains from Saint-Césaire. *Nature* 351, 737–739.

Meriläinen, J., Huttunen, P. and Battarbee, R.W. (eds) 1983. Paleolimnology. *Developments in Hydrobiology* 15, 1–318.

Merrilees, D. 1968. Man the destroyer: late Quaternary changes in the Australian marsupial fauna. *Journal of the Royal Society of Western Australia* 51, 1–24.

Merritts, D.J., Vincent, K.R. and Wohl, E.E. 1994. Long river profiles, tectonism, and eustasy: a guide to interpreting fluvial terraces. *Journal of Geophysical Research* 99(B7), 14031–14050.

Mesolella, K.J., Matthews, R.K., Broecker, W.S. and Thurber, D.L. 1969. The astronomical theory of climatic change: Barbados data. *Journal of Geology* 77, 250–274.

Messerli, B., Winiger, M. and Rognon, P. 1980. The Saharan and East African uplands during the Quaternary. In Williams, M.A.J. and Faure, H. (eds), *The Sahara and the Nile*. Balkema, Rotterdam, 87–132.

Middleton, N. 1995. *The Global Casino. An Introduction to Environmental Issues*. Edward Arnold, London.

Mikolajewicz, U., Maier-Reimer, E., Crowley, T.J. and Kim, K.-Y. 1993. Effects of Drake and Panamanian gateways on the circulation of an ocean model. *Paleoceanography* 8, 409–426.

Mikolajewicz, U., Santer, B.D. and Maier-Reimer, E. 1990. Ocean response to greenhouse warming. *Nature* 345, 589–593.

Milankovitch, M. 1941. *Canon of Insolation and the Ice-Age Problem*. Royal Serbian Academy Special Pub. No. 132. (Translated from the German by the Israel Program for Scientific Translations, Jerusalem, 1969.)

Miller, G.H. and de Vernal, A. 1992. Will greenhouse warming lead to Northern hemisphere ice-sheet growth? *Nature* 355, 244–246.

Miller, G.H. and Kaufman, D.S. 1991. Ice-sheet/ocean interaction at the mouth of Hudson Strait, Canada, as a trigger for Younger Dryas cooling. *Norsk Geologisk Tidsskrift* 71, 149–151.

Miller, G.H., Magee, J.W. and Jull, A.J.T. 1997. Low-latitude glacial cooling in the southern hemisphere from amino-acid racemisation in emu eggshells. *Nature* 385, 241–244.

Miller, G.H. and Mangerud, J. 1986. Aminostratigraphy of European marine interglacial deposits. *Quaternary Science Reviews* 4, 215–278.

Milliman, J.D. 1997. Blessed dams or damned dams? *Nature* 386, 325–327.

Milliman, J.D. and Haq, B.U. (eds) 1996. *Sea-Level Rise and Coastal Subsidence. Causes, Consequences, and Strategies*. Kluwer, Dordrecht.

Mitchell, J.F.B., Grahame, N.S. and Needham, K.J. 1988. Climate simulations for 9000 years before present: seasonal variations and effect of Laurentide ice sheet. *Journal of Geophysical Research* 93(D7), 8283–8303.

Mitrovica, J.X. and Peltier, W.R. 1991. On postglacial geoid subsidence over the equatorial oceans. *Journal of Geophysical Research* 96(B12), 20053–20071.

Mix, A.C. 1987. The oxygen-isotope record of glaciation. In Ruddiman, W.F. and Wright H.E. Jr (eds), *North America and Adjacent Oceans during the Last Deglaciation*. Geological Society of America, The Geology of North America Vol. K-3. Boulder, Colorado, 111–135.

Mix, A.C. and Ruddiman, W.F. 1985. Structure and timing of the last deglaciation: oxygen-isotope evidence. *Quaternary Science Reviews* 4, 59–108.

Molles, M.C. Jr and Dahm, C.N. 1990. A perspective on El Niño and La Niña: Global implications for stream ecology. *Journal of the North American Benthological Society* 9, 68–76.

Molnar, P. and England, P. 1990. Late Cenozoic uplift of mountain ranges and global climate change: chicken or egg? *Nature* 346, 29–34.

Monod, T. 1963. The Late Tertiary and Pleistocene in the Sahara. In Howell, F.C. and Bourlière, F. (eds), *African Ecology and Human Evolution*. Aldine, Chicago, 117–229.

Moore, P.D., Chaloner, B. and Stott, P. 1996. *Global Environmental Change*. Blackwell, Oxford.

Moore, T.C., Pisias, N.G. and Dunn, D.A. 1982. Carbonate time series of the Quaternary and late Miocene sediments in the Pacific Ocean: a spectral comparison. *Marine Geology* 46, 217–233.

Moore, W.S. 1996. Large groudwater inputs to coastal waters revealed by 226Ra enrichments. *Nature* 380, 612–614.

Morales, C. (ed.) 1979. *Saharan Dust – Mobilization, Transport, Deposition*. Wiley, New York.

Morell, V. 1995. *Ancestral passions. The Leakey Family and the Quest for Humankind's Beginnings*. Simon and Schuster, New York.

Moritz, C., Joseph, L., Cunningham, M. and Schneider, C. 1997. Molecular perspectives on historical fragmentation of Australian tropical and subtropical rainforests: implications for conservation. In Lawrence, W.F. and Bierregaard, R.O. Jr (eds), *Tropical Rainforest Remnants*. University of Chicago Press, Chicago, 57–59.

Morley, J.J. and Heusser, L.E. 1997. Role of orbital forcing in east Asian monsoon climates during the last 350 kyr: evidence from terrestrial and marine climate proxies from core RC14–99. *Paleoceanography* 12, 483–493.

Mörner, N.-A. 1976. Eustasy and geoid changes. *Journal of Geology* 84, 123–151.

Mörner, N.-A. 1978. Low sea levels, droughts and mammalian extinctions. *Nature* 271, 738–739.

Mörner, N.-A. 1980a. Eustasy and geoid changes as a function of core/mantle changes. In Mörner, N.-A. (ed.), *Earth Rheology, Isostasy and Eustasy*. Chichester, Wiley, 535–553.

Mörner, N.-A. 1980b. The Fennoscandian uplift: geological data and their geodynamical implication. In Mörner, N.-A. (ed.), *Earth Rheology, Isostasy and Eustasy*. Wiley, Chichester: 251–284.

Mörner, N.-A. 1987. Pre-Quaternary long-term changes in sea level. In Devoy, R.J.N. (ed.), *Sea Surface Studies: a Global View.* Croom Helm, London, 233–241.

Mörner, N.-A. 1989. Global changes: the lithosphere: internal processes and earth's dynamicity in view of Quaternary observational data. *Quaternary International* 2, 341–352.

Mörner, N.-A. 1994. Internal response to orbital forcing and external cycle sedimentary sequences. *Special Publication of the International Association of Sedimentology* 19, 25–33.

Moyà-solà, S. and Köhler, M. 1996. A *Dryopithecus* skeleton and the origins of great-apes locomotion. *Nature* 379, 156–159.

Mulder T. and Syvitski J.P.M. 1996. Climatic and morphologic relationships of rivers: implications of sea-level fluctuations on river loads. *Journal of Geology* 104, 509–523.

Muller, R.A. and MacDonald, G.J. 1995. Glacial cycles and orbital inclination. *Nature* 377, 107–108.

Muller, R.A. and MacDonald, G.J. 1997a: Glacial cycles and astronomical forcing. *Science* 277, 215–218.

Muller, R.A. and MacDonald, G.J. 1997b: Simultaneous presence of orbital inclination and eccentricity in proxy records from Ocean Drilling Program site 806. *Geology* 25, 3–6.

Mulvaney, D.J. 1975. *The Prehistory of Australia.* Dominion Press, Victoria.

Mungall, C. and McLaren, D.J. (eds) 1991. *Planet Under Stress. The Challenge of Global Change.* Oxford University Press, Toronto.

Munhoven, G. and François, L.M. 1996. Glacial-interglacial variability of atmospheric CO_2 due to changing continental silicate rock weathering: a model study. *Journal of Geophysical Research* 101(D16), 21423–21437,

Munson, P.J. 1976. Archaeological data on the origins of cultivation in the south-western Sahara and its implications for West Africa. In Harlan, J.R., de Wet, J.M.J. and Stemler, A.B.L.(eds), *Origins of African Plant Domestication.* Mouton, The Hague, 187–210.

Murdock, T.Q., Weaver, A.J. and Fanning, A.F. 1997. Paleoclimatic response of the closing of the Isthmus of Panama in a coupled ocean–atmosphere model. *Geophysical Research Letters* 24, 253–256.

Murray, J.W., Barber, R.T., Roman, M.R., Bacon, M.P. and Feely, R.A. 1994. Physical and biological controls on carbon cycling in the equatorial Pacific. *Science* 266, 58–65.

Muzzolini, A. 1995. *Les Images Rupestres du Sahara.* Alfred Muzzolini, Toulouse.

Myneni, R.B., Keeling, C.D., Tucker, C.J., Asrar, G. and Nemani, R.R. 1997. Increased plant growth in the northern high latitudes from 1981 to 1991. *Nature* 386, 698–702.

Nakada, M. and Lambeck, K. 1988. The melting history of the late Pleistocene Antarctic ice sheet. *Nature* 333, 36–40.

Nakada, M. and Lambeck, K. 1989. Late Pleistocene and Holocene sea-level change in the Australian region and mantle rheology. *Geophysical Journal* 96, 497–517.

Nanson, G.C., Chen, X.Y. and Price, D.M. 1995. Aeolian and fluvial evidence of changing climate and wind patterns during the past 100 ka in the western Simpson Desert, Australia. *Palaeogeography, Palaeoclimatology, Palaeoecology* 113, 87–102.

Nanson, G.C., Price, D.M. and Short, S.A. 1992. Wetting and drying of Australia over the past 300 ka. *Geology* 20, 791–794.

National Academy of Sciences 1975. *Understanding Climatic Change: A Program for Action.* National Academy of Sciences, Washington, DC.

Neftel, A., Moor, E., Oeschger, H. and Stauffer, B. 1985. Evidence from polar ice cores for the increase in atmospheric CO_2 in the past two centuries. *Nature* 315, 45–57.

Neumann, A.C. and Moore, W.S. 1975. Sea level events and Pleistocene coral ages in the northern Bahamas. *Quaternary Research* 5, 215–224.

Newby, J.E. 1984. Large mammals. In Cloudsley-Thompson, J.L. (ed.), *Sahara Desert.* Pergamon, Oxford, 277–290.

Newell, R.E. 1974. Changes in the poleward energy flux by the atmosphere and ocean as a possible cause for ice ages. *Quaternary Research* 4, 117–127.

Newman, W.S. and Fairbridge, R.W. 1986. The management of sea level rise. *Nature* 320, 319–321.

Nicholls, N. 1997. Increased Australian wheat yield due to recent climate trends. *Nature* 387, 484–485.

Nicholson, S.E. 1989. African drought: characteristics, causal theories and global teleconnections. In Berger, A., Dickinson, R.E. and Kidson, J.W. (eds) *Understanding Climatic Change.* American Geophysical Union, Geophysical Monograph 52, 79–100.

Nicholson, S.E. and Flohn, H. 1980. African environmental and climatic changes and the general atmospheric circulation in Late Pleistocene and Holocene. *Climatic Change* 2, 313–348.

Nilsson, T. 1983. *The Pleistocene. Geology and Life in the Quaternary Ice Age.* Reidel, Dordrecht.

Nisbet, E.G. 1992. Sources of atmospheric CH_4 in early postglacial time. *Journal of Geophysical Research* 97(D12), 12859–12867.

North, G.R. and Coakley, J.A. 1979. Differences between seasonal and mean annual energy balance model calculations of climate and climate sensitivity. *Journal of the Atmospheric Sciences* 36, 1189–1204.

Nott, J.F., Price, D.M. and Bryant, E.A. 1996. A 30,000 year record of extreme floods in tropical Australia from relict plunge-pool deposits: implications for future climate change. *Geophysical Research Letters* 23, 379–382.

Nowak, C.L., Nowak, R.S., Tausch, R.J. and Wigand, P.E. 1994. Tree and shrub dynamics in northwestern great basin woodland and shrub steppe during the late-Pleistocene and Holocene. *American Journal of Botany* 81, 265–277.

Oakley, K.P. 1980. Relative dating of fossil hominids of Europe. *Bulletin of the British Museum Natural History Geology Series* 34, 1–63.

Oerlemans, J. 1980. Model experiments on the 100,000-year glacial cycle. *Nature* 287, 430–432.

Oerlemans, J. 1989. A projection of future sea level. *Climatic Change* 15, 151–174.

Oerlemans, J. 1991. The role of ice sheets in the Pleistocene climate. *Norsk Geologisk Tidsskrift* 71, 155–161.

Oerlemans, J. 1993a. Evaluating the role of climate cooling in iceberg production and the Heinrich events. *Nature* 364, 783–786.

Oerlemans, J. 1993b. Possible changes in the mass balance of the Greenland and Antarctic ice sheets and their effects on sea level. In Warrick, R.A., Barrow, E.M. and Wigley, T.M.L. (eds), *Climate and Sea Level Change: Observations, Projections and Implications*. Cambridge University Press, Cambridge, 144–161.

Oerlemans, J. and Van Der Veen, C.J. 1984. *Ice Sheets and Climate*. Reidel, Dordrecht.

Ohmura, A., Wild, M. and Bengtsson, L. 1996. A possible change in mass balance of Greenland and Antarctic ice sheets in the coming century. *Journal of Climate* 9, 2124–2135.

Oliver, J.S., Sikes, N.E. and Stewart, K.M. 1994. Introduction to early hominid behavioural ecology: new looks at old questions. *Journal of Human Evolution* 27, 1–5.

Olsen, P.E. 1986. A 40-million-year lake record of Early Mesozoic orbital climatic forcing. *Science* 234, 842–848.

Olsen, P.E. and Kent, D.V. 1996. Milankovitch climate forcing in the tropics of Pangaea during the late Triassic. *Paleogeography, Paleoclimatology, Paleoecology* 122, 1–26.

Open University Course Team 1989a. *The Ocean Basins: Their Structure and Evolution*. Pergamon Press, Oxford.

Open University Course Team 1989b. *Ocean Chemistry and Deep-Sea Sediments*. Pergamon Press, Oxford.

Open University Course Team 1989c. *Ocean Circulation*. Pergamon Press, Oxford.

Ostlund, H.G., Craig, H., Broecker, W.S. and Spencer, D. 1987. *GEOSECS Atlantic, Pacific, and Indian Ocean Expeditions, Vol. 7*. National Science Foundation, Washington DC.

Otto-Bliesner, B.L. 1996. Initiation of a continental ice sheet in a global climate model (GENESIS). *Journal of Geophysical Research* 101(D12), 16909–16920.

Otto-Bliesner, B.L. and Upchurch, G.R. 1997. Vegetation-induced warming of high-latitude regions during the Late Cretaceous period. *Nature* 385, 804–807.

Overpeck, J., Rind, D., Lacis, A. and Healy, R. 1996. Possible role of dust-induced regional warming in abrupt climate change during the last glacial period. *Nature* 384, 447–449.

Overpeck, J.T., Prentice, I.C. and Webb, T. III. 1985. Quantitative interpretation of fossil pollen spectra: dissimilarity coefficients and the method of modern analogues. *Quaternary Research* 23, 87–108.

Owen, R. 1870. On the fossil mammals of Australia – Part III. *Diprotodon australis. Philosophical Transactions of the Royal Society* 162, 241–258.

Pachur, H.-J. and Altmann, N. 1997. The Quaternary (Holocene, ca. 8000 BP). In Schandelmeier, H. and Reynolds, P.-O.(eds), *Palaeogeographic-Palaeotectonic Atlas of North-Eastern Africa, Arabia, and Adjacent Areas. Late Neoproterozoic to Holocene*. Balkema, Rotterdam, 111–125.

Paillard, D. and Labeyrie, L. 1994. Role of the thermohaline circulation in the abrupt warming after Heinrich events. *Nature* 372, 162–164.

Panagoulia, D. and Dimou, G. 1997. Sensitivity of flood events to global climate change. *Journal of Hydrology* 191, 208–222.

Park, J. 1992. Envelope estimation for quasi-periodic geophysical signals in noise: a multitaper approach. In Walden, A.T. and Guttorp, P. (eds), *Statistics in the Environmental and Earth Sciences*. Arnold, London, 189–219.

Parker, B.B. 1996. Sea level as an indicator of climate and global change. In Pirie, R.G. (ed.), *Oceanography: Contemporary Readings in Ocean Sciences*. Oxford University Press, New York, 115–129.

Parkin, D.W. 1974. Trade-winds during the glacial cycles. *Proceedings of the Royal Society of London* A337, 73–100.

Parkin, D. and Shackleton, N.J. 1973. Trade-winds and temperature correlations down a deep-sea core off the Saharan coast. *Nature* 245, 455–457.

Parmenter, C. and Folger, D.W. 1974. Eolian biogenic detritus in deep sea sediments: a possible index of equatorial Ice Age aridity. *Science* 185, 695–698.

Parry, M. and Duncan, R. (eds) 1995. *The Economic Implications of Climate Change in Britain*. Earthscan, London.

Parsons, B. and Sclater, J.G. 1977. An analysis of the variation of ocean floor bathymetry and heat flow with age. *Journal of Geophysical Research* 82, 803–827.

Partridge, T.C., Bond, G.C., Hartnady, C.J.H., DeMenocal, P.B. and Ruddiman, W.F. 1995a. Climatic effects of Late Neogene tectonism and volcanism. In Vrba, E.S., Denton, G.H., Partridge, T.C. and Burckle, L.H.(eds), *Paleoclimate and Evolution*. Yale University Press, New Haven, 8–23.

Partridge, T.C., Wood, B.A. and DeMenocal, P.B. 1995b. The influence of global climate change and regional uplift on large-mammalian evolution in east and southern Africa. In Vrba, E.S., Denton, G.H., Partridge, T.C. and Burckle, L.H. (eds), *Paleoclimate and Evolution*. Yale University Press, New Haven, 331–355.

Pastouret, L., Charmley, H., Delibrias, G., Duplessy, J.C. and Thiede, J. 1978. Late Quaternary climatic changes in western tropical Africa deduced from deep-sea sedimentation off the Niger Delta. *Oceanologica Acta* 1, 217–232.

Paterson, W.S.B. and Hammer, C.U. 1987. Ice core and other glaciological data. In Ruddiman W.F. and Wright H.E. Jr (eds), *North America and Adjacent Oceans During the Last Deglaciation*. Geological Society of America, The Geology of North America Vol. K-3, 91–109.

Paton, T.R. 1978. *The Formation of Soil Material*. George Allen and Unwin, London.

Paton, T.R., Humphreys, G.S. and Mitchell, P.B. 1995. *Soils: A New Global View*. UCL Press, London.

Patterson, W.A. III, Edwards, K.J. and Maguire, D.J. 1987. Microscopic charcoal as a fossil indicator of fire. *Quaternary Science Reviews* 6, 3–23.

Paytan, A., Kastner, M. and Chavez, F.P. 1996. Glacial to interglacial fluctuations in productivity in the equatorial Pacific as indicated by marine barite. *Science* 274, 1355–1357.

Pearce, R.H. and Barbetti, M. 1981. A 38,000 year-old site at Upper Swan, W.A. *Archaeology in Oceania* 16, 173–178.

Pearman, G. (ed.) 1988. *Greenhouse: Planning for Climate Change.* CSIRO Publications, Melbourne.

Peel, D.A. 1995. Profiles of the past. *Nature* 378, 234–235.

Peixoto, J.P. and Oort, A.H. 1992. *Physics of Climate.* American Institute of Physics, New York.

Peltier, W.R. 1982. Dynamics of the ice age earth. *Advances in Geophysics* 24, 1–144.

Peltier, W.R. 1987. Glacial isostasy, mantle viscosity and Pleistocene climatic change. In Ruddiman, W.F. and Wright, H.E. (eds), *North America and Adjacent Oceans During the Last Deglaciation.* Geological Society of America, The Geology of North America Vol. K-3, 155–182.

Peltier, W.R. 1988. Lithospheric thickness, Antarctic deglaciation history, and ocean basin discretization effects in a global model of postglacial sea level change: a summary of some sources of nonuniquenenss. *Quaternary Research* 29, 93–112.

Peltier, W.R. 1994. Ice age paleotopography. *Science* 265, 195–201.

Peltier, W.R. and Hyde, W. 1984. A model of the ice age cycle. In Berger, A., Imbrie, J., Hays, J., Kukla, G. and Saltzman, B. (eds), *Milankovitch and Climate. Understanding the Response to Astronomical Forcing.* NATO ISI Series C, Volume 126. Reidel, Dordrecht, 565–580.

Peltier, W.R. and Tushingham, A.M. 1989. Global sea level rise and the Greenhouse effect: might they be connected? *Science* 244, 806–810.

Peltier, W.R. and Tushingham, A.M. 1991. Influence of glacial isostatic adjustment on tide gauge measurements of secular sea level change. *Journal of Geophysical Research* 96(B4), 6779–6796.

Penck, A. and Brückner, E. 1909. *Die Alpen im Eiszeitalter.* Tauchnitz, Leipzig.

Peng, C.H., Guiot, J., van Campo, E. and Cheddadi, R. 1994. The vegetation carbon storage variation in Europe since 6000 BP: reconstruction from pollen. *Journal of Biogeography* 21, 19–31.

Pennington, W. 1986. Lags in adjustment of vegetation to climate caused by the pace of soil development: evidence from Britain. *Vegetatio* 67, 105–118.

Perlmutter, M.A. and Matthews, M.D. 1990. Global cyclostratigraphy – a model. In Cross, T.A. (ed.), *Quantitative Dynamic Stratigraphy.* Prentice-Hall, New Jersey, 233–260.

Pestiaux, P., van der Mersch, I. and Berger, A. 1988. Paleoclimatic variability at frequencies ranging from 1 cycle per 10000 years to 1 cycle per 1000 years: evidence for nonlinear behaviour of the climate system. *Climatic Change* 12, 9–37.

Petit, J.-R., Briat, M. and Royer, A. 1981. Ice age aerosol content from East Antarctic ice core samples and past wind strength. *Nature* 293, 391–394.

Petit, J.-R., Mounier, L., Jouzel, J., Korotkevich, Y.S., Kotlyakov, V.I. and Lorius, C. 1990. Palaeoclimatological and chronological implications of the Vostok core dust record. *Nature* 343, 56–58.

Petit-Maire, N. 1990a. Will greenhouse green the Sahara? *Episodes* 13, 103–107.

Petit-Maire, N. 1990b. Natural aridification or man-made desertification? A question for the future. In Paepe, R., Fairbridge, R.W. and Jelgersma, S. (eds), *Greenhouse Effect, Sea Level and Drought.* Kluwer Academic, Dordrecht, 281–285.

Petit-Maire, N. 1994. Natural variability of the Asian, Indian and African monsoons over the last 130 ka. In Desbois, M. and Désalmand, F. (eds), *Global Precipitations and Climate Change.* NATO ASI Series I, 26. Springer, Berlin, 3–26.

Petit-Maire, N., Fontugne, M. and Rouland, C. 1991. Atmospheric methane ratio and environmental changes in the Sahara and Sahel during the last 130 kyrs. *Palaeogeography, Palaeoclimatology, Palaeoecology* 86, 197–204.

Petit-Maire, N., Sanlaville, P. and Yan, Z.W. 1995. Oscillations de la limite nord du domaine des moussons africaine, indienne, et asiatique, au cours du dernier cycle climatique. *Bulletin Société Géologique de France* 166, 213–220.

Petschick, R., Kuhn, G. and Gingele, F. 1996. Clay mineral distribution in surface sediments of the South Atlantic: sources, transport, and relation to oceanography. *Marine Geology* 130, 203–229.

Petts, G.E., Möller, H. and Roux, A.L. (eds) 1989. *Historical Change of Large Alluvial Rivers – Western Europe.* Wiley, Chichester.

Péwé, T.L. 1981. Desert dust: an overview. *Geological Society of America Special Paper* 186, 1–10.

Péwé, T.L. 1983a: Alpine permafrost in the contiguous United States: a review. *Arctic and Alpine Research* 15, 145–156.

Péwé, T.L. 1983b: The periglacial environment in North America during Wisconsin time. In Porter, S.C. (ed.), *Late Quaternary Environments of the United States. Volume 1. The Late Pleistocene.* Longman, Essex, 157–189.

Pflaumann, U., Duprat, J., Pujol, C. and Labeyrie, L. 1996. SIMMAX: a modern analog technique to deduce Atlantic sea surface temperatures from planktonic foraminifera in deep-sea sediments. *Paleoceanography* 11, 15–35.

Phillipps, P.J. and Held, I.M. 1994. The response to orbital perturbations in an atmospheric model coupled to a slab ocean. *Journal of Climate* 7, 767–782.

Pilbeam, D. and Gould, S.J. 1974. Size and scaling in human evolution. *Science* 186, 892–901.

Pimentel, D., Houser, J., Preiss, E., White, O., Fang, H., Mesnick, L., Barsky, T., Tariche, S., Schreck, J. and Alpert, S. 1997. Water resources: agriculture, the environment, and society. *BioScience* 47, 97–106.

Piperno, D.R. 1987. *Phytolith Analysis: An Archaeological and Geological Perspective.* Academic Press, San Diego.

Pirazzoli, P.A. 1978. High stands of Holocene sea levels in the northwest Pacific. *Quaternary Research* 10, 1–29.

Pisias, N.G., Martinson, D.G., Moore, T.C., Shackleton, N.J., Prell, W., Hays, J. and Boden, G. 1984. High resolution stratigraphic correlation of benthic oxygen isotope records spanning the last 300,000 years. *Marine Geology* 56, 119–136.

Pisias, N.G. and Moore, T.C. 1981. The evolution of Pleistocene climate: a time series approach. *Earth and Planetary Science Letters* 52, 450–458.

Pisias, N.G. and Shackleton, N.J. 1984. Modelling the

global climate response to orbital forcing and atmospheric carbon dioxide changes. *Nature* 310, 757–759.

Pittock, A.B. and Nix, H.A. 1986. The effect of changing climate on Australian biomass production – a preliminary study. *Climatic Change* 8, 243–255.

Pokras, E.M. and Mix, A.C. 1985. Eolian evidence for spatial variability of late Quaternary climates in tropical Africa. *Quaternary Research* 24, 137–149.

Pokras, E.M. and Mix, A.C. 1987. Earth's precession cycle and Quaternary climatic change in tropical Africa. *Nature* 326, 486–487.

Pollard, D. 1982. A simple ice sheet model yields realistic 100 kyr glacial cycles. *Nature* 296, 334–338.

Pollard, D., Ingersoll, A.P. and Lockwood, J.G. 1980. Response of a zonal climate–ice sheet model to the orbital perturbations during the Quaternary ice ages. *Tellus* 32, 301–319.

Pollock, D.E. 1997. The role of diatoms, dissolved silicate and Antarctic glaciation in glacial/interglacial climatic change: a hypothesis. *Global and Planetary Change* 14, 113–125.

Polyak, L., Lehman, S.J., Gataullin, V. and Jull, A.J.T. 1995. Two-step deglaciation of the southeastern Barents Sea. *Geology* 23, 567–571.

Ponting, C. 1991. *A Green History of the World*. Sinclair-Stevenson, London.

Porter, S.C. and An, Z. 1995. Correlation between climate events in the North Atlantic and China during the last glaciation. *Nature* 375, 305–308.

Porter, S.C., Stuiver, M. and Heusser, C.J. 1984. Holocene sea-level changes along the Strait of Magellan and Beagle Channel, southernmost South America. *Quaternary Research* 22, 59–67.

Postel, S.L., Daily, G.C. and Ehrlich, P.R. 1996. Human appropriation of renewable fresh water. *Science* 271, 785–788.

Potts, R. and Shipman, P. 1981. Cutmarks made by stone tools on bones from Olduvai Gorge, Tanzania. *Nature* 291, 577–580.

Prell, W.L. 1984. Monsoonal climate of the Arabian Sea during the Late Quaternary: a response to changing solar radiation. In Berger, A.L. and Labeyrie, L. (eds), *Milankovitch and climate*. Riedel, Dordrecht, 349–366.

Prell, W. L. 1985. *The Stability of Low Latitude Sea Surface Temperatures: An Evaluation of the CLIMAP Reconstruction with Emphasis on the Positive SST Anomalies*. US Department of Energy, Report TR025, 1–60.

Prell, W.L. and Kutzbach, J.E. 1992. Sensitivity of the Indian monsoon to forcing parameters and implications for its evolution. *Nature* 360, 647–652.

Prell, W.L., Imbrie, J., Martinson, D.G., Morley, J.J., Pisias, N.G., Shackleton, N.J. and Streeter, H.F. 1986. Graphic correlation of oxygen isotope stratigraphy application to the late Quaternary. *Paleoceanography* 1, 137–162.

Prell, W.L., Hutson, W.H., Williams, D.F., Be, A.W.H., Geitzenauer, K. and Molfino, B. 1980. Surface circulation of the Indian Ocean during the last glacial maximum, approximately 18,000 yr B.P. *Quaternary Research* 14, 309–336.

Prentice, C., Cramer, W., Harrison, S.P., Leemans, R., Monserud, R.A. and Solomon, A.M. 1992. A global biome model based on plant physiology and dominance, soil properties and climate. *Journal of Biogeography* 19, 117–134.

Prentice, I.C., Guiot, J., Huntley, B., Jolly, D. and Cheddadi, R. 1996. Reconstructing biomes from palaeoecological data: a general method and its application to European pollen data at 0 and 6 ka. *Climate Dynamics* 12, 185–194.

Prentice, I.C., Sykes, M.T., Lautenschlager, M., Harrison, S.P., Denissenko, O. and Bartlein, P.J. 1993. Modelling global vegetation patterns and terrestrial carbon storage at the last glacial maximum. *Global Ecology and Biogeography Letters* 3, 67–76.

Prentice, I.C. and Fung, I.Y. 1990. The sensitivity of terrestrial carbon storage to climate change. *Nature* 346, 48–51.

Prentice, M.L. and Denton, G.H. 1988. The deep-sea oxygen isotope record, the global ice sheet system and hominid evolution. In Grine, F.E. (ed.), *Evolutionary History of the 'Robust' Australopithecines*. Aldine de Gruyter, New York, 383–403.

Prentice, M.L. and Matthews, R.K. 1991. Tertiary ice sheet dynamics: the snow gun hypothesis. *Journal of Geophysical Research* 96(B4), 6811–6827.

Press, F. and Siever, R. 1986. *Earth*. Freeman, San Francisco.

Prest, V.K. 1969. *Retreat of Wisconsin and Recent Ice in North America: Speculative Ice Marginal Positions During Recessions of the Last Ice Sheet Complex*. Map 1257A, Geological Survey of Canada.

Prestwich, J. 1860. On the occurrence of flint-implements, associated with remains of animals of extinct species in beds of a late geological period, in France at Amiens, and in England at Hoxne. *Philosophical Transactions of the Royal Society* 150, 277–317.

Pye, K. 1984. Loess. *Progress in Physical Geography* 8, 176–217.

Pye, K. 1987. *Aeolian Dust and Dust Deposits*. Academic Press, London.

Pye, K. and Tsoar, H. 1990. *Aeolian Sand and Sand Dunes*. Unwin Hyman, London.

Quade, J., Cater, J.M.L., Ojha, T.P., Adam, J. and Harrison, T.M. 1995. Late Miocene environmental change in Nepal and the northern Indian subcontinent: stable isotopic evidence from paleosols. *Bulletin of the Geological Society of America* 107, 1381–1397.

Quade, J., Cerling, T.E. and Bowman, J.R. 1989. Development of Asian monsoon revealed by marked ecological shift during the latest Miocene in northern Pakistan. *Nature* 342, 163–166.

Rahmstorf, S. 1994. Rapid climate transitions in a coupled ocean–atmosphere model. *Nature* 372, 82–85.

Rahmstorf, S. 1995. Bifurcation of the Atlantic thermohaline circulation in response to changes in the hydrological cycle. *Nature* 378, 145–149.

Ramanathan, V., Barkstrom, B.R. and Harrison, E.F. 1992. Climate and the Earth's radiation budget. In Levi, B.G., Hafemeister, D. and Scribner, R. (eds), *Global warming: Physics and facts*. American Institute of Physics, New York, 55–77.

Ramaswamy, V. 1992. Explosive start to last ice age. *Nature* 359, 14.

Rampino, M.R. and Self, S. 1992. Volcanic winter and accelerated glaciation following the Toba super-eruption. *Nature* 359, 50–52.

Ramstein, G., Fluteau, F., Besse, J. and Joussaume, S. 1997. Effect of orogeny, plate motion and land–sea distribution on Eurasian climate change over the past 30 million years. *Nature* 386, 788–795.

Ranov, V.A., Carbonell, E. and Rodríguez, X.P. 1995. Kuldara: earliest human occupation in central Asia in its Afro-Asian context. *Current Anthropology* 36, 337–346.

Raper, S.C.B., Wigley, T.M.L. and Warrick, R.A. 1996. Global sea-level rise: past and future. In Milliman, J.D and Haq, B.U. (eds), *Sea-Level Rise and Coastal Subsidence. Causes, Consequences, and Strategies.* Kluwer, Dordrecht, 11–45.

Rasmusson, E.M. 1987. Global climate change and variability: effects on drought and desertification in Africa. In Glantz, M.H. (ed.), *Drought and Hunger in Africa: Denying Famine a Future.* Cambridge University Press, Cambridge, 3–22.

Raval, A. and Ramanathan, V. 1989. Observational determination of the greenhouse effect. *Nature* 342, 758–761.

Raymo, M.E. 1992. Global climate change: a three million year perspective. In Kukla, G.J. and Went, E. (eds), *Start of a Glacial.* NATO ISI Series, Vol. 3. Springer, Berlin, 207–223.

Raymo, M.E. 1994. The initiation of northern hemisphere glaciation. *Annual Reviews of Earth and Planetary Sciences* 22, 353–383.

Raymo, M.E. 1997. The timing of major climate terminations. *Paleoceanography* 12, 577–585.

Raymo, M.E. and Ruddiman, W.F. 1992. Tectonic forcing of late Cenozoic climate. *Nature* 359, 117–122.

Raymo, M.E., Ruddiman, W.F. and Froelich, P.N. 1988. Influence of late Cenozoic mountain building on ocean geochemical cycles. *Geology* 16, 649–653.

Raynaud, D., Jouzel, J., Barnola, J.M., Chappellaz, J., Delams, R.J. and Lorius, C. 1993. The ice record of Greenhouse gases. *Science* 259, 926–934.

Reeh, N. 1989. Dynamic and climatic history of the Greenland ice sheet. In Fulton, R.J. (ed.), *Quaternary Geology of Canada and Greenland.* Geological Survey of Canada, Geology of Canada. No.1/Geology Society of America, Geology of North America K1. Canadian Government Publishing Centre, Ottawa, 793–822.

Retallack, G.J. 1990. *Soils of the Past. An Introduction to Paleopedology.* Unwin Hyman, Boston.

Revel, M., Sinko, J.A., Grousset, F.E., and Biscaye, P.E. 1996. Sr and Nd isotopes as tracers of North Atlantic lithic particles: paleoclimatic implications. *Paleoceanography* 11, 95–113.

Rice, R.A. and Vandermeer, J. 1990. Climate and the geography of agriculture. In Carroll, C.R., Vandermeer, J.H. and Rosset, P. (eds), *Agroecology.* McGraw-Hill, New York, 21–63.

Rind, D., Chiou, E.-W., Chu, W., Larsen, J., Oltmans, S., Lerner, J., McCormick, M.P. and McMaster, L. 1991. Positive water vapour feedback in climate models confirmed by satellite data. *Nature* 349, 500–503.

Rind, D. and Peeet, D. 1985. Terrestrial conditions at the last glacial maximum and CLIMAP sea surface temperature estimates: are they consistent? *Quaternary Research* 24, 1–22.

Rind, D., Peteet, D. and Kukla, G. 1989. Can Milankovitch orbital variations initiate the growth of ice sheets in a general circulation model? *Journal of Geophysical Research* 94(D10), 12851–12871.

Roberts, L. 1988. Is there life after climate change? *Science* 242, 1010–1012.

Roberts, N. 1989. *The Holocene: An Environmental History.* Blackwell, Oxford.

Roberts, N. (ed.) 1994. *The Changing Global Environment.* Blackwell Publishers, Cambridge, MA.

Roberts, R.G., Jones, R. and Smith, M.A. 1990. Thermoluminescence dating of a 50,000–year old human occupation site in northern Australia. *Nature* 345, 153–156

Robock, A. 1983. Global mean sea level: indicator of climate change? *Science* 219, 996.

Roche, H. 1980. *Premiers Outils Taillés d'Afrique.* Société d'Ethnographie, Paris.

Rodbell, D.T. 1993. The timing of the last deglaciation in Cordillera Oriental, northern Peru, based on glacial geology and lake sedimentology. *Geological Society of America Bulletin* 105, 923–934.

Rogers, A.R. 1995. Genetic evidence for a Pleistocene population explosion. *Evolution* 49, 608–615.

Rognon, P. 1967. *Le Massif de l'Atakor et ses Bordures (Sahara Central): Etude Géomorphologique.* CNRS and CRZA, Paris.

Rognon, P. 1987. Late Quaternary reconstruction for the Maghreb (North Africa). *Palaeogeography, Palaeoclimatology, Palaeoecology* 58, 11–34.

Rognon, P. 1989. *Biographie d'un Désert.* Plon, Paris.

Rognon, P. and Coudé-Gaussen, G. 1996. Paleoclimates off northwest Africa (28°–35°N) about 18,000 yr B.P. based on continental eolian deposits. *Quaternary Research* 46, 118–126.

Rognon, P. and Williams, M.A.J. 1977. Late Quaternary climatic change in Australia and North Africa: a preliminary interpretation. *Palaeogeography, Palaeoclimatology, Palaeoecology* 21, 285–327.

Rooth, C. 1982. Hydrology and ocean circulation. *Progress in Oceanography* 11, 131–149.

Roset, J.-P. 1984. The prehistoric rock paintings of the Sahara. *Endeavour* 8, 75–84.

Rossignol-Strick, M. 1983. African monsoons, an immediate climate response to orbital insolation. *Nature* 304, 46–49.

Rossignol-Strick, M., Nesteroff, W., Olive, P. and Vergnaud-Grazzini, C. 1982. After the deluge: Mediterranean stagnation and sapropel formation. *Nature* 295, 105–110.

Rotnicki, K. 1991. Retrodiction of palaeodischarges of meandering and sinuous alluvial rivers and its palaeohydroclimatic implications. In Starkel, L., Gregory, K.J. and Thornes, J.B. (eds), *Temperate Palaeohydrology: Fluvial Processes in the Temperate Zone during the Last 15000 Years.* Wiley, Chichester, 431–471.

Rousseau, D. and Wu, N. 1997. A new molluscan record of the monsoon variability over the past 130,000 yr in the Luochuan loess sequence, China. *Geology* 25, 275–278.

Royal, P.D. Delcourt, P.A. and Delcourt, H.R. 1991. Late Quaternary paleoecology and paleoenvironments of the central Mississippi alluvial valley. *Bulletin of the Geological Society of America* 103, 157–170.

Ruddiman, W.F. 1984. The last interglacial ocean. CLIMAP Project Members. *Quaternary Research* 21, 123–224.

Ruddiman, W.F. and Kutzbach, J.E. 1989. Forcing of late Cenozoic northern hemisphere climate by plateau uplift in southern Asia and the American west. *Journal of Geophysical Research* 94(D15), 18409–18427.

Ruddiman, W.F. and Kutzbach, J.E. 1991. Plateau uplift and climatic change. *Scientific American* 264, 42–50.

Ruddiman, W.F. and McIntyre, A. 1979. Warmth of the subpolar Atlantic Ocean during northern hemisphere ice-sheet growth. *Science* 204, 173–175.

Ruddiman, W.F. and McIntyre, A. 1981. The mode and mechanism of the last deglaciation: oceanic evidence. *Quaternary Research* 16, 125–134.

Ruddiman, W.F., McIntyre, A., Niebler-Hunt, V. and Durazzi, J.T. 1980. Oceanic evidence for the mechanism of rapid northern hemisphere glaciation. *Quaternary Research* 13, 33–64.

Ruddiman, W.F. and Raymo, M.E. 1988. Northern hemisphere climate regimes during the past 3 Ma: possible tectonic connections. *Philosophical Transactions of the Royal Society of London* B318, 411–430.

Ruddiman, W.F., Raymo, M. and McIntyre, A. 1986. Matuyama 41,000-year cycles: north Atlantic Ocean and northern hemisphere ice sheets. *Earth and Planetary Science Letters* 80, 117–129.

Ruddiman, W.F., Raymo, M.E., Martinson, D.G., Clement, B.M. and Backman, J. 1989. Pleistocene evolution: northern hemisphere ice sheets and north Atlantic Ocean. *Paleoceanography* 4, 353–412.

Ruff, C.B., Trinkaus, E. and Holliday, T.W. 1997. Body mass and encephalization in Pleistocene *Homo*. *Nature* 387, 173–176.

Rust, B.R. and Nanson, G.C. 1986. Contemporary and palaeochannel patterns and the Late Quaternary stratigraphy of Cooper Creek, southwest Queensland, Australia. *Earth Surface Processes and Landforms* 11, 581–590.

Rutter, N. and Ding, Z. 1993. Paleoclimates and monsoon variations interpreted from micromorphogenic features of the Baoji paleosols, China. *Quaternary Science Reviews* 12, 853–862.

Rutter, N., Ding, Z., Evans, M.E. and Liu, T.S. 1991. Baoji-type pedostratigraphic section, Loess Plateau, north-central China. *Quaternary Science Reviews* 10, 1–22.

Ryan, W.B.F. 1973. Geodynamic implications of the Messinian crisis of salinity. In Drooger, C.W. (ed.), *Messinian Events in the Mediterranean*. North Holland, Amsterdam, 26–38.

Sabadini, R., Yuen, D.A. and Boschi, E. 1982. Interaction of cryospheric forcings with rotational dynamics has consequences for ice ages. *Nature* 296, 338–341.

Sagan, C., Toon, O.B. and Pollack, J.B. 1979. Anthropogenic albedo changes and the Earth's climate. *Science* 206, 1363–1368.

Sageman, B.B., Rich, J., Arthur, M.A., Birchfield, G.E. and Dean, W.E. 1997. Evidence for Milankovitch periodicities in Cenomanian–Turonian lithologic and geochemical cycles, western interior USA. *Journal of Sedimentary Research* 67, 286–302.

Sakai, K. and Peltier, W.R. 1996. A multibasin reduced model of the global thermohaline circulation: paleoceanographic analyses of the origins of ice-age climate variability. *Journal of Geophysical Research* 101(C10), 22535–22562.

Saltzman, B. 1977. Global mass and energy requirements for glacial oscillations and their implications for mean ocean temperature oscillations. *Tellus* 29, 205–212.

Saltzman, B. 1983. Climatic systems analysis. *Advances in Geophysics* 25, 173–233.

Saltzman, B., Hansen, A.R. and Maasch, K.A. 1984. The late Quaternary glaciations as the response of a three-component feedback system to earth-orbital forcing. *Journal of the Atmospheric Sciences* 41, 3380–3389.

Saltzman, B. and Moritz, R.E. 1980. A time-dependent climatic feedback system involving sea-ice extent, ocean temperature, and CO_2. *Tellus* 32, 93–118.

Saltzman, B. and Sutera, A. 1984. A model of the internal feedback system involved in late Quaternary climatic variations. *Journal of the Atmospheric Sciences* 41, 736–745.

Saltzman, B. and Sutera, A. 1987. The mid-Quaternary climatic transition as the free response of a three-variable dynamical model. *Journal of the Atmospheric Sciences* 44, 236–241.

Saltzman, B., Sutera, A. and Hansen, A.R. 1982. A possible marine mechanism for internally generated long-period climate cycles. *Journal of the Atmospheric Sciences* 39, 2634–2637.

Saltzman, B. and Verbitsky, M. 1994. CO_2 and glacial cycles. *Nature* 367, 419.

Saltzman, B. and Vernekar, A.D. 1971. Note on the effect of earth orbital radiation variations on climate. *Journal of Geophysical Research* 76, 4195–4197.

Saltzman, B. and Vernekar, A.D. 1975. A solution for the northern hemisphere climatic zonation during a glacial maximum. *Quaternary Research* 5, 307–320.

Samuels, R. and Prasad, D.K. (eds) 1994. *Global Warming and the Built Environment*. Spon, London.

Sancetta, C. 1992. Primary production in the glacial North Atlantic and north Pacific Oceans. *Nature* 360, 249–251.

Sanlaville, P. 1995. Marge septentrionale du désert arabique: la zone sud-levantine au Pléistocène supérieur et à l'Holocène. *Mémoires Société Géologique de France* 167, 57–66.

Sanson, G.D., Riley, S.J. and Williams, M.A.J. 1980. A late Quaternary Procoptodon fossil from Lake George, New South Wales. *Search* 11, 39–40.

Sanyal, A., Hemming, N.G., Broecker, W.S., Lea, D.W., Spero, H.J. and Hanson, G.N. 1996. Oceanic pH control on the boron isotopic composition of foraminifera: evidence from culture experiments. *Paleoceanography* 11, 513–518.

Sarich, V.M. and Wilson, A.C. 1967. Immunological time scale for hominid evolution. *Science* 158, 1200.

Sarmiento, J.L. and Toggweiler, J.R. 1984. A new model for the role of the oceans in determining atmospheric PCO_2. *Nature* 308, 621–624.

Sarnthein, M. 1978. Sand deserts during glacial maximum and climatic optimum. *Nature* 272, 43–46.

Sarnthein, M., Tetzlaff, G., Koopmann, B., Wolter, K. and Pflaumann, U. 1981. Glacial and interglacial wind regimes over the eastern subtropical Atlantic and North-West Africa. *Nature* 293, 193–196.

Sarnthein, M., Thiede, J., Pflaumann, U., Erlenkeuser, H., Fütterer, D., Koopmann, B., Lange, H. and Seibold, E. 1982. Atmospheric and oceanic circulation patterns off North West Africa during the past 25 million years. In von Rad, U., Hinz, K., Sarnthein, M. and Seibold, E. (eds), *Geology of the Northwest African Continental Margin.* Springer-Verlag, Berlin, 545–604.

Sarnthein, M. and Tiedemann, R. 1990. Younger Dryas-style cooling events at glacial terminations I–VI at ODP site 658. associated benthic $\delta^{13}C$ anomalies constrain meltwater hypothesis. *Paleoceanography* 5, 1041–1055.

Savin, S.M. and Yeh, H.-W. 1981. Stable isotopes in ocean sediments. In Emiliani, C. (ed.), *The Sea, Vol. 7. The Oceanic Lithosphere.* Wiley, New York, 1521–1554.

Scarre, C., Bray, W., Cook, J., Daniel, G.E. and Sabloff, J.A. (eds) 1988. *Past Worlds: The Times Atlas of Archaeology.* Times Books, London.

Schenk, P.E. 1991. Events and sea-level changes on Gondwana's margin: the Meguma Zone (Cambrian to Devonian) of Nova Scotia, Canada. *Bulletin of the Geological Society of America* 103, 512–521.

Schick, K.D. and Toth, N. 1995. *Making Silent Stones Speak. Human Evolution and the Dawn of Technology.* Phoenix, London.

Schild, R. and Wendorf, F. 1989. The Late Pleistocene Nile in Wadi Kubbaniya. In Wendorf, F., Schild, R. and Close, A.E. (eds), *The Prehistory of Wadi Kubbaniya. Vol. 2. Stratigraphy, Paleoeconomy and Environment.* Southern Methodist University Press, Dallas, 15–100.

Schmitz, W. J. Jr 1995. On the interbasin-scale thermohaline circulation. *Reviews of Geophysics* 33, 151–173.

Schneider, S.H. 1989. *Global Warming: Are We Entering the Greenhouse Century?* Sierra Club, San Francisco.

Schneider, S.H. and Thompson, S.L. 1979. Ice ages and orbital variations: some simple theory and modelling. *Quaternary Research* 12, 188–203.

Schnitker, D. 1980. Global paleoceanography and its deepwater linkage to the Antarctic glaciation. *Earth Science Reviews* 16, 1–20.

Scholtz, C.A. and Rosendahl, B.R. 1988. Low lake stands in Lakes Malawi and Tanganyika, East Africa, delineated with multifold seismic data. *Science* 240, 1645–1648.

Schönwiese, C.D., Ullrich, R., Beck, F. and Rapp, J. 1994. Solar signals in global climatic change. *Climatic Change* 27, 259–281.

Schumm, S.A. 1977. *The Fluvial System.* Wiley, New York.

Schumm, S.A. and Winkley, B.R. (eds) 1994. *The Variability of Large Alluvial Rivers.* American Society of Civil Engineers, New York.

Schumm, S.A. and Parker, R.S. 1973. Implications of complex response of drainage systems for Quaternary alluvial stratigraphy. *Nature* 243, 99–100.

Schumm, S.A., Erskine, W.D. and Tilleard, J.W. 1996. Morphology, hydrology, and evolution of the anastomosing Ovens and King Rivers, Victoria, Australia. *Bulletin of the Geological Society of America* 108, 1212–1224.

Schwarcz, H.P. and Skoflek, I. 1982. New dates for the Tata, Hungary archaeological site. *Nature* 295, 590–591.

Schwarcz, H.P. and Grün, R. 1992. Electron spin resonance (ESR) dating of the origin of modern man. *Philosophical Transactions of the Royal Society of London* 337, 145–149.

Schwarzacher, W. 1993. *Cyclostratigraphy and the Milankovitch Theory.* Developments in Sedimentology No. 52, Elsevier, Amsterdam.

Schweingruber, F.H. 1988. *Tree Rings: Basics and Applications of Dendrochronology.* Reidel, Dordrecht.

Scott, L. and Bousman, C. 1990. Palynological analysis of hyrax middens from southern Africa. *Palaeogeography, Palaeoclimatology, Palaeoecology* 79, 367–379.

Sefton, C.E.M. and Boorman, D.B. 1997. A regional investigation of climate change impacts on UK streamflows. *Journal of Hydrology* 195, 26–44.

Seibold, E. and Berger, W.H. 1996. *The Sea Floor. An Introduction to Marine Geology.* Springer, Berlin.

Sellers, P.J., Dickinson, R.E., Randall, D.A., Betts, A.K., Hall, F.G., Berry, J.A., Collatz, G.J., Denning, A.S., Mooney, H.A., Nobre, C.A., Sato, N., Field, C.B. and Henderson-Sellers, A. 1997. Modelling the exchanges of energy, water, and carbon between continents and the atmosphere. *Science* 275, 502–509.

Sellers, W.D. 1970. The effect of changes in the earth's obliquity on the distribution of mean annual sea-level temperatures. *Journal of Applied Meteorology* 9, 960–961.

Semaw, S., Renne, P., Harris, J.W.K., Feibel, C.S., Bernor, R.L., Fesseha, N. and Mowbray, K. 1997. 2.5–million-year-old stone tools from Gona, Ethiopia. *Nature* 385, 333–336.

Servant, M. 1973. *Séquences Continentales et Variations Climatiques: Évolution du Bassin du Tchad au Cénozoique Supérieur.* DSc thesis, University of Paris.

Servant, M. and Servant-Vildary, S. 1980. L'environnement quaternaire du bassin du Tchad. In Williams, M.A.J. and Faure, H. (eds), *The Sahara and the Nile.* Balkema, Rotterdam, 133–162.

Shackleton, N. 1967. Oxygen isotope analyses and Pleistocene temperatures re-assessed. *Nature* 215, 15–17.

Shackleton, N.J. 1977. The oxygen isotope stratigraphic record of the late Pleistocene. *Philosophical Transactions of the Royal Society* 280, 169–179.

Shackleton, N.J. 1987. Oxygen isotopes, ice volume and sea level. *Quaternary Science Reviews* 6, 183–190.

Shackleton, N.J. 1993. The climate system in the recent geological past. *Philosophical Transactions of the Royal Society of London Series B* 341, 209–213.

Shackleton, N.J. 1995. New data on the evolution of Pliocene climatic variability. In Vrba, E.S., Denton, G.H., Partridge, T.C. and Burckle, L.H.(eds), *Paleoclimate and Evolution, with Emphasis on Human Origins.* Yale University Press, New Haven, 242–248.

Shackleton, N.J., Backman, J., Zimmerman, H., Kent, D.V., Hall, M.A., Roberts, D.G., Schnitker, D., Baldauf, J.G., Desprairies, A., Homrighausen, R., Huddlestun, P., Keene, J.B., Kaltenback, A.J., Krumsiek, K.A.O., Morton, A.C., Murray, J.W. and Westberg-Smith, J. 1984. Oxygen isotope calibration of the onset of ice-rafting and history of glaciation in the North Atlantic region. *Nature* 307, 620–623.

Shackleton, N.J., Berger, A. and Peltier, W.R. 1990. An alternative astronomical calibration of the lower Pleistocene time scale based on ODP site 677. *Transactions of the Royal Society of Edinburgh, Earth Sciences* 81, 251–261.

Shackleton, N.J. and Kennett, J.P. 1975. Paleotemperature history of the Cenozoic and the initiation of Antarctic glaciation: oxygen and carbon analyses of DSDP sites 277, 279, 281. *Initial Reports of the Deep Sea Drilling Project 29.* US Government Printing Office, Washington, DC, 743–755.

Shackleton, N.J. and Opdyke, N.D. 1973. Oxygen isotope and paleomagnetic stratigraphy of equatorial Pacific core V28–238: oxygen isotope temperatures and ice volumes on a 10^5 year and a 10^6 scale. *Quaternary Research* 3, 39–55.

Shackleton, N.J. and Opdyke, N.D. 1977. Oxygen isotope and palaeomagnetic evidence for early northern hemisphere glaciation. *Nature* 270, 216–219.

Shaomeng, L., Yulou, Q. and Walker, D. 1986. Late Pleistocene and Holocene vegetation history at Xi Hu, Er Yuan, Yunnan Province, southwest China. *Journal of Biogeography* 13, 419–440.

Sheets-Johnstone, M. 1989. Hominid bipedality and sexual-selection theory. *Evolutionary Theory* 9, 57–70.

Short, D.A. and Mengel, J.G. 1986. Tropical climate phase lags and Earth's precession cycle. *Nature* 323, 48–50.

Short, D.A., Mengel, J.G., Crowley, T.J., Hyde, W.T. and North, G.R. 1991. Filtering of Milankovitch cycles by Earth's geography. *Quaternary Research* 35, 157–173.

Shreeve, J. 1996. *The Neandertal Enigma. Solving the Mystery of Modern Human Origins.* The Softback Preview, New York.

Siegenthaler, U. and Oeschger, H. 1980. Correlation of ^{18}O in precipitation with temperature and altitude. *Nature* 285, 314–317.

Siegert, M.J. and Dowdeswell, J.A. 1995. Numerical modeling of the Late Weichselian Svalbard–Barents Sea ice sheet. *Quaternary Research* 43, 1–13.

Sillen, A., Hall, G. and Armstrong, R. 1995. Strontium calcium ratios (Sr/Ca) and strontium isotopic ratios ($^{87}Sr/^{86}Sr$) of *Australopithecus robustus* and *Homo* sp. from Swartkrans. *Journal of Human Evolution* 28, 277–285.

Simmons, I.G. 1989. *Changing the Face of the Earth: Culture, Environment, History.* Blackwell, Oxford.

Simmons, I.G. 1996. *Changing the Face of the Earth. Culture, Environment, History.* (2nd edn). Blackwell Publishers, Oxford.

Singh, G. 1988. History of arid land vegetation and climate: a global perspective. *Biological Review* 63, 159–195.

Singh, G. and Geissler, E.A. 1985. Late Cainozoic history of fire, lake levels and climate at Lake George, New South Wales, Australia. *Philosophical Transactions of the Royal Society of London* 311, 379–447.

Singh, G., Joshi, R.D., Chopra, S.K. and Singh, A.B. 1974. Late Quaternary history of vegetation and climate of the Rajasthan Desert, India. *Philosophical Transactions of the Royal Society of London* 267B, 467–501.

Sioli, H. 1984. The Amazon and its main effluents: hydrography, morphology of the river courses, and river types. In Sioli, H. (ed.), *The Amazon: Limnology and Landscape Ecology of a Mighty Tropical River and its Basin.* Junk, Dordrecht, 127–165.

Sirocko, F., Garbe-Schönberg, D., McIntyre, A. and Molfino, B. 1996. Teleconnections between the subtropical monsoons and high-latitude climates during the last deglaciation. *Science* 272, 526–529.

Sirocko, F., Sarnthein, M., Erlenkeuser, H., Lange, H., Arnold, M. and Duplessy, J.C. 1993. Century-scale events in monsoonal climate over the past 24, 000 years. *Nature* 364, 322–324.

Sissons, J.B. 1983. Shorelines and isostasy in Scotland. In Smith, D.E. and Dawson, A.G. (eds) *Shorelines and isostasy.* Institute of British Geographers Special Publication 16, 209–225.

Slowey, N.C., Henderson, G.M. and Curry, W.B. 1996. Direct U–Th dating of marine sediments from the two most recent interglacial periods. *Nature* 383, 242–244.

Smart, J. 1977. Late Quaternary sea-level changes, Gulf of Carpentaria, Australia. *Geology* 5, 755–759.

Smart, P.L. and Richards, D.A. 1992. Age estimates for the late Quaternary high sea-stands. *Quaternary Science Reviews* 11, 687–696.

Smil, V. 1990. Planetary warming: realities and responses. *Population and Development Review* 16,1–29.

Smith, A.B. 1984. The origins of food production in northeast Africa. *Palaeoecology of Africa* 16, 317–324.

Smith, B.D. 1995. *The Emergence of Agriculture.* Scientific American Library, New York.

Smith, D.E. and Dawson, A.G. (eds) 1983. *Shorelines and Isostasy.* Institute of British Geographers Special Publication Number 16. Academic Press, London.

Smith, H.J., Wahlen, M., Mastroianni, D. and Taylor, K.C. 1997. The CO_2 concentration of air trapped in GISP2 ice from the last glacial maximum–Holocene transition. *Geophysical Research Letters* 24, 1–4.

Smith, J.E., Risk, M.J., Schwarcz, H.P. and McConnaughey, T.A. 1997. Rapid climate change in the North Atlantic during the Younger Dryas recorded by deep-sea corals. *Nature* 386, 818–820.

Smythe, F.W., Ruddiman, W.F. and Lumsden, D.N. 1985. Ice-rafted evidence of long-term North Atlantic circulation. *Marine Geology* 64, 131–141.

Snead, R.E. 1980. *World Atlas of Geomorphic Features.* Krieger, New York.

Sowers, T., Bender, M., Raynaud, D., Korotkevich, Y.S. and Orchardo, J. 1991. The $\delta^{18}O$ of atmospheric CO_2 from air inclusions in the Vostok ice core: timing of CO_2 and ice volume changes during the penultimate glaciation. *Paleoceanography* 6, 679–696.

Sparks, B.W. and West, R.G 1972. *The Ice Age in Britain.* Methuen, London.

Spencer, R.J., Baldecker, M.J., Eugster, H.P., Forester, R.M., Goldhaber, M.B., Jones, B.F., Kelts, K., McKenzie, J., Madsen, D.B., Rettig, S.L., Rubin, M. and Bowser, C.J. 1984. Great Salt Lake, and precursors. Utah: the last 30,000 years. *Contributions to Mineralogy and Petrology* 86, 321–334.

Staffelbach, T., Stauffer, B. and Sigg, A. 1991. CO_2 measurements from polar ice cores: more data from different sites. *Tellus* 43B, 91–96.

Stanley, E.H., Fisher, S.G. and Grimm, N.B. 1997. Ecosystem expansion and contraction in streams. *Bioscience* 47, 427–435.

Stanley, D.J. and Warne, A.G. 1993a. Sea level and initiation of Predynastic culture in the Nile delta. *Nature* 363, 435–438.

Stanley, D.J. and Warne, A.G. 1993b: Nile delta: recent geological evolution and human impact. *Science* 260, 628–634.

Stein, R., Nam, S.-I., Schubert, C., Vogt, C., Fütterer, D. and Heinemeier, J. 1994. The last deglaciation in the eastern central Arctic Ocean. *Science* 264, 692–696.

Steinen, R.P., Harrison, R.S. and Matthews, R.K. 1973. Eustatic low stand of sea level between 125 000 and 105 000 BP: evidence from the subsurface of Barbados, West Indies. *Geological Society of America Bulletin* 84, 63–70.

Stemler, A.B.L. 1980. Origins of plant domestication in the Sahara and Nile Valley. In Williams, M.A.J. and Faure, H. (eds), *The Sahara and the Nile*. Rotterdam, Balkema, 503–526.

Stocker, T.F. and Schmittner, A. 1997. Influence of CO_2 emission rates on the stability of the thermohaline circulation. *Nature* 388, 862–865.

Stokes, S., Thomas, D.S. and Washington, R. 1997. Multiple episodes of aridity in southern Africa since the last interglacial period. *Nature* 388, 154–158.

Stommel, H. 1957. A survey of ocean current theory. *Deep Sea Research* 4, 149–184.

Stoneking, M. and Cann, R.L. 1989. African origin of human mitochondrial DNA. In Mellars, P.A. and Stringer, C.B. (eds), *The Human Revolution: Behavioural and Biological Perspectives on the Origins of Modern Humans*. Princeton University Press, New Jersey.

Stoneking, M., Sherry, S.T., Redd, A.J. and Vigilant, L. 1992. New approaches to dating suggest a recent change for the human mtDNA ancestor. *Philosophical Transactions of the Royal Society of London* 337, 167–177.

Stothers, R.B. 1987. Beat relationships between orbital periodicities in insolation theory. *Journal of the Atmospheric Sciences* 44, 1875–1876.

Strahler, A.N. and Strahler, A.H. 1987. *Modern Physical Geography*. Wiley, Brisbane.

Straus, L.G. 1985. Stone Age prehistory of northern Spain. *Science* 2130, 501–507.

Street, F.A. and Grove, A.T. 1979. Global maps of lake-level fluctuations since 30,000 year BP. *Quaternary Research* 10, 83–118.

Streeter, S.S. and Shackleton, N.J. 1979. Paleocirculation of the deep North Atlantic: 150,000–year record of benthic foraminifera and oxygen-18. *Science* 203, 168–171.

Street-Perrott, F.A. 1993. Ancient tropical methane. *Nature* 366, 411–412.

Street-Perrott, F.A. and Perrott, R.A. 1990. Abrupt climate fluctuations in the tropics: the influence of Atlantic Ocean circulations. *Nature* 343, 607–612.

Street-Perrott, F.A., Roberts, N. and Metcalfe, S. 1985. Geomorphic implications of late Quaternary hydrological and climatic changes in the Northern Hemisphere tropics. In Douglas, I. and Spencer T. (eds), *Environmental Change and Tropical Geomorphology*. George Allen and Unwin, London, 165–183.

Street-Perrott, F.A., Marchand, D.S., Roberts, N. and Harrison, S.P. 1989. *Global Lake-Level Variations from 180 000 to 0 Years Ago: A Paleoclimatic Analysis*. US Department of Energy Technical Report (TRO46), 1–213.

Stringer, C.B. and Grün, R. 1991. Time for the last Neanderthals. *Nature* 351, 701–702.

Stringer, C.B., Grün, R., Schwarcz, H.P. and Goldberg, P. 1989. ESR dates for the hominid burial site of Es Skhul in Israel. *Nature* 338, 756–758.

Stringer, C. and McKie, R. 1996. African Exodus. *The Origins of Modern Humanity*. Pimlico, London.

Stuiver, M., Grootes, P.M. and Braziunas, T.F. 1995. The GRIP 2 $\delta^{18}O$ climate record of the past 16,500 years and the role of the sun, ocean, and volcanoes. *Quaternary Research* 44, 341–354.

Stute, M., Forster, M., Frischkorn H., Serejo, A., Clark, J.F., Schlosser, P., Broecker, W.S. and Bonani, G. 1995. Cooling of tropical Brazil (5°C) during the last glacial maximum. *Science* 269, 379–383.

Stute, M., Schlosser, P., Clark, J.F. and Broecker, W.S. 1992. Paleotemperatures in the southwestern United States derived from noble gases in ground water. *Science* 256, 1000–1003.

Suarez, M.J. and Held, I.M. 1979. The sensitivity of an energy balance climate model to variations in the orbital parameters. *Journal of Geophysical Research* 84(C8), 4825–4836.

Suc, J.-P 1984. Origin and evolution of the Mediterranean vegetation and climate in Europe. *Nature* 307, 429–432.

Suc, J.-P., Bertini, A., Leroy, S.A.G. and Suballyova, D. 1997. Towards the lowering of the Pliocene/Pleistocene boundary to the Gauss–Matuyama Reversal. *Quaternary International* 40, 37–42.

Suc, J.-P. and Zagwijn, W. 1983. Plio-Pleistocene correlations between the northwestern Mediterranean region and northwestern Europe according to recent biostratigraphic and palaeoclimatic data. *Boreas* 12, 153–166.

Sugden, D. and John, B. 1976. *Glaciers and Landscape: a Geomorphological Approach*. Edward Arnold, London.

Summerfield, M.A. and Hulton, N.J. 1994. Natural controls of fluvial denudation in major world drainage basins. *Journal of Geophysical Research* 99(B7), 13871–13883.

Susman, R.L. 1994. Fossil evidence for early hominid tools. *Science* 265, 1570–1573.

Sutton, J.E.G. 1977. The African Aqualithic. *Antiquity* 51, 25–34.

Suwa, G., Asfaw, B., Beyene, Y., White, T.D., Katoh, S., Nagaoka, S., Nakaya, H., Uzawa, K., Renne, P. and Woldegabriel, G. 1997. The first skull of *Australopithecus boisei*. *Nature* 389, 489–492.

Suzuki, D. 1990. *Inventing the Future. Reflections on Science, Technology and Nature*. Allen and Unwin, Sydney.

Swineford, A. and Frye, J.C. 1945. A mechanical analysis of wind-blown dust compared with analyses of loess. *American Journal of Science* 243, 249–255.

Swisher, C.C. III, Curtis, G.H., Jacob, T., Getty, A.G., Suprijo, A. and Widiasmoro 1994. Age of the earliest known hominids in Java, Indonesia. *Science* 263, 1118–1121.

Szabo, B.J., Haynes, C.V. Jr and Maxwell, T.A. 1995. Ages of Quaternary pluvial episodes determined by uranium-series and radiocarbon dating of lacustrine deposits of Eastern Sahara. *Palaeogeography, Palaeoclimatology, Palaeoecology* 113, 227–242.

Szilder, K. and Lozowski, E.P. 1996. The influence of greenhouse warming on the atmospheric component of the hydrologic cycle. *Hydrological Processes* 10, 1317–1327.

Takahashi, T. 1975. Carbonate chemistry of seawater and the calcium carbonate compensation depth in the oceans. *Journal of Foraminiferal Research, Special Publication* 13, 11–26.

Talbot, M.R. 1980. Environmental responses to climatic change in the West African Sahel over the past 20,000 years. In Williams, M.A.J. and Faure, H. (eds), *The Sahara and the Nile*. Balkema, Rotterdam, 37–62.

Tanke, M. and Gulik, J. van, 1989. *The Global Climate*. Mirage Publishing, Amsterdam.

Tappen, M. 1995. Savanna ecology and natural bone deposition. Implications for early hominid site formation, hunting, and scavenging. *Current Anthropology* 36, 223–260.

Tattersall, I. 1995. *The Fossil Trail. How We Know What We Think We Know About Human Evolution*. Oxford University Press, New York.

Tauber, H. 1967. Investigations of the mode of pollen transfer in forested areas. *Review of Palaeobotany and Palynology* 3, 277–286.

Tauxe, L., Herbert, T., Shackleton, N.J. and Kok, Y.S. 1996. Astronomical calibration of the Matuyama–Brunhes boundary: consequences for magnetic remanence acquisition in marine carbonates and the Asian loess sequence. *Earth and Planetary Science Letters* 140, 133–146.

Taylor, K.C., Lamorey, G.W., Doyle, G.A., Alley, R.B., Grootes, P.M., Mayewski, P.A., White, J.W.C. and Barlow, L.K. 1993. The 'flickering switch' of late Pleistocene climate change. *Nature* 361, 432–436.

Taylor, K.E. and Penner, J.E. 1994. Response of the climate system to atmospheric aerosols and greenhouse gases. *Nature* 369, 734–737.

Tchernia, P. 1980. *Descriptive Regional Oceanography*. Pergamon Press, Oxford.

Tchernov, E. 1987. The Age of the 'Ubeidiya Formation, an early Pleistocene hominid site in the Jordan Valley, Israel. *Israel Journal of Earth Sciences* 36, 3–30.

Tegen, I., Lacis, A.A. and Fung, I. 1996. The influence on climate forcing of mineral aerosols from disturbed soils. *Nature* 380, 419–422.

Teller, J.T. 1990. Volume and routing of late-glacial runoff from the southern Laurentide ice sheet. *Quaternary Research* 34, 12–23.

Teller, J.T. and Kehew, A.E. 1994. Introduction to the late glacial history of large proglacial lakes and meltwater runoff along the Laurentide ice sheet. *Quaternary Science Reviews* 13, 795–799.

Ten Brink, N.W. and Weidick, A. 1974. Greenland ice sheet history since the last glaciation. *Quaternary Research* 4, 429–440.

Thieme, H. 1997. Lower Palaeolithic hunting spears from Germany. *Nature* 385, 807–810.

Thom, B.G. and Roy, P.S. 1985. Relative sea levels and coastal sedimentation in southeast Australia in the Holocene. *Journal of Sedimentary Petrology* 55, 257–264.

Thomas, D.S.G. (ed.) 1997. *Arid Zone Geomorphology: Process, Form and Change in Drylands*. (2nd edn). Wiley, New York.

Thomas, D.S.G. 1989a. Aeolian sand deposits. In Thomas, D.S.G. (ed.), *Arid Zone Geomorphology*. Belhaven, London, 232–261.

Thomas, D.S.G. 1989b. Reconstructing ancient arid environments. In Thomas, D.S.G. (ed.), *Arid Zone Geomorphology*. Belhaven, London, 311–334.

Thomas, D.S.G. and Middleton, N.J. 1994. *Desertification: Exploding the Myth*. Wiley, Chichester.

Thomas, K.D. 1993. Molecular biology and archaeology: a prospectus for inter-disciplinary research. *World Archaeology* 25, 1–17.

Thomas, M.F. and Thorp, M.B. 1995. Geomorphic response to rapid climatic and hydrologic change during the late Pleistocene and early Holocene in the humid and sub-humid tropics. *Quaternary Science Reviews* 14, 193–207.

Thomas, R.H. and Bentley, C.R. 1978. A model for Holocene retreat of the West Antarctic ice sheet. *Quaternary Research* 10, 150–170.

Thompson, L.G. and Mosley-Thompson, E. 1992. *Tropical Ice Core Paleoclimatic Records, Quelccaya Ice Cap, Peru A.D. 470 to 1984*. Byrd Polar Research Centre Miscellaneous Publication 321. Ohio State University, Printing Services, Columbus.

Thompson, L.G., Mosley-Thompson, E., Dansgaard, W. and Grootes, P.M. 1986. The Little Ice Age as recorded in stratigraphy of the tropical Quelccaya Ice Cap. *Science* 234, 361–364.

Thompson, L.G., Mosley-Thompson, E., Davis M., Lin, P.-N., Dyurgerov, M. and Dai, J. 1993. 'Recent warming': ice core evidence from tropical ice cores with emphasis on Central Asia. *Global and Planetary Change* 7, 145–156.

Thompson, L.G., Mosley-Thompson, E., Davis, M.E., Lin, P.-N., Henderson, K.A., Cole-Dai, J., Bolzan, J.F. and Liu, K.-B. 1995. Late Glacial stage and Holocene tropical ice core records from Huascarán, Peru. *Science* 269, 46–50.

Thompson, R.S. 1988. Western North America. In Huntley, B. and Webb, T. (eds), *Vegetation History*. Kluwer, Dordrecht, 415–458.

Thompson, R.S. and Krautz, R.R. 1983. Pollen analysis. *Anthropological Papers of the American Museum of Natural History* 59, 136–151.

Thorpe, R.B., Law, K.S., Bekki, S., Pyle, J.A. and Nisbet, E.G. 1996. Is methane-driven deglaciation consistent with the ice-core record? *Journal of Geophysical Research* 101(D22), 28627–28635.

Thunell, R.C. and Mortyn, P.G. 1995. Glacial climatic instability in the northeast Pacific Ocean. *Nature* 376, 504–506.

Tiedemann, R., Sarnthein, M., and Shackleton, N.J. 1994. Astronomic timescale for the Pliocene Atlantic $\delta^{18}O$ and dust flux records of ODP Site 659. *Paleoceanography* 9, 619–638.

Tiedemann, R., Sarnthein, M. and Stein, R. 1989. Climatic changes in the western Sahara: aeolio-marine sediment record of the last 8 million years (Sites 657–661). In Ruddiman, W.F. and Sarnthein, M. *et al.* (eds), *Proceedings of the Ocean Drilling Progam, Scientific Results* 108. Ocean Drilling Progam, College Station, TX, 241–261.

Tobias, P.L.V. 1979. Men, minds and hands: cultural awakenings over two million years of humanity. *South African Archaeological Bulletin* 34, 92–95.

Tobias, P.V. 1994. The changing face of hominid evolution in the twentieth century. *South African Journal of Science* 90, 205–209.

Tolba, M.K. 1992. *Saving our Planet. Challenges and Hopes.* Chapman and Hall, London.

Tooley, M.J. 1993. Long term changes in eustatic sea level. In Warrick, R.A., Barrow, E.M. and Wigley, T.M.L. (eds), *Climate and Sea Level Change: Observations, Projections and Implications.* Cambridge University Press, Cambridge, 81–107.

Toscano, M.A. and York, L.L. 1992. Quaternary stratigraphy and sea-level history of the U.S. middle Atlantic coastal plain. *Quaternary Science Reviews* 11, 301–328.

Tricart, J. and Cailleux, A. 1972. *Introduction to climatic geomorphology.* Longman, London.

Tricot, Ch. and Berger, A. 1988. Sensitivity of present-day climate to astronomical forcing. In Wanner, H. and Siegenthaler, U. (eds), *Long and Short Term Variability of Climate.* Lecture Notes in Earth Science No. 16, Springer, Berlin, 132–152.

Trinkhaus, E. 1983. *The Shanidar Neandertals.* Academic Press, New York.

Trinkhaus, E. 1986. The Neandertals and modern human origins. *Annual Review of Anthropology* 15, 193–218.

Trinkhaus, E. and Howells, W.W. 1979. The Neandertals. *Scientific American* 241, 118–133.

Troels-Smith, J. 1955. Characterisation of unconsolidated sediments. *Geology Survey of Denmark* IV, 3(10), 1–73.

Troelstra, S.R., van Hinte, J.E. and Ganssen, G.M. (eds) 1995. *The Younger Dryas.* North-Holland, Amsterdam.

Turner, A. and Wood, B. 1993. Taxonomic and geographic diversity in robust *australopithecines* and other African Plio-Pleistocene larger mammals. *Journal of Human Evolution* 24, 147–168.

Turner II, B.L., Clark, W.C., Kates, R.W., Richards, J.F., Mathews, J.T. and Meyer, W.B. (eds) 1990. *The Earth as Transformed by Human Action. Global and Regional Changes in the Biosphere over the Past 300 Years.* Cambridge University Press, Cambridge, with Clark University.

Turner, J. and Peglar, S.M. 1988. Temporally-precise studies of vegetation history. In Huntley, B. and Webb, T. (eds), *Vegetation History.* Kluwer, Dordrecht, 753–777.

Tushingham, A.M. and Peltier, W.R. 1991. Ice-3G: a new global model of late Pleistocene deglaciation based upon geophysical predictions of post-glacial relative sea level change. *Journal of Geophysical Research* 96(B3), 4497–4523.

Tuttle, R.H. 1988. What's new in African paleoanthropology? *Annual Review of Anthropology* 17, 391–426.

Tziperman, E. 1997. Inherently unstable climate behaviour due to weak thermohaline ocean circulation. *Nature* 386, 592–595.

van Andel, T.H. 1985. *New Views on an Old Planet.* Cambridge University Press, Cambridge.

van Campo, E., Duplessy, J.C., Prell, W.L., Barratt, N. and Sabathier, R. 1990. Comparison of terrestrial and marine temperature estimates for the past 135 kyr off southeast Africa: a test for GCM simulations of palaeoclimate. *Nature* 348, 209–212.

Vandenberghe, J. 1995. Timescales, climate and river development. *Quaternary Science Reviews* 14, 631–638.

van der Hammen, T. and Absy, M.L. 1994. Amazonia during the last glacial. *Palaeogeography, Palaeoclimatology, Palaeoecology* 109, 247–261.

van der Kaars, W.A. and Dam, M.A.C. 1995. A 135,000–year record of vegetational and climatic change from the Bandung area, west-Java, Indonesia. *Palaeogeography, Palaeoclimatology, Palaeoecology* 117, 55–72.

van der Merwe, N.J. 1982. Carbon isotopes and archaeology. *South African Journal of Science* 78, 14–16.

van Devender, T.R., Burgess, T.L., Piper, J.C. and Turner, R.M. 1994. Paleoclimatic implications of Holocene plant remains from the Sierra Bacha, Sonora, Mexico. *Quaternary Research* 41, 99–108.

van Woerkom, A.J.J. 1953. The astronomical theory of climate changes. In Shapley, H. (ed.), *Climate Change. Evidence, Causes, and Effects.* Harvard University Press, Cambridge, MA, 147–157.

Van Zinderen Bakker, E.M. and Mercer, J.H. 1986. Major late Cainozoic climatic events and palaeoenvironmental changes in Africa viewed in a world wide context. *Palaeogeography, Palaeoclimatology, Palaeoecology* 56, 217–235.

Varney, M. 1996. The marine carbonate system. In Summerhayes, C.P. and Thorpe, S.A. (eds), *Oceanography. An Illustrated Guide.* Manson Publishing, London, 182–194.

Vaughan, D.G. and Doake, S.M. 1996. Recent atmospheric warming and retreat of ice shelves on the Antarctic Peninsula. *Nature* 379, 328–331.

Veeh, H.H. and Chappell, J.M.A. 1970. Astronomical theory of climatic change: support from New Guinea. *Science* 167, 862–865.

Vermeij, G.J. 1993. The biological history of a seaway. *Science* 260, 1603–1604.

Vernekar, A.D. 1971. Long-period global variations of incoming solar radiation. *Meteorological Monographs* 12, 1–11.

Vernet, R. 1995. *Climats Anciens Du Nord De L'Afrique.* L'Harmattan, Paris.

Veum, T., Jansen, E., Arnold, M., Beyer, I. and Duplessy, J.-C. 1992. Water mass exchange between the North Atlantic and the Norwegian Sea during the past 28,000 years. *Nature* 356, 783–785.

Vincent, E. and Berger, W.H. 1981. Planktonic foraminifera and their use in paleoceanography. In Emiliani, C. (ed.), *The Sea, Vol. 7; The Oceanic Lithosphere.* Wiley, New York, 1025–1119.

Vita-Finzi, C. 1973. *Recent Earth History.* Macmillan, London.

Vogel, J.C. 1978. Isotopic assessment of the dietary habits of ungulates. *South African Journal of Science* 74, 298–301.

Vogel, J.C., Fuls, A. and Ellis, R.P. 1978. The geographical distribution of Kranz grasses in South Africa. *South African Journal of Science* 74, 209–215.

Volk, T. 1989. Rise of angiosperms as a factor in long-term climatic cooling. *Geology* 17, 107–110.

von Grafenstein, U., Erlenkeuser, H., Müller, J. and Kleinmann-Eisenmann, A. 1992. Oxygen isotope records of benthic ostracods in Bavarian lake sediments. *Naturwissenschaften* 79, 145–152.

von Grafenstein, U., Erlenkeuser, H., Müller, J., Jouzel, J. and Johnsen, S. 1998. The cold event 8,200 years ago documented in oxygen isotope records of precipitation in Europe and Greenland. *Climate Dynamics* 14, 73–81.

Vrba, E.S. 1988. Late Pliocene climatic events and hominid evolution. In Grine, F.E. (ed.), *Evolutionary History of the 'Robust' Australopithecines.* Aldine de Gruyter, New York, 405–426.

Vrba, E.S., Denton, G.H. and Prentice, M.L. 1989. Climatic influences on early hominid behavior. *Ossa* 14, 127–156.

Vrba, E.S. 1995. The fossil record of African antelopes (Mammalia, Bovidae) in relation to human evolution and paleoclimate. In Vrba, E.S., Denton, G.H., Partridge, T.C. and Burckle, L.H. (eds), *Paleoclimate and Evolution.* Yale University Press, New Haven, 385–424.

Wadia, S., Korisettar, R. and Kale, V.S. (eds) 1995. *Quaternary environments and geoarchaeology of India.* (Essays in Honour of Professor S.N. Rajaguru). Memoir 32, Geological Society of India, Bangalore.

Waitt, R.B. 1985. Case for periodic, colossal jökulhlaups from Pleistocene glacial Lake Missoula. *Bulletin of the Geological Society of America* 96, 1271–1286.

Walker, A., Leakey, R.E., Harris, J.M. and Brown, F.H. 1986. 2.5 Myr *Australopithecus boisei* from west of Lake Turkana, Kenya. *Nature* 322, 517–522.

Walker, D. and Chen, Y. 1987. Palynological light on tropical rainforest dynamics. *Quaternary Science Reviews* 6, 77–92.

Walter, R.C. 1994. Age of Lucy and the first family: single-crystal ^{40}Ar/^{39}Ar dating of the Denen Dora and lower Kada Hadar members of the Hadar Formation, Ethiopia. *Geology* 22, 6–10.

Walther, J. 1924. *Das Gesetz der Wüstenbildung in Gegenwart und Vorzeit* (3rd edn). Reimer, Berlin.

Wang, H., Ambrose, S.H., Liu, C.J. and Follmer, L.R. 1997. Paleosol stable isotope evidence for early hominid occupation of east Asian temperate environments. *Quaternary Research* 48, 228–238.

Ward, W.R. 1982. Comments on the long-term stability of the Earth's obliquity. *Icarus* 50, 444–448.

Warrick, R.A. and Ahmad, Q.K. (eds) 1996. *The Implications of Climate and Sea Level Change for Bangladesh.* Kluwer, Dordrecht.

Warrick, R.A., Barrow, E.M. and Wigley, T.M.L. (eds) 1993. *Climate and Sea Level Change. Observations, Projections and Implications.* Cambridge University Press, Cambridge.

Warrick, R.A., Le Provost, C., Meier, M.F., Oerlemans, J. and Woodworth, P.L. 1996. Changes in sea level. In Houghton, J.T., Meira Filho, L.G., Callander, B.A., Harris, N., Kattenberg, A. and Maskell, K. (eds), *Climate Change 1995. The Science of Climate Change.* Cambridge University Press for the Intergovernmental Panel on Climate Change, Cambridge, 359–405.

Warrick, R. and Oerlemans, J. 1990. Sea level rise. In Houghton, J.T., Jenkins, G.J. and Ephraums, J.J. (eds), *Climatic Change: The IPCC Assessment.* Cambridge University Press for The Intergovernmental Panel on Climatic Change, Cambridge, 257–281.

Washburn, A.L. 1973. *Periglacial Processes and Environments.* Edward Arnold, London.

Washburn, A.L. 1979. *Geocryology – a Survey of Periglacial Processes and Environments.* Edward Arnold, London.

Wasson, R.J. and Hyde, R. 1983. Factors determining desert dune type. *Nature* 304, 337–339.

Wasson, R.J. and Nanninga, P.M. 1986. Estimating wind transport of sand on vegetated surfaces. *Earth Surface Processes and Landforms* 11, 505–514.

Wasson, R.J., Rajaguru, S.N., Misra, V.N., Agrawal, D.P., Ohir, R.P., Singhvi, A.K. and Rao, K.K. 1983. Geomorphology, Late Quaternary stratigraphy and palaeoclimatology of the Thar Desert. *Zeitschrift für Geomorphologie, Supplementband* 45, 117–152.

Watson, A.J. 1997. Volcanic iron, CO_2, ocean productivity and climate. *Nature* 385, 587–588.

Watts, R.G. and Hayder, M.E. 1983. The origin of the 100–kiloyear ice sheet cycle in the Pleistocene. *Journal of Geophysical Research* 88(C9), 5163–5166.

Watts, W.A. 1978. Plant macrofossils and Quaternary palaeoecology. In Walker, D. and Guppy, J.C. (eds) *Biology and Quaternary Environments.* Australian Academy of Sciences, Canberra, 53–67.

Weaver, K.F. 1985. Stones, bones and early man. The search for our ancestors. *National Geographic Magazine* 168, 560–623.

Webb, L.J. and Tracey, J.G. 1981. Australian rainforests: patterns and change. In Keast, A. (ed.), *Ecological Biogeography in Australia.* Junk, The Hague, 607–694.

Webb, R.S., Rind, D.H., Lehman, S.J., Healy, R.J. and Sigman, D. 1997. Influence of ocean heat transport on the climate of the last glacial maximum. *Nature* 385, 695–699.

Webb, S.D. 1995. Biological implications of the Middle Miocene Amazon seaway. *Science* 269, 361–362.

Webb, T. 1986. Is vegetation in equilibrium with climate? How to interpret late-Quaternary pollen data. *Vegetatio* 67, 75–91.

Weertman, J. 1976. Milankovitch solar radiation variations and ice age ice sheet sizes. *Nature* 261, 17–20.

Wendorf, F. and Schild, R. 1980. *Prehistory of the Eastern Sahara.* Academic Press, New York.

Wendorf, F. and Schild, R. 1989. Summary and synthesis. In Wendorf, F., Schild, R. and Close, A.E. (eds), *The Prehistory of Wadi Kubbaniya*, Volume 3. *Late Paleolithic Archaeology.* Southern Methodist University Press, Dallas, 768–824.

West, R.G. and Sparks, B.W. 1977. *Pleistocene Geology and Biology, with Special Reference to the British Isles* (2nd edn). Longman, London.

Wheeler, D.A. 1985. An analysis of the aeolian dustfall on eastern Britain. November 1984. *Proceedings of the Yorkshire Geological Society* 45, 307–310.

Wheeler, P.E. 1993. The influence of stature and body form on hominid energy and water budgets: a comparison of *Australopithecus* and early *Homo* physiques. *Journal of Human Evolution* 24, 13–28.

Wheeler, P.E. 1994. The thermoregulatory advantages of

heat storage and shade-seeking behaviour to hominids foraging in equatorial savannah environments. *Journal of Human Evolution* 26, 339–350.

White, J.P. and O'Connell, J.F. 1978. Australian prehistory: new aspects of antiquity. *Science* 203, 21–28.

White, J.P. and O'Connell, J.F. 1982. *A prehistory of Australia, New Guinea and Sahul*. Academic Press, Sydney.

White, T.D. 1984. Pliocene hominids from the Middle Awash, Ethiopia. *Courier Forschungsinstitut Senckenberg* 69, 57–68.

White, T.D. 1986. Cutmarks on the Bodo cranium: a case of prehistoric defleshing. *American Journal of Physical Anthropology* 69, 503–509.

White, T.D. 1995. African omnivores: global climatic change and Plio-Pleistocene hominids and suids. In Vrba, E.S., Denton, G.H., Partridge, T.C. and Burckle, L.H. (eds), *Paleoclimate and Evolution*. Yale University Press, New Haven, 369–384.

White, T.D. and Harris, J.M. 1977. Suid evolution and correlation of African hominid localities. *Science* 198, 13–21.

White, T.D., Johanson, D.C. and Kimbel, W.H. 1981. *Australopithecus afarensis*: its phyletic position reconsidered. *South African Journal of Science* 77, 445–470.

White, T.D. and Suwa, G. 1987. Hominid footprints at Laetoli: facts and interpretations. *American Journal of Physical Anthropology* 72, 485–514.

White, T.D., Suwa, G. and Asfaw, B. 1994. *Australopithecus ramidus*, a new species of early hominid from Aramis, Ethiopia. *Nature* 371, 306–312.

Whitlock, C., Bartlein, P.J. and Watts, W.A. 1993. Vegetation history of Elk lake. In Bradbury, J.P. and Dean, W.E. (eds), *Elk Lake, Minnesota: Evidence for Rapid Climatic Change in the North-Central United States*. Geological Society of America Special Paper 276, 251–274.

Whitmore, T.C. and Prance, G.T. (eds) 1987. *Biogeography and Quaternary History in tropical America*. Clarendon Press, Oxford.

Wiggs, G.F.S., Thomas, D.S.G. and Bullard, J. E. 1995. Dune mobility and vegetation cover in the southwest Kalahari Desert. *Earth Surface Processes and Landforms* 20, 515–529.

Wigley, T.M.L. 1976. Spectral analysis and the astronomical theory of climatic change. *Nature* 264, 629–631.

Wigley, T.M.L. 1995. Global-mean temperature and sea level consequences of greenhouse gas concentration stabilization. *Geophysical Research Letters* 22, 45–48.

Wigley, T.M.L. and Raper, S.C.B. 1987. Thermal expansion of sea water associated with global warming. *Nature* 330, 127–131.

Wigley, T.M.L. and Raper, S.C.B. 1990. Climatic change due to solar irradiance changes. *Geophysical Research Letters* 17, 2169–2172.

Wigley, T.M.L. and Raper, S.C.B. 1992. Implications for climate and sea level of revised IPCC emissions scenarios. *Nature* 357, 293–300.

Wigley, T.M.L. and Raper, S.C.B. 1993. Future changes in global mean temperature and sea level. In Warrick, R.A., Barrow, E.M. and Wigley, T.M.L. (eds), *Climate and Sea Level Change: Observations, Projections and Implications*. Cambridge University Press, Cambridge, 111–133.

Wigley, T.M.L., Richels, R. and Edmonds, J.A. 1996. Economic and environmental choices in the stabilization of atmospheric CO_2 concentrations. *Nature* 379, 240–243.

Wilgus, C.K., Hastings, B.S., Kendall, C.G., St., C., Posamentier, H., Ross, C.A. and van Wagoner, J. (eds) 1988. *Sea-Level Changes: An Integrated Approach*. Society of Economic Paleontologists and Mineralogists, Tulsa, Special Publication No. 42.

Williams, C.A. 1986. An oceanwide view of Palaeogene plate tectonic events. *Palaeogeography, Palaeoclimatology, Palaeoecology* 57, 3–25.

Williams, D.F., Moore, W.S. and Fillon, R.H. 1981. Role of glacial Arctic Ocean ice sheets in Pleistocene oxygen isotope and sea level records. *Earth and Planetary Science Letters* 56, 157–166.

Williams, G.E. 1993. History of the Earth's obliquity. *Earth-Science Reviews* 34, 1–45.

Williams, G.P. 1984. Palaeohydrologic equations for rivers. In Costa, J.E. and Fleisher, P.J. (eds), *Developments and Applications of Geomorphology*. Springer-Verlag, Berlin, 343–367.

Williams, M.A.J. 1966. Age of alluvial clays in the western Gezira, Republic of the Sudan. *Nature* 211, 270–271.

Williams, M.A.J. 1968. Termites and soil development near Brocks Creek, Northern Territory. *Australian Journal of Science* 31, 153–154.

Williams, M.A.J. 1975. Late Quaternary tropical aridity synchronous in both hemispheres? *Nature* 253, 617–618.

Williams, M.A.J. 1982. Quaternary environments in North Africa. In Williams, M.A.J. and Adamson, D.A. (eds), *A Land Between Two Niles. Quaternary Geology and Biology of the Central Sudan*. Balkema, Rotterdam, 13–22.

Williams, M.A.J. 1984a: Geology. In Cloudsley-Thompson, J.L. (ed.), *Sahara Desert*. Pergamon, Oxford, 31–39.

Williams, M.A.J. 1984b: Late Quaternary environments in the Sahara. In Clark, J.D. and Brandt, S.A. (eds), *The Causes and Consequences of Food Production in Africa*. University of California Press, Berkeley, 74–83.

Williams, M.A.J. 1984c: Cenozoic evolution of arid Australia. In Cogger, H.G. and Cameron E.E. (eds), *Arid Australia*. Australia Museum, Sydney, 59–78.

Williams, M.A.J. 1985a: Pleistocene aridity in Africa. Australia and Asia. In Douglas, I. and Spencer, T. (eds), *Environmental Change and Tropical Geomorphology*. Allen and Unwin, London, 219–233.

Williams, M.A.J. 1985b. On becoming human: geographical background to cultural evolution. 11th Griffith Taylor Memorial Lecture. *Australian Geographer* 16, 175–184.

Williams, M.A.J. 1986. The creeping desert: what can be done? *Current Affairs Bulletin* 63, 24–31.

Williams, M.A.J. 1994a: Cenozoic climate changes in deserts: a synthesis. In Abrahams, A.D. and Parsons, A.J. (eds) *Geomorphology of Desert Environments*. Chapman and Hall, London, 644–670.

Williams, M.A.J. 1994b: Some implications of past climatic changes in Australia. *Transactions of the Royal Society of South Australia* 118, 17–25.

Williams, M.A.J. Abell, P.I. and Sparks, B.W. 1987. Quaternary landforms, sediments, depositional environments and gastropod isotope ratios at Adrar Bous,

Tenere Desert of Niger, south-central Sahara. In Frostick, L. and Reid, I. (eds), *Desert Sediments: Ancient and Modern*. Geological Society Special Publication, London, 35, 105–125.

Williams, M.A.J., Adamson, D.A., Williams, F.M., Morton, W.H. and Parry, D.E. 1980. Jebel Marra volcano: a link between the Nile Valley, the Sahara and Central Africa. In Williams, M.A.J. and Faure, H. (eds), *The Sahara and the Nile*. Balkema, Rotterdam, 305–337.

Williams, M.A.J. and Balling, R.C. Jr 1996. *Interactions of Desertification and Climate*. Edward Arnold, London.

Williams, M.A.J. and Clarke, M.F. 1984. Late Quaternary environments in north central India. *Nature* 308, 633–635.

Williams, M.A.J. and Clarke, M.F. 1995. Quaternary geology and prehistoric environments in the Son and Belan Valleys, north central India. *Memoirs Geological Society of India* 32, 282–308.

Williams, M.A.J., De Deckker, P. and Kershaw, A.P. (eds) 1991. *The Cainozoic in Australia: A Re-appraisal of the Evidence*. Geological Society of Australia Special Publication 18.

Williams, M.A.J. and Royce, K. 1982. Quaternary geology of the Middle Son Valley, north central India: implications for prehistoric archaeology. *Palaeogeography, Palaeoclimatology, Palaeoecology* 38, 139–162.

Williams, M.A.J., Williams, F.H., Gasse, F., Curtis, G.H. and Adamson, D.A. 1979. Pleistocene environments at Gadeb prehistoric site, Ethiopia. *Nature* 282, 29–33.

Williams, R.B.G. 1969. Permafrost and temperature conditions in England during the last glacial period. In Péwé, T.L. (ed.), *The Periglacial Environment: Past and Present*. McGill-Queen's University Press, Montreal, 399–410.

Williamson, P. and Oeschger, H. 1993. Climate instability in a warmer world. *Ambio* 22, 411.

Wind, H.G. (ed.) 1987. *Impact of Sea Level Rise on Society*. Balkema, Rotterdam.

Woldegabriel, G., White, T.D., Suwa, G., Renne, P., DE Heinzelin, J., Hart, W.K. and Heiken, G. 1994. Ecological and temporal placement of early Pliocene hominids at Aramis, Ethiopia. *Nature* 371, 330–333.

Wolters, H.J., Shaffer, J.A., Cerveny, R.S. and Barnhill, R.E. 1996. Visualization of Milankovitch climate-change theory. *Journal of Geoscience Education* 44, 7–12.

Wood, B. 1989. Hominid diversity in the Plio-Pleistocene. *Ossa* 14, 19–31.

Wood, B. 1994. The oldest hominid yet. *Nature* 371, 280–281.

Woodwell, G.M., Hobbie, J.E., Houghton, R.A., Melillo, J.M., Moore, B., Peterson, B.J. and Shaver, G.R. 1983. Global deforestation: contribution to atmospheric carbon dioxide. *Science* 222, 1081–1086.

Worsley, T.R., Nance, D. and Moody, J.B. 1984. Global tectonics and eustasy for the past 2 billion years. *Marine Geology* 58, 373–400.

Wright, H.E. Jr 1996. Global climatic changes since the last glacial maximum: evidence from paleolimnology and paleoclimate modeling. *Journal of Paleolimnology* 15, 119–127.

Wright, H.E. Jnr., Kutzbatch, J.E., Webb III, T., Ruddiman, W.E.F., Street-Perrott, F.A. and Bartlein, P.J. (eds) 1993.

Global Climates Since the Last Glacial Maximum. University of Minnesota Press, Minneapolis.

Wright, V.P. (ed.) 1986. *Paleosols: Their Recognition and Interpretation*. Princeton University Press, Princeton, New Jersey.

Wu, R. and Lin, S.H. 1983. Peking Man. *Scientific American* 248, 78–86.

Wyatt, S.P. 1977. *Principles of Astronomy*. (3rd edn). Allyn and Bacon, Boston.

Wyrtki, K. 1990. Sea level rise: the facts and the future. *Pacific Science* 44, 1–16.

Xue Yongkang 1996. The impact of desertification in the Mongolian and the Inner Mongolian grassland on the regional climate. *Journal of Climate* 9, 2173–2189.

Yaalon, D.H. (ed.) 1971. *Paleopedology: Origin, Nature and Dating of Paleosols*. Papers of the Symposium on the Age of Parent Materials and Soils, Amsterdam, Netherlands, August 10–15, 1970. International Society of Soil Science and Israel Universities Press, Jerusalem.

Yao, T., Thompson, L.G., Mosley-Thompson, E., Zhihong, Y., Xingping, Z. and Lin P.-N. 1996. Climatological significance of $\partial^{18}O$ in north Tibetan ice cores. *Journal of Geophysical Research* 101(D23), 29531–29537.

Yasuda, Y. 1995. Climatic changes and the development of Jomon culture in Japan. In Ito, S. and Yasuda, Y. (eds), *Nature and Humankind in the Age of Environmental Crisis*. International Research Centre for Japanese Studies, Kyoto, 57–77.

Yellen, J.E., Brooks, A.S., Cornelissen, E., Mehlman, M.J. and Stewart, K. 1995. A Middle Stone Age worked bone industry from Katanda, Upper Semliki River (Kivu), Zaire. *Science* 268, 553–556.

Yemane, K., Taieb, M. and Faure, H. 1987. Limnogeologic studies on an intertrappean continental deposit from the northern Ethiopian Plateau (37°03′E, 12°25′N). *Journal of African Earth Sciences* 6, 91–101.

Yiou, F., Raisbeck, G.M., Bourles, D., Lorius, C. and Barkov, N.I. 1985. ^{10}Be in ice at Vostok Antarctica during the last climatic cycle. *Nature* 316, 616–617.

Yiou, P., Genthon, C., Ghil, M., Jouzel, J., le Treut, H., Barnola, J.M., Lorius, C. and Korotkevictch, Y.N. 1991. High-frequency paleovariability in climate and CO_2 levels from Vostok ice core records. *Journal of Geophysical Research* 96(B12), 20365–20378.

Young, M.A. and Bradley, R.S. 1984. Insolation gradients and the paleoclimatic record. In Berger, A., Imbrie, J., Hays, J., Kukla, G. and Saltzman, B. (eds), *Milankovitch and Climate*. Reidel, Dordrecht, 707–713.

Yu E.-F. François R. and Bacon M.P. 1996. Similar rates of modern and last-glacial ocean thermohaline circulation inferred from radiochemical data. *Nature* 379, 689–694.

Yu, Z., McAndrews, J.H. and Eicher, U. 1997. Middle Holocene dry climate caused by change in atmospheric circulation patterns: evidence from lake levels and stable isotopes. *Geology* 25, 251–254.

Yung, Y.L., Lee, T., Wang, C.-H. and Shieh, Y.-T. 1996. Dust: a diagnostic of the hydrologic cycle during the Last Glacial Maximum. *Science* 271, 962–963.

Zachos, J.C., Flower, B.P. and Paul, H. 1997. Orbitally paced climate oscillations across the Oligocene/Miocene boundary. *Nature* 388, 567–570.

Zagwijn, W.H. 1957. Vegetation, climate and time-correlations in the Early Pleistocene of Europe. *Geologie en Mijnbouw* 19, 233–244.

Zahn, R. 1994. Core correlations. *Nature* 371, 289–290.

Zarate, M.A. and Fasano, J.L. 1989. The Plio-Pleistocene record of the central eastern Pampas, Buenos Aires province, Argentina: the Chapadmalal case study. *Palaeogeography, Palaeoclimatology, Palaeoecology* 72. 27–52.

Zeng, N., Dickinson, R.E. and Zeng, X. 1996. Climatic impact of Amazonian deforestation – a mechanistic model study *Journal of Climate* 9, 859–883.

Zerbini, S., Plag, H.-P., Baker, T., Becker, M., Billiris, H., Bürki, B., Kahle, H.-G., Marson, I., Pezzoli, L., Richter, B., Romagnoli, C., Sztobryn, M., Tomasi, P., Tsimplis, M., Veis, G. and Verrone, G. 1996. Sea level in the Mediterranean: a first step towards separating crustal movements and absolute sea-level variations. *Global and Planetary Change* 14, 1–48.

Zeuner, F.E. 1950. *Dating the Past. An Introduction to Geochronology* (2nd edn). Methuen, London.

Zhang, J. 1995. The vulnerability of socioeconomical developments on climatic change in China. In Ito, S. and Yasuda, Y. (eds), *Nature and Humankind in the Age of Environmental Crisis.* International Research Centre for Japanese Studies, Kyoto, 161–191.

Zhang, X.Y., An, Z.S., Chen, T. and Zhang, G.Y. 1994. Late Quaternary records of the atmospheric input of eolian dust to the center of the Chinese loess plateau. *Quaternary Research* 41, 35–43.

Zheng, H.B., Rolph, T., Shaw, J. and An, Z.S. 1995. A detailed palaeomagnetic record for the last interglacial period. *Earth and Planetary Science Letters* 133, 339–351.

Zielinski, G.A., Mayewski, P.A., Meeker, L.D., Whitlow, S., Twickler, M.S., Morrison, M., Meese, D.A., Gow, A.J. and Alley, R.B. 1994. Record of volcanism since 7000 BC from the GISP2 Greenland ice core and implications for the volcano-climate system. *Nature* 264: 948–952.

Zinderen Bakker, E.M. van (ed.), 1978. *Antarctic Glacial History and World Palaeoenvironments.* Balkema, Rotterdam.

INDEX